The Virginia Regimental
Histories Series

52nd Virginia Infantry

2nd Edition

Robert J. Driver, Jr.

HERITAGE BOOKS
2024

HERITAGE BOOKS
AN IMPRINT OF HERITAGE BOOKS, INC.

Books, CDs, and more—Worldwide

For our listing of thousands of titles see our website
at
www.HeritageBooks.com

Published 2024 by
HERITAGE BOOKS, INC.
Publishing Division
5810 Ruatan Street
Berwyn Heights, MD 20740

Copyright © 1986 H. E. Howard, Inc.

All rights reserved. No part of this book may be reproduced or transmitted in any form or by any means, electronic or mechanical, including photocopying, recording or by any information storage and retrieval system without written permission from the author, except for the inclusion of brief quotations in a review.

International Standard Book Number
Paperbound: 978-0-7884-3024-4

This book is dedicated to the unknown Confederate soldiers from Virginia who lie in unmarked graves throughout the state. Lest we forget.

"NOT FOR FAME OR REWARD, NOT FOR PLACE OR RANK, NOT LURED BY AMBITION, OR GOADED BY NECESSITY, BUT IN SIMPLE OBEDIENCE TO DUTY, AS THEY UNDERSTOOD IT, THESE MEN SUFFERED ALL, SACRIFICED ALL, DARED ALL — AND DIED! NO STATELY ABBEY WILL EVER COVER THEIR REMAINS, THEIR DUST WILL NEVER REPOSE BENEATH FRETTED OR FRESCOED ROOF. NO COSTLY BRONZE WILL EVER BLAZON THEIR NAMES FOR POSTERITY TO HONOR — BUT THE POTOMAC AND THE RAPPAHANNOCK, THE JAMES AND THE CHICKAHOMINY, THE CUMBERLAND AND THE TENNESSEE, THE MISSISSIPPI, AND THE RIO GRANDE, AS THEY RUN THEIR LONG RACE FROM THE MOUNTAINS TO THE SEA, WILL SING OF THEIR PROWESS FOREVERMORE."

> Bronze plaque in memory of the Confederate soldiers from Staunton and Augusta County, Thornrose Cemetery, Staunton, Virginia.

1861

The Fifty-second Regiment of Virginia Infantry was organized at Staunton from the militia units of Augusta and Rockbridge Counties. This followed Governor John Letcher's proclamation of July 3 calling for 3,000 additional volunteers from the state. General Robert E. Lee, Commander of the Forces of Virginia, had, in turn, ordered the commanding officers of the militia regiments of the two counties to rendezvous at Staunton at 11 a.m. on July 16. "All who have arms will bring them, with ammunition, ready to do service. Each man will bring a blanket and change of clothes."

The companies organized from the militia were:

Company A — Augusta Fencibles (orginally Confederate Guards). Enlisted Staunton July 9, Captain James H. Skinner, a prominent lawyer and colonel of militia, commanding. 80 officers and men.

Company B — (1st) — Fairfield McDowell Guards, Fairfield, Rockbridge County. Enlisted Fairfield July 10. Captain John Miller, Princeton graduate and Minister of Oxford Presbyterian Church, commanding. 120 officers and men.

Company B — (2nd) — Waynesboro Guards, Waynesboro, Augusta County. Enlisted July 15. Captain William Long, Mexican War Veteran and saddler, commanding. 75 officers and men. This company may have served as Company K originally until the Fairfield McDowell Guards were redesignated an artillery company September 28, 1861.

Company C — Letcher Guard, Mossy Creek area, Augusta County. Enlisted Staunton July 16, Captain Edward M. Dabney, VMI '62, commanding. 66 officers and men.

Company D — Harper Guard, Mount Solon area, Augusta County. Enlisted Staunton July 16. Captain Joseph F. Hottell, constable and militia officer, commanding. 46 officers and men.

Company E — Captain Thomas H. Watkin's Company, Colliers Creek, Buffalo Creek, Broad Creek and Natural Bridge area, Rockbridge County. Enlisted Staunton August 1. Captain Watkins, manager of Glenwood Furnace, commanding. 64 officers and men.

Company F — Captain Joseph E. Cline's Company, Mount Sidney area, Augusta County. Enlisted Staunton July 31. Captain Cline, successful farmer and militia officer, commanding. 56 officers and men.

Company G — Veteran Guards, Fishersville area, Augusta County. Enlisted Staunton Augusta 2. Captain Samuel H. McCune, prominent farmer and militia colonel, commanding. 50 officers and men.

Company H — Staunton Pioneers, Augusta Pioneers or Mountain Pioneers, Staunton, Augusta County. Enlisted Staunton July 23. Captain Claiborne R. Mason, railroad engineer, commanding. 50 officers and men.

Company I — Men of West Augusta, Middlebrook and Greenville sections, Augusta County. Enlisted Staunton July 16. Captain Samuel A. Lambert, successful farmer, commanding. 42 officers and men.

Company K — Bath Rifles, Bath County. Enlisted Shenandoah Mountain April 9, 1862. Captain Benjamin J. Walton, successful farmer and militiaman, commanding. 59 officers and men.

Company K — Captain Morrison's Company, Rockbridge County. Listed in unofficial sources only. Became Company G, 58th Virginia Infantry.

The 52nd was officially organized for state service on August 16, 1861. The new soldiers were camped on the grounds of the Deaf, Dumb and Blind Asylum in Staunton. The inmates of the asylum had been moved to the smaller Staunton Female Academy. Many of the men and boys would later return to the asylum under direr circumstances when it became a Confederate hospital.

Governor Letcher appointed John B. Baldwin of Staunton as Colonel of the Regiment. Baldwin, a graduate of the University of Virginia, was a lawyer and politician, had served as Captain of the pre-war Staunton Light Infantry and as a Colonel in the state militia. He had been appointed Colonel and Inspector General of Virginia troops earlier, and had been responsible for the successful organizing and equipping of state forces at Staunton.

Michael G. "Mike" Harman was named Lieutenant Colonel. A self-made man of considerable talents, he owned a hotel in Staunton and ran a stage line throughout western Virginia and Tennessee. His knowledge of local resources and organizational ability had also been put to use by Governor Letcher. He was serving as Quarter Master of Virginia forces at Staunton, with the rank of Major, in charge of supplying the thousands of troops serving in western Virginia. These duties put him at cross purposes with his assignment as second in command of the 52nd. Harman did not resign his position as Quarter Master until January 1862, and was acting Quartermaster for General Thomas J. Jackson as late as February 1862.

The Major's billet was given to John D. Ross, a recent VMI graduate, and an engineer officer on the staff of General William W. Loring in northwestern Virginia.

These appointments did not set well with the new soldiers from Rockbridge County. They believed one of the field officers should have been from their county. Ross, a native of Culpeper County, and an instructor at VMI when the war began, was not considered by them to represent Rockbridge. The matter apparently blew over, as nothing more appeared in the Lexington newspaper.

The Adjutant's slot went to John W. Lewis, a VMI classmate of Ross's, and a descendant of the founder of Bath County.

George M. Cochran, Jr., a University of Virginia graduate and Staunton lawyer, was appointed Captain and Quartermaster.

Staunton lawyer and state senator Bolivar Christian, a Washington College graduate, was made Captain and Commissary of Subsistence.

Richard H. Phillips, President of the Female Institute in Staunton, was appointed Chaplain.

The Surgeon's assignment went to elderly Livingston Waddell, a University of Pennsylvania Medical School graduate, and prominent Waynesboro physician. He was to be assisted by John Lewis, a graduate of the University of Virginia Medical School and Charlottesville medic, and William S. McChesney, who attended Washington College and was a Staunton doctor.

The regiment was fortunate to have in its ranks a number of VMI alumni, most of whom had served as drillmasters of other Confederate units in Richmond. William "Willie" Galt and Richard S. Kinney in Company A, Eugene Curry and Lewis Harman, son of the Lieutenant Colonel, in Company C, Cyrus B. "Bent" Coiner in Company G, John D. Lilley in Company H, and John A. Stuart, were to instill discipline and efficiency in the rank and file of the 52nd. Many of these men would provide the leadership for the regiment in the next four years.

Staunton had resembled an armed camp almost from the beginning of the war, as numerous Confederate units arrived by train and marched west and north, but the local populace had not seen some of their "own" in training. The other companies organized in Staunton and Augusta County had marched off immediately to the seats of war at Harper's Ferry and Manassas. Lieutenant William Knick wrote to a friend in Rockbridge County that "over 100 girls came every evening for the dress parade." This was certain to inspire all the young men in the regiment.

The 52nd remained in Staunton until September 10, adding numerous recruits to the smaller companies, and probably due to a lack of arms and equipment. Orders came for the regiment to march for Greenbrier River and join other Confederate units in the defense of western Virginia.

Leaving Staunton at 2 p.m. on the 10th the fledgling soldiers began the first of many long marches. The citizens of Staunton lined the streets and roads cheering and waving as the 52nd marched past. The day's trek brought them to Buffalo Gap about dark, and to the first of many hasty camps.

The second day's tramp of fifteen miles brought them to Ryan's on the Calfpasture River. Continuing their march westward on the morning of the 12th, they reached Hull's on the Bullpasture River. They marched the next day to Hevener's Store at Crab Bottom in Highland County where the waters from two springs flow in different directions, one sending water into the James River and the other into the Potomac. Here a report was received that "about one thousand yankees" were near Petersburg in Hardy County. The 52nd immediately struck its tents and marched northward in that direction.

Arriving at Ottobine Chapel on Straight Creek, the men rested a few days awaiting reinforcements. According to Private Adam Kersh, "all members of his company F that had made the march were well and the regiment was in good health and generally in good spirits." The fact that the men had their knapsacks hauled in wagons over the mountains helped. Some few members of the regiments had become exhausted and had to be hauled in wagons too.

Major Ross wrote from "Camp at Fork of Waters", Highland County, on September 21, describing the camp surrounded by mountains; "with streams full of trout on each side and the soldiers busy catching them." The move toward Petersburg had been delayed because the expected reinforcements had been diverted back to Greenbrier River by the report of a Union advance from Cheat Mountain. General Henry R. Jackson was in command of the Hardy line of operations. Shumaker's battery of artillery had now joined the 52nd and Ross was critical of the plan to post the guns halfway between Fork of the Waters and Greenbrier River.

On the 28th Ross wrote to his future father-in-law during a terrible storm, describing his feeling about a cold, damp camp with no fire. They had camped below a milldam and bedding and camp equipment had been washed away. Private John H. Stover commented in his diary that "their first baptism was not of fire but of water." The camp was moved to a nearby hillside.

This was also the day the Fairfield McDowell Guards reorganized as the 2nd Rockbridge Artillery and separated from the regiment, although they would continue to serve together for some time.

Private John S. Robson of Company D recalled in later years: "the perfect autumn weather, buckwheat cakes, honey, cakes, pies, roast beef and wild turkey."

The 52nd received orders to reinforce the troops at Greenbrier River, in anticipation of another Federal move from Cheat Mountain. When the top of the Alleghanies was reached on September 30, the troops were halted and went into camp. This proved to be their home for the next six months.

On the morning of October 3, the men of the 52nd were alarmed by the sound of cannon fire from Greenbrier River, eight miles away. The officers immediately ordered them under arms, issued extra ammunition, and quickly had them ready to march to support General Jackson's forces. A messenger from the general arrived advising Colonel Baldwin to guard the road by which the enemy might get in his rear. Baldwin and Ross led two companies and placed them in ambush, and sent scouts six miles down the road. Another messenger arrived informing the Colonel to be ready to move immediately. A scout returned reporting no enemy in sight on the road in question. Soon another courier dashed up with orders to march and protect Jackson's rear. The troops marched off cheering and singing, with drums beating. Rushing on for two miles, the 52nd met General Jackson's aide who informed them that the battle was over, the enemy defeated, and its services no longer needed. Major Ross commented in a letter home that "the troops were greatly disappointed." The regiment camped where they had halted — with orders to march toward the river if more firing was heard.

The 52nd returned to "Camp Baldwin" on top of the mountain the next day.

The rest of October was devoted to building log cabins for winter quarters and entrenching the camp. Adam Kersh wrote home on October 20 reporting they had "received blankets and were expecting overcoats any day." He only desired boots and gloves from home. Many of the men were sick with typhoid fever, mumps, measles, colds, and bowel complaints. The water there was not good. Over 150 acres of timber on the Greenbank Road had been cut down for cabins. Major Ross wanted the cabins finished quickly as "the wind was whistling through the tents like greased lightning through a crab orchard." Water had frozen in buckets left outside. Delays had been caused by the regiment's wagons being used to haul logs for units stationed at Greenbrier River which were scheduled to winter at "Camp Tip Top." Ross was also kept busy keeping the roads to Monterey open, as they were becoming impassable.

The only military activities were scouting parties to keep the Union troops from surprising the Confederates and from driving off horses and cattle. Corporal John A. Wilkinson wrote: "We are now...about 17 miles from the top of Cheat (Cheat Mountain) where the yankees are. We can see their tents...."

Wilkinson commented on camp life:

I drew an overcoat the other day and by covering with a blanket and 2 overcoats I can sleep pretty comfortable. We get plenty of beef, sugar, coffee, rice, salt, we don't get quite enough flour but we can make out. I have not got much money but I don't need much out here there is nothing to buy only pumpkins....This is a disagreeable place to stay but if the yankees can stand it on ahead we can stand it here....There are a great many sick here caused by standing guard duty in bad weather. We have built a hospital and it is full besides a great many that have gone home....It has been raining and snowing about every other day for about 3 weeks which goes very hard about cooking. Our tent is good and it leaks but very little. A great many of us have fireplace in our tents and by building on a big fire it makes it very comfortable. We can make out to bake our bread by the fire but there is no room for anything else....We have to work every day at our cabins Sunday not excepted. We have got them up and are now building chimneys and putting on roofs and daubing them. Each company is to have six cabins and about 12 men to each one....

Lieutenant Colonel Harman and his wife visited the 52nd to see their son Lewis who had been ill. Mrs. Baldwin came to visit the Colonel who had been sick with typhoid fever for some time. The effects of this illness so debilitated Baldwin he was forced to go home to recuperate. This led to his being assigned to duty in Staunton for the rest of the winter. The command of the regiment fell entirely on Major Ross.

In November Colonel Baldwin announced his candidacy for the Confederate Congress.

The troops stationed at Greenbrier River soon pulled back to Camp Alleghany and Colonel Edward Johnson of the 12th Georgia Infantry took command of all the units posted there. Under his direction additional trenches were dug around the camp facing Greenbrier River and the most likely avenues of approach by Union forces.

On December 7 Johnson reported to General Lee that the enemy troops on Cheat Mountain had built winter quarters and a strongly entrenched camp. They were also actively scouting the old Confederate positions on Greenbrier River and near Green Bank, a small hamlet nearby. Johnson believed his position on top of Alleghany Mt. should be held; otherwise the Federals would advance further toward Staunton and occupy his present strong position.

Major Ross, with two companies of the 52nd, was sent on December 11 to ambush a reported Union advance on the Parkersburg Turnpike, near the Greenbrier River. Led by a local civilian, the small force lay in ambush at the narrows of Greenbrier River all day. When night fell they withdrew into a recess in the mountains near the ambush site, and spent a cold night without fires. Early the next morning Ross marched them back to the narrows and concealed them in a pine thicket. Adam Kersh described the action:

> Next morning they seen about 17 yankees coming down the river. Our men ambushed until they come within shot of them. Fired on them. Killed about seven of them. The rest run. One run behind a tree and commenced fireing a pistol at our men. He was commanded to surrender by Major Ross. He would not. Major ordered our men to charge on him. He undertook to run when Lieut. Woodel (William H. Woddell of Company D) fired and killed him. No sooner had he fired when he spied about one thousand yankees coming double quick after them. Our men had to take up the mountain through the bushes. The yankees fired after our men but hurt no one. Our men all reached camp safe the evening of the 13th.

December 13 was to prove a busy day for the men of the 52nd. About 4 a.m. the camp was awakened by the pickets of the regiment firing on the Union advance down the pike from their position. Sergeant Major Thomas D. Ranson had difficulty rousing drummer boy Kene Morris to beat the long roll without the benefit of his "toilet." The men were rapidly formed in the cold darkness and ordered into the trenches. The fighting raged on either flank but the 52nd was not actively engaged. Only companies A and F, on the extreme left flank, were able to get a shot at the enemy, who were defeated after a six hour fight. Led by Union sympathizers from the area, the Federal troops had avoided the Confederate entrenchments and attacked on the left flank and the right rear, only to be repulsed by other Rebel units. There were no casualties among the 52nd in the ditches, but the pickets had suffered three wounded and seven captured. Their actions had saved the entire command from a surprise attack. Privates William D. Baskins and Henry D. Welch, both of Company H, were mortally wounded. Both died on the same day, becoming the first mortalities from combat in the regiment.

John A. Wilkinson wrote his father concerning the action:

> We had a powerful battle last Friday here. I was on picket when they came on us on the outside post. . . . It was about 3 o'clock Thursday night. I heard them when they were half a mile off. When they came up I fired on them & killed two, then all the picket fell back within 2 or 3 hundred yards of the camp where we were reinforced by part of the 12th Georgia regiment. We lay in the brush waiting for them to come up the road, but they didn't come that way. They came around us & made the attack about daylight. There were 4,000 yankees against about 1300 of our men. The fight lasted seven hours. Our men charged on them and run them. We

have buried 17 yankees and found five more yesterday & got 7 prisoners. I have not heard how many of our men were lost. The yankees were carrying off their killed & wounded all the time the fight was going on. The yankees got my blanket and canteen & haversack that I had to leave where I was standing picket. They came up so close on us that I didn't have time to get them....

Ross, who commanded the regiment during the assault, wrote home that he was not surprised that the 52nd had not received much mention in the newspapers.

Owing to the position of the trenches they did very capital service, and were in an account I saw, as was also the whole Regiment in another account. The 52nd, however, I am glad to say, does not fight for newspaper reputation. It is sufficient for us to know that we did our duty, and hold ourselves 'always ready' to do that, and even more than that. When the time comes, and we are called upon, I feel sure from what I saw today, that the boys from Rockbridge & Augusta will do honor to the counties which they represent. At all events, we desire devoutly an opportunity to show our metal.

Some of the 52nd helped bury the enemy dead the next day. Robson found greenbacks on the corpses, the first he had ever seen.

The victory at Alleghany Mountain encouraged the Confederate high command to hold the position there, and reinforcements and additional supplies were sent to Johnson. He earned the sobriquet "Alleghany" to distinguish him from the other officers of high rank with that surname and also received a promotion to Brigadier General. "Alleghany" and his troops received the thanks of the Confederate Congress for their feat.

A company from the 52nd and one from the 12th Georgia were sent out on December 21 to capture the traitors who had acted as guides for the Union forces. They were spotted across Greenbrier River but immediately fled to the mountains upon seeing the Confederates.

Kersh wrote his brother on Christmas day that Colonel Johnson was taking no chances on being surprised again:

We have been trimming & burning brush ever since the fight. We had no fortifications for the cannon the day of the fight. We have five now. We have been blockading some roads also and extending our trenches around further. We will have a better chance at the yanks if they come now. They had a better chance than we the day of the battle. They could lay behind the brush and logs concealed and when we raise up to look for them they would fire at us....The health of our regiment is improving....We have a dry Christmas here. The boys had to go to work today as well as any other day. We work here on Sundays too....

Wilkinson wrote his sister on Christmas day:

As this is Christmas day I thought I could not celebrate it in a better way than by writing a letter. . . .We have to work every day on the fortifications but I suppose we will get done in a short time. An alarm was given day before yesterday morning that the yankees were coming. So we went to the brestwork and waited two or three hours but it turned out to be a false alarm. We are looking for another attack most every day. I received everything you sent me by Entsminger but the whiskey there was not more than a quart in the jug. I dont know what became of it. . . .

1862

The regiment remained camped on top of Alleghany Mountain the rest of the winter, with little to disturb the monotony of camp life. Captain Hottell wrote his son describing the conditions:

> Our cabins are very comfortable. We have two rows of bunks on each side of the room...one row above the other...are just as wide as our house at home and nearly as long...18 feet. Thousands of families are raised in worse houses....The men are all busy, some fortifying, some chopping down timber, some building cabins....Our bakery will start this week & then we will have good bread.

Private Gerald E. Crist agreed: "We have good times here now. All are in good spirits, have plenty to eat and good cabins to stay in and not much to do...4 on guard per day....A great quantity of whiskey is being used here ...about 25 men in the guard house. Have been running the 'Blockade' as we call it, going home without permission."

Lieutenant Coiner recalled: "we had a Glee Club in our Regt. vocal and instrumental and when in winter quarters and on the march the Club was a pleasure for the members and to others." "Have men playing the fiddle all the time and the men have great sport dancing after him through the camp," wrote Private Patrick Loyd. Later he added: "Marched 20 miles in 7 hours without food." He also complained: "the Captain has not been with the company 2 months since entering service....rained 5 days and snowed 4....52nd had snowball fight with (12th) Ga Rgt. 3,000 men engaged...but our Regiment locked them out and drove them back home....have spelling schools 3 times a week and prayer meeting 3 times a week which affords us great pleasure." Another commented: "We have a regular post office here now-moved up from 'Traveler's Repose' (Greenbrier River).

Kersh wrote in January: "We had snow here on the 24th, about 4 inches deep. It stormed and blowed like the Dickens here. It blowed through our clapboard roofs & sifted it over us some....The roads were alfull here then but in the evening turned cold and the ground froze hard....Since commenced this letter it has commenced to rain. You may judge from this how uncertain the weather in these old mountains." He added on February 4:...."We had snow on the 2nd about 5 inches deep. It drifted some. The snow being very fine and the wind blowing kept our cabins pretty wet inside....We fare pretty well here much better than I expected. We get plenty to eat....The boys are lively and kicking around. All our company are well except two. Soldiers do not have any objection to a box of provisions as they sometimes get tired of Bull Beef...."

Private Stover looked back and recalled: "This camp proved to be very unhealthy — a great many dying of fever. Co. F lost 3 men here. . . . The Co. as well as the whole Army suffered greatly on account of the inclimency of the weather, and excessive picket duty." Stover was correct. More than 50 men in the regiment died of disease during the winter of 1861-62. Many men were absent sick for long periods, and the health of others was so impaired they had to be discharged.

Private Marion Koiner remembered: "We had comfortable log cabins and good rations and fuel."

The 52nd remained at "Camp Baldwin" until April 1, 1862. On that date the fortified camp was burned, and Johnson marched his "Army of the North-west" to Shenandoah Mountain. This move was prompted by the defeat of "Stonewall" Jackson at Kernstown on March 23 and his subsequent retirement up the Valley. General Lee had suggested Shenandoah Mountain as a possible defensive position for Johnson to protect Staunton and yet be within supporting distance of Jackson.

Major Ross, who had left the regiment to return to VMI for a short period, as well as to marry, rejoined the regiment there and reported on conditions:

> I am sitting on a rail pile, before the fire with smoke nearly blinding me, writing on my knee. The sick and wounded are ordered to Buffalo Gap so I think some portion . . .of the Command will be left at Shenandoah. I hope our Regiment will follow Jackson, as it is better to be marching around . . .than to be seated on top of Shenandoah . . .you have to sleep on the side of the mountain so steep that you have to stick your spurs in the ground to keep from sliding down.

Ross would soon get his wish.

The company from Bath County joined the regiment at Shenandoah Mountain as Co. K, giving it a full complement of ten companies for the first time since September.

Johnson soon pulled his army back through Buffalo Gap and camped at Valley Mills and West View. Private "Doc" Yount of Company A wrote his sister describing the move:

> I tell you the retreat from the Shenandoah was the hardest time I ever saw in my life. I fell down in the mud two or three times. We left the mountain about 3 o'clock PM and marched. 12 in the night when we stopped. I went to an old stable and got a plank and laid down on it & slept very well til morning & then took up the line of march to Staunton but did not get there. . . .It is the talk in camp that we are to cross the Blue Ridge but I hope that day may never come that we will go across the mountains and leave this Beautiful Valley behind. . . .

During this period many volunteers and conscripts joined the regiment, replacing losses caused by wounds and disease. On May 1 the 52nd re-enlisted and reorganized for the war. Colonel Baldwin, who had rejoined and was commanding a brigade of two regiments, declined to stand for reelection, having been elected to the Confederate Congress and being in poor health. Lieutenant Colonel Harman was elected Colonel, Captain James H. Skinner of Company A was elected Lieutenant Colonel, and Ross was reelected Major. There were many changes among the company officers; younger and more active men replaced older ones in most cases.

The morning of May 7 dawned with the news that the regiment was to march toward the enemy. Jackson's army had arrived at Staunton and he directed Johnson to lead the advance against General Milroy's Union forces, which had crossed Shenandoah Mountain. With the 52nd in the lead and Lieutenant "Bent" Coiner commanding the advance guard, Federal cavalry were encountered and driven from the junction of the Harrisonburg-Warm Springs and Staunton-Parkersburg turnpikes and over Shenandoah Mountain. The Union troops were forced to abandon their baggage at Rogers' at the eastern base of the mountain. Coiner continued his advance, driving the Yankees off the mountain and did not stop until the western base was reached. The exhausted 52nd camped at Shaw's Fork for the night.

Jackson ordered the advance to continue the next day, and Johnson's army, preceded by Coiner and his company, drove the enemy cavalry until the top of Bull Pasture Mountain was reached. Jackson and Johnson, with members of their staffs, and accompanied by Coiner and 30 of his men, moved to Sitlington's Hill. This was a large spur of the Bull Pasture, on the left of the turnpike, overlooking the village of McDowell and the surrounding valley. There they found Milroy's army camped around the town. Two enemy regiments were sighted on a hill to their right front, across the turnpike, but out of range. A battery supported by infantry was spotted about a mile to the front. Milroy's men, sighting the Confederates, immediately sent forward skirmishers which Coiner and his men drove back.

Jackson ordered Johnson to move his troops to Sitlington's Hill. Colonel Harman led the way as the rest of the 52nd climbed the steep and rugged incline. Moving to the top of the hill, the regiment was quickly positioned by Johnson on the extreme left as skirmishers. The men, loading and firing rapidly, drove the advancing Union skirmishers down the hill in their front. During the action Colonel Harman was wounded by a minie ball in the right arm, but refused to leave the field. General Johnson was also wounded. Lieutenant Colonel Skinner commanded the left wing of the regiment and Ross the right, as they repelled the enemy from their front. More Union troops advanced and though hotly engaged, Johnson ordered Harman to move the 52nd around the left flank of the foe. The Federals were driven back, but the fighting continued to rage until 8:30 p.m., when darkness fell. The battle had commenced at 4:30 p.m.

Private Robson recounted the details of the battle:

When the enemy advanced to the attack we received the full assult of their first line and repulsed it, thus giving time for the arrival of the other regiments. The enemy, after being driven back, opened on us with their artillery a rapid and incessant fire of case shot and shell but 'us boys' laid low among the rocks and trees, which afored us ample protection, and also the angle of the elevation of their guns being so great, no damage, except to the timber, resulted from the cannonade, and the noise was all on the Yankees' side, we having no artillery in position. About 5 o'clock General Milroy . . .advanced his whole force of 8,000 men, and the battle roared and raged along the side of the hill with terrific force for a long time, but our two little brigades held them back until Jackson got his flank movement worked out, and then the Federals gave way, as a matter of course. In the final closing up of the business, just as Taliferro's Brigade reached the field, the 52nd, backed by the 10th Virginia, made a charge which drove them headlong down the hill, and the battle ended at 8 o'clock p.m. It seemed to me we had been at it about a week, but the other boys spoke as though it was a very short half day. . . .We found 103 dead on the mountain side the next morning; and during the night Milroy set the woods on fire behind him, and retreated toward Franklin, whither General Jackson followed the next day. Many wounded in Company D.

The Confederates suffered many head and shoulder wounds because they silhouetted themselves against the skyline in order to shoot down on the Federals below.

Harman wrote of the 52nd: "My officers and men behaved with great courage, but it is not necessary to particularize them by name, as they fought under your (Johnson's) eye." The greatest loss to the regiment was the death of Captain William Long of Company B. John A. Carson, a new Lieutenant in Company D, was also killed. Captains Dabney of Company C, Watkins of Company E and John M. Humphreys of Company I were the officers wounded. The casualty list for the regiment was heavy:

MCDOWELL May 8, 1862

	Co A	Co B	Co C	Co D	Co E	Co F	Co G	Co H	Co I	Co K	F&S	Total
KIA	0	3	1	3	3	2	0	0	0	2	0	14
WIA	1	18	5	4	7	3	5	0	11	3	2	59

KIA figure include eight who died of wounds.

Private G. Marion K. Coiner of Company B was extremely fortunate. He "would have been killed instantly, but for his diary and bible which he had in his coat pocket; these arrested the ball which struck through the diary and penetrated to the middle of the bible."

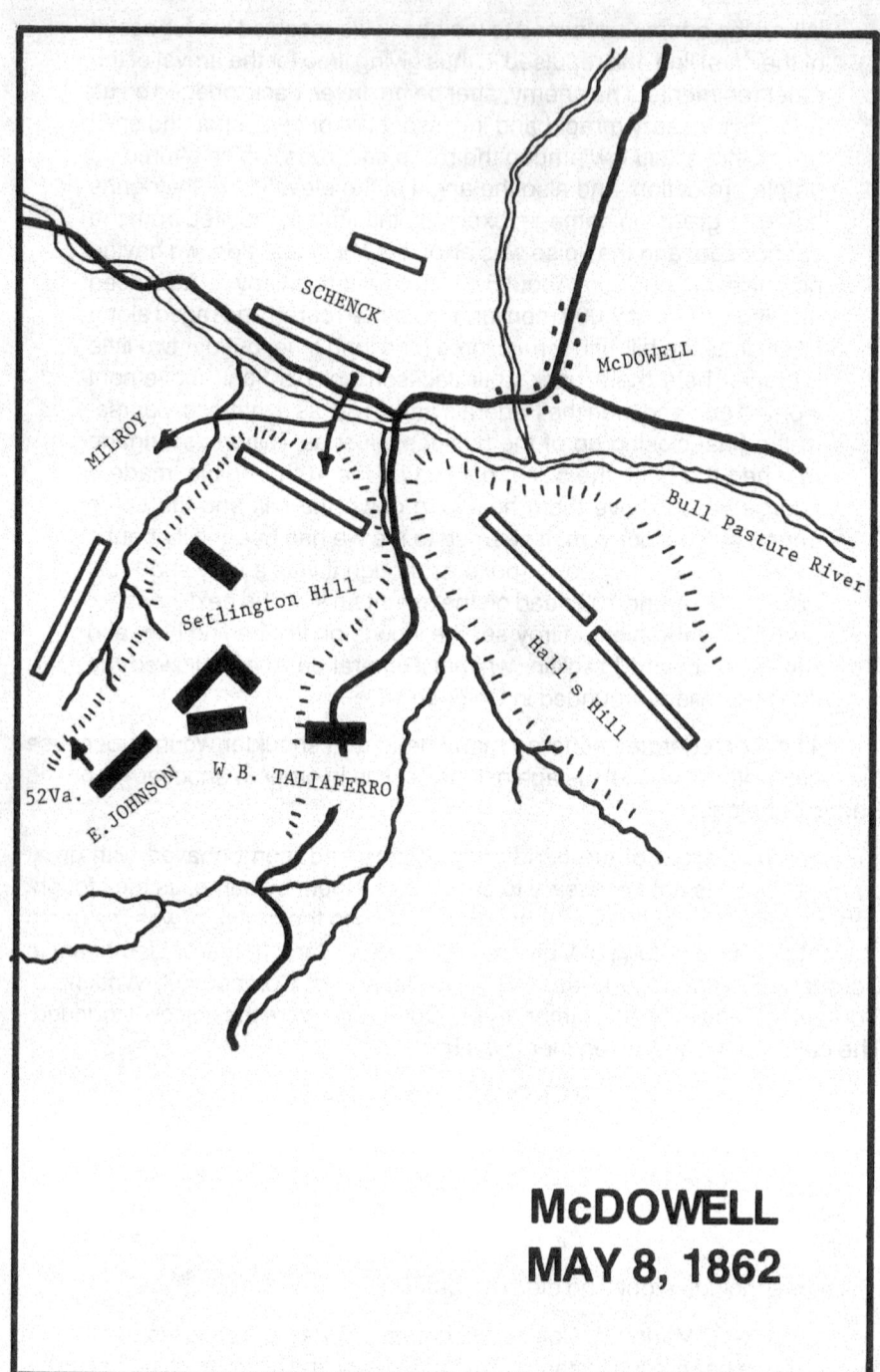

Jedediah Hotchkiss of Jackson's staff, an eyewitness, wrote in his diary: "Colonel Michael G. Harman moved his regiment...gallantly up to the fight."

Ross was singled out by Colonel Scott of the 44th Virginia, commanding the Second Brigade in Johnson's Army: "I derived considerable assistance from Major Ross...who acted with great gallantry." Ross was reported in Lexington as having been killed, and he was much exercised by the event: "It must have been carried by some miscreant rascal who deserted the battlefield, too much scared to know anything, and too much knave to tell the truth if he knew it." He wrote with high regards for the Union soldiers:

> Their discipline is immensely superior to ours. I watch them well during the fight, or rather at the beginning of it and I never saw Cadets at drill march with greater precision and more regularity than did the Yankee skirmishers under our fire. Not a man shrank from his position but they all marched all alike, true soldiers to the attack. And they fight well, too, as the mortality on our side proves. Southern pluck, however, was too much for them, and they had finally to give up. If they had not attacked us that evening, but had waited 'til next morning, we would have killed or captured the whole army. Captain Miller (2nd Rockbridge Artillery) had orders to go by a road which came in two miles in rear of McDowell so as to operate upon their retreating columns, but the smart rascals took time by the forelock, pitched into us in the evening, and that night pulled up stakes and cut for home.

Milroy, who had been reinforced by General Schenck's forces, retreated to Franklin after destroying his ammunition and supplies. Jackson ordered an immediate pursuit the next morning, and Confederate cavalry captured a few Federal stragglers and supplies. The Union forces effectively used the mountainous terrain and defended each hilltop in their rear. The Southerners, channeled into the narrow road through the steep mountains, were unable to bring Milroy's Army to bay.

Ross wrote his new bride concerning the pursuit on May 13 from "Camp Simmons," 10 miles from Franklin:

> We left our position in front of Franklin yesterday about dawn and fell back eight miles, in the hope that the enemy would be tempted from his strong position and follow us — thus allowing us an opportunity of falling upon him at any point. Gen. Jackson pleased. We are still here and I am uninformed whether Gen. Milroy has the livelihood to pursue. Cannon have been firing at intervals all the morning but I can't tell whether from the same position as yesterday or not. Am inclined to think from the report of the Artillery, they have not come any closer. Report said that Milroy had been reinforced by Fremont with 4,000 men, and I suppose it is true. That, however, will not give them advantage over us, if any, and we only await an occassion when we can fall upon them with anything like equal

ground to make our front attack.

The next day he added from "Camp at foot of Shennandoah":

I have just gotten into camp and am wet and cold as a rat....We had a very unpleasant march this morning owing to the heavy fall of rain as we have to sleep on the ground without tents, expect to have a right damp snooze tonight. Last night I slept under a wagon with Adj. Lewis and Dr. Lewis, and kept dry whilst both of them got wet as could be. We had beautiful weather during our pursuit of the enemy, and in any other country could have come up with him and given him his death stroke, but every night he would post himself in one of the thousand and one narrow passes, when one man is equal to three or four, and there defy us. Our advance guard was constantly skirmishing with their rear and we took twenty odd prisoners, but could not bring them to any general engagement...by taking a position upon Shenandoah we can rub off nearly any number....I rather think...Johnson's forces will be left here and Jackson will go down the Valley to help Ewell in his operations against Banks. We ought to be able to clear the Valley entirely....

Ross proved to be an excellent prophet concerning Banks' army.

Jackson moved his army from Shenandoah Mountain and camped for several days near Lebanon White Sulphur Springs, giving the troops a well earned rest.

The Confederates continued on toward Banks' army on May 17. The Southern forces marched through Harrisonburg and down the Valley Turnpike to New Market, near where General Ewell's command linked up. Jackson changed his line of march and moved through Luray toward Front Royal, hoping to flank Banks out of his strong position at Strasburg.

On May 23 the Confederates attacked the Union force near Front Royal, routing them, and capturing more than 700 men, several pieces of artillery, and large amounts of food and ammunition. The 52nd was not engaged in the battle.

Jackson continued his advance the next day, sending his troops in a two pronged movement toward Banks' exposed flank at Newtown and Winchester. The Rebels only succeeded in attacking and destroying a portion of the Federal troops. The 52nd, in Colonel W. C. Scott's Brigade, advanced to assault the Federals around Winchester, when the Yankees retreated. Lieutenant Coiner believed the regiment was the first unit to enter the town. Robson recalled how the people lined the streets giving out good things to eat. Lieutenant John D. Summers of Company I noted in his diary: "routed them ran them through the streets." The regiment had one man wounded during the day.

Private James M. McClintic wrote his wife:

The Yankees were completely routed and they left in double quick, passed through Winchester with our boys at their heel, firing on them at every turn. We made a great many prisoners, at least 5,000, how many were killed and wounded I cannot say but I think it was heavy. Our loss was trifling but I cannot say how many. They succeeded in burning the quartermaster and commissary stores but we saved a great deal of army supplies, they having no time to destroy them.

Much was saved out of the burning houses by the citizens and if they could have had more help they could have saved much more, as it was our victory was complete and our spoils were great. In arms and ammunition we captured many hundreds of thousands worth (perhaps a million dollars worth) of hay, flour, bacon and great quantities of stores for the army.

Our infantry followed at double quick for 6 miles and the Yankees strewed the road with knapsacks, canteens, overcoats, everything but their guns and now and then a gun was cast aside. Every here and there an officer's wagon loaded with baggage, which seemed to be the only wagons they had left at the end of 6 miles, was captured. We captured a train of cars from Baltimore together with 700 soldiers coming in to reinforce Banks. The infantry halted here and gave over the pursuit and are here yet. The cavalry kept up the pursuit until night making a great many prisoners which passed by every hour in squads of 10 to 90 in number till after dark.

McClintic added a footnote the next day:

Since I began to write there has passed toward Winchester another list of prisoners, 100 in number. . . .Some were officers. . . .

We observed yesterday as a thanksgiving day. We had an excellent sermon by the Rev. McElwain, our chaplain. He preached Sunday evening, he held services after the battle yesterday and is to preach again this evening so you see we do not lack for preaching. I think it is not without its effect. There is more morality in our regiment than is usual in camp. There is still plenty of vice yet I seldom hear profanity which they say is common in the army, but of course there is always room to mend.

On May 28th, the 52nd, now part of General Ewell's command, moved north to Charlestown in support of the Stonewall Brigade's operations against Harper's Ferry.

Jackson, learning that General Shields was advancing from the east toward Front Royal and General Fremont from the west from the South Branch

Valley, rapidly redeployed his troops to Winchester. Meanwhile, Shields had attacked a small Confederate force left at Front Royal, resulting in the loss of the town and some supplies. The 52nd was not engaged, but lost ten sick captured in the hospital there.

Early on the morning of May 31, Jackson moved his units south to Strasburg to avoid the Federal pincer movement. The next day Ewell's division engaged Fremont's advance and the 52nd had one man wounded and one man captured. Six more of the regiment's sick were captured and paroled in the hospital at Winchester.

Jackson continued his successful retrograde movement to Harrisonburg, arriving on June 5. Shields and Fremont followed rapidly. Captain Samuel B. Coyner of Ashby's cavalry, formerly a private in the 52nd, slowed Shields' advance by burning the bridge over the South Fork of the Shenandoah River at Conrad's Store.

General Ashby, commanding Jackson's cavalry, met Fremont's advance on the Port Republic Road near Harrisonburg on June 6. Jackson had moved the rest of the army toward Port Republic. General George H. Steuart was sent by Jackson to support Ashby and his men. Steuart led the 1st Maryland and the 52nd and 58th Virginia regiments of his new command into action to relieve the hard pressed Confederate cavalrymen. The 1st Maryland and the 58th Virginia attacked and drove off the Federal troops. Lieutenant Colonel Skinner and the 52nd, on the extreme right flank in dense woods, failed to receive the order to charge and were not engaged. General Ashby was killed during the engagement. Robson reported that the loss of Ashby "cast a gloom over the whole army." The Confederates continued to move toward Port Republic unmolested and rested quietly in camp on June 7.

Fremont advanced his army toward the Confederate position near Cross Keys on June 8. The 52nd was sent by Colonel Scott, commanding the brigade, to support the artillery batteries of Ewell's division on the left flank of the Rebel lines. Lying down under a hot cannonade, the regiment suffered few casualties.

Later in the day Skinner was ordered farther to the left and the 52nd was actively engaged to the end of the day. The stiff Southern resistance stopped Fremont's army in its tracks and prevented it from joining forces with Shields. Fremont retreated to Mount Jackson. Robson remembered that the regiment received four distinct charges from Union troops during the battle. The 52nd was located in the woods and fought from behind trees. The regiment was withdrawn at midnight and marched toward Port Republic before camping.

CROSS KEYS JUNE 8, 1862

	Co A	Co B	Co C	Co D	Co E	Co F	Co G	Co H	Co I	Co K	F&S	Total
KIA	0	1	0	0	0	0	1	1	0	2	0	5
WIA	1	5	0	2	6	1	16	1	4	3	3	42

Lieutenant Clinton M. King of Company B was killed. Major Ross, Adjutant Lewis and Assistant Surgeon Lewis of the staff were wounded, as were Lieutenants William Galt of Company A, Thomas D. Ranson of Company I and Samuel Paul of Company D, who lost a leg.

The following day the regiment was engaged in the bloody battle of Port Republic. The 52nd was again detached from the brigade and led by Skinner to the support of the Stonewall Brigade's assault on the Federal position. Winder's command was pinned down near the Lewis House. The regiment advanced into a storm of fire from the Union artillery and infantry posted in the woods, and both units were driven back with heavy casualties. Lieutenant Coiner ordered the men of his company behind the bank of a small stream at the foot of the hill and had only two men wounded. The other companies were not so fortunate.

General Richard Taylor's Louisiana Brigade, supported by the remainder of Scott's troops, soon hit the Federals' left flank and drove them from the hill after a sharp fight. This defeat forced Shields to withdraw from the Valley.

The 52nd incurred high losses in the aborted assault. The leadership of the regiment suffered heavily again. Captain Benjamin T. Walton of Company K was killed, and Lieutenants Samuel B. Brown and Lewis Harman of Company A, John N. Hanna of Company D and James A. White of Company H, were wounded.

PORT REPUBLIC JUNE 9, 1862

	Co A	Co B	Co C	Co D	Co E	Co F	Co G	Co H	Co I	Co K	F&S	Total
KIA	3	0	5	3	0	3	0	3	1	2	0	20
WIA	9	2	16	4	3	3	3	9	5	7	3	64

Jackson camped his army at Mount Meridan after the battle. The Confederate cavalry kept an eye on Fremont's forces at Harrisonburg. "Old Blue Light" moved his infantry near Weyer's Cave for a well deserved rest. Banks' army still lay in camp at Winchester.

During its stay near Weyer's Cave the 52nd was placed in the Fourth Brigade of the Valley District under General Arnold Elzey, a Marylander and a West Point graduate. The new brigade now consisted of the 12th Georgia and the 13th, 25th, 31st, 44th, 52nd and 58th Virginia regiments, all greatly reduced in numbers from battle and illness.

General Lee reinforced Jackson and requested "Stonewall" to move near Richmond to help destroy General McClellan's Union army, which was closing in on the city. Starting on June 18, Jackson marched his troops over the Blue Ridge through Jarman's Gap to the railroad at Mechum's River. Robson remembered the men "stuffing themselves with just ripened apples in Albermarle County." Jackson's quartermaster used the available railroad equipment to move the infantry to Frederick's Hall and Beaver Dam Station north-west of Richmond. The artillery, cavalry, and wagon train moved by road.

Lee wanted Jackson to attack McClellan's army north of the Chickahominy River. "Stonewall" marched his troops over the hot, narrow, dusty roads to Shady Grove Church on June 26. General J. E. B. Stuart's cavalry drove the opposing Union troopers from the banks of Totopotomoy Creek and crossed. Lee's plan had been for Jackson to roll up the Federal flank in coordination with General A. P. Hill's division, but "Old Blue Light" was unable to meet the planned timetable. Hill's troops had assaulted alone and driven the Yankees across Beaver Dam Creek.

Jackson's men, marching through unfamiliar terrain in the parching heat and dust were finally put into action on the evening of the 27th. Ewell ordered Elzey and Taylor to take their brigades and support Hill's men, stalled on Boatswain's Swamp near Gaines' Mill by Federal resistance from the far side. Passing over Hill's troops, the two units splashed through the swamp and up the far slope. They were met by a hurricane of artillery and musket fire and, after suffering numerous casualties, the attack bogged down. Finally at about 7 p.m., additional troops were thrown into battle by Jackson. The Southerners rushed forward yelling and screaming, and despite stubborn resistance overwhelmed the Union forces.

The 52nd under Skinner joined in the assault with the Stonewall Brigade, through the swamp and up the hill near the McGhee house. The men of Augusta, Bath and Rockbridge again proved their mettle, capturing an enemy battery, including a rapid-fire cannon. Lieutenant Coiner's attention was called to it before the charge: "...lying on the ground...seeing the little shells exploding above us it was then getting dark. I only saw it as I passed by it stopping a fraction of a minute." The gallant Confederates pursued the Federals about 300 yards beyond the McGhee house. General Winder stopped them because of darkness; it was about 9 p.m. The exhausted men sank down in relief and slept undisturbed during the night. Captain Watkins of Company E captured Major Henry B. Clitz of the 12th U.S. Regulars. He took him to Jackson, who recognized him as a former acquaintance of West Point days.

During the night, while searching for a spring, Watkins came upon nine Union soldiers looking for their regiment. He tricked them into following him into the Confederate lines where they were captured, much to their surprise.

The 52nd had, as usual, paid a heavy price for its success.

GAINES' MILL JUNE 27, 1862

	Co A	Co B	Co C	Co D	Co E	Co F	Co G	Co H	Co I	Co K	F&S	Total
KIA	3	0	0	0	0	1	1	0	1	0	0	6
WIA	0	2	2	1	15	3	1	2	1	5	0	32

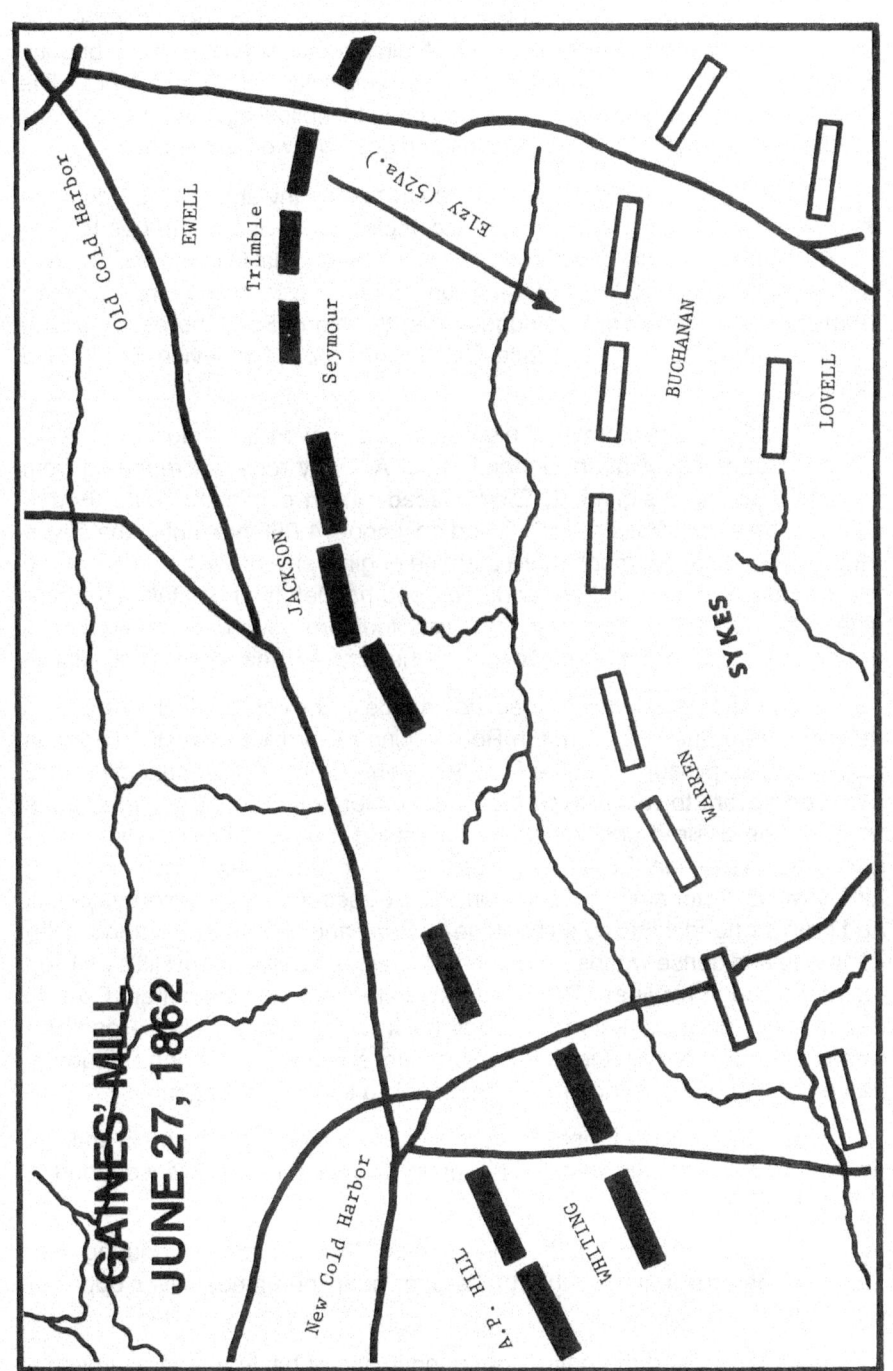

Captain Watkins and Lieutenant William V. Knick of Company E were the only officers wounded. General Elzey had been wounded late in the day and Colonel James A. Walker of the 13th Virginia took command of the brigade during the night. He camped three of his regiments at Beulah Church. The other three, including the 52nd, rejoined them the following day. The 28th was spent drawing ammunition and rations and getting a well-earned rest.

Ewell advanced his division to the Chickahominy at Bottom's Bridge on the 29th, but finding the bridge still incomplete, returned the troops to their former camps. Captain Claiborne R. Mason, the original Captain of Company H, now an Engineer Officer on Jackson's staff, arrived with a force of negro laborers and completed the bridge during the night. Early the next morning Ewell marched his men to White Oak Swamp near Grapevine Bridge and camped.

On July 1, Ewell crossed his troops and moved them down the Willis Church Road. About noon General Jubal A. Early rode up, reporting from General Lee for assignment. Ewell placed him in command of the Fourth Brigade, relieving Walker. Early found only about 1,050 present in the seven regiments, a total not much more than the original size of each of them. Hard fought battles, long marches, and poor food had left the hospitals and roadsides filled with the wounded, sick and footsore. During all these rapid marches more than 60 men had deserted the 52nd and few recruits had joined.

In the afternoon Early formed the brigade in line of battle in the woods on the left side of the Willis Church Road facing toward Malvern Hill. The men rested under sporadic shell fire from the enemy artillery. Near sundown Early received orders to move to the support of General D. H. Hill. The brigade reached the desired position after undergoing a heavy shelling from Union gunboats. Lieutenant Coiner had a shell explode almost in his face, knocking him down, and stunning him for a moment. He escaped with only his eyelashes and brows singed. Little could be done by Early and his small command in the darkness and dense woods, with enemy fire and stragglers from Hill's Division running through his lines. The 52nd had gotten lost, and Lieutenant Colonel Skinner led the fragments of the unit back to the Willis Church Road and remained until morning. Lieutenant Summers summed up the fight: "fought a battle under the command of Gen. Early — in a tight place, got out."

The regiment was more fortunate in the battle of Malvern Hill than at Gaines' Mill, losing only one man killed in Company H and two wounded in Company C.

On July 2 Early rested the brigade near Willis Church in a pouring rain. The next day he marched them to a spot near the James River opposite Westover

Jackson moved the entire command in line of battle to near Harrison's Landing, but McClellan's position was too strong and only a little skirmishing occurred. The 52nd and the rest of the troops remained camped there until

July 10, when they moved to within four miles of Richmond. The regiment reported 752 present and absent, with an aggregate of 797 men on that date.

On July 13 the 52nd, with the rest of Jackson's Corps, marched to Richmond and the next morning took cars for Gordonsville. This move was caused by the arrival of General John Pope's Union army near Fredericksburg. The 52nd and the rest of Early's Brigade moved by easy marches to Liberty Mills. While camped there, Commissary Sergeant "Tom" McClure wrote Skinner concerning the meat rations being issued the regiment:

> The proportion of fresh beef issued to the Regt (and — I believe to each of the Regts of this brigade) is one pound to each man per diem including neck & shankbones. The shankbone averages, I am told not far from 8 pds. to each quarter of beef; thus occasioning a dead loss of 8 rations to each company or mess receiving a quarter for its portion. This particular bone cannot from the nature of the case be itself distributed among individuals, hence to some asking for meat I am compelled to give only a worthless bone. (The soldiers are impertenate as Shylocks in demanding their pound of flesh!)
>
> If the supplemental rations — rice, sugar, coffee, flour — were regularly issued, the loss by neck & shank bone might, for the once, occasion no special complaint from a mess — but it becomes a matter of consequence when these supplemental items are so rarely recd. The men are undergoing meanwhile the active physical exercise like that of the last two months. It was about two months ago that the prescribed legal ration was reduced, by some authority, to 1 pd. of beef(20 pr. ct.less) and ½ pd. of bacon (33⅓pr. ct. less) and 18 ounces(from 20 oz.)of flour to the present quantum.

The last part of this letter is missing, as well as any conclusion about what Skinner did to change the policy of issuing bones.

The brigade moved near Orange Court House during early August. On August 8, Early's troops splashed across Barnett's Ford on the Rapidan, leading Jackson's army toward Culpeper Court House. Confederate cavalry screened the advance the next morning and pushed the opposing Federal troopers beyond Major's School House, near Cedar Run. The Rebel riders came to a halt in the face of Union artillery fire and increased resistance by their counterparts on the other side.

Early, after a quick reconnaissance, advanced his brigade into a meadow on the left side of the road. With the 13th Virginia acting as skirmishers, the brigade, now 1,500 strong, moved obliquely across the road and on the flank of the Federal cavalry. The regiments advanced to the crest of a long ridge and were met by heavy artillery fire. Sending for reinforcements and artillery, Early withdrew his troops behind the crest of the hill. When the Rebel batteries arrived and unlimbered, Union skirmishers advanced to capture them. Early spotted this, and after a volley sent his troops charging and yelling to drive them back. After this success, the tired Confederates rested and watched the artillery duel. General Banks, commanding the Union Corps, sent troops to assail the Confederate left flank. Attacking through the woods, the Yankees succeeded in breaking the leftward Southern brigades and then in getting behind Early's men, where they poured a destructive fire into them. Several regiments broke and ran, but the 52nd and the 12th Georgia and part of the 58th Virginia stood their ground and prevented the Federals from rolling up the Confederate flank and capturing the batteries.

The Stonewall Brigade and the 13th Virginia assailed the breakthrough, but the situation was desperate until Branch's North Carolina Brigade assaulted the Union flank and drove them off the field. All the Southern units now advanced but pursuit was called off because of darkness. The day had been oppressively hot, and the weary troops fell down by the roadside and slept. Lieutenant Summers recapped the battle: "fight three hours whip them very badly-follow in the night."

The 52nd had done its usual hard fighting and Lieutenant Colonel Skinner was singled out by Early for his gallant conduct. Six companies had been absent picketing roads but the regiment still had significant casualties.

CEDAR RUN AUGUST 9, 1862

	Co A	Co B	Co C	Co D	Co E	Co F	Co G	Co H	Co I	Co K	F&S	Total
KIA	0	0	0	1	1	0	0	0	0	2	0	4
WIA	0	0	2	0	5	2	0	0	1	2	0	12

Lieutenant William V. Knick of Company E received his second wound of the war.

August 10 was spent burying the dead of both armies, sending the wounded to the rear, and collecting arms and the spoils of war. Early's men picked up six wagon loads of rifles.

Jackson soon advanced to the Rappahannock River. Finding the crossings undefended, he threw Early's Brigade across near Warrenton Springs on August 22. The 52nd waded the river on a mill dam and advanced into the woods beyond. A terrible rain storm came up and Early's command spent two days in the damp forest without rations, trapped on the north side of the swollen stream. Union troops probed the position but never attacked and Early was finally able to recross the river on a temporary bridge.

Lee next directed Jackson to try to get to the rear of Pope's army, believing Pope would be forced to retreat. Jackson soon had his men on the march to Manassas Junction, Pope's supply depot. Covering 54 miles in two days, through the heat of an August sun and clouds of dust, the Confederates successfully captured Manassas Junction.

Early's Brigade acted as rear guard for Jackson's Corps and had a brief skirmish at Broad Run near Bristoe Station on August 27, resulting in two men wounded in the 52nd. Arriving last at the junction, Early's men were only able to get hard bread and salt meat from the abundant supplies captured.

On the night of August 27-28 Early was ordered to march to near Groveton and form his line of battle in front of a railroad cut. Here they were exposed to a heavy fire of canister and shrapnel from the Union artillery during the fight on the evening of August 28. Lieutenant Colonel Skinner was hit in the head by a shell fragment and was lost to the regiment. Thomas H. Watkins, the youthful Captain of Company E, took command. The same shell killed and wounded 18 men, mostly in Company C. Seven men were killed outright. Lieutenant John M. Lambert of Company I was among the wounded. The 52nd lay in this exposed position during the night.

The next morning Early moved his brigade back into a railroad cut. The 52nd lay behind the embankment with the fighting going on around them until about 3:30 p.m. The Federal troops gained a foothold in the woods and railroad cut in from of A. P. Hill's division and Hill's men were almost out of ammunition. Early led the brigade in a charge to seal off the Union penetration and secure the Southern lines. Lieutenant Coiner described the assault: "we charged a railroad cut where two other brigades had failed to take. We killed wounded and captured nearly all of them. Our loss was considerable." Private Robson added:

> the boys in Blue...attacked our position...only to be driven back with slaughter...General Early found they had gotten...in the railroad cut in his front. We promptly attacked them, drove them out. As we pressed across the railroad bank, where lay numbers of dead and wounded Federals, I inadvertently stepped on the foot of a wounded man, which brought a groan of pain, and I asked his pardon for the accident. After our line halted — which was in a short distance I returned to the poor fellow, gave him water, and asked if I could do anything for him. He was very grateful, but thought nothing could be done then; however, I asked my Captain Airhart, a noble-souled christian gentleman, to assist me, and we moved the man to a more comfortable position under a tree, where Captain Airhart, who had considerable knowledge of surgery, dressed his wounds, and I did what I could to make him comfortable, and, after exchanging slips of paper with our names written on them, I rejoined my Company, and in the busy scenes then and afterwards being enacted, almost forgot the incident.

Robson later met the Union soldier he had aided, in 1885. This ended the fighting for the brigade for the day.

Three regiments of the brigade, including the 52nd, were engaged in the Union repulse on August 29, but suffered no casualties.

The 52nd paid its usual high price for its success. Captain Edward V.

Garber of Company A was killed and Captains John D. Lilley of Company H and John M. Humphreys of Company I were wounded.

2nd MANASSAS 28-29 AUGUST 1862

	Co A	Co B	Co C	Co D	Co E	Co F	Co G	Co H	Co I	Co K	F&S	Total
KIA	3	1	4	1	0	0	1	2	1	0	0	13
WIA	8	2	10	5	2	2	1	4	7	1	1	43

Lee sent his victorious columns in pursuit of Pope on August 31. The 52nd was not actively engaged in the confusing battle in the rainstorm at Chantilly the next day. Company H had one man killed and Company I lost one man wounded.

Early paid high tribute to his troops for their recent exploits: "My own brigade, which has never been broken or compelled to fall back or left one of its dead to be buried by the enemy, but has invariably driven the enemy when opposed to him and slept upon the ground on which it has fought in every action. . . ."

During an earlier reorganization the 49th Virginia, under ex-governor William "Extra Billy" Smith had been assigned to the brigade. The redoubtable 12th Georgia, which had fought side by side with the 52nd since Alleghany Mountain, was transferred to another brigade. Early now commanded the 13th, 25th, 31st, 44th, 49th, 52nd and 58th Virginia regiments.

The Confederate army marched north, forded the Potomac, and moved to Frederick, Maryland. There, Jackson was directed to attack the Federal garrison at Harper's Ferry, thus opening the Shenandoah Valley as a supply line to Lee's army. Subsisting his men on green corn and apples, Jackson led his corps back across the Potomac and lay seige to the town. After compelling the trapped Union forces to surrender on September 15 without the infantry becoming engaged, Jackson and his men were soon on their way to reinforce the remainder of the army near Sharpsburg, Maryland. Wading the Potomac at Boteler's Ford on September 16, Ewell's division, now under General A. R. Lawton, took position on the left of the Southern lines, next to General Hood's division. Early's Brigade was posted on the extreme left flank in support of General Stuart's cavalry and a few batteries.

On the morning of the 17th the Confederates were subjected to a heavy shelling by the Union artillery. This was followed by an infantry assault on the thin Rebel lines. General Lawton was wounded early in the action, and Early took command of the division and "Extra Billy" the brigade.

The fighting raged back and forth for some time and the decimated ranks of Jackson's Corps were driven back. The brigade, leaving the 13th Virginia to support the batteries, was ordered to the relief of its beleagured comrades. Early succeeded in maneuvering his troops behind the Union forces who had penetrated the woods west of the Hagerstown-Sharpsburg pike near the Dunker Church. The brigade was formed in line of battle and drove the Yankees from the woods. This successful assault had exposed the brigade's own left flank to another advancing Union column. Early quickly ordered the regiments to fall back, changed front parallel to the road, and faced the new onslaught. The Federals made a heavy attack on the small brigade, but with the help of some reinforcements, were repulsed. Confederate units were driven back on Early's right, and "Old Jube" found a large Union force between him and the remainder of the army. Early rapidly redeployed his men to face this new threat. The embattled units, seeing reinforcements nearing the Federal front, assailed them from the flank. Driving the Yankees back across the pike, Early placed his weary band of warriors in position for yet another attack. The men remained in this alignment until the evening of the 18th, with only shellfire and sharpshooting to contend with. The bloody battle was over and Captain Watkins' small command was small indeed.

SHARPSBURG SEPTEMBER 17, 1862

	Co A	Co B	Co C	Co D	Co E	Co F	Co G	Co H	Co I	Co K	F&S	Total
KIA	2	1	0	0	1	1	2	1	0	0	0	8
WIA	4	5	0	0	7	4	2	1	6	1	0	30
POW	0	0	0	0	1	0	0	0	0	1	0	2

KIA's include 4 who died of wounds.

Of the wounded, six were so badly injured they had to be left behind when the Confederates recrossed the Potomac on the night of the 18th. Lieutenant William F. Dold of Company H was killed and Lieutenant James A. Burns of Company B was wounded. The brigade had lost 18 killed and 166 wounded. The 65-year-old Colonel Smith had been wounded in three places.

Stuart complimented Early and Smith, stating that Early had "behaved with great coolness and good judgement...." Smith "was conspicuously brave and self possessed." Jackson reported: "Early attacked with great vigor and gallantry...."

Lieutenant Coiner wrote: "I believe our Brigade did the best fighting and acomplished greater things in the battle of Sharpsburg than any battle we were ever in. I heard General Early remark more than once that his brigade saved Gen. Lee's Army in the two charges we made one right after the other....We charged first a line advancing in our front and drove them back over a hill and then changed front to the left and charged three lines charging on our left wing."

Private G. Marion Koiner wrote to his sister:

We marched around Williamsport and recrossed the river-came by Martinsburg to Harper's Ferry. There we captured 10,000 prisoners, 25,000 stand of arms, 40 or 50 pieces of artillery-horses, wagons, etc.....Here on the 17th we had another terrible battle which closes with no considerable advantages on either side, I don't think....I am told that Gen. Lee said the army did not do as good fighting as they have done or he would not have had to retreat to the Va. side. This is true, I think, and that ain't the worst of it. If they drag us across the river a few more times we won't have many to fight at all. I hear a great many say they don't intend to cross again.

Early's brigade was among the last Confederate units to wade back across the Potomac.

On September 20, Union troops attacked the Southern rear guard and the brigade was ordered back to Mohler's Crossroads near Shepherdstown. Although they were not engaged, the troops had to undego an artillery bombardment.

The regiment camped near Martinsburg on September 22 and reported 755 present and absent, aggregate 792.

Moving back to Bunker Hill, the 52nd received a respite of nearly five weeks. Recruits and conscripts joined; wounded, sick, and lame returned to duty and the regiment grew rapidly in strength. The brigade ranks increased from 1,221 men on September 22 to 1,522 on the 30th. On October 10 the 52nd had 790 present and absent, aggregate 832. Colonel Harman, still not recovered from his wound, returned and stayed a few weeks. Lieutenant Colonel Skinner was able to return in November and take command of the regiment. These events had been clouded by the knowledge that Lieutenant James A. White had deserted while on the march from Shepherdstown to Bunker Hill. Lieutenant James A. Dold wrote the Staunton Vindicator expressing his disbelief and stated that at both 2nd Manassas and Sharpsburg White had "laid aside his sword and used a musket believing he could do better service that way."

Adam Kersh had another matter on his mind when he wrote his brother from a camp near Cedarville on November 22. He wanted a substitute for himself: "Our Captain won't except none but boys from sixteen to seventeen or men born in the South unless the Colonel is agreed. Then he will take them in. I will go fifteen hundred. If you can get one for that bring him, if not don't." The experience of the officers was that foreigners hired as substitutes deserted immediately and older men could not keep up on the march.

Kersh described the 52nd's recent activities: "We have been around Martinsburg tearing up the railroad and marching around right smart....We are now near Front Royal. We have been tearing up railroad here on the Manassas railroad for a couple of days now. We quit this evening. We go on picket tomorrow." Kersh also requested that his overcoat, blanket, shirt, a pair of pants and mittens be sent to him because of the cold weather.

The 52nd and the brigade left the Valley and marched via New Market across Massanutten Mountain and the Blue Ridge to Madison Court House, Guinea Depot, and finally to Skinker's Neck, on the Rappahannock between Port Royal and Fredericksburg. Colonel James A. Walker of the 13th Virginia now commanded the brigade, in the absence of Colonel Smith from wounds; Early still commanded the division. Many of the men were nearly barefoot from the long march, but great efforts were made to supply their needs. Blankets and overcoats were issued also. The muster rolls for Company G for October-December showed discipline, instruction, military appearance, arms and accoutrements marked "good." Clothing was "good" also, except for shoes. This was probably true for the rest of the regiment as well.

During the morning of December 13, Walker moved the brigade to Hamilton's Crossing, near Fredericksburg. The Union army, now under General Ambrose P. Burnside, had forced a river crossing and was attempting to drive Lee's army from the city.

Federal troops penetrated the Rebel lines on December 13, near where Walker's regiments were stationed in reserve. Walker led the men to the breakthrough, formed them into line of battle, and the brigade drove the Yankees from the woods and across a railroad into the open fields next to the river. Some of the men advanced so far that they became casualties while returning across the open plain to the Confederate lines behind the railroad bed. Captain Edward M. Dabney, acting major of the 52nd, dismounted from his horse and was hit by a canister ball which went through both hips. He died in a Richmond hospital later in the month. The tallest man in the regiment, and a brave and gallant soldier, he had only just recovered from his wounds received at McDowell. His chances for promotion had been very promising.

The brigade remained behind the railroad embankment the rest of the day, with only skirmishers in front of the line firing occasionally. After dark the main body moved back into the woods and camped in the bitter cold.

The 52nd had again suffered moderately heavy casualties.

FREDERICKSBURG DECEMBER 13, 1862

	Co A	Co B	Co C	Co D	Co E	Co F	Co G	Co H	Co I	Co K	F&S	Total
KIA	0	0	1	0	0	0	4	0	0	0	0	5
WIA	3	0	2	0	1	1	1	0	5	0	1	14

Sergeant Major Charles L. Weller had been wounded. Lieutenant Colonel Skinner had commanded the regiment during the battle, Colonel Harman still being bothered by his wound.

Walker complimented the brigade on "the gallantry and bravery exhibited by both officers and men. I had no trouble getting them to fight, but a good deal to get them to stop, when, in my opinion, it was imprudent to go farther."

For two day the Confederates waited in the freezing weather for the Union army to advance but Burnside retreated back across the river on the night of December 15. On the 16th the 52nd Virginia and the rest of the brigade moved back to the vicinity of Port Royal and went into winter quarters.

Colonel Harman, convalescing at his home in Staunton, had printed in the Staunton <u>Vindicator</u> that the Christmas meal consisted of "Bull Beef and patty bread" and whatever the men had received in boxes from home. He also published this request:

> The families and friends of any of the men belonging to my Regiment who have left their overcoats and blankets at home will please forward them to me to send to the Regiment, as they are greatly in need of them.
>
> I appeal to the citizens of Bath, Augusta and Rockbridge to furnish 200 pairs of socks for my Rgt, for which I will pay the highest cash price on delivery. I also appeal to all good citizens of Bath, Rockbridge and Augusta, to give information in regard to any members of my Rgt who are at home without leave, as every man is needed now and should be at his post.

1863

The winter of 1862-63 was spent like the last one, building cabins cutting and hauling wood, erecting fortifications and maintaining roads. The 52nd rotated with the other regiments of the brigade in standing picket along the Rappahannock. The regiment had one sentinel slightly wounded on January 3.

Robson remembered the roads becoming so bad and the mud so deep that they were almost impassable for a man on foot. Kersh wrote to his brother.

> We had some right rough weather here a few days back. It snowed several snows. The deepest was about six inches deep. . . . We got our houses done before it set in cold. We have as comfortable quarters as we had on the Alleghany Mountain now. . . . It appears to be more healthy here than on the Alleghany Mts. At least the health of the regiment is better now. . . We go on picket once in eight days. We go ten miles. Our posts are in sight of Port Royal. Our whole regiment goes at a time and stays three days at a time. We have pickets fifteen or twenty miles along the Rappahannock River. We go on post right on the riverbank. Some times can see the yankees riding and walking on the other side. Our reserve is left about one mile and a half in the rear. We have about fifty prisoners in the guard house now and they are still coming in(men who had been absent without leave).

Lieutenant Colonel Skinner wrote the Vindicator requesting additional clothing for the regiment: "We have recently received fourty pair of excellent socks, presented by kind ladies of Augusta, which were given to men entirely destitute, and yet there are others altogether without socks, and many insufficently supplied. . . ."

The appeal met with an outpouring of clothing and blankets. Captain Humphreys wrote the newspaper in February thanking the ladies of the Greenville "Soldiers Aid Society" for 8 blankets, 19 pr socks, 33 pr drawers, and 33 cotton shirts; the Greenville and Mint Spring society for 2 blankets, 4 pr drawers, 15 pr socks, 8 shirts and 10 pr gloves; the Gospel Hill society for 3 pr socks, and 10 pr gloves. This left "every man in Company supplied with socks and gloves."

Lieutenant William R. Gillett wrote thanking the ladies of Bath County for their donation of socks, drawers and blankets for Company K.

Captain Watkins wrote to the Lexington Gazette: "Permit me through the columns of your paper to acknowledge the receipt of, and to return the thanks of my company and myself, for 17 pair of socks, and 14 pair of gloves, from the

Ladies Soldiers aid Society of the Natural Bridge District, also a pair of excellent shoes, the gift of Thomas P. Paxton Esq. of Rockbridge Co."

Skinner again wrote the Staunton paper thanking Mr. Albert G. Garber of Augusta for donating 42 pairs of socks, 12 pairs of pants, 3 blankets, 2 overcoats, 2 pairs of shoes and underclothing. He noted that government issued shoes had leather soles but canvas tops. Skinner added that they were unsatisfactory for the rain, snow, mud and cold of northern Virginia during the winter months, but said they might suffice during the summer.

The muster rolls of Company G were marked "good" in all catagories except clothing for January-February 1863. Clothing was shown as "tolerable."

The appeals for clothing continued to be answered throughout the winter. Harman wrote his hometown paper on March 6 thanking Miss. E. V. Hanger and the Soldiers Aid Society of Waynesboro for a barrel of socks and shirts.

Chaplain John McGill also wrote thanking Miss Hanger for 55 pairs of socks, 22 pairs of drawers, 10 pairs of gloves, 22 shirts and 2 neckties. "These articles were distributed among the various companies according to their several wants and we can testify that they were most acceptable to the men, and were received with feelings of gratitude. Will the ladies accept, therefore, the heartfelt thanks of the Regiment."

The members of the 52nd showed their concern for the destitute citizens of Fredericksburg who had been driven from their homes with only the clothes on their backs. Their homes lay in ruins. They were living in the nearby woods in tents and cabins in distressful conditions. A member of the 52nd wrote to the Vindicator that the officers and men of the regiment had donated $439 toward their relief. Other Confederate units had done likewise.

The brigade had moved closer to Fredericksburg during March. Private Koiner wrote his sister on March 12: "We are now camped about 4 miles from Fredericksburg near on the railroad & about a short mile from Hamilton's crossing. We have a very good camp here. The Regt. is camped above the road about two miles." Rations seemed to have improved, as he added: "We have been drawing a good deal of sugar and rice since we have been up here. We draw 10 or 12 pounds a week of sugar for 5 of us, so we have everything sweetened up to nature. You ought to see the pies we bake. We have plenty of shortning but sometimes we get it in the rong way. I have not been over the battleground yet. It's about a mile to where our brigade fought. We can see Yankey tents, plenty of them, from the Xing.

A few days later Koiner wrote: "We have had some very rough weather for the last 30 days. It snowed for several days & then rained a day so it still keeps the road in a mudy condition." He wrote again on April 6: "it snowed all Saturday night and Sunday till the middle of the day (six inches). It was also very cold and windy. . . . They went on picket yesterday. . . .Colonel Harman is in command of the regt now. . . .Old Jack says that he will have to get among the yanks again before long for he said that he never lived so poor in his life as he did this winter. We hear cannon nearly every day but I don't recon its anything but the yanks trying their guns. Their balloon is up nearly every day or two."

Colonel Harman wrote to the editor of the Vindicator on April 17 announcing his reason for declining to run for Congress: "The condition of our country demands the services of every man able to be in the field. . . .My duty and the men of the Rgt, who have so nobly endured the hardships of war, urge me to stand by them, as they have always stood by me."

Chaplain McGill wrote the Staunton paper on April 28 reporting that the 52nd had suffered 54 killed or died of wounds, 68 deaths from disease and 15 from causes unknown. McGill's list was far too modest, as the records show 181 deaths from all causes up to that date. Nearly the equivalent of two companies had paid the supreme sacrifice for the Confederacy.

The chaplain lost his horse during this period and the men raised four or five hundred dollars and bought him a new one. Captain James Bumgardner presented it to McGill in front of the regiment.

Colonel Smith, recovered from his wounds, was promoted to brigadier general and resumed command of the brigade.

The lull of camp life and picketing came to an end on April 29 when it was reported that Union forces were crossing the Rappahannock under cover of a dense fog. Early, still leading the division, was left to defend Fredericksburg, while Lee moved the rest of the army westward to oppose General Joseph Hooker's advance.

On May 3, Smith led the brigade in a repulse of a Federal penetration to the railroad near Deep Run. The Southerners quickly drove the enemy back across the plain to the river. The 52nd reported no loss in this action. Meanwhile, back in Fredericksburg, Union forces under General John Sedgwick had finally overcome stubborn Confederate resistance on the heights back of the town. This forced Early to withdraw his troops down the Telegraph Road.

Lee, having defeated Hooker at Chancellorsville, sent General Lafayette McLaws with his division to aid Early. After considerable confusion and misunderstanding of orders, Early advanced his units up the Telegraph Road on May 4 with General John B. Gordon's Georgia Brigade leading and General William Barksdale's Mississippi Brigade and Smith's Virginians in support. Gordon successfully reoccupied Marye's Heights. Early sent "Extra Billy's Boys" down Lee's Hill and across Hazel Run to capitalize on Gordon's success. Moving across the open terrain the brigade was subjected to a heavy shelling by Union batteries on Stafford Heights on the other side of the Rappahannock. Finding the Federals in strong force on this hill mass and seeing the effects of the cannon fire, Early ordered Smith to withdraw his regiments to the trenches along the Plank Road. Smith formed his men along the crest of these hills from the road toward Taylor's Hill, facing the enemy to the west. Late on May 4, three Confederate brigades assailed the Union position from the west and south. After much confusion and delay, the Confederates were finally successful. The 52nd and the 49th Virginia, which had been sent back to Marye's Hill to support Barksdale, were ordered to turn and support the attack. Arriving as darkness set in, they were placed in the trenches along the Plank Road overlooking the canal. Sedgwick retired across the river during the night.

Finding the Federals gone the next day, Early moved his weary troops to near Salem Church. When he found that Hooker had also retreated, Lee ordered Early's command back to Fredericksburg. While crossing Hazel Run on the Telegraph Road, the brigade underwent another bombardment from the Union cannoneers.

Private Kersh wrote a description of the fighting:

About the fighting on 29th of April. The yankees commenced crossing the river about a half mile below Fredericksburg. They drive our pickets — company of the 13th Georgians and killed some few — but nothing like the number we killed of theirs. We were ordered down there about day light and put in the ditches in line of battle. Skirmishers were out and some fireing was done and shelling during the day but not many were hurt. On the 30th some fireing amongst the skirmishers and some shelling done not much damage done. May 1st every thing quiet except a shott occassionally amongst the skirmishers. 2nd — Our regiment went on skirmish the yankees recrossed the river. Some shelling done the yankees left a good many things behind.

We (were) then ordered up the river about ten miles where Gen Lee was fighting. We marched about 8 miles and was ordered back to Fredericksburg to the ditches. The yankees were crossing back again. We got back about 12 o'clock in the night. May the 3rd Our batteries opened this morning about 7 o'clock and a hot shelling was kept up for about 3 hours and then ceased. They poured it into our batteries like thunder but our artillarymen held their position nobly and gave them as good as they sent. While this was going on the yankees were flanking us...on the left. Our force was not strong enough to keep them back. They took about eight hundred prisoners Alabamians and Mississippians and about eight pieces of artillary. The artillary is on the other side of the river amost of it and heavy pieces rangeing with every road — we bring our troops in we were ordered back this eaving in the rear on our left to reinforce our left where we lay in line of battle all night May 4th. The next morning we advanced toward the enemy (ex)posed to Shells which they threw at us in abundance not many were hurt in our brigade. We were held in reserve. About 5o'clock the general fight commenced. Our men charged them and drove them like chaff before the wind. During the action one man in our regiment . . .was killed and one wounded. . . .About sunset we were ordered to the field to reinforce but when we got there the fireing had ceased so we were not in the fight atall. Our men drove them back about 4 miles next to the river. Got the heights back. We took about one thousand prisoners and killed a great many of them more than they killed of ours. They said we had one division against thirty thousand in this fight. They crossed back during the night on the other side. May the 5th we were ordered to fall back. They threw shells at us like forty. Got about a mile and was ordered back that the yankees were about crossing below Fredericksburg. They kept throwing shells at us. One fell in our regiment and killed four men and wounded seven. . . .None of our company hurt.

Sergeant Adam G. Cleek added this description: "We had two killed and two wounded in Company (K) John A. Lindays (Lindsay) was badly wounded in the arm. Andrew Gillett was slightly wounded-lost two of his toes. Brown Archy (Archie) was killed dead and Cornelius T. McClung was also killed done by the same shell which was thrown a distance of at least three miles. Lindsay had just gotten back a few days before. . . ."

Lieutenant John A. Lindsay and a couple of his friends were sitting on a pile of rails eating their rations when they saw a cannon ball rolling in front of them. They thought it was a solid shot and ignored it. It exploded and a fragment of it struck Lindsay in the shoulder. The fragment was removed from his back, as were the pieces of clothing that were carried in with it. In the hospital, a young doctor asked him what was wrong. Lindsay told him what the problem was, and the doctor took hold of his arm, raised it up in the air then dropped it. Needless to say, the pain was tremendous. Lindsay had his pistol under his pillow, and he took it out, and asked an orderly to put a cap on it for him, as he intended to shoot the doctor. The orderly put all six caps on it; as this took so much time, that the doctor was able to escape according to a family tradition recounted by Lindsay's son.

Kersh wrote this further account: "Our regiment did not get to fire a gun the whole eight days. The yankees had all went back on the other side of the river when we left and said they did not intend to fire on us now for a while. They did not throw any shells at us when we left. We arrived in our old camp on the 7th. The hardes fight was above us where Lee and Jackson fought them. They (said) it was the (hardest) fight thats been fought yet. The ground is blue with yankees. They had about nine thousand prisoners at Guinea station taken in the fight."

FREDERICKSBURG MAY 1-4, 1863

	Co A	Co B	Co C	Co D	Co E	Co F	Co G	Co H	Co I	Co K	F&S	Total
KIA	2	0	1	0	2	0	0	1	0	2	0	8
WIA	2	0	1	0	0	0	0	0	2	3	0	8

The leadership of the regiment continued to take a high proportion of the casualties. Lieutenant William V. Knick of Company E received his third and fatal wound of the war. Lieutenants Richard S. Kinney of Company A, Charles L. Weller of Company C and Lindsay were wounded.

Sergeant Cleek wrote to his wife: "Oh Ann I am so sorry. I have just heard that General T. J. Jackson is dead. He died 25 minutes past 3 o'clock this morning at Guinea Station. He died of Pneumonia. Oh what a loss we have met with to lose such a man as General Jackson at this time his place can not (be) filled. I do think that he was the greatest man in the Confederacy. It is certainly the greatest loss that we have ever met with since the commencement of the war."

Later, Private "Dil" Koiner added:

We have a good many troops camped around here now. It is thought that the yankees intend to try us again here since they found out Jackson is dead. It is said that they appear keen to try us again. It is true we have lost a good General but we still have a few more smart men left let them come. I don't believe that they are as anxious to fight us as they let on to be. Just mere talk. They have the advantage of us at Fredericksburg on account of the big number of seage guns they have planted there. As long as they are under shelter of them they can talk large. Just let them advance and we are ready for them. We got a great many guns knapsacks shovels picks and such things in the last fight. They have been sending them off from the depot — Hamilton's Crossing ever since the fight. It is a detail made in our regiment of about twenty five men every day to help onload and load the cars. It is also details made out of other regiments in our Brigade to stand guard and fatiegue duty. It is a good many suttlers at the Station. They sell very high too beans one dollar a quart one dollar for about a quart of irish potatoes and other things in proportion. It is awful warm here now (May 24) has been for several days. We have no shade. The timber has all been cut for firewood. We expect to move out of our cabbins in a few days about a half mile in the woods in the shade.

We have tolerable good water here but not as good and cool as in Augusta. Our Mess got ourselves good little tents and knapsacks and blankets in the last fight. It was thousands of old united States blankets and little tents scattered over the battlefield. . . . Fredricksburg is pretty much torn to pieces with shells.

The most of the citizens have moved away. A great many houses have been entirely destroyed.

Me and some of the boys took a walk over the battlefield at Frederickburg before the last fight. We could see plenty of old half decayed yankees some with their heads sticking out hands and feet sticking out. . . . It was an awful sight to behold.

The 52nd collected $481 toward the proposed statue of General Jackson and sent it to Governor Letcher.

A great loss to the regiment occurred on June 1. Colonel Harman, who had commanded the 52nd in the battle of Fredericksburg, resigned for disability caused by his old wounds. Skinner was elected Colonel, Ross became Lieutenant Colonel, and Captain Watkins was promoted to Major.

The 52nd Virginia remained camped along the Rappahannock until June 4, when Lee began a flanking movement around Hooker's army. On the night before their departure the glee club serenaded General Ewell, who had just returned to duty. One man was wounded in Company G in a skirmish on June 6. Moving across Raccoon Ford the brigade reached Culpeper Court House on June 7. On the 9th the brigade was ordered out in support of Stuart's cavalry at Brandy Station but soon returned to camp. Marching northward the next day, the troops crossed the Blue Ridge at Chester Gap and reached Front Royal on June 12. The Shenandoah River being low, the men waded across and went into camp. As they were covered with dust and dirt from the long march, the soldiers returned to the river for a bath and to wash their clothes.

Moving to Newtown and along the familiar Valley turnpike the next day, the Confederates drove a small enemy force back toward Winchester. Reaching Kernstown, Early moved his veterans around Winchester to a position facing Bower's Hill. Smith maneuvered the brigade to the extreme left of the army, under shell fire from Union guns. Positioned southwest of the enemy fortifications, two regiments were sent forward as skirmishers. The troops rested in line of battle during the rest of the day. From their location they were able to see the flag flying over the main fort in Winchester. It rained hard during the night and the men got soaked.

Early's units remained in position until 2 p.m. on the 14th. Hays' Louisiana, Hoke's North Carolina and Smith's Brigades, accompanied by Jones' battalion of artillery, marched farther northward around Little North Mountain to the vicinity of Round Hill and then turned eastward, hidden from view of the Federals troops. Jones' artillerymen moved their pieces silently by hand to the edge of the woods facing the enemy position. It was about 5 p.m. before Early had everything ready and gave the gunners permission to open fire on the entrenched Federals. Out rolled twenty cannon into an apple orchard and cornfield and the Rebel cannoneers blasted away at the west fort of the Union lines. The Yankee artillery soon responded and the gun duel lasted for about forty-five minutes before the Union guns almost ceased firing.

Early ordered Hays' Brigade, supported by Smith's, to assault the west fort. Hays' veterans charged with the Rebel yell, through the abatis in front of the enemy lines and into the trenches. The Louisianians overwhelmed the Federal troops and sent them flying to the other forts nearby. Hays' men turned captured cannon on the retreating foe. Smith's Virginians helped occupy the breastworks and turn them toward the other Union forces, but the glory went to the brave "Pelicans." The 52nd had only one casualty, a man wounded in Company G.

Lieutenant Coiner called this affair "a sharp little battle. Gen. Early got his artillery in position and opened on their breastworks. Gen. Hayes Brig advanced together (with Smith's) captured works, and the next morning nearly all of the enemy was captured below the town."

More than three thousand prisoners and most of General Milroy's artillery were captured. That morning Smith's men were up at dawn and Early sent out skirmishers toward the nearby forts surrounding Winchester and found them abandoned. The Union flag was taken down and a Confederate one raised in its place, amid the cheers of the men. Early soon marched his division in pursuit northward down the pike. Finding he was too late to assist in the capture, "Old Jube" ordered his men into camp near where General "Alleghany" Johnson's troops had been engaged. Smith's soldiers were in high spirits as they cooked their rations and savored the easily won battle.

The next day Smith moved the brigade to a campsite near Stephenson's Depot, where it remained for three days. Some of the men were issued badly needed items of clothing and equipment from the captured stocks in Winchester. The 13th Virginia was detached from the brigade and sent to Winchester to act as provost guard. The 58th Virginia was sent to take charge of more than 3,000 prisoners and march them to Staunton.

Winchester and the defeat of Milroy's army had not been Lee's goal. Confederate units had already advanced into Maryland. On June 18 the brigade marched northward through the heat and dust via Leetown and camped near Kearneyville, on the Shepherdstown pike. Here they remained in camp for several days.

At daybreak on June 22, the brigade resumed the march and passed through Shepherdstown, where the ladies and pretty girls waved Confederate flags at the Southerners as they passed by. The Rebels cheered them and were in excellent spirits. After the men waded the Potomac at Boteler's Ford, the reception for the gray-clad troops in Sharpsburg and the other Maryland villages was far from warm. Moving through Boonsboro, the tired men went into camp at Benerola Chapel, near Hagerstown.

On the march at dawn the next day, Early's division moved through Ringgold into Pennsylvania and camped at Waynesboro, worn out by the heat and dust. The half-starved Confederates were in the land of milk and honey, untouched by war, and feasted on all the good things to be found on the nearby farms. Rebel quartermasters were busy gathering horses, cattle and other badly needed supplies.

The march continued on June 24, through Quincy and Funkstown to near Greenwood, where the exhausted troops bivouaced. The next day all wagons, except the ones carrying ammunition, were sent to the rear and three days rations were cooked in preparation for another advance.

Leaving Greenwood in a pouring rain on June 26, the brigade was directed to destroy the iron works of Thaddeus Stevens, a Union Congressman. Marching along the now-muddy roads, Smith's troops crossed the Blue Ridge and went into camp at Mummasburg. Because of the rain the Confederates were allowed to burn fence rails for the first time since crossing the Potomac.

Moving at 7 a.m. the next day, the trek continued through Hunterstown, New Chester, Hampton and into camp near Berlin.

"Extra Billy's" boys marched at sunrise on June 28, trudging through the oppressive heat and dust to York. Nearing the town, Smith ordered the brigade band to lead them into the city, playing "Yankee Doodle." Smith and his staff bowed and scraped to everyone and were all smiles. The astonished and amused citizens cheered the music and General Smith. The people invited General Smith to make a speech and he halted the brigade in the town square, had them stack arms, and gave an eloquent talk to all. Early, disturbed by the delay, pushed his way though the crowd and demanded of "Extra Billy" the reason for the delay. When the old politician replied he was only "having a little fun" and also trying to win the cooperation of the people, "Old Jube" stopped cursing and left. Smith finished his speech and led the brigade into camp north of town.

The next morning many of Smith's men were drunk on whiskey procured in York. Early seized all the stores and confiscated large numbers of shoes, boots, hats and other articles of clothing for his men. Some of the soldiers managed to get items they needed "on their own hook." All weapons were inspected in the afternoon and three days rations were issued. Orders were given to march at daylight.

Moving at the prescribed time on June 30, Early's troops retraced their steps through Berlin and back to near Hampton. The march was a rapid one, covering twenty-four miles. Smith's Brigade acted as rear guard and had to eat the dust of the rest of the division and the wagon trains. Many of the men suffered from blistered heels and toes from the new shoes and boots. This had been the hardest march of the campaign.

At dawn on July 1, Smith led the brigade down the Cantersburg Road. Soon cannon fire was heard. Hastening through Heidlersburg, the Virginians reached the scene of action two miles north of Gettysburg at 3 p.m. Early ordered Smith to support a battery going into action. The rest of the division Early led toward Gettysburg, in line of battle. Early's assault was successful and the Union troops retreated through the town and onto the heights above. About 5 p.m. Early directed Smith farther to the left to guard that flank of the army. The 49th Virginia was advanced as skirmishers and moved toward Culp's Hill, about three-fourths of a mile away, with the 52nd and 31st regiments in support. While making this movement the gray-clad soldiers were shelled by the enemy's cannon and harrassed by sharpshooters. Relieved by "Alleghany" Johnson's men at dusk, the 49th and the other two regiments rejoined the brigade on the York turnpike and camped for the night. Only the 49th had suffered casualties.

Lieutenant Coiner recalled the action:

My command...reached Gettysburg, July 1st, about 3 p.m. and was placed in position on the extreme left. As we were told to guard the flank and support the cavalry we had there. After the troops of Hoke Gordon and Hays had entered the town Gen. Smith from some cause I do not know what, unless he mistook a fence with a growth of small trees for a line of troops, was so impressed with the idea the enemy was advancing upon him that he called his son (Lt. Smith) and sent him to Gen. Early with the report that the enemy was advancing and that he needed reinforcements at once. I heard what passed between the Gen. and remembered seeing Lt. Smith dash off toward town at a furious gallop. By the time he could get to town and make his report I saw Gordon's Brigade coming out of town towards us at a double quick. By this time I suppose General Smith had found out his mistake and saw that there was no flanking or the movement of the enemy upon us so Gordon's Brigade was halted before it reached our position. There was no further movement of our Brigade that afternoon or night....

Smith's men rested in bivouac until noon on July 2, when Confederate cavalry pickets reported an enemy advance down the York Pike. The 49th was sent out as skirmishers and the gray-clad men were able to repulse their blue clad counterparts. Starting at 1 p.m., the 52nd witnessed the four-hour bombardment of the Federal positions by the Confederate artillery. Colonel Skinner's men also watched the assault on Culp's Hill by Johnson's division. The brave Southern infantrymen fought for nearly four hours trying to take the position, without success. "Alleghany's" units suffered severe losses.

Awakened in the middle of the night, Smith was ordered to move the brigade at 3 a.m. across Rocky Run to the extreme left flank of Johnson's line. Attacking at dawn on July 3, Johnson's troops, aided by Smith's small unit, assailed the entrenched enemy's fortifications. While advancing in support, Smith's men received a severe shelling. Johnson ordered Smith farther to the left, nearly one half mile, the 49th leading and the 52nd following closely behind. The leading elements soon became engaged with a Federal force advancing toward Johnson's flank. Deploying rapidly, the 49th dispersed these Yankees after a sharp fight. The 52nd acted in support of its sister regiment. The 31st Virginia, which had been left behind on picket, then rejoined the brigade. The 52nd and 31st were pulled back across the stream while the 49th remained watching the enemy. About noon "Extra Billy" was directed to join in Johnson's attempts to take the Union bastion.

Sergeant W. O. Johnson of the 49th described the assault:

Our two regiments (49th and 52nd)...charged and drove everything before us until we came to a small field at the foot of Culp's Hill on the bank of Rocky Run where the Yankees opened on us with grape and canister from a 20 gun battery and in less than 5 minutes killed and wounded 150 of our two regiments, which caused a halt. . . .We held our position on the edge of the woods until noon. During the whole time the fighting on the right of Johnson's pos. was furious and incessant. There was a momentary stillness and a cessation of fire. The Stonewall Brigade had given way and the enemy regained the ground. Gen.'l Johnson then ordered back his troops, leaving a line of skirmishers around the hill. Our Brigade was ordered back out of fire to rest and took pos. on the right of Gen's Johnson's lines & commenced to skirmish with the enemy. . . .We were terribly exposed to the enemy's shells until night closed the scene. We were withdrawn and lined up by the side of an old fence where we fell down and went to sleep.

Major R. W. Hunter of Johnson's staff witnessed General Smith's charge to the relief of the 2nd Virginia of Jackson's old brigade:

At the supreme moment was heard the voices of Smith and his men dashing forward to the rescue.

They stood not upon the order of their coming, but came with a rush, the old Governor in the lead, his voice rising above the din of battle. . . .Taking the highest pos. he could find, reckless of shot and shell, with bare head & sword in hand, pointing to the enemy, he harangued each regiment as it double quicked past into the area of blood & fire. . . .the old Governor either could not recall the "orders" as laid down in the books on Tactics . . .however . . .there was not a moment hesitation as to what it meant. His "boys,"as he affectionately called them, knew him & understood him, and off they dashed with a spirit and a vim that soon drove back the enemy.

It was done so handsomely; the old Governor's bearing was so superbly gallant; his voice so ringing and inspiring; the reinforcement he brought so opportune, so welcome and so effective, that the troops in that quarter, rejoicing in their deliverance, in hearfelt tribute to that "good gray head that all men knew," and with a spontaneous impulse such as only soldiers, in such a plight can feel, with one accord raised the shout: "Hurrah for Governor Smith," which went along the lines like an electric current, mingling with the sullen roar of the enemy's cannon.

"Bent" Coiner remembered it differently: "Gen. Smith took us into battle with out orders and had us badly cut up and we accomplished nothing. The enemy was hidden behind the rocks & trees on Culp's Hill. LtCol Ross asked that we be permitted to charge (Col Skinner was wounded earlier) or be permitted to drop back a few paces where we might be protected. I have always believed we could have taken them (the Union position)."

The 52nd again suffered severe casualties but not nearly as many as the 49th which lost 100 men out of about 250 engaged.

GETTYSBURG JULY 3, 1863

	Co A	Co B	Co C	Co D	Co E	Co F	Co G	Co H	Co I	Co K	F&S	Total
KIA	0	0	2	0	2	1	2	0	1	1	0	9
WIA	0	3	1	1	3	3	4	5	2	3	1	26
POW	4	3	2	1	0	2	0	0	3	4	0	19

KIA figures include those who died of wounds; POW totals include those lost during the campaign including those left behind with the wounded at Gettysburg. Seven of the wounded were left behind.

Captain Andrew J. Thompson of Company B and Lieutenant Erasmus S. Trout of Company H were wounded.

Awakened near midnight, the brigade moved at 1 a.m. on July 4 around Gettysburg into a reserve position behind A. P. Hill's corps. At dawn it commenced raining. Soon the whole of Lee's army fell back about one half mile and occupied a defensive position on a range of hills. Smith's exhausted men lay down in an orchard near Gettysburg College. They camped there in the pouring rain without rations.

Alerted at 9 p.m., the worn out soldiers were ordered to pack up, load weapons, and prepare to march. Smith's troops waited in the falling rain until midnight before taking the road. Because they were delayed by mud and wagons blocking the road, Smith's men made little progress. Lieutenant Robert D. Funkhouser of the 49th wrote: "our bivouack that night was mostly standing and sitting along the roadside." Burdened with miles of wagons, vast herds of cattle and horses, and ambulances full of sick and wounded, Lee's army slowly retreated toward the Potomac.

Lieutenant Colonel Ross, commanding the regiment, wrote to his wife from a camp near Hagerstown on July 8:

> Here we are again in Maryland, after quite an exciting campaign in Pennsylvania. We came from Waynesboro, Penn. yesterday to this place and from the present indications are likely to remain here some days. We do not seem in the least desirous of getting away from the enemy. . . . They did not follow us across the Mountain, and I do not know exactly where their army is. We are very anxious to give them battle whenever they come after us, and I hope most sincerely we will not cross the Potomac without having another trial of strength with them upon fair ground. Gen. Lee's Adjt. Gen. says our whole loss in killed, wounded and missing is a little over 7,000, while we captured a good many more than that from them, not counting their killed and wounded. I think they were severely punished. Certainly they don't seem anxious to renew the engagement. Last night we had a most unpleasant time sleeping. It rained in torrents, and I did not have a tent. I got right damp. Dick,

Uncle Cameron and I were sleeping under a tree, upon gum blankets, but it rained so hard, the water got under them, and we had to lie so during the night. A soldier's life is not pleasant at any time, but when he has to contend with limited transportation, and such elements as mine had since the battle of Gettysburg, it is boring in the extreme. If we had succeeded in dislodging the enemy from his stronghold there I think the war would have been near its close. As it is, the time is probably postponed, but I hope the next battle may be a decisive one, but men with one voice want to fight them again before we leave Maryland, and I cannot think the contest will be doubtful if there is any sort of equality of positions. Our rations during the three days of the fight were very limited, as we did not have an opportunity to cook, I don't think, in the three days of the battle and two days after it, I ate enough for one good meal. . . .

Lieutenant Coiner recalled the ordeal: "In the retreat our Brig. was the rear guard and was engaged several times. When we got back to the Potomac Gen. Lee stoped for a day to give the enemy an opportunity to attack us (the river was high) but they declined."

Continuing the retrograde movement, Smith's Brigade waded the waist-deep river on July 14, during a pouring rain. Moving into camp near Martinsburg, the exhausted men lay down and slept despite the storm.

The battle-weary Army of Northern Virginia lay camped in the northern Shenandoah Valley for several weeks. When Smith's Brigade marched through Winchester on July 23, the 13th Virginia rejoined. They were the subject of much abuse from the members of the other regiments, being nattily dressed with paper collars and Yankee uniforms. The 58th Virginia soon was reunited with the brigade also.

"Extra Billy's boys" were soon back on the familiar terrain along the Rapidan. Lieutenant Funkhouser calculated that the brigade had tramped 540 miles in 54 days and fought one major battle and numerous skirmishes.

During this period the 52nd, along with the rest of Lee's units suffered from numerous absences without leave and desertions. Eighteen of the regiment left during July never to return, despite a Presidential amnesty. This was the largest number to leave permanently since May 1862, when twenty had left because of the Conscript Act, which ordered them to be held in service. Desertion would continue to be a problem to the end of the war. Major Watkins reported only 235 men present on August 8, 1863.

Smith was promoted to Major General and sent on a speaking tour of the state to boost the morale of the people and encourage the return of absentees. Officers and men from all the regiments were sent on detail to their respective counties to arrest and return those absent from their commands. Colonel Hoffman of the 31st Virginia assumed temporary command of the brigade.

While they were camped near Pisgah Church in Orange County in the late summer of 1863, the men of the 52nd, and the rest of Lee's army, were swept up in a religious revival that continued through the entire Confederate army during the fall and winter of 1863-64. Chapels were built by every brigade. The chapel built by Hoffman's Brigade was also used as a school for the less educated men.

The troops were immediately involved in erecting quarters, cooking, drilling, cutting fire wood, building roads, erecting fortifications and picketing Sommerville Ford. Trading with the Yankees was a common practice, too. Some of the men received well-earned furloughs. There were several alarms, requiring the troops to break camp and march to different positions and remain in muddy trenches, but no serious action resulted. Ewell's entire corps was reviewed by Lee on September 9.

Adam Kersh wrote on September 29:

We are still camped on the rappidan river at Summerville Ford. We guard the ford and picket along the river. The river is the line between Orange and Cupepper us and the enemy. The yankees advanced their picket lines. The distance between us is about five hundred yards at places. They have also reinforced their picket. The picket dont fore on each other now. They hollow at each other sometimes and sometimes exchange news papers. It is against orders but they will do it. Our brigade and Hokes guard Summerville Ford. Hays brigade guards Raccoon Ford a mile below here. Gordons Brigade is about a mile above here. Camp is about a half mile from the river behind a big hill. If we want to see blue bellies all we have to do is to walk on top of the hill and we have a fair view of the yankees picket lines Camps and wagon trains. We have a good position along the river here. We have the heights. We have ditches and rife pits thrown up here. I dont think they will attact us here as we have a good position. . . .

On October 8, Colonel "Wagonmaster" Hoffman's Brigade crossed the Rapidan and Robertson Rivers, as part of Lee's flanking movement around General George G. Meade's Union army. While the maneuver was successful, the Confederates were unable to destroy the Federal forces or capture any appreciable amount of supplies. The 52nd was only subjected to hard marching and scant rations. The men became so hungry they resorted to roasting acorns to ease their pangs. Returning across the Rappahannock, the Valley soldiers moved through a waste land; all vacant houses and all mills and barns had been burned. The rest of the month was spent in protecting working parties removing the iron rails from the railroad. The only loss of the campaign was one man captured in Company B.

Brigadier General John Pegram arrived on November 1 and took command of the brigade. He was a West Point graduate who had given up the command of a cavalry brigade in the Army of Tennessee to return and defend his native state.

Colonel Skinner, who had recovered from his wounds at Gettysburg and resumed command of the regiment, wrote the Staunton paper that he was sending two men to Staunton and they would receive packages for the men at the Quartermaster's office.

Early had left two of his brigades entrenched across the Rappahannock, with only one pontoon bridge as an escape route. On November 9 these units were overrun and Pegram's Brigade was sent to their relief but arrived too late. The new commander had volunteers burn the bridge, preventing the Federals from crossing. The next day Lee's army recrossed the Rapidan and Pegram's Brigade resumed its old position near Sommerville Ford and resumed the routine of camp life, broken only by periods of picket duty.

The lull did not last long. General Meade sent his army across the Rapidan on November 27 in an attempt to turn Lee's right flank. Early moved his division toward the Wilderness, leaving Pegram to defend Sommerville Ford until relieved by cavalry units. The gray-clad troopers arrived later in the day and the new brigade commander marched his men rapidly eastward. When Pegram's men rejoined the division near Locust Grove, Early placed them in reserve behind Mine Run. Colonel Skinner had the men construct breastworks. Two regiments were sent out as skirmishers. A stray round struck "Bent" Coiner on the thigh but the thickness of his old Cadet overcoat prevented a serious wound. The young Lieutenant described the action: "We fought the enemy standing in line until the men had exhausted there ammunition and then charged them and ran them." Finding the Federals had retreated on December 2, Lee sent his army in pursuit but was unable to overtake them before they recrossed the river. Pegram's Brigade returned to Mine Run and spent the night and moved back to Sommerville Ford the next day.

MINE RUN NOVEMBER 27-29, 1863

	Co A	Co B	Co C	Co D	Co E	Co F	Co G	Co H	Co I	Co K	F&S	Total
WIA	1	0	0	1	0	0	1	1	2	0	0	6
POW	0	0	0	0	1	0	0	0	0	0	0	1

Winter quarters became the prime concern of the regiment. The construction of cabins and fireplaces went forward as rapidly as military duties would permit. On December 5 the 52nd was ordered across the Rappahannock to gather forage for the brigade animals. Some of the skirmishers got between a Yankee picket of the 2nd New Jersey Cavalry and his reserve and captured him. The wagons were quickly loaded with hay and returned safely across the river.

Private Silas Jones wrote home on December 19:

> Christmas is near at hand & today finds me in the cold —. We are building winter huts . . . I am afraid that I will not have a cabin to go in

to at Christmas — working hard to get it done. . . .I had to go on picket the next morning (after) I got here — on the river towards Fredericksburg. They awaited the enemy in rifle pits but instead of advancing they retired-as usual. . . .these are fifing times with us. All the boys have their fifes but myself. . . .All are cheerful and gay with their fifes. . . .

It snowed on December 23 and the men of the 52nd spent Christmas in their huts to escape the extreme cold weather.

During December, Lieutenant Colonel Ross resigned because of his wounds and because he had brought charges of incompetence against Colonel Skinner and they had been dropped by a Court of Inquiry. Major Watkins was promoted Lieutenant Colonel and Captain John D. Lilley of Company H was elected Major.

1864

The new year started with a continuation of cold weather and snow. The monotony of camp life was broken by the inevitable picket duty on the Rappahannock. Rations were much better during this season than the previous one. Coffee, sugar, molasses, desiccated vegetables and rice were often issued to supplement the usual fare of bacon, beef, pork, corn meal and flour. The brigade sutler even had whiskey for sale—for $2 a drink. Inspections and drills were held when the weather would permit. On January 19 the whole brigade was marched out to witness the execution of a deserter from the 49th, a sight they would never forget.

During January the 52nd was issued a new flag. The old one, full of bullet holes and badly worn and faded, was delivered to Governor Letcher by Colonel Skinner.

In February an effort was made to reenlist the men for the war. General Pegram, Colonel Terrill of the 13th, and Colonel Skinner spoke to the brigade. The colors of the regiments were then marched to the front and the men willing to reenlist were told to march to the colors. About one half of the 52nd expressed their willingness to serve to the end of the war at that time. Company D held a meeting on the subject and wrote to the Staunton Spectator to announce that they had reenlisted to the man. The other companies soon did likewise.

Chaplain McGill wrote to the newspapers requesting books for the men to read in their spare time.

Returning from leave on March 12, Private Kersh found the regiment in a new camp:

> I found the boys in our new camp cabbins all complete and comfortable. They moved in them five or six days before I arrived. I found the boys in F all well and in fine spirits. And about the first question asked after how do you do how did you enjoy yourself at home was have you got your old fiddle along. I told them I had. I had to get it right out and commence playing right off. We had a stag dance in one of our cabbins every night most. . . .We had several big rains here lately which raised the Rapidan very much. It is a river that raises very quick any way and is vey rapid. . . .I went fishing the other day, We caught a good mess. Duty is heavyer than ever on us. A heavy picket is kept up along here and a good many in the guard house to guard. Keeps us on duty nearly all the time. Role call four times a day and drill twice a day when the weather will permit. . . .

On March 24 the 52nd challenged the 13th Virginia to a snowball fight. Private H. J. Mugler of the 13th described the two hour mock battle as "Both parties making and receiving charges."

Writing his brother on April 3, Kersh described the bad weather: "We still picket along the rapidan. We had some very rough weather last month. On the 22nd the show fell 11 inches deep here. We had a good deal of rain to boot. On the 30th the rapidan was very high. It was level with the dam...."

Mugler, an ex-U.S. Army soldier, described General Pegram: "Pegram is a very neat looking officer and shows the West Point training very plain.... (He) is too fond of drilling to be a good General."

Inspections and drills increased as Spring arrived. On May 1 a corps of sharpshooters was organized in each brigade in Lee's army. Lieutenant Coiner explained:

> An order was given that every regiment must organize a Corps of Sharpshooters. When I heard of it I concluded I would like to command the sharpshooters of our Regt. I went to the Col tent to apply for the position & when I entered his tent he remarked that he had just sent his orderly to tell me that he wanted to see me about organizing the sharp shooter Corps. I received what I wanted. I was permitted to select proportionately from all the Companies forty six men & four Non Com officers & one Lieut-and by that time I knew all the men in the Regt well enough to select the best soldiers. I first drilled them throughly in skirmish drill. Then I got permission to practice shooting at a target. And when the spring opened and Gen. Grant began his move on Richmond my little Corps was ready for business.

The sharpshooters and the rest of the 52nd didn't have long to wait. General Grant sent his army across the Rapidan on May 4. The Yankee pickets across the river opened fire on their Rebel counterparts. Once Lee was aware of where Grant was crossing, Pegram's Brigade was relieved by cavalry and then marched to Locust Grove and camped for the night in a grove of pines.

Aroused at dawn on May 5, Pegram led his men down the Orange Turnpike. The roar of musketry ahead hurried their footsteps. Loaded ambulances and walking wounded soon came past in increasing numbers. Gray-clad dead also lay along the road. Moving to within 200 yards of the Confederate lines, the men were greeted with the sound of a Federal charge and a terrific roar of musketry to their front. Pegram's men ducked behind logs, trees and rocks, as bullets flew around them. After about 20 minutes the firing ceased and the brigade continued moving to the extreme left flank of the Southern army.

Lieutenant Coiner remembered the first day's action in the Wilderness:

Our Brig. did not arrive on the battle-field until nearly night. We were place in line on the left of Gen. Lee's line of battle. I was immediately ordered to the front. When I had advanced as far as I was ordered to go my men knew what to do, to pick the best place near at hand for protection. It was not long that we had to wait, for the enemy skirmish line came in sight of us, but they did not see any thing to stop them until they were near enough that every one of my men could pick out his man. Then I told the sergeant to fire which was the signal for the whole line to fire, and if any of the enemy were not killed or wounded he did not show himself. Then it began to get dark. I could only hear commands given by officers and I could judge that they were forming to advance upon our line of battle & I believe that they thought that there was nothing in front of them but skirmishers. I sent one of my men back to tell Gen Pegram what I believed he might expect. I heard the command given to advance and on they came. I started another of my men back to tell the line to be ready and when the first time one came to the top of a little rise in the ground I ordered my men to fire and to fall back without firing any more untill in the line of battle. The brigad (e) was ready for them behind breastworks and when within twenty paces the whole line fired at once. I have never heard any thing like it except at the battle of Gain's Mill. The enemy did not advance a step after we gave them the first volley and when we could no longer see any more flashes of there guns the order came; Sharpshooters to the front and frankly that was not a welcome command-to me only for an instant. I remembered I was in the position that I had and that the propper place for me to give my first Command was out in front of the breastworks. My Command knew my voice and forward we went. We could easily tell by the line of dead & wounded the distance the enemies line was from the breastworks when the Brig fired the first volley, And I found out what I expected that some of the enemy had not retreated but laying down behind logs and large trees. One a Lieut-told me that he was in the third line in the advance.

Captain Jame Bumgardner wrote of the days action:

The brigade reached the extreme left of the position taken by the Confederate army about 4 o'clock in the evening, and was placed on the left of the army, on the prolongation of the line. The line was through a wood and the ground in front was covered by trees and by dense undergrowth. Immediately after taking position, the troops were ordered to erect breastworks. The men worked very reluctantly, they were tired and grumbled a good deal. They remarked to one another that the work was useless, that they had been throwing up breastworks ever since the war began, and had never fired a shot behind one. That they had always whipped the Yankees on open ground and that the Yankees were not such fools, as to attempt to drive them out of breastworks. By about sundown, a pile of logs three or four feet high, with some dirt thrown up against it had been raised along the front of the brigade, and between sundown and dark the skirmishers came in, announcing that they were followed by the Federal forces.

The timber and undergrowth in front was so dense, that the attacking force could not have been seen until it appeared within twenty or thiry yards of the Confederate line.

The attacking force moved in a close column with a front of one regiment, and with a depth of eight or ten regiments, about eight or ten steps apart.

The Fifty-second Regiment was in the center of the brigade, and the attacking force covered the entire front of the Fifty-second regiment, and about half of the regiment on the right, I think the Thirteenth Battalion (Regiment), I am not sure.

As soon as the foremost regiment of the attacking force came within sight, the brigade opened fire upon it. The range was very short and the Federals disappeared from sight in the jungle in a very few minutes after the firing commenced.

A desultory fire was kept up by the Federal troops back in the jungle out of sight for about twenty minutes, which was replied to by us. And then the firing ceased, and the engagement ended. It was quite dark when the firing ceased.

Some time, an hour or more, perhaps, after the close of the engagement, a skirmisher came in from the front, and as he crossed th breastworks, I heard him say to his comrades: "Boys, you just ought to see how the Yankees are momicked up out there in the brush."

This remark excited my curiosity. My supposition was that the Yankees had retired too soon to have suffered a heavy loss. The moon had risen by that time, and I went out into the thicket, and was astonished at the number of dead and wounded lying on the ground. I never saw dead and wounded lying more thickly anywhere. Evidently the rear lines stood, after the front lines fell back out of sight, and our fire was into a packed mass which retired slowly under it.

The casualties on our side were very few. In the Fifty-second only two or three were wounded.

General Pegram, who was sitting on his horse behind the breastworks, and therefore exposed, was wounded.

The result of the engagement reconciled the men to the labor of erecting works.

They went to work vigorously, and by morning, a substancial breastwork was completed.

A space of perhaps forty yards in width in front of the works was cleared of brush. At my suggestion trees along the edge of the clearing were blazed, to guide the men in the elevation of their guns, and the men were instructed to aim below the blaze marks.

The next morning between daylight and sunrise, the attack was repeated, with the same formation, and with apparently about the same number. The company next to mine, located in the center of the regiment, happened to have no officers present, and was under my charge, and thinking it possible, that as the distance from the edge of the clearing to the works was so short, that the Yankees might attempt, after the first volley was fired, to rush over us before we could reload our guns, I ordered the two companies to hold their fire until I gave the word.

When the first line came into view, a crushing volley was poured into it, and it immediately fell back out of sight, and in a minute or two afterwards, the second line moved into the open. I ordered my two companies to fire, and every gun cracked at once. The Federal line staggered back. Our firing into the jungle continued for perhaps twenty minutes, and the engagement ended, Captain (William H.) Burns of the Fifty-second was killed, Captain (Robert C.) Davis was struck in the head, as he was in the act of firing a gun which he had taken from a private. As he fell backwards, his gun went off killing Sergeant (Lieutenant George W.) Moore and wounded a private whose name I have forgotten.

During the two engagements not a straggler left the line. And it was told us, during the day in the works that General Early had ordered the rearguard to be taken away as a useless formality and an unmerited imputation upon the men of the brigade.

General Early reported on the action in part that "a very heavy attack was made on the front occupied by Pegram's brigade (now under the command of Colonel John S. Hoffman of the 31st Virginia Regiment); but it was handsomely repulsed, as were several subsequent attacks on the same point. These attacks were so persistant, that two regiments of Johnson's division were moved to the rear of Pegram's brigade, for the purpose of supporting it; and when an offer was made to relieve it, under the apprehension that its ammunition might be exhausted, the men of the gallant brigade begged that they might be allowed to retain their position, stating that they were getting along very well indeed and wanted no help."

Captain James Bumgardner's account continues the regiment's story:

I was told a day or two afterwards, as coming from the officer who superintended the removal of the dead and wounded, that thirty-six hundred dead and wounded Yankees were buried and taken to the hospital from the front of the brigade. My recollection is, that Captain Garber of the Staunton Battery, made the statement to me; and that he stated that the information was given him by the officer under whose charge the dead and wounded were removed.

The brigade remained in the works until toward sundown in the evening of May 6th, when General John B. Gordon and his command moved to strike the Federal Army to its right flank and rear, and Early's old brigade (the brigade continued to be called Early's Brigade in the army after General Early's promotion), moved out of the works to cooperate with Gordon. There appears from the beginning to have been some confusion among the staff officers as to the direction in which the brigade was to move. One idea appears to have been, that the regiment on the left was in connection with Gordon's right and the others that the regiment on the right was in contact with our line of breastworks.

The Fifty-second Regiment was in the center of the brigade. In moving the Fifty-second had to cross a marsh, covered with fallen trees, and a dense growth of underwood and brush, interlaced with vines and briers, and the mud and water in the place was half-leg deep. In struggling through this difficult ground, moving as rapidly as possible in order to keep abreast of the line, the men necessarily became much scattered. Conflicting orders, were repeatidly given, one order was to dress and close to the right, the other order was to close and dress to the left. I heard both orders several times while forcing my way through the swamp. After getting through the swamp the ground was still very difficult; there was no time at least none was given, to reform the regiment. On reaching the immediate front of the Federal works the men were in scattered groups, the rearmost ones twenty or thirty yards behind the foremost ones, and no company under effective control of its captain. This place was probably far to the right of the position the regiment should have occupied if it had kept close up towards Gordon.

The scattered, it would be an exaggeration to say, the disorganized regiment, confronted at short range, the Federal troops in their works, who as yet, were unaffected by the attack of Gordon, who was sweeping along the Federal flank and rear, way off on its left. In fact some of the men had gone clear away to the right and others to the left.

A short line was hastily formed of the foremost men, which moved towards the Federal works. One or two men were wounded by the fire of the men in the rear. There was no support in sight, on either right or left. The line after moving towards the Federal works fell back. The men knew nothing of Gordon's progress and success on their left; they evidently felt that a charge by them against the Federals was useless. A few officers urged the line forward until it dissolved.

The brigade was collected in the night in the Wilderness, between the Confederate and Federal lines of breastworks, and moved back into the works early the next morning. The upshot of the matter as I think, was that the extent of the success of Gordon's movement, brilliant and successful as it was, was checked and restricted in its result by the failing of combined and concentrated coordination of the part of Pegram's or Early's old brigade. And that failing resulted from the seperation of the brigade and conflicting orders. I am sure that the disorganization of the Fifty-second Regiment, the center regiment of the brigade, practically cut the brigade in two at the very important moment when it was most needed and this disorganization resulted from conflicting orders, combined with very difficult ground over which the regiment had to move.

"In Gen Gordons flank attack the sharpshooters advanced when he advanced but Pegrams brigad(e) was not engaged but were fired into by some of our men by mistake in the dark. There was no more fighting on this part of the line by the sharpshooters," added Coiner.

WILDERNESS MAY 5-6 1864

	Co A	Co B	Co C	Co D	Co E	Co F	Co G	Co H	Co I	Co K	F&S	Total
KIA	0	3	0	1	0	1	0	1	1	0	0	7
WIA	2	2	0	1	4	6	2	0	2	1	0	20

KIA's include 4 who died of wounds.

Lieutenant Adam F. Craun of Company F died of wounds received on May 5. Company B lost Lieutenants William H. Burns and Moore killed on May 6. In addition to Captain Davis of Company A, Captains John M. Humphreys of Company I and Elijah Bateman of Company G were wounded, Bateman losing an arm.

Pickets came in on the morning of May 7 and reported the Yankees had abandoned their position. Sergeant "Sandy" Coffman of Company G wrote in his diary: "The Yanks left their ditches during last night. We marched over to their ditches and stayed there until 10 o'clock that night, when we left and marched toward Fredericksburg. Marched until near daylight distance 10 miles. Left there soon after sunup and marched toward Fredericksburg, it was a very hot day and we marched very hard. There was a great many men give out, and fell down in the road. We have no rations and very little water...."

Coffman continued:

9th — Left here in the morning and marched about 10 miles and

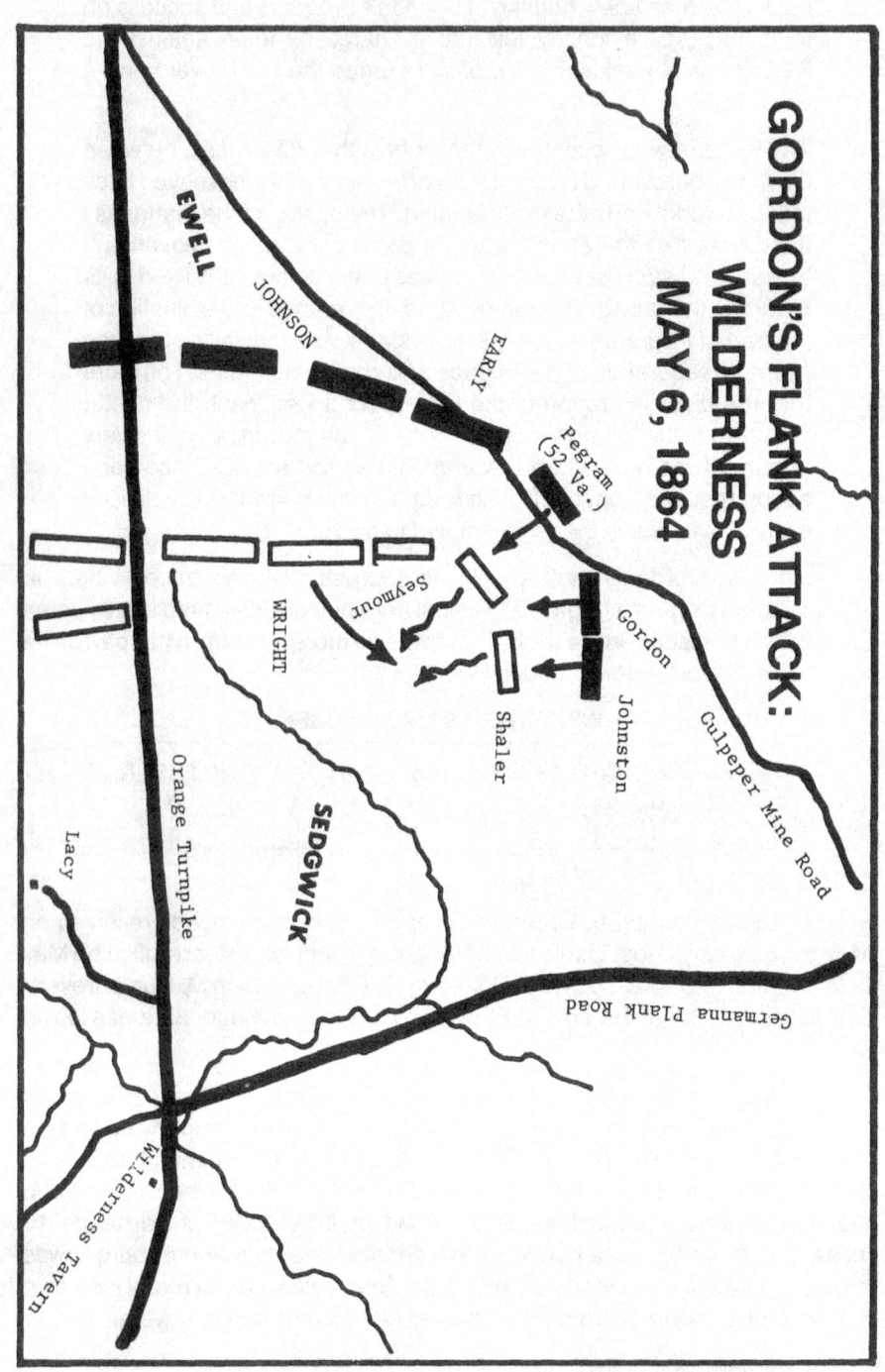

dug a ditch, and stayed there until nearly dark. We then marched around in the rear of some Yanks. When it got too dark for us to execute and then we went back to our ditch and stayed there all night.

10th — Laid there until 10 o'clock in the morning. Then the Yanks got to shelling us, then we marched down behind a hill and laid there a short time; then we marched on down our lines in a piece of pines. Fighting very hard. Then we marched up to the left of our line to support Longstreet's Corps. The Yanks were shelling the front line and killed our assistant surgeon. The Yanks then charged our right wing and run Gen. Doles brigade out of the ditches. Then our men reinforced and run them back to their lines; back to the right and was now dark. We marched back 5 mi. We got there about 12 in the night and laid there until morning.

11th — We stayed there until eight o'clock in the morning and marched to the front ditches. Then at night we were ordered back to the rear again.

Spottsylvania May 12 — Left there before daylight and marched to the ditches again. Soon after we got there the Yanks charged our ditches and run our men out. Then our brigade fell back, formed a line and moved forward on the Yanks. Gen. Lee rode out in front of our line but some of the officers told him to go back to the rear for he was in too much danger, and took his horse by the bridle and told him we could drive back the enemy without him coming in so much danger.

General Gordon commanded the 3,000 men who were about to attempt to retake the "Mule Shoe" salient which had been overrun by the Union troops. Captain Bumgardner stated:

The effect of General Lee was wonderful. General Lee offered to lead the troops. There at once burst out from 3,000 lips the cry 'General Lee to the rear', and not a foot would stir until Lee was led back through a gap in the line, and the word was given and the line moved forward without pause, or waver, or break, right on, up to the very face of the solid opposing mass; on till sabres clashed and bayonets crossed; on 'till the first line was diven back in confusion upon the second, and the 1st & 2nd upon the 3rd; on, and into the angle of the salient, where batteries, massed on right, massed on left, poured a storm of shot and shell upon either flank, and still on, pressing back the stubborn heavy mass, covering the earth in piles with the slain, 'till the enemy, his organization lost in confusion, retired from the dreadful carnage, yielded back the captured works, and the crisis passed and the field was saved.

"Bent" Coiner also recorded an account of that day's fighting at Spotsylvania:

It was our Brig. & Gen Gordon on the 12 of May that recaptured our breastworks at the Bloody Angle. The troops that Gen Lee wanted to lead in the Charge Gen Lee passed to the rear through the 52nd Regt. We all had some new experiences in that charge Officers fought with sword & pistols & the men their bayonets. Our Regt went into that charge with a new flag & it was simply torn in shreds and the staff was shot intoo three times in the hands of the Ensign (Lieut. Lackey of Rockbridge Co.). My Brother who Commanded CoC was killed after we had retaken the works. That was my first real trouble and it was terrible. Our Regt captured three battle flags from the yanks & one of our flags captured when the works were taken. The fighting was terrifick all the day. The sharpshooters fought with their Co in that Charge.

Sergeant Coffman continued: "Our brigade soon drove them back but lost a great many men. In our Company had 1 killed and 5 wounded. That was the hardest fight had fought since the war. We stayed in the ditch until nearly daylight the next morning. Then we fell back one mile and dug more ditches. It rained all day. The Yankees tried to run us back but could not."

Adam Kersh wrote home:

Thursday the 12th. This morning the yankees made a great charge on our breastworks. Jones Brigade gave away and they got through. Our brigade was in reserve. We charged them and drove them back. They got through at several other places but were driven back again with great slaughter. By Jones giving away others had to give away to keep from getting flanked. Wounded in our company in this charge David Snell in wrist but not very badly and Sergeant C. Walker in the face not serious. And myself very slightly on the head with spent ball is all in our comp. Our loss in officers is greater than privates according. Col Skinner was badly wounded in the face. Captain Airheart (Abraham Airhart) was wounded, Lieut. (Charles L.) Weller wounded, Lieut Fry (Joseph Frey), and two Paxton lieuts (Captain Samuel W. and Lieutenant Joseph S.) in the Rockbridge Company. Captain Humphreys wounded. Lieut (Joseph F.) Coiner killed. Our Dr was killed by a shell while we were supporting our lines Dr Edwinson (William H. Edmondson). We have two Drs with our regt yet Gibson and Ewings (Ewing). 111 killed wounded and missing in our regt 78 total 18 com officers 4 killed 10 privates killed dead ballance wounded —.

Captain Airhart later died from his wounds. Captain William F. Gillett of Company K, Lieutenants William S. Kinney of Company A, John D. Summers and John M. Lambert of Company I were also wounded. Lieutenant Colonel Watkins had taken command of the 52nd during the battle.

SPOTSYLVANIA COURT HOUSE MAY 10-12, 1864

	Co A	Co B	Co C	Co D	Co E	Co F	Co G	Co H	Co I	Co K	F&S	Total
KIA	0	0	3	2	2	1	3	0	0	1	1	13
WIA	6	3	3	2	7	3	4	3	4	2	2	39
POW	0	0	0	0	0	0	0	1	2	0	0	3

Those who died of wounds are included in the KIA figures. Kersh's figures cannot be resolved except for slightly wounded men such as himself or those who disappeared from the muster rolls during this period.

"Nothing but sharpshooting for the past three or four days here," wrote Kersh later. "We are lying behind our breastworks waiting for the enemy to advance. Some think that old Grant is crossing back but I dont believe it. They are fortifying in our front still." Coffman added on May 15; "...it has been raining nearly all day. It is very muddy...we had one man wounded by a stray ball. It was D. M. Coiner."

Coffman continued in his diary:

16th — Laid there all day everything was quiet in fore part of the day. But in the afternoon the Yanks advanced their skirmishers. But our skirmishers kept them back. Our batteries fired at them a great deal.

17th — Skirmishing all day and after dark the Yanks beat their drums and played their bands a great deal, and so did ours.

18th — The Yankees attacked us early in the morning and drove our skirmishers in but their line of battle did not get close enough to our ditch for us to fire on them with our small arms. But our artillery drove them back and killed a great many of them. Lieutenant C. B. (Bent) Coiner was wounded by a stray ball.

Coiner knew better: "I was wounded on the 18th of May one of the enemies sharpshooters climbed out onto a large tree where he could see my head over the little pits which we had and he was about one thousand yards away his first shot missed my face about an inch. We had not discovered him then & (I supposed the ball had hit a limb and glanced) but the second shot passed through my cheek out of my mouth wounding my wrist and wounding one of my men that was in the pit with me."

A grandson recalled more of the story: "He was sitting with his back to a tree directing fire when a Yankee sniper shot him through both cheeks. He lost some teeth and a chew of tobacco. One of his men saw the puff of smoke when the Yankee shot. He was up a tree and said, 'Captain, watch me bring him out,' and when the gun cracked, out came the Yankee head first. I guess that was why Grandfather always wore a beard to hide the scars on his face."

Coffman went on to describe the Battle of Harris Farm on May 19: "19th — Left there about the middle of the day and marched in direction of Fredericksburg. After going 7 miles we met the Yanks and our regiment was engaged from sun down till ten o'clock in the night. It was a very hot place we were in an open field. We fell back from there. About 12 o'clock marched back to our same old place, our Co. lost 2 killed, 1 wounded, and 1 prisoner. It commenced raining about 10'oclock and rained all night and was very muddy."

General Stephen D. Ramseur reported that at Harris Farm, "fortunately Pegram's gallant brigade came in on my left in elegant style just as the enemy was about to turn me there." General Ewell wrote: "I was about to retire when he attacked me. Part of my line was shaken, but Pegram's brigade of Early's division (Colonel Hoffman Commanding) . . .held their ground so firmly that I maintained my position till night-fall, then withdrew unmolested. . . .".

Lieutenant Coiner noted: " . . .in the battle of the 19th of May and were opposed by Colored troops in this battle, and at Bethesda Church." Coffman added "Caroline Co. May 20th — Everything is quiet in front today and no Yanks to be seen."

SPOTSYLVANIA MAY 16-20, 1864

	Co A	Co B	Co C	Co D	Co E	Co F	Co G	Co H	Co I	Co K	F&S	Total
KIA	0	1	0	2	3	0	3	0	0	1	0	10
WIA	2	3	1	0	4	1	4	1	4	1	1	22
POW	0	0	0	2	2	0	3	0	0	0	0	7

Major John D. Lilley was wounded on May 19.

Coffman's diary detailed the regiments movements the rest of the week:

2st — Left here at daylight and marched in the direction of Hanover. I mentioned it was a very warm day and we had very bad water to drink. This is a very bad country for there is no houses to see. The country is nearly all pine bushes. We passed a very few neat places after dark. We stopped about 11 o'clock in the night. Distance that day was 20 miles or more.

22nd — Left there before daylight, the country is a great deal better than it was. We reached Hanover about 1 o'clock in the morning and laid there until nearly morning. Then we marched one mile to a place to spend the night. (Two men were taken prisoner during this march).

May 23rd — Left next day marched toward the river (North Anna), and dug ditches. Distance 1 mile. Left there after dark, and marched back where we were and stayed all night.

24th — Left there before daylight and marched up our line, laid behind the V.C.R.R. (Virginia Cental Railroad) but soon left there and marched further up. Then in the evening marched back to the Junction and laid there in the front line of battle until our men could dig ditches. It rained very hard and we had a very muddy place to stay. (Lieutenant Oscar C. Lipscomb of Company A was wounded during the day).

25th — Left there early in the morning and marched to the rear and laid there nearly all day. Then we marched to the right of our lines, and dug another ditch. We had to work all night, and it rained nearly all the times. We spent a bad night.

26th — Left there before daylight and marched to the Junction. Laid there until after sun up and marched near Catholic Station on the V.C.R.R. 10 miles from Richmond. We passed by the residence of General Wickham. It was the neatest place I ever saw. We passed some very Bad water. We marched 15 miles that day.

28th — Left there at three o'clock and marched toward Gaines' Mill where we fought 2 years ago in the Chickahominy Swamp. We stopped before night and spent the night here.

29th — Laid there until nearly night then we marched to the front to see how many Yanks were there in the marshes. Drove the Yankees into their line of battle. . . .We dug a ditch during the day. We spent the night where we were last night.

30th — Stayed there until twelve o'clock, then we marched to the right and went to the front and turned . . . to the left flank. Our brigade had to charge the Yanks ditch but they had a very high fence about 50 yards in front of their ditch.

Six or eight pages unfortunately have been lost from his diary but Sergeant Cleek's diary picks up the story: "The skirmishing was very heavy all along the line until late in the evening, when Pegram's brigade was ordered to charge on the enemy's works, which they did without any support and was badly cut up."

Lieutenant Colonel C. B. Christian of the 49th Virginia described the attack:

The enemy on our front shifted their position and threw up earthworks lower down the road, and parallel to it. Orders came to Early's old brigade . . . to march down the road and make a reconnoissance. . . . The brigade was marched down the road, the Forty-ninth at its head, for some distance and halted, General Ramseur "bossing the job."

I then heard a single piece of artillery firing at intervals in a strip of woods on the left, and being at the head of the column, I heard General Ramseur say to General Early: "Let me take that gun out of the way." General Early vigorously advised and protested against it, but Ramseur insisting, he finally acquiesed in the move.

The brigade was fronted to the left and the advance started. The gun immediately retired to the works as a decoy and no resistance was made to our advance then.

Presently we came to a level, open field, one half-mile across and could see on the opposite side at the edge of another strip of timber behind which artillery was massed-heavier than I had ever seen, unless it was at Malvern Hill...and bayonets bristling as thick as "leaves of Vallambrose," supported by three distinct line of battle....

They had evidently taken the exact range to the edge of the woods. As soon as the brigade was well into the open fields the enemy opened with the heaviest and most murderous fire I had ever seen with grape, canister and musketry. Our veterans of a hundred fights knew at a glance that they were marching to die, but like the old guard under Cambranne at Waterloo they preferred to die, rather than to waver. Our line melted away as if by magic-every brigade, staff, and field officer was cut down (mostly killed outright) in an incredibly short time.

...The men who usually charged with the 'rebel yell' rushed on in silence. At each successive fire, great gaps were made in our ranks, but immediately closed up. We crossed that field of carnage and mounted the parapet of the enemy's works and poured a volley in their faces. They gave way but two lines of battle, close in their rear, rose and each delivered a volley into our ranks in rapid succession. Some of our killed and wounded fell forward into the enemy's trenches-some backward outside the parapet. Our line, already decimated, was now almost annihilated. The remnants... formed and sheltered behind a fence partly thrown down just outside the parapet, and continued the unequal struggle, hoping for support that never came.

This was the bloodiest fight of our Civil War considering the number engaged on our side. The per cent killed and wounded was three times as great at that of the French at Waterloo. The loss of officers was fully ninety per cent of all engaged....The dashing Colonel Edward Willis of the 12th Georgia (in temporary command of the brigade, Colonel Hoffman being sick from a spider bite) was killed ...also Major (Lieutenant Colonel) Watkins, the brave soldier of the Fifty-second....

Captain James Bumgardner of Staunton, who was an officer in the Fifty-second Va., told me that his regiment also had only three officers and eighteen men left....Thus and there at Bethesda Church well nigh perished one of the grandest corps of men the world has ever known-made up of the best young blood of Virginia, fighting for their "Lares and Penates...." The absent wounded returned, the ranks were recruited by conscription, but the heroic old Fourth Virginia Brigade died then and there at Bethesda Church.

Lieutenant Coiner added: "...and at Bethesda Church they made a famous charge. Gen-Willis was killed (he had just been promoted & put in command of the Brig.). All of the field officers in the Brig. were killed or wounded. There was but one commissioned officer in the 52nd Regt. that was not killed or wounded. (Capt James Bumgardner)...."

BETHESDA CHURCH MAY 30, 1864

	Co A	Co B	Co C	Co D	Co E	Co F	Co G	Co H	Co I	Co K	F&S	Total
KIA	3	5	5	0	3	4	4	1	3	9	1	38
WIA	10	5	5	2	6	10	5	3	5	1	0	52
POW	5	3	2	1	0	2	4	4	4	1	0	26

KIAs include those who died of wounds.

In addition to Lieutenant Colonel Watkins, who was shot from his horse and killed while gallantly leading the regiment, Captain James A. Dold of Company H, Lieutenants Robert B. McFarland of Company I and Walter Boon of Company K were killed, and Lieutenant William A. Ross of Company C died of wounds. Captain John M. Humphreys of Company G, John E. Hamilton of Company B, Lieutenants John N. Hanna of Company D, Joseph S. Paxton of Company E and John M. Lambert of Company I were wounded. There were only 500 present in the brigade and 298 of them were killed, wounded and captured. Private Mugler described the action as a "complete blunder and failure."

Neither the survivors of the 52nd nor the rest of Lee's army were given any respite from Grant's relentless pressure. The two armies were locked in a war of attrition and there were few replacements for the embattled Rebels. Cooks, teamsters, musicians and other detailed men and the slightly wounded were given muskets and returned to the fray.

Colonel Robert D. Lilley of the 25th Virginia, brother of recently promoted Lieutenant Colonel John D. Lilley of the 52nd, was promoted Brigadier General and assigned to command the brigade on June 2.

During its stay in the trenches around Cold Harbor the regiment had three more men killed. Captain Humphreys, who returned to duty on June 4, was wounded again on June 6 while commanding the unit. Two men were also taken prisoner on the 6th. Unknown to the members of the regiment, four of their number had been captured in the defense of Staunton on June 4.

Sergeant Coffman continued his diary:

June 9th — Stayed there until late in the morning then we marched to the right — about 2½ miles and then stopped in some pines. We are now near Gaines Mill where we fought two years ago....We got orders after dark to be ready to march at 3 o'clock in the morning. It was very warm.

June 10th — Stayed there all day everything quiet —. But late in the evening there was some skirmishing in front and some artillery firing all day and night. It is reported the Yanks are at Staunton. . . .It has been warm all day long.

11th — Laid there all day long everything quiet in front only a shot fired once in a while and is reported the Yankeees has fallen back from Augusta.

12th — Laid there all day, everything quiet in front fore part of the day but in the evening the Yankees throw some shells over us. That night we got orders to march at 2 o'clock in the morning.

When he learned that General Hunter had defeated the small Confederate force in the Valley and was moving toward Lexington, Lee ordered Early's Second Corps to Lynchburg.

The troops marched over the dusty country roads of central Virginia in the summer heat and reached the railroad near Charlottesville on June 16. Taking a train ride the next day, most of Early's units reached Lynchburg in time to save the city. On arrival, General Lilley marched his small brigade immediately to the front, near the outskirts of town, and watched as General Hoke's North Carolina Brigade attacked and drove back the enemy's advance. Lilley then marched his regiments out the New London Road two miles, formed a line of battle, and spent the night.

The morning of June 18 was spent building earthworks and burying the dead cavalrymen of both armies. The Confederate artillery was engaged during the day, and about 4 p.m. the brigade sharpshooters advanced and drove their Yankee counterparts back to their line of battle. They in turn drove back the Confederates, with a few casualties on both sides. The 52nd had one man mortally wounded and one captured.

The Southerners had been without food for two days as the wagons had not arrived. Early found a supply of hard bread in their city, confiscated it, and issued it to the half-starved men. The bread was full of worms, but the hungry troops ate it anyway.

Early was ready to attack Hunter on the morning of the 19th but found that the Union forces had retreated during the night. The Confederates immediately gave pursuit. Catching up with the Federal rear guard near Liberty, Lilley's Brigade was deployed as skirmishers. Advancing toward the Union position on a high hill, the gray-clad soldiers watched the Federals abandon the hill and flee to the rear. The brigade captured a few prisoners without firing a shot. Sergeant Coffman wrote: "We overtook the enemy at Liberty and fought them and drove them back, it was a hard day's march---27 miles. . . ." No rations were issued to Lilley's men that day.

Having led the advance on the 19th, Lilley's Brigade brought up the rear of Early's corps on June 20. They marched down the railroad following

Hunter's army, which had split into three columns. Private Peyton of the 13th Virginia noted that the Federals "killed and destroyed all food they couldn't carry off." Pursuing the enemy to Buford's Gap, the weary Southerners again brought Hunter to bay on a mountain top. After a sharp engagement the Yankees continued their retreat. The 52nd had one man killed, one mortally wounded and one slightly wounded in the fray. The exhausted and famished troops were finally issued rations that night.

Early's men continued the chase for several more days but finally left it up to the cavalry. "Old Jube" started his corps down the Valley toward Winchester.

While camping near Lexington on June 24, seven men from Company E deserted. The next day, as they passed through Lexington, Early's men marched at "reverse arms" and with bared heads past Jackson's grave.

Moving on to Staunton, several more men from the other companies took advantage of being near home to desert. Twelve men had deserted in May and a total of sixteen left during June. Some of these men were recent conscripts but many were veterans of three years service. Several days were spent near Staunton equipping the men. More than half needed new shoes and other items of clothing and equipment.

When Early reached Winchester on July 2, he received orders from Lee to remain in the lower Valley to force Grant to send reinforcements from his army to defend the area, and to destroy the vital Baltimore and Ohio Railroad and the canal system used to supply Washingon and the Federal forces.

Moving to Bolivar Heights the next day, Lilley's sharpshooters drove the enemy through Harper's Ferry and across the Potomac.

On July 6, Early commenced crossing the river at Shepherdstown. Unable to maneuver the Federals off Maryland Heights, Early moved his corps northward through Boonsboro Gap and on to Frederick, Maryland.

Early's advance reached Monocacy Junction on July 9. Finding the Monocacy River defended, Early sent Gordon's division to the right and cavalry units to the left, while Lilley's Brigade was part of the corps demonstrating in front. Gordon's flanking movement proved successful and after a hot fight the Yankes were forced to flee. Private Peyton wrote: "Gordon cried like a child seeing so many of his officers and men dead." Sergeant Coffman added: "...met with the Yankees about 2 miles from the city (Frederick) on the Washington Pike. We left, marching in around there until nearly night when Gordon flanked the Yankees out of their positon, then we marched to the front and crossed the river and followed after for a piece. Then we can back and went into camp. The Yanks left a great deal of plunder behind....We marched about 10 miles altogether."

Coffman continued in his diary:

Montgomery, Md. July 10th — Left there at 12 o'clock in the day and marched towards Washington City, the roads very rough and dusty. we passed through Hyattstown and through Darksburg, it was a very hard days march. We stopped about 1 o'clock in the night. I was very tired. Distance 20 miles.

Washington July 11th — Left there at sunup and marched on towards Washington, passed through a small place called Rockville. It was nearly as large as Staunton. Today was the hottest day that I ever felt. I came very near giving out about the middle of the day. We reached the forts, 3 miles of Washington about 3 o'clock. The Yankees threw some few shells at us. We stopped and waited for further orders. We marched 10 miles. We will spend the night here. Heavy skirmishing in front.

Peyton claimed that "over one half of the men fell out. There were only 100 present in the brigade when we went into camp."

Coffman's diary went on:

12th — Laid there all day, there was some shells thrown over where we were. Very heavy skirmishing in front. Montgomery, Md. July 13 — Left here last night at 11o'clock and marched back the same road that we came, as far as Rockville, then we took the road leading to Leesburg. We marched the balance of the night and the next day until 1 o'clock. Then we went into camp to stay until dark. We have marched near 20 miles.

July 14th — Left there at sundown and marched on towards Leesburg, did not march very fast in the fore part of the night. Went into camp soon after sunup. We came 12 miles, camped near the river. We passed through a little town called Coolsville (Poolsville). We soon left camp and crossed the Potomac River and marched towards Leesburg. Camped 4 miles from Leesburg. The Yankees threw a few shells in our camp. Distance we marched today 7 miles. (Three men were wounded.)

15th — Stayed there all day. Everything quiet in front.

16th — Left there after sunup and marched through Leesburg. Took the Pike that leads to Winchester. Passed through a place called Hamilton, there was some Yankee Cavalry there; made a dash on our wagon train, burning and braking some few things, had to leave. We camped in the forks of the Blue Ridge at a place called Smokestill (Smoketown). We come about 23 miles.

17th — Sunday — Left there about 8 o'clock and marched over the Blue Ridge. After we crossed the mountain we waded the Shenandoah River then we marched down towards Charlestown. Went into camp about 1 o'clock. We only came 12 miles. We have marched 585 miles up to the 18th of July.

Returning to the Valley, Early maneuvered his small force between Winchester and the Potomac. On July 20, General Ramseur, commanding Early's old division, received word from the cavalry of a Federal unit on the Martinsburg Pike, near Winchester. Ramseur, disbelieving that a force of any size was in the area, marched his troops in column out the pike. General Averell's Union force attacked the leading brigade, Johnston's North Carolina, in the flank and routed them, in a battle called, among other names, Rutherford's Farm. Private Peyton wrote humorously: "In a little while here came the artillery and all of North Carolina running like sheep — lost 4 guns and never fired a shot. We stood our ground awhile but so many ran through us that our line was broken up & we took off after the tar heels. I met Ramseur & he was crying, he stopped me & ordered me back, but I didn't go far & went on again. They killed some of our men & broke Col. (General) Lilley's arm so it had to be amputated."

Lilley was also captured but he was rescued a few days later. The Confederates fell back to the large forts around Winchester and Averell's men didn't pursue any farther. While resting there Peyton "met old Jube (General Early). He stopped & said Hello! What Brig. is that? We said 'Pegram's.' 'What the hell did you let Yankees run you for this evening?' We said Gen. Ramseur was the blame for it. We heard he gave Ramseur hail Columbia for it." "Wagonmaster" Hoffman again took command of the brigade.

Coffman's diary reported:

July 20th — Stayed there (near Winchester) until 3 o'clock, then we went down the road about 3 miles and met the enemy. Before we had our line formed the Yanks charged us and flanked us on the right. Our whole division broke and run back to our forts near Winchester. Our loss is heavy. We stayed at Winchester until dark. Then we marched back as far as Kernstown and laid down but was soon wakened up and marched a little piece further. Then we laid down there until 9 o'clock, the we marched on up the Valley. Passed through Middletown, went into camp 2 miles below Strasburg at the river. Distance 10 miles.

RUTHERFORD'S FARM JULY 20, 1864

	Co A	Co B	Co C	Co D	Co E	Co F	Co G	Co H	Co I	Co K	F&S	Total
KIA	1	0	0	0	2	0	0	0	0	0	0	3
WIA	1	0	0	0	1	2	0	0	1	2	0	7
POW	2	0	1	0	0	0	0	1	0	0	0	4

Lieutenant Joseph B. Fauver of Company I was the only officer injured.

Ramseur and his men had the satisfaction of seeing the Union army driven from the field at Kernstown on July 24. Hoffman's Brigade and the rest of the division were held in reserve. "Sandy" Coffman wrote: "...marched towards Winchester, met the Yankees 5 miles on this side of town but soon drove them back with heavy loss. We camped below Winchester, marched 25 miles."

Early continued to vex the larger Federal forces by constant maneuvering. The troops were continually on the go, tearing up railroad track, skirmishing and scouting. Hard marches and short rations were the order of the day. Captain Bumgardner's 52nd Virginia had men wounded on August 11, 18, and 23.

General Sheridan took command of the Union forces opposing Early and began moving his vastly superior army to attack Early's outnumbered men.

The morning of September 19 found Hoffman's Brigade stationed northeast of Winchester between the Millwood and Berryville roads. At daybreak the cavalry videttes in front of the lines were driven in and the camp shelled by the Union artillery. Ramseur's thin division was called upon to stop two Federal army corps. Only a line of cavalry pickets protected the flanks. By 11 o'clock these 1,700 veterans were being forced back by the overwhelming numerical superiority of Sheridan's units. Lieutenant Coiner, who had rejoined the regiment after Monocacy, gave this account: "Our Brig. was the first troops engaged in the second battle on both flanks so we had to change our position a little to the rear (that was I think the first time I ever ran from the Yankeys). We were ordered to fall back. We reformed and charged the enemy and drove them back (Cadet Galt) our Adjutant was killed in this battle." Just as the division was forced to retreat again Rodes' and Gordon's divisions arrived on either flank and assaulted the Federal columns. Private Peyton described Rodes' attack on the enemy's flank:

> When Rhodes started into the woods it sounded like every tree was falling down that a terriffic thunderstorm was raging in the woods. When our men heard all the noise & saw the Yankees running they got over their panic and started back....A Lt of the 12th NC grabbed a flag & rushed ahead crying "forward boys, forward...."

> Our line was soon moving forward. <u>We held the line all day</u>. Rhodes drove enemy back to where they had started & then he was killed & his men stopped & that ended the drive. We held this line all day but did not advance any more.

When the division was flanked out of this position late in the day, Peyton continued: "Fell back to a brick house about 1 mile from Winchester & faced about & fired at the Yankees. They stopped 400 yards off & so spent the balance of the day." In this position the brigade remained under heavy artillery and musket fire. When men began running out of ammunition, volunteers, including Peyton, ran a gauntlet of fire for a resupply.

Late in the day, hearing heavy firing on their left flank and seeing the Confederate troops retiring, Ramseur's division fell back to the Berryville Pike. General Pegram, who had just returned to duty, gathered men to act as sharpshooters to harass the solid lines of bluecoats advancing toward them. A few pieces of artillery arrived and together with the marksmen held the Union forces in check. The main body of the exhausted Confederates retreated through the fields to the Valley pike. Pegram used his brigade as the rear guard covering Early's retirement through Kernstown to Newtown. Reaching their assigned position, Pegram's men fell down and slept in line of battle during the night.

Private John Palmer wrote the parents of James Wiseman concerning his wound and capture:

> he was wounded...and left in the hands of the Yankeys. We had one of the hardest days fitings yesterday that has ever been in the valley yet....he was stricken in the side...and thear was no chance to get a man away. We had a very hard days fitings. We commenced just after daybreak and never ceased till after dark and we had to fall back till what is called Fishers hill. I look to have it hear again....I hope we will drive the Yankeys back recapture the place where our wounded is and get them all back again....

Young Wiseman later died of his wounds.

The all-day fight had cost the small 52nd dearly.

WINCHESTER SEPTEMBER 19, 1864

	Co A	Co B	Co C	Co D	Co E	Co F	Co G	Co H	Co I	Co K	F&S	Total
KIA	0	2	0	0	0	0	1	2	1	0	1	7
WIA	2	3	1	2	5	1	0	0	1	3	0	18
POW	2	0	0	4	0	1	1	0	1	1	1	11

KIA's include 6 who died of wounds.

Adjutant William Galt led the long list of officers casualties in the regiment. His was a particularly hard-felt loss. Lieutenants Joseph Frey of Company G and James A. Burns of Company B died from wounds received in the battle. James J. Wright, a Union soldier from Boston, found Burns with a mortal stomach wound on the battlefield. "I had him carried to a large oak tree out of the sun," Wright wrote. "I did all I could to relieve his suffering. Before he died, he requested me to take his diary and letters to his father near Mt. Jackson or Roseland. I told him I would carry out his wishes." Wright was true to his word.

Captain William R. Gillett of Company K and Lieutenant Oscar C. Lipscomb of Company A were wounded and captured. Captain James Bumgardner of Company F, Lieutenant John N. Hanna of Company D and Assistant Surgeon William D. Ewing were captured. Ewing had been left behind to attend to the wounded.

Captain Humphreys led the survivors of the 52nd to Fisher's Hill the next day. Entrenching at this strong position, the Confederates didn't have long to wait as Sheridan's skirmishers and the brigade sharpshooters were engaged on September 21.

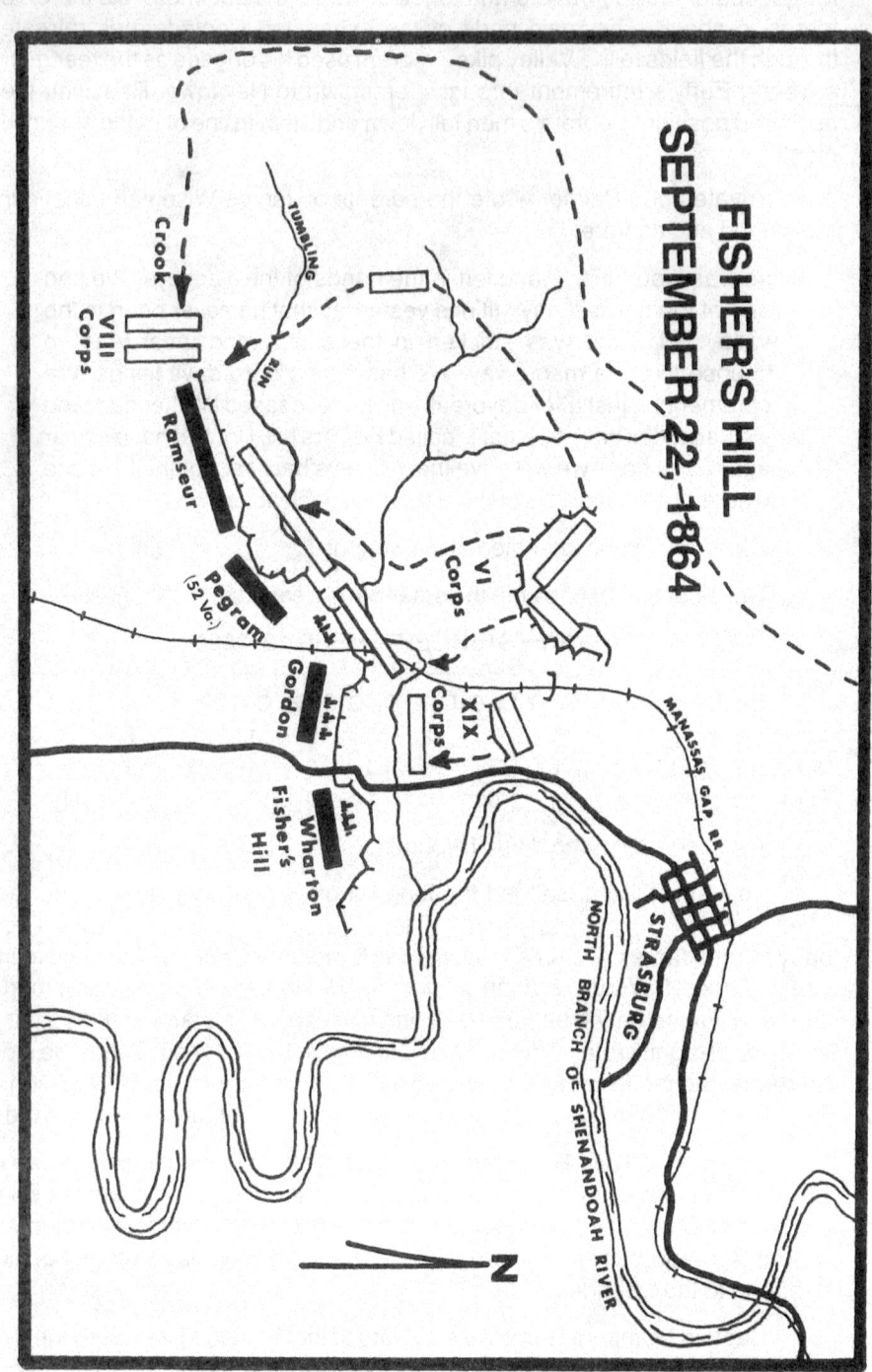

Fighting of a more serious nature broke out on September 22. Early's forces, stretched over a four-mile front, presented little more than a skirmish line to Sheridan's masses. The Southerners' left flank, held by a thin line of cavalry, was turned late in the day. Pegram, now commanding the division, pulled his brigades out of the trenches and tried to form them against the advancing foe, but the other troops ran through and broke his lines and he was forced to retreat also. Upon getting clear of the stragglers, Pegram was able to march his troops off in good order to Woodstock, where they camped. Gordon reportedly interfered with Pegram's plans and caused some of his men to be captured unnecessarily by ordering them back into the works. The 52nd had two men wounded and nine captured.

Retiring up the Valley, Early's worn out troops halted near Waynesboro on September 29. Pegram's Brigade, again under Colonel Hoffman, had only 663 present, with but 409 effectives. More than 500 officers and men were absent as prisoners of war. The 52nd had lost three men captured on the 26th, one wounded and two captured on the 29th, all the results of minor skirmishing and shelling on the retreat.

Early, reinforced by Kershaw's division of infantry and Rosser's Brigade of cavalry, started north on October 5 to drive Sheridan out of the Valley. This was "the burning time" as horrified residents of the Shenandoah called it. Sheridan burned houses, barns, mills, straw stacks, and killed, destroyed or carried off all livestock and food supplies. It was with grim determination and anger, as well as empty stomachs, that the embattled Confederates marched northward through the scene of destruction and ruin.

Arriving at Fisher's Hill on October 12, Early found Sheridan entrenched in a strong position across Cedar Creek. Finding the mere presence of his army would not force the Federals to retire, and unable to gather food supplies for men or animals from the now-barren farmlands, Early had to attack or retreat.

Gordon and Topographical Engineer Jed Hotchkiss made a reconnaissance and found a route that would place Early's men on the flank of the Union forces. Leaving canteens and other noisy equipment behind, Gordon's division and Pegram's Brigade, still under Hoffman, moved to Bowman's Ford on the north fork of the Shenandoah River. Most of the night of October 18-19 was spent moving in single file over a mountain path getting in position for the assault.

Led by troopers of Payne's Brigade of cavalry, the redoubtable Confederates splashed across the river at 5 a.m. the next morning. The gray-clad riders easily routed the Union videttes and Gordon's troops moved to flank the Federal lines. Pegram's division followed in support of Evans' and Ramseur's as they overran the Yankee encampments. Early had led other Southern infantry in a successful attack on the other entrenched camps. Pegram's division was soon ordered to the front as additional resistance was met near Belle Grove. Captain Samuel Buck of the 13th Virginia best describes the action; "We moved forward in beautiful style, driving the skirmishers as dust before the wind until we rose upon the hill in our front where we struck a solid line and much artillery. Passing over a hill, swept by artillery and infantry tries mens' souls and to my horror the brigade stopped. . . .The whole line was heavily engaged and it was all-important to charge and pass this point."

Buck quickly ordered the 13th Virginia to charge and the whole brigade moved forward to Marsh Run. "Col. Hoffman came up and ordered the brigade to halt and reform for another charge," Buck added. Buck saw the Union artillery, which had fallen back, going into battery to open on the Virginians and he pleaded with Hoffman to charge and take the guns. Hoffman finally relented to Buck's insistence and again the young Captain moved to the front of the 13th and called on them to take the artillery. The 13th moved forward with a shout and " . . .they followed across the run and up the hill and we were upon the guns and in a moment they were turned upon the enemy who were in full retreat, "Buck wrote. "Bent" Coiner recalled the 52nd's role in this feat: "The enemy made a stand and our Regt was ordered to charge a battery, which we did, the infantry run before we came very close to them but the gunners fought there guns until they were knocked down or bayoneted. One of my men killed a Major with the butt of his gun."

Buck went on: "The brigade now being in trim we moved forward, driving everything and halting for nothing until we passed through on to left of Middletown where we formed with our right on the turnpike at the tollgate and here we stayed all day waiting for orders or to be attacked."

Sheridan, absent in Winchester when the battle commenced, returned and rallied the Federal troops. Maneuvering his much larger command against the triumphant Rebels, he began to force them back. Early directed Pegram to march his division back from Middletown. Coiner remembered: "Our Brig. was the only Brig. that marched off the battlefield in the evening in order fighting as we fell back with a Battery with the (gun) trails dragging the ground." Most of the Confederate units had become discouraged and disorganized in the face of the overpowering Union advance, and fear of being captured caused many otherwise brave men to break to the rear. Peyton said the "retreat became a mob." "When we got to Cedar Creek Gen. Pegram halted the Brigade until stragglers got by them, " Buck continued. "At Stickley's Shop Gen. Pegram rallied about a hundred men who we thought could stop the cavalry but they came at us in such force we had to break."

Coiner continued:

After crossing Cedar Creek on the hill this side we were asked to make one more effort (it getting dark hoping to save the Artillery & wagon trains.) We formed across the pike, the enemys cavalry had crossed above where we crossed and came thundering upon us. I do not believe they knew we were there until we fired upon them, which did not stop them (we killed a lot of them). I took in the situation quickly, the yanks were ordering our men to surrender etc. There was a rock fence on one side of the (road) and some woods about two hundred yards away, I jumped the fence regardless of the consequences. I could hear the pistol balls passing. Once I got to the woods and made my escape.

Buck also escaped but Peyton caught his knapsack on the fence and was captured.

The diminutive 52nd had taken more casualties. Captains Humphreys, who commanded the regiment, and Robert C. Davis of Company A were captured and Lieutenant Benjamin F. Hildebrand of Company A was wounded.

CEDAR CREEK OCTOBER 15, 1864

	Co A	Co B	Co C	Co D	Co E	Co F	Co G	Co H	Co I	Co K	F&S	Total
WIA	1	1	1	1	2	3	1	0	3	1	0	14
POW	2	0	1	1	1	1	1	0	1	0	0	8

Lieutenant Coiner was proud of his unit's part in the battle: "Our Brig. & Rgt distinguished themselves...(where we whipped Sheridan's Army so badly in the early morning and our Army acted so shamefully in the evening). There was a report current in the Brig. for some time that we would receive a badge of Honor for our conduct in that battle but it never materialized."

Withdrawing to New Market, Early went into camp to reorganize and refit his exhausted men. Some conscripts were assigned to the 52nd and a number of men returned to duty from detail and wounds but the regiment was still pitifully small.

Early, ever aggressive, advanced his little army to Newtown and Fisher's Hill, inviting Sheridan with 40,000 men to attack his resolute 8,500, but no engagement took place. Unable to subsist his men in the lower Valley, Early was forced to withdraw toward Staunton. Lt. Col. Lilley, who returned on October 27, had his hand and arm accidentally crushed the next day while building fortifications. Despite this painful injury he remained in command of the brigade during this operation.

The Confederates went into camp near Fishersville and the railroad, where supplies could be obtained.

Moving to Waynesboro on on December 8, the members of the 52nd boarded trains and left the Valley, many for the last time, to rejoin Lee's army at Petersburg. Arriving there, Lilley marched the regiment to winter quarters near Burgess' Mill, south of the city. Buck commented on the camp; "Col. Hoffman had it all done by rule and a very nice camp it was. All the huts were the same size and the grounds were well kept, fenced in by a brush fence all around the Brigade. We also had a large chapel built in which we would attend prayer meetings every Wednesday evening and prayer meetings were held every night in some of the huts."

Some slight skirmishing took place on the picket lines — the regiment had two men wounded on December 12 and one on Christmas eve. The numbers of the 52nd were again bolstered by new conscripts and the return of sick and wounded men. A liberal leave policy helped morale as many were able to visit their homes and obtain much-needed food and clothing. Rations were meager but boxes from home helped the men through the Christmas season.

1865

The winter of 1864-65 proved to be a very severe one with deep snows and extreme temperatures. A member of the 52nd took time to write to the Staunton papers thanking the ladies of Churchville for the dinner sent for the new year. Good huts and abundance of fire wood helped fend off the cold while the men were in camp, but thin and wornout clothing and shoes did little to ease the suffering when on picket.

The lull in active campaigning was broken on February 5 when Grant attempted another turning movement on Lee's right flank to cut the Southside Railroad around Hatcher's Run. Captain Buck described the brigade's part in the action: "We were ordered in great haste to the line and had a terribly severe engagement. Gen. Pegram put his Division in position....The fighting commenced about 10 o'clock and lasted all day and while I have been in more severe engagements....I never had so many shots fired directly at me. Almost every officer showed himself was shot...the enemy came up and made a dash at us but we repulsed them but did not keep them from renewing their efforts, still they could not move us."

General Pegram tried to advance his division and fell mortally wounded. Colonel Hoffman, commanding the brigade, also received a wound. General Gordon brought up reinforcements and the Confederates finally drove the enemy back.

HATCHER'S RUN FEBRUARY 5-6, 1865

	Co A	Co B	Co C	Co D	Co E	Co F	Co G	Co H	Co I	Co K	F&S	Total
WIA	1	2	0	0	0	1	0	5	1	0	1	11

Lieutenant Colonel John D. Lilley and Lieutenant John B. Lipscomb of Company B were the only officers injured.

"Bent" Coiner had finally been promoted to Captain and commanded the regiment in the battle. He also wrote the preamble and resolutions adopted by the Fifty-second Virginia Regiment at about this time:

> Whereas, the Southern States never bound themselves to make the union that they formed with those of the North a permanent one; and whereas the war waged against our wills is injust, be it therefore,
>
> 1. Resolved, That we have as good a right to be free and to govern ourselves as our enemies have, and that we are unwilling to forefeit that right by accepting at their hands any terms short of independence.

2. That we are firmly convinced that a valorous zeal and steady adherence to determination to conquer by force of arms cannot fail to reward us with honorable peace; that with our resources and an inflexible resolve to be free, subjugation is impossible, and that should we be conquered, it will not be by the stern valor of our foes, but by a want of firmness and determination in ourselves.

3. That we ever acknowledge the hand of Providence in our past successes, and that we trust in the righteousness of our cause and in the justice of our God, and lean upon his strong right arm for deliverance in time to come.

The above preamble and resolutions were enthusiastically adopted by this Rgt without a dissending vote.

The two opposing armies settled down into the seige operations that had gone on before. Coiner reported: "Our Brig. was moved from the right around and placed in the breastworks across the City Point RR & spent the rest of the winter under ground in bomb proofs."

On March 8 a number of men deserted to the Union army, the only such occurrence of record. Desertions from the 52nd had been frequent, but the men had usually gone home to look after their families and farms. While some of the deserters were recent conscripts, others had served faithfully throughout the war. Poor and inadequate food and clothing, and the seemingly hopeless situation of the Confederacy had apparently driven them to take this course of action.

Lee was desperate to break Grant's stranglehold on his army and decided on an audacious plan to destroy as much of the Union army as possible, hoping to gain freedom to maneuver his troops out of their hopeless situation. At 2 a.m. on March 25, the 52nd was aroused and marched by Colonel John G. Kasey of the 58th Virginia, now commanding the brigade, to participate in the attack on Fort Stedman. Captain Coiner was sent for to command the sharpshooters in the assault, but was fortunately absent on leave. The officer chosen was killed in the attack. Led by fifty men armed with sharp axes to cut away obstructions, the brave Confederates approached the Federal sentinels. When challenged the Rebels told their challenger that they were looking for corn in a nearby field and for him not to fire.

When all was ready, General Gordon, commanding the sortie, fired three rounds from his pistol, and the 10,000 men moved forward in their forlorn hope. The axemen rapidly cut away the barriers and the sharpshooters, despite casualties while waiting for an opening, rushed into the first redoubt and gave the 'Rebel yell', which was the signal for Gordon and the main body to charge the enemy works. Rushing forward in two ranks the resolute Confederates captured many of the Federals still in their bunks. Moving forward to the next line of entrenchments, the gray-clads stormed them. Taking this line of works after a short fight, the Rebels found themselves being attacked from both flanks by the fully aroused Yankees. Daylight had come and the Union forces had recovered from their initial surprise. Kasey found himself and the brigade in an isolated position, and sent a messenger to Gordon requesting permission to fall back. Seeing the hoplessness of the situation. Kasey did not wait for an answer, but ordered the men back. Under a severe fire from front and both flanks, the brigade attempted to return to friendly lines. Confederate fire added to the confusion. The gauntlet of over 400 yards was too much for many of the soldiers, and the Federals had reoccupied some of the first line of works. Many of the Southerners were killed and wounded and more than 1,900 were captured, including Colonel Kasey. Major Henry Kyd Douglas of General Walker's staff took command of the remnants of the brigade.

Quarter Master Sergeant "Tom" McClure wrote his parents the day of the battle concerning the fate of his brothers: "Jimmy was wounded today in the thigh . . . he was left in the enemy lines. George was taken prisoner he stopped with Jimmy for a few minutes and the enemy came upon him. . . . The entire line of Yankee works were taken but on account of some of our troops behaving badly we were compelled to fall back to our original lines. Jimmy was wounded while coming back. . . . I can't give you the particulars at present but we have gained nothing. Captured one thousand prisoners. We lost about five hundred." Jimmy's death were reported to Sergeant McClure on March 27 by the surgeon of the 7th Rhode Island Infantry.

Captain Samuel W. Paxton of Company E, who commanded the 52nd, was wounded and Captain Charles L. Weller of Company C was taken prisoner.

FORT STEDMAN MARCH 25, 18, 1865

	Co A	Co B	Co C	Co D	Co E	Co F	Co G	Co H	Co I	Co K	F&S	Total
KIA	0	0	0	1	0	1	0	1	1	0	0	4
WIA	0	4	4	1	2	1	0	2	2	1	0	17
POW	0	2	2	0	0	0	0	2	1	3	0	10

KIA's include 3 who died of wounds 3 of the wounded were also captured.

Grant now was convinced the time was ripe to destroy Lee's army and engage the gray-clads all along the line, night and day. The Confederates were stretched so thin that men were posted 15 feet apart. Captain Coiner returned from furlough and took command of the remains of the 52nd. The young Captain described the action on April 1 and 2: "When Gen. Grant began his attack on the right wing of our army our Brig. was attack(ed) early in the morning but we repulsed the enemy easily although we were not more than a skirmish line." The known casualties for the two days were two wounded and three captured.

Major Douglas led the battle weary brigade on the retreat. He recalled:

While my Brigade was waiting for its order to move....shells were bursting...over the town and the air was...illumined by the baleful light of mortars....(The next morning) I was directed by General Walker to fall back with my brigade and march it until relieved as the infantry rear guard of the army. They impossed on me a dreary and ceaseless vigil for more than twenty-four hours — sleepless, foodless, cheerless. At Amelia Court House another trial was added to our many: the rations for the army expected to be there had been run on to Richmond. We had nothing to eat; hunger became an ally of the enemy....The little army was willing to march and fight, but starvation made stragglers of them; commands were reduced to skeletons, as their men had been some time before.

Coiner added: "On our retreat one day we received one ear of corn for a days rations."

The alternate marching all night and fighting all day rapidly took its toll of the tired and starved soldiers. Moving through Farmville, where some rations were finally issued, the Confederates reached Sayler's Creek, and were sharply engaged. This fight and the other skirmishes had cost the 52nd at least one wounded and 19 prisoners. Douglas described the action at Sayler's Creek: "...We were roughly handled and the loss in my brigade was very heavy for its size. We were driven back but I had held the hill assigned to me until the trains and artillery of the division had crossed the stream in my rear....After this battle my brigade did not number over 500 men and we moved on to High Bridge."

The exhausted and worn out veterans plodded on to Appomattox Court House where Union troops blocked their retreat toward Lynchburg.

Here, on the morning of April 9, the 52nd participated in what it claimed was the last infantry charge of the Army of Northern Virginia. Major Douglas led the brigade and Captain Paxton the regiment, in conjunction with General Fitzhugh Lee's cavalry, in an attempt to reopen the avenue of retreat. The Confederates advanced with the "Rebel yell" and the blue-clad cavalry men were driven back upon a solid line of Union infantry, which did not advance. Major Hunter of Gordon's staff arrived and notified Douglas that General Lee had surrendered the army. Douglas commented: "This accounted for the peculiar conduct of the enemy. But my little brigade had fired the last shot from Lee's army, and I sadly moved it back to its place in the division."

Coiner would have none of this surrender business; "We did some fighting on the morning of the surrender. At the last moment when the white flags we see along our lines an assistant Surgeon came riding along in rear of our lines leading a horse. I hailed him & asked where he was going & he said that he intend to try to get out the rear and not surrender. I asked permission to ride his extra horse. He told me I could do so if I would stay with him. We got through safely to Lynchburg . . .I arrived home safely."

Major Douglas recounted the actual surrender:

> When the time came to march out and give up our guns and flags, in surrender, I asked General Gordon to let my brigade — as it had fired the last shot — be the last to stack arms. This he readily granted. . . .A heavy line of Union soldiers stood opposite us in absolute silence. As my decimated and ragged band with their bullet torn banner marched to its place, someone in the blue line broke the silence and called for three cheers for the last brigade to surrender. It was taken up all about him by those who knew what it meant. But for us this soldierly generosity was more than we could bear. Many of the grizzled veterans wept like women, and my own eyes were as blind as my voice was dumb. . . .That line of blue broke its respectful silence to pay such tribute, at Appomattox, to the little line in grey that had fought them to the finish and only surrendered because it was destroyed.

Musician Daniel I. Bush "hung his drum on a gun stock at Appomattox" and the 52nd Virginia Infantry Regiment was no more.

Captain Samuel W. Paxton surrendered 8 officers and 53 men. Only 14 men were armed on April 9, 1865.

Pvt. John J. McCoun

Pvt. David Franklin Rosen

1st. Lt. John Moffett Brown

Pvt. Alexander F. Staubus

Officers of "West Augusta Guards" 5th Va. Inf. Capt. James Bumgardner, Jr. Co. F, 52nd Va. Inf. on far left.

Pvt. Peter Lucas, Jr.

1st. Lt. Lewis Harman

Charles W. Allen
Lincoln U. S. General Hospital
Ward 20, Bed 47

Adj. John William Lewis

THE MUSTER ROLLS

"HERE ARE THEY WHO MARCHED AWAY
FOLLOWED BY OUR HOPES AND FEARS,
NOBLER NEVER WENT THAN THEY
TO A BLOODIER MADDER FRAY,
IN THIS LAPSE OF ALL THE YEARS
GARLANDS STILL SHALL WREATH THE SWORDS
THAT THEY DREW AMID OUR CHEERS,
CHILDREN'S LISPINGS, WOMEN'S WORDS,
SUNSHINE AND THE SONGS OF BIRDS
GREET THEM HERE THROUGH ALL THE YEARS"

From Monument to Confederate Dead — Thornrose Cemetery

This listing of 1506 men who served in the 52nd Virginia Infantry was made up primarily from the Compiled Service Records in the National Archives. Original muster rolls for one company were found in the Virginia State Library. Several post war rosters were located and also used. This information presents a fairly accurate account of these soldiers through December 31, 1864. No muster rolls exist after that time. The rolls of Company K for the year 1862 exist for the month of April only. Therefore, many individuals records end for that unit on April 30, 1862. There are missing rolls for all companies at various times, especially September-October 1861, May-June and September-December 1862, and January-April 1864. Pension applications, county death registers, county histories, UDC and CV records, cemetery listings, family genealogies and papers have provided additional names and dates.

Each company was organized with a Captain, First, Second, and a Third or Junior Second Lieutenant, a First through Fifth Sergeant, and a First through Fourth Corporal. Due to so many absentees in 1864 and 1865 many vacancies were not filled. Most companies operated with only three officers during this period.

The 1860, 1870 and 1910 censuses were used extensively to verify ages, prewar and postwar occupations and places of residence. A man's age has been given as indicated on the censuses, despite some inaccuracies.

Numerous individuals claimed service in the regiment in the postwar era. All men who did so have been included in the roster and are identified as such. Several soldiers have been identified in their obituary, on their tombstone or in family records, as having served in the 52nd. They have been included except where it has been clearly proved that they served in another unit.

The surnames presented a difficult problem, especially in the Coiner, Coyner, Koiner family. An individual's record often contained all three spellings. Even within a family in the Koiner family history individuals spelled their names differently. In these cases the family history or the most frequent spelling was used.

Each man's record of service is arranged chronologically from place and date of birth to date and cause of death. Where dates of death differ tombstone data was used when available. Every attempt has been made to give a complete record of each individual's service.

The rank indicated is the highest held while a member of the regiment. Promotions and demotions have been shown when they occurred. Promotions outside the 52nd have been indicated also. Temporary assignment in higher positions has also been noted.

Company designation is the unit the man served with in the regiment. Those soldiers who were assigned to the field and staff of the 52nd from a company have that data indicated in their record.

The date of enlistment is the date sworn into Confederate service. Some of the companies were organized before this date, in state service. Many were serving in the militia until entering the Confederate States Army.

The place of enlistment again indicates where the man entered Confederate service. Company E, for example, enlisted in Rockbridge County but was sworn in at Staunton on August 1, 1861.

The dates have been abbreviated. "Present 4-6/64" means the man was present during the period April through June 1864 specific dates unknown. The muster rolls were generally compiled for pay purposes in two month increments, often after the period had expired. For example, some rolls for the period April 30-December 31, 1864 were actually completed in January 1865, therefore the presence or absence of a soldier during that period is uncertain. Often, if present at the beginning of the period, the man was marked present. Particular attention was paid to periods of absence without leave (AWOL), but short periods of leave and hospitalization were often not noted. The term "present through" means the man was not reported absent for an extended period during that time. Officers' presence or absence frequently were not indicated on the rolls. Unless other evidence showed them absent they have been indicated as present through a specific date.

Entries for dates wounded and captured, especially in May 1864, often conflict with Federal records, and the Federal records disagree among themselves. The term "WIA and captured 5/18-19/64 indicates the events occurred on one of the two days involved. The soldier may have been wounded on May 18 and captured and removed from the battlefield on May 19.

Periods of confinement as a prisoner of war are not shown specifically but run from the date captured until exchanged, released, took the oath of allegience or died. Transfer dates between Union prisons are not shown.

Hospitalization periods are shown as "in hospital Richmond 1/1-4/3/65" and "furloughed for 60 days 3/4/65". The soldier may have been in several hospitals in the Richmond area during this period. The fact that he was granted leave for 60 days shows that he was well enough to travel but it would be at least 60 days before he could return to duty. The exact hospital is shown when death occurred there.

Most of the men enlisting in the 52nd volunteered from the militia units of the three counties. Therefore, the term "conscript" is not used in 1861 and early 1862, although it appears often on the muster rolls. The regiment reenlisted on May 1, 1862. The term "reenlisted 5/1/62" was used in each case where it was a matter of record. When a man failed to reenlist or whether he reenlisted or not was not shown on the rolls, but service continued, the term "conscript" was not used. "Non conscript" was copied from the man's record which indicated he was too young or too old for conscription, or not a resident of the Confederate States. "Conscript" was used from May, 1862 on because of the large number of desertions by this class of individuals and also to distinguish those soldiers who "volunteered" for whatever reason.

The term "no further record" (NFR) has been used to show when a man's service ended and no reason is shown.

Periods of unauthorized absence have been shown as absent without leave (AWOL) and desertion based on the soldier's record. All periods of AWOL and desertion have been indicated. Extenuating circumstances, such as hospitalization to justify the absence have been shown also. Where punishment was not indicated, the offense was probably handled by extra duty or some other minor punishment. All forfeitures of pay seem to have been carefully documented. The fact that no records exist for 1865 does not mean a man AWOL in 1864 did not return and serve during the period. However, individuals absent for an extended period showed their intent not to return.

No descriptive lists exist, except for age in the case of a few companies. Descriptive lists, therefore, are taken from hospital, pay, discharge, leave, prisoner of war and death records. Pension applications, postwar rosters, UDC and CV applications, family records, genealogies, tombstone inscriptions, and the 1860, 1870 and 1910 censuses present data and ages that conflict with the roll. Specific dates of birth were used where found. Otherwise, a soldier's age is a matter of conjecture in cases of conflicting information.

All physical descriptions available were used. This was done to show the difference of opinion or agreement on the appearance of the man. The original spellings were used. The description of those soldiers being released from prison in 1865 may differ greatly after the hardships of prison life.

Periods of hospitalization were shown to indicate the degree of illness sustained by the individual soldier as well as the loss of strength to the regiment. Some medical terms have been converted to layman's terms for clarity. Medical details pertaining to wounds and or cause of death were used to show the degree of suffering and horrors these men experienced because of lack of medicine, medical knowledge or indifference.

All locations are as indicated on the muster rolls. Virginia locations are not followed by the state name. Locations outside Virginia are followed by the state name the first time they are used, including those in W. Va.

Quoted material appears in its original form, including spelling in parentheses.

Certain standard abbreviations have been used throughout the roster. These are listed below:

Absent - Ab.
Absent without leave - AWOL
Assistant Adjutant General - AAG
Born - B.
Brigadier General - Brig. Gen.
Buried - Bur.
Cavalry - Cav.
Captured - Capt.
Church - Ch.
College - Col.
Corporal - Cpl.
Court House - CH
Died - D.
Died of wounds - DOW
Enlisted - Enl.
First Sergeant - 1stSgt.
General - Gen.
Hospital - Hosp.
Infantry - Inf.
Lieutenant - Lt.
Major - Maj.
Missing in action - MIA
Mount - Mt.
Ordnance - Ord.
Private - Pvt.
Quarter Master - QM
Second Lieutenant - 2ndLt.
Sergeant - Sgt.
Third Lieutenant - 3rdLt.
Wounded in action - WIA

Adjutant - Adj.
Artillery - Arty.
Brigade - Brig.
Captain - Capt.
Cemetery - Cem.
Colonel - Col.
Company or County - Co.
Commissary - Comm.
Court Martial - CM
Deserted - Des.
Division - Div.
First Lieutenant - 1stLt.
Fort - Ft.
General Courts Martial - GCM
Killed in action - KIA
Lieutenant Colonel - LtCol.
Major General - Maj. Gen.
Obituary - Obit.
Post Office - PO
Provost Marshal - PM
Reenlisted - Reenl.
Regiment - Regt.
Route - Rt.
Sergeant Major - SgtMaj.
Transferred - Transf.
United Daughters of the Confederacy - UDC

PRISONERS OF WAR CAMPS

Camp Chase, Ohio - Camp Chase
Camp Morton, Indiana - Camp Morton
Elmira, New York - Elmira
Fort Delaware, Delaware - Ft. Delaware
Fort Henry, Maryland - Ft. Henry
Johnson's Island, Ohio - Johnson's Island

Newport News Prison, Virginia - Newport News
Old Capitol Prison, Washington, D.C. - Old Capitol
Point Lookout, Maryland - Point Lookout
Rock Island, Illinois - Rock Island
Wheeling, West Virginia - Wheeling

OCCUPATIONS

Farm hands - 412	Cook - 1	Wagoners - 5
Laborers - 103	Gardner - 1	Machinists - 4
Students - 30	Joiner - 1	Surveyors - 3
Shoemakers - 27	Mail Carrier - 1	Saddlers - 3
Teachers - 10	Paper maker - 1	Distillers - 2
Wagonmakers - 9	Proprietor - 1	Coachmakers - 2
Cabinetmakers - 9	Stonecutter - 1	Trader - 2
Merchants - 6	Stage Driver - 1	Baker - 1
Mechanics - 5	Planter - 1	Coachpainter - 1
Preachers - 4	Farmers - 224	Constable - 1
Clerks - 5	Carpenters - 67	Engineer - 1
Plasterers - 5	Blacksmiths - 28	Grocer - 1
Miners - 5	Doctors - 11	Bricklayer - 1
Tailors - 3	Millers - 9	Potter - 1
Millwrights - 4	Coopers - 6	Pedler - 1
Tanners - 3	Lawyers - 6	Sheriff - 1
Brickmoulder - 1	Stonemasons - 5	Lumberman - 1
Chairmaker - 1	Ironmakers - 5	
Confectioneer - 1	Painters - 4	

Of the 1506 men who served in the 52nd, only twenty-three are identified as foreign born. Fifteen were born in Ireland, six in Germany and two in Italy. All but nineteen of the remainder were born in Virginia. Seven were natives of Maryland, three of Ohio, two of Kentucky, Missouri, and Tennessee and one each in Indiana, New York and Pennsylvania.

The oldest man was Drummer John B. Rankin, Sr. who enlisted at age sixty-two. Private William G. Miller was sixty. Drummer Boy Kene Morris, age eleven, enlisted with his father's consent. His father later served for a short time in the regiment. The younger Morris was also partially if not totally blind. Private L. Knopp enlisted at age twelve but apparently did not like military life and deserted five months later.

AGES AT ENLISTMENT

11 - 1	12 - 1	21 - 73	22 - 63
13 - 1	14 - 1	23 - 80	24 - 65
15 - 7	16 - 18	25 - 43	26 - 53
17 - 66	18 - 99	27 - 33	28 - 44
19 - 84	20 - 68	29 - 44	30 - 25

31 - 40	32 - 31	47 - 5	48 - 3
33 - 26	34 - 33	49 - 0	50 - 1
35 - 20	36 - 26	51 - 2	52 - 0
37 - 23	38 - 24	53 - 1	54 - 1
39 - 21	40 - 13	55 - 2	56 - 1
41 - 17	42 - 11	57 - 1	58 - 2
43 - 8	44 - 13	59 - 1	60 - 1
45 - 11	46 - 6	61 - 0	62 - 1

From the sparse descriptive lists, Captain Edward M. Dabney was probably the tallest man in the regiment. He was over six feet, four inches. Private Richard M. Hayslett of Company E was the shortest at five feet.

HEIGHTS

6' 4" - 5	6' 3" - 4
6' 2" - 7	6' 1" - 11
6' - 45	5' 11" - 37
5' 10" - 65	5' 9" - 62
5' 8" - 56	5' 7" - 33
5' 6" - 28	5' 5" - 18
5' 4" - 7	5' 3" - 3
5' 2" - 1	5' 1" - 2
5' - 1	

Privates William D. Baskins and Henry D. Welch were both born in 1835, had the same middle initial, were both farmhands in Augusta County, enlisted in Staunton in the same company on the same day, were both wounded on picket on Alleghany Mountain the same day, died of wounds the same day and were buried in the same cemetery. They were the first of 184 deaths from battle. Private William B. Curry was the youngest to die in battle. He was only fifteen when struck down at Second Manassas. The last known deaths from battle were the four men who fell at Fort Stedman on March 25, 1865. Lieutenant John B. Eskridge was the first of 150 to die of disease. He was stricken on August 14, 1861, only thirty five days after he enlisted. The 52nd conducted a dress parade in his honor. Private William C. Hemp survived over three months in prison only to die of typhoid fever contracted while there. He died at home on June 24, 1865, the last man to die from disease.

DEATHS FROM DISEASE

1861 - 40
1862 - 61 3 as POWs
1863 - 12 2 as POWs
1864 - 18 8 as POWs
1865 - 10 7 as POWs
Date unknown - 9 All POWs

Twenty-nine died while prisoners of war. George A. Rogers was the last man in the 52nd to die. He survived until June, 1940, when he succumbed at age 95.

Captain William Long and Privates John B. Gardner, John M. Heflin and John A. Smith were the only men identified as Mexican War veterans.

DESERTIONS

1861 - 11	1862 - 97
1863 - 95	1864 - 91
1865 - 30	Total - 324

The regiment had 308 men discharged. Some of these soldiers served in other units. Transfers accounted for the loss of 120 men, nearly all to the cavalry. Another 101 were transferred with Co B (1st). One man was killed while a deserter and one was killed accidentally. 184 were killed in action or died of wounds. 559 were known to be wounded. 158 were captured, some several times.

FIELD AND STAFF OFFICERS

John Brown Baldwin - Colonel
Michael Garber Harman - Lieutenant Colonel
James H. Skinner - Lieutenant Colonel, Colonel
John DeHart Ross - Major, Lieutenant Colonel
Thomas H. Watkins - Major, Lieutenant Colonel
John Doak Lilley - Major, Lieutenant Colonel
Livingston Waddell - Surgeon
Briscoe Baldwin Donaghe - Surgeon
William Steele McChesney - Surgeon
James Edgar Chancellor - Surgeon
John St. P. Gibson - Assistant Surgeon, Surgeon
John Lewis - Assistant Surgeon
William Davis Ewing - Hospital Steward, Assistant Surgeon
William H. Edmundson - Assistant Surgeon
William Mabry Strickler - Assistant Surgeon
Richard Henry Phillips - Chaplain
John McGill - Chaplain
Bolivar Christian - Commissary
Matthew Thompson McClure - Commissary Sergeant
Claiborne Rice Mason - Quarter Master
George Moffett Cochran, Jr. - Quarter Master
Patrick Maloy - Quarter Master Sergeant
George Pilson Lightner - Quarter Master Sergeant
Alexander Givens McCune - Adjutant
John William Lewis - Adjutant
William Galt - Adjutant
Alexander Fisher Kinney - Adjutant

William Harvey Lackey - Ensign
John Andrew Stuart - Drillmaster
Thomas Davis Ranson - Sergeant Major
Erasmus Stribling Trout - Sergeant Major
William Alexander Ross - Sergeant Major
William L. Black - Sergeant Major
John Freeman Parrish - Sergeant Major
Adam Given Cleek - Ordnance Sergeant
Daniel L. Bush - Musician
Jesse L. Harris - Musician
Kene Morris, Sr. - Musician
Kene Morris, Jr. - Musician
John B. Rankin, Sr. - Musician
John Wesley Booz - Musician

REGIMENTAL STRENGTH DURING THE WAR

COMPANY A	158
COMPANY B (1st)	123
COMPANY B (2nd)	157
COMPANY C	138
COMPANY D	106
COMPANY E	180
COMPANY F	106
COMPANY G	120
COMPANY H	129
COMPANY I	115
COMPANY K	128
F&S	29
COMPANY UNKNOWNS	17
TOTAL	1506

COMPANY OFFICERS

Company A - Captains James H. Skinner, Edward Valentine Garber, Robert Coleman Davis. Lieutenants - John Lynn Cochran, John B. Eskridge, William Galt, Richard Stevenson Kinney, James Bumgarner, Jr., Thomas Hall Neilson, Benjamin Franklin Hildebrand, Oscar C. Lipscomb, Jacob Vaiden.

Comapny B (1st) - Captain John Miller. Lieutenants Samuel Wallace, John Andrew Montgomery Lusk, John Carter Dickinson.

Company B (2nd) - Captains William Long and Andrew James Thompson. Lieutenants Clinton M. King, Thomas Houston Antrim, John S. Myers, James A. Burns, William H. Burns, George William Moore and John Sinclair Lipscomb.

Company C - Captains Edward Moon Dabney, John S. Byers and Charles Lastran Weller. Lieutenants Samuel Bradford Brown, Jr., Lewis Harman and William Alexander Ross.

Company D - Captains Joseph Franklin Hottell and Abraham Airhart. Lieutenants Phillip Airhart, William Harrison Wooddell, John Addison Carson, Samuel Paul, John Newton Hanna and Benjamin Franklin Wooddell.

Company F - Captains Joseph E. Cline and James Bumgardner, Jr. Lieutenants Cyrus W. Snapp, Jonathan B. Nash, John Alexander Fauver, George M. Crist, Adam H. Craun.

Company G - Captains Samuel Houston McCune, Elijah Bateman and Cyrus Benton Coiner. Lieutenants James Western, David W. Coiner, Joseph Smith Coiner, John Edward Hamilton and Joseph Frey.

Company H - Captains Claiborne Rice Mason, John Doak Lilley, James Addison Dold and Erasmus Stribling Trout. Lieutenants Isaac A. Bushong, William F. Dold and James A. White, Issac W. Airy, Mathew Thompson McClure.

Company I - Captains Samuel A. Lambert and John Moore Humphreys. Lieutenants John D. Summers, John Moore Lambert, Robert P. McFarland and Thomas Davis Ranson.

Company K - Captains Benjamin T. Walton, William Crawford Burger, Samuel Bradford Brown, Jr. and William Rofe Gillett. Lieutenants John Alexander Lindsay, Sr., William Wallace Byrd, Walter Boon and J. E. Mayer.

STRENGTH REPORTS

DATE	AGGREGATE PRESENT	AGGREGATE ABSENT	AGGREGATE PRESENT & ABSENT
10 April, 1862	692	143	835
16 April, 1862	681	162	843
1 May, 1862	703	182	885
5 May, 1862	693	180	873
10 July, 1862	286	511	797
22 Sept., 1862	235	557	792
10 Oct., 1862	370	462	832
9 Nov., 1862	386	436	822
10 Dec., 1862	310	482	792
1 Jan., 1863	330	509	839
31 Jan., 1863	373	419	792
28 Feb., 1863	358	416	774
31 Mar., 1863	383	381	764
18 April, 1863	391	371	762
8 May, 1863	372	385	757

31 May 1863	398	353	751
9 June, 1863	369	390	759
19 June, 1863	341	407	748
10 July, 1863	289	454	743
30 July, 1863	249	489	738
31 Aug., 1863	291	436	727
30 Sept., 1863	318	392	710
20 Oct., 1863	317	395	712
20 Nov., 1863	349	337	686
31 Dec., 1863	345	341	686

No reports found for 1861 and 1864-1865.

ABLE, MATHIAS: Pvt. Co. G. B. Germany 1834? Farmer, age 25, Burkes Mill District, Augusta County, 1860 census. Enl. Staunton 8/2/61. Capt. Allehangy Mt. 12/13/61 while on picket. Sent to Camp Chase. Exch. Vicksburg, Miss. 9/62. Present 12/25/62-7/24/63. AWOL 7/25/63-8/26/63. Fined $11.36. Present 8/26/63 until WIA (hip) Bethesda Ch. 5/30/64. Returned to duty 11/9/64. Present until WIA (buttock) near Petersburg 12/24/64. Furloughed from hospital Petersburg 1/6/65 for 60 days. NFR. Farmer, age 37, Mt. Sidney PO, Augusta Co. 1870 census. Laborer, Waynesboro 1878.

ACKERLY, GEORGE W.: 3rdSgt. Co. E. B. Rockbridge Co. 3/30/32. Enl. Staunton 8/1/61. Ab. on leave 11-12/61. Ab. sick leave 1-2/62. Present 3-5/11/62. Ab. 11/6/62-2/28/63 in hosp. Staunton 11/20/62-12/31/62 with pneumonia. Present 3/63-6/2/63. Ab. sick 6/3/63-12/31/64 in hospital Staunton. NFR. Farmhand, age 39, Buffalo Township, Rockbridge Co. 1870 census. Stabbed and killed by a negro on Buffalo Creek, Rockbridge Co. 6/3/71. Bur. Stonewall Jackson Cem., Lexington.

ACORD, GEORGE F.: Pvt. Co. K. B. Augusta Co. 1835? Carpenter, age 25, Hot Springs, Bath County 1860 census. Enl. Shepherdstown, 9/23/62. Present 9/23/62-1/8/63. In hosp. Charlottesville 1/9/63-2/17/63 "Morbai Cutis". Transf. to Co. F, 11th Va. Cav. 3/31/63. KIA Wilderness 5/8/64. Age 28. Left widow.

ADAMS, JACOB HARVEY: Pvt. Co. A. B. Va. 1840? Farmhand, age 19, Burkes Mill District, Augusta Co. 1860 census. Enl. Staunton 7/17/61. Ab. sick Staunton 9/10/61-12/31/62. Present 1/63 until WIA (hand) Bethesda Ch. 5/30/64. Ab. wounded through 12/31/64. NFR. Farmer Madison Co., Va. D. Madison Co. 2/25/95.

ADAMS, WILLIAM.: Pvt., Co. K. Present 5/26/62. Over 35. NFR.

ADAMS, WILLIAM W.: Pvt. Co. K. B. Va. 1837? Farmhand, age 25, Green Valley Dist., Bath Co. 1860 census. Enl. Shenandoah Mt. 4/9/62. Present 4/9/62 until WIA (arm) Bethesda Ch. 5/30/64. DOWs at Camp Winder General Hosp., Richmond 7/23/64. Bur. Hollywood Cem., Richmond.

ADOLPH, WILLIAM: Pvt. Co. A. Resident of Richmond. Enl. 5/30/62 Camp Shenandoah as a substitute. Des. 5/30/62. NFR.

AILER, JOHN L.: Pvt. Co. A. B. Va. 1840? Shinglemaker, age 19, Deerfield, Augusta Co., 1860 census. Enl. Staunton 7/16/61. Present 7/16/61-11/15/61. Ab. sick 11/16/61-3/5/62. Des. 4/2/62. NFR. Alive 1903.

AIRHART, ABRAHAM: Capt. Co. D. B. Augusta Co. 2/23/28. Papermaker, age 32, Mt. Solon, Augusta Co., 1860 census. Enl. Staunton 7/16/61 as 2nd Lt. Present 7/16/61-10/31/61. Presence or absence not stated 11/61-2/62. Present 3/62-4/62. Elected Capt. 5/1/62. Commanded the Co. until WIA (right thigh) at Spotsylvania CH 5/12/64. DOWs in hosp. at Spotsylvania CH 6/5/64. Left widow and 3 children. Bur. Old Methodist Ch. Cem., Mt. Solon, Augusta Co.

AIRHART, PHILLIP: 2ndLt. Co. D. B. Rockingham Co. 12/13/21. Farmer, age 38, Burkes Mill Dist., Augusta Co., 1860 census. Enl. Staunton 7/16/61. Present 7/16/61-11/10/61. Ab. sick 11/11/61-2/62. Not reelected 5/1/62. Farmer, age 47, Mt. Sidney, Augusta Co. 1870 census. D. near Sangersville, Augusta Co., 9/6/02. Bur. Emanuel Ch. Cem., Mt. Solon, Augusta Co.

AIRY, GEORGE W.: Pvt., Co. F. WIA Gaines Mill 6/27/62 (head). NFR d. Centreville, Augusta Co. 2/25/66.

AIRY, ISAAC W.: 2ndLt., Co. H. B. Staunton 1826? Carpenter, age 33, Long Glade, Augusta, Co., 1860 census. Enl. Staunton 7/23/61, age 33. Present 7/23/61-2/23/62. Ab. on leave 2/24/62-2/28/62. Present 3-4/62. Not reelected 5/1/62. Reenlisted as Pvt. at Bunker Hill 10/3/62. AWOL 11/10/62-12/31/63. Serving in McClanahan's Va. Battery 3/18/64. Carpenter, age 82, Staunton 1910 census. D. Staunton 3/27/15. Bur. Thornrose Cem., Staunton.

ALDHIZER, JAMES HENRY: 4thSgt., Co. G. B. Augusta Co. 1832? Farmer New Hope, Augusta Co. Enl. Staunton 8/2/61 as Pvt. Ab. sick 1/1-3/6/62. Present 3/7/62-4/62. Reenlisted and elected 4thCpl. 5/1/62. Present 5/62-2/63. Appointed 4thSgt. 3/15/63. Present 3/63 until KIA Bethesda Ch. 5/30/64. Left widow.

ALEXANDER, JOHN FRANKLIN: Pvt., Co. G. B. on Walker's Creek, Rockbridge Co., 1833, Shoemaker, Walker's Creek Dist., Rockbridge Co. Not on muster rolls. Paroled Staunton 5/23/65. Shoemaker, Walker's Creek, Rockbridge Co., 1870 census. Died Zack, Rockbridge Co., 6/16/22. Bur. United Brethren Ch. Cem., Walker's Creek.

ALEXANDER, JOHN W. (1st): Pvt., Co. B (2nd). B. Augusta Co. 1824. Merchant, Deerfield, Augusta Co. On postwar roster. Served as Provost Guard part of time. (Probably a late conscript.) Stockraiser Bath Co., and then McDowell, Highland Co., Va. D. 1914.

ALEXANDER, JOHN W. (2nd): Pvt., Co. B (2nd). B. Waynesboro, Augusta Co., 6/22/37. Farmer, age 22, Sherando, Augusta Co., 1860 census. Served in Capt. Hall's Co., 32nd Va. Militia. Enl. West View, Augusta Co., 5/1/62. AWOL 5/10/62-10/1/63. In arrest Orange CH 10/1-31/63. In hospital Richmond 11/3/63 "debilitas". Sent to Castle Thunder 11/5/63, "accidental pistol wound in groin". AWOL 12/3/63. Discharged 1/27/64 at Staunton Gen. Hosp. for "Organic disease of the heart and thoratic aneurism of many years standing". 5' 9", fair complexion, grey eyes, light hair. Farmer, age 37, Fishersville PO, 1870 census. Died Lynhurst, Augusta Co., 3/16/20. Buried Bethlehem Ch. Cem., Ladd, Augusta Co.

ALEXANDER, JOHN WILLIAM: Pvt., Co. A. B. King William Co., Va. 1819. On postwar roster. D. Malvern Hill, Henrico Co., Va., 1872.

ALEXANDER, THOMAS WOODWARD: Pvt. Co. B (2nd). B. Augusta Co. 10/13/45. Cabinetmaker, Waynesboro. Not on roster. Served 4 yrs. Capt. Bethesda Ch. 5/30/64. Sent to Ft. Delaware. Released 5/65. Owner, Alexander Church Furniture Co., Waynesboro; Fire Chief and Member of City Council, Waynesboro. D. Waynesboro 1/1/24. Bur. Riverview Cem., Waynesboro.

ALEXANDER, WILLIAM POWHATAN: Pvt. Co. B (1st). B. Rockbridge Co. 9/7/27. Laborer, age 30, South River Dist., Rockbridge Co. 1860 census. Enl. Fairfield 7/10/61. Present until transf. 9/28/61.

ALLEN, CHARLES W.: Pvt. Co. F. B. Augusta Co. 6/26/35. Farmhand age 24, Burkes Mill Dist., Augusta Co. 1860 census. Enl. Staunton 7/31/61. Present 7/31/61-11/4/61. Ab. sick leave 11/5/61-12/31/61. Present 1-4/62. Reenlisted 5/1/62. WIA (head) Port Republic 6/9/62. Present 11/5-23/62. AWOL 11/24/62-2/10/63. In arrest 2/28/63. Present 4/30/63-6/15/63. Ab. in hosp. 7/16/63-12/31/63. Present 2/64 until WIA (head) Wilderness 5/6/64. WIA (right foot) and capt. Bethesda Ch. 5/30/64. Admitted to Lincoln USA Hosp., Wash., D.C. 6/4/64. Sent to Old Capitol Prison 2/5/65. Transf. to Elmira. Released 5/15/65, 5' 9", fair complexion, grey eyes, light hair. Farmer, age 34, Mt. Sidney PO, Augusta Co. 1870 census. Murdered by poison 8/23/73 near Mt. Sidney. "A brave and faithful Confederate soldier".

ALLEN, JAMES GEORGE: Pvt. Co. B (1st). B. Fluvanna Co. 1839? Cooper, age 21, South River Dist., Rockbridge Co. 1860 census. Enl. Fairfield 7/10/61. Ab. on detail Staunton when transf. 9/28/61.

ALLEN, WILLIAM RICHARDSON: Pvt. Co. B (1st). B. Fluvanna Co. 1841? Enl. Fairfield 7/17/61. D. of measles in Rockbridge 9/9/61, age 20.

ALLISON, J.H.: Pvt. Co. A. Not on muster rolls. D. 1/3/62. Bur. Thornrose Cem., Staunton. Grave #874.

ALMARODE, GEORGE HENRY: 3rdCpl. Co. I. B. Augusta Co., 7/26/44. Student age 18, 1st District, Augusta Co. 1860 census. Enl. Staunton 7/16/61 as Pvt. Present 7/16/61-7/15/62. AWOL 7/16/62-8/13/62. Present 8/14/62-9/14/62. AWOL 9/15/62-10/25/62. Fined $24.56 for both periods. Present 10/25/62 until WIA (thigh) Spotsylvania CH 5/12/64. Returned to duty 5/17/64. WIA (thigh bone shattered and leg shortened 4 in.) and capt. Winchester 9/19/64. In USA Hosp. Winchester until transf. to another Fed. hosp. 10/23/64. Paroled Staunton 5/13/65. 5' 5", fair complexion, light hair, blue eyes, Farmer, Middlebrook, Augusta Co. Mem. Stonewall Jackson Camp Confederate Veterans, Staunton. D. near Middlebrook 7/12/22. Bur. Oakland Meth. Ch. Cem., Augusta Co. Brother of Jacob C. and John W. Almarode.

ALMARODE, JACOB CHRISTIAN: Cpl., Co. H. B. Augusta Co., 7/20/38. Carpenter, age 22, 1st Dist., Augusta Co. 1860 census. Enl. Staunton 7/23/61 as Pvt., age 23. Present until WIA (hand) Cross Keys 6/8/62. Ab. wounded through 8/31/62. Present 1-4/63. Promoted Cpl. 4/15/63. Present 5/63 until des. near Winchester 9/12/64. Took oath and sent to Frederick, Md. 9/21/64. NFR. Brother of George H. and John W. Almarode.

ALMARODE, JOHN WILLIAM: Pvt., Co. D and Co. I. B. Augusta Co. 5/23/40. Carpenter, age 20, 1st Dist., Augusta Co. 1860 census. Enl. Mt. Jackson 3/23/62. Transf. to Co. I 2/1/62. Present 13/62-12/64 In hosp. Richmond 1/26-27/65 and 2/8-10/65 and returned to duty. Paroled Staunton 5/13/65. 5' 8", dark eyes, dark hair. Brother of George H. and Jacob C. Almarode.

ALVIS, WILLIAM H.: Pvt., Co. C. B. Augusta 1839. Farmhand, age 20, Laurel Hill, Augusta Co. 1860 census. Enl. Staunton 7/16/61, age 20. Present 7/16/61-4/62. Reenlisted 5/1/62. Present 5/1/62 until WIA 2nd Manassas 8/25/62. Returned to duty 9/1/62. Present until KIA Fredericksburg 5/5/63. Left widow and child.

ANDERSON, CHARLES MOORE: Pvt., Co. D. B. Va. 1826? Tailor, age 33. Mt. Solon, Augusta Co. 1860 census. Enl. Staunton 7/16/61. Present 7/16/61-4/62. Capt. Strasburg 6/24/62. Sent to Wheeling and Camp Chase. Exchanged 8/25/62. 5' 10", brown hair, grey eyes, brown whiskers. Present 1/63-4/24/63. Ab. sick in hosp. 4/25/63-12/31/63. Capt. Fisher's Hill 9/24/64. Sent to Pt. Lookout 10/1/64. D. there 1/2/65 of acute dysentery. Bur. in POW Cem. there. Grave #757. Left widow and 2 children.

ANDERSON, EDWARD MANOR: 3rdSgt., Co. F. B. "Burnt Cabin Place" near Schutterlee's Mill, Augusta Co. 6/4/37. Farmer, Augusta Co. Enl. Staunton 7/31/61. Present 11/61-4/62. Reenlisted 5/1/62. WIA (head) Pt. Republic 6/9/62. WIA (hip) 2nd Manassas 8/29/62 or 8/31/62. Returned to duty 6/63. Transf. to Co. C, 14th Va. Cav. 5/63. WIA Gettysburg 7/3/63. Capt. Fall '64. POW Pt. Lookout to end of war. Farmer near Anderson PO, Augusta Co. D. 8/18/99. Bur. Union Pres. Ch. Cem. near Churchville, Augusta, Co.

ANDERSON, GEORGE HARVEY: Pvt., Co. C. B. Augusta Co. 4/17/44. Not on muster roll. Transf. Co. C, 14th Va. Cav. D. 9/16/62. Bur. Augusta Stone Ch. Cem., Augusta Co.

ANTRIM, THOMAS HOUSTON: 2ndLt., Co. B (2nd). B. Charlottesville, Albemarle Co., 4/2/32. Merchant, age 28, Waynesboro, Augusta Co. 1860 census. Capt. Hall's Co., 32nd Va. Militia. Enl. Waynesboro 7/15/61. Resigned 1/8/61 ill health and disease of the lung. Appointed Tax In Kind Assessor 54th Dist. of Va. and Clerk to Post QM 11th Dist. of Va., Staunton, to end of war. Merchant, age 37, Waynesboro 1870 census. Member, Waynesboro City Council. D. Waynesboro 1/14/16. Bur. Riverview Cem., Waynesboro.

ARCHIE, HENRY BROWN: Pvt., Co. K. B. Bath Co. Farmer, Lexington, Rockbridge Co. Married Lexington 8/3/62. Enl. Sheperdstown 9/23/62. Present 1/63 until KIA Fredericksburg 5/5/63 by explosion of a shell. Could be "H.A." Va. Bur. Fredericksburg Confederate Cem. Widow lived at Kerr's Creek, Rockbridge Co.

ARCHIE, STEPHEN P.: Pvt., Co. K.B. Bath Co. 1843? Laborer, age 17, Cleek's Mill PO, Bath Co. 1860 census. Enl. Shenandoah Mt. 4/9/62. Present 4/9/62-4/30/62. Present 1/63-6/30/63. Ab. on sick leave 7/1/63. Capt. Waterloo, Pa. 7/5/63. Sent to Ft. Delaware. Exchanged 11/1/63. Issued clothing 11/23/64. Listed as POW at Elmira in "Staunton Vindicator" 12/1/64. NFR.

AREY, GEORGE F.: Pvt., Co. F. B. Va. 7/43. Farmhand, age 17, Burkes Mill Dist., Augusta Co. 1860 census. Enl. Staunton 7/31/61. Ab. sick leave 11/5/61-12/31/61. Present 1/62-4/62. Reenlisted 5/1/62. WIA (hand) Gaines Mill 6/27/62. WIA 2nd Manassas 8/28/62. Present 8/31/62-2/16/63. Transf. to Carpenter's "Alleghany Artillery" 2/17/63, in exchange for John Pannell. WIA 3 times including saber wound to the head. Grocer, Staunton, 1897. Farmer, Bridgewater, Rockingham Co. 1914. D. there 1916. Bur. Oakland Cem., Bridgewater.

ARGENBRIGHT, JAMES SAMUEL: Pvt., Co. F. B. Augusta Co. 11/15/35. Farmer, Augusta Co. Enl. Mt. Meridian 6/15/62. Present 8/31/62 until WIA (lost left eye) Carter's Farm near Winchester 7/20/64 (lost left eye). Ab. wounded through 12/31/62 (Roll dated 2/15/65). Paroled Staunton 5/1/65. 5' 10", fair complexion, brown hair, grey eyes. Farmer and carpenter, Long Glade, Augusta Co. D. Western State Hosp., Staunton 2/19/07.

ARGENBRIGHT, JOHN ROBERTSON: Pvt., Co. A. B. Augusta, Va. 4/29/44. Resident, age 15, Jennings Gap, Augusta Co. 1860 census. Enl. Camp Alleghany 3/27/62. Present 3/27/62-11/22/62. AWOL 11/23/62-1/20/63. Present in arrest 1/21/63-4/63. Sentenced by GCM to forfeit 6 mos. pay. Present 5/1/63-12/22/64. Ab. on leave 12/23/64-12/31/64. Surrendered Appomattox 4/9/65. Farmer, age 26, Pastures Dist., Augusta Co., 1870 census. D. near Jerusalem Chapel, Augusta Co. 12/21/20. Bur Jerusalem Chapel Cem., near Churchville.

ARMENTROUT, ROBERT N.: Pvt., Co. G. B. Va. 1841. Farmhand, age 18, Burkes Mill Dist., Augusta Co. 1860 census. Enl. Staunton 8/2/61. Present 11/61-4/62. Present 5/1/62 until capt. Front Royal 6/12/62. 5' 8", age 20. Exchanged date unknown. WIA and capt. Sharpsburg 9/17/62. DOWs.

ARMISTEAD, ROBERT: Pvt., Co. G. Listed on post war roster.

ARMSTRONG, JOHN C.: Pvt., Co. E. B. Va. 1835? Farmer, age 24, Collierstown, Rockbridge Co. 1860 census. Enl. Staunton 8/1/61. Present 11/61-8/15/62. AWOL 8/16/61-12/31/63. Dropped as a des. Laborer, age 34, Buffalo Township, Rockbridge Co. 1870 census. Farmer. D. on upper Colliers Creek 2/11/04. Bur. Collierstown Cem.

ARTHUR, JOSEPH D.: Pvt., Co. E. B. on Sinking Creek, Botetourt Co. 1819? Farmer, Lexington PO, Rockbridge Co. 1860 census. Enl. Staunton 8/1/61. Present 11/61-10/62. Paid 11/24/62. Transf. to 14th Va. Cav. Laborer, age 48, Buffalo Township, Rockbridge Co. 1870 census. D. on North Buffalo Creek 2/7/03. Bur. in private cem. near John A. Hickman's (obit.) Rockbridge Co.

ATKINS, ALEXANDER B.: Pvt., Co. C. B. Va. 1845? Living Burkes Mill Dist., age 14, Augusta Co. 1860 census. Enl. Staunton 4/1/63. Present 4/1-5/1/63 when transf. to 5th Va. Inf. in exchange for Samuel J. Byers. Paroled Staunton 5/10/65. Living in Augusta Co. 1926.

AYERS, JOHN W.: Pvt., Co. E. B. Rockbridge Co. 5/1/47. Enl. Rockbridge Co. 12/9/64. Present until des. to Army of the Potomac near Petersburg 3/8/65. Took oath and transportation furnished to Greene Co., Ind. Farmhand, age 24, Walker's Creek Dist., Rockbridge Co. 1870 census. D. Rockbridge Co., 5/18/24. Bur. Broad Creek Ch. Cem.

AYERS, STEPHEN P.: Pvt., Co. K. B. Bath Co. 1835? Laborer, age 25, Millboro Springs PO, 1860 census. Enl. Shenandoah Mt. 4/18/62. Present 4/18/62-4/30/62. AWOL 12/17-31/63. Dropped as a deserter. NFR.

AYRES, ALFRED GRAHAM: Pvt., Co. E. B. Rockbridge Co. 12/27/25. Millwright, Rockbridge Co. Served in Co. E. 8th Va. Militia and rejected for service 3/62 because of consumption. Enl. 27th Va. Inf. and discharged. Enl. Rockbridge Co. 10/28/64. Present 11-12/64. WIA (scalp) and capt. Ft. Stedman 3/25/65. Adm. to Lincoln Gen. Hosp., Wash., D C. 3/28/65. Released 6/12/65. 5' 8", dk. complexion, dk. hair, grey eyes. Millwright, age 46, Lexington PO, 1870 census. D. near Zollman PO, Rockbridge Co., 1/11/93. Bur. Oxford Pres. Ch. Cem., Rockbridge Co.

AYRES, JAMES: Pvt., Co. E. Farmer, age 35, Kerrs Creek Dist., Rockbridge Co., 1860 census. Not on muster rolls. WIA and capt. Gettysburg 7/3/63. Rec. at DeCamp Gen. Hosp, David's Island, N.Y. 7/17-24/63. NFR. Believed to have DOWs.

BAILEY, BENJAMIN F.: Pvt., Co. C. B. Va. 1835? Farmhand, age 20, Burkes Mill Dist., Augusta Co. 1860 census. Enl. Staunton 7/16/61 age 21. Present 11/61-4/62. Reenlisted 5/1/62. WIA (arm) 2nd Manassas 8/28/62. Present 8/31/62-12/31/63. WIA near Lynchburg 7/18/64. DOWs at Pratt Hosp., Lynchburg 7/20/64. Bur. Conf. Sec., No 8, Row 3, Lynchburg City Cem.

BAILEY, GEORGE W.: Pvt., Co.D. B. Va. 1827? Farmhand, age 32, Sangersville, Augusta Co. 1860 census. Enl. Staunton 7/16/61. Present 7/16/61-10/31/61. D. in hosp. at Yagar's Alleghany Mt., of fever 11/5/61.

BAKER, GEORGE FREDERICK: Pvt., Co. D. B. Hesse Darnstadt, Germany 1831? Shoemaker, age 28, Burkes Mill Dist., Augusta Co. 1860 census. Enl. Staunton 7/16/61. Present 7/16/61-11/9/61. AWOL 11/10/61-12/23/61. Present 12/24/61-4/1/62. Reenlisted 5/1/62. AWOL 6/2/62-6/15/62, fined $4.76. Ab. sick Gordonsville 7/20/62-2/17/63. Present 2/18/63-7/14/63. AWOL 7/15/63-8/5/63, fined $8.06. Present 8/6/63-11/7/63. Des. 11/8/63. NFR. Farmer, age 43, Mt. Sidney PO, Augusta Co. 1870 census. D. North River Dist., Augusta Co. 10/15/92. Bur. St. Paul's Lutheran Ch. Cem.

BALDWIN, JOHN BROWN: Col., F&S. B. "Spring Farm" near Staunton 1/11/20. Grad. Staunton Academy and UVa. '38. Lawyer, Staunton. Capt. of "Staunton Light Infantry" and Col. of Militia. Speaker in Va. House of Delegates 1848-61. Member of State Conv. 1861. Sent by Conv. to visit Pres. Lincoln. Appt. Inspector Gen. of State Forces by Gov. Letcher. Appt. Col. of 52nd 8/19/61. Health was broken at Alleghany Mtn. and he was not reelected 5/1/62. Member of Conf. Congress 1862-65. Col. of Reserves in Augusta Co. and commanded the "Augusta Raid Guard" 1863. Member Va. Legislature 1865-67. Bd. of Visitors UVa 1856-64. Organizer of Augusta Co. Fair (later called Baldwin Co. Fair in His Honor). Attorney for Valley RR. Supt. of Elections 1870. D. Staunton 9/30/73. Bur. Thornrose Cem. "of striking physique, tall and well proportioned. A man remarkable for the purity and integrity of his character both in public and private life. No man ever enjoyed a larger share of the confidence and affection of his fellow citizens...As a man of intellect, he was almost unrivaled. His talents were of the most varied character. As a logical reasoner, he had no superior in the State."

BALDWIN, PETER: Pvt., Co. K. B. Rockbridge Co. 9/10/27. Farmer. age 32, Bath CH PO, 1860 census. Capt. Hamilton's Co., 81st Va. Militia. Not on muster rolls. Enl. Co. F. 11th Va. Cav. 12/2/63. Farmer, age 45, Warm Springs Dist, Bath Co. 1870 census D. near Cleeks Mill, Bath Co., 5/15/04.

BALSLEY, ELIJAH G.: Pvt., Co. B (2nd). B. Va. 1833? Carpenter, age 26, 1st. Dist., Augusta Co. 1860 census. Enl. Waynesboro 7/15/61. Ab. sick 8/21/61-10/63. Des. Clarksburg, W. Va. 10/31/63. Took oath and sent north, age 30, 5' 8". Millwright, age 37, Fishersville PO, Augusta Co. 1870 census.

BANE, ABNER McC.: Pvt., Co. E. Not on muster rolls. D. in camp of diphtheria in 1862. Left widow.

BARGER, DAVID W.: Pvt., Co. G. B. Augusta Co. 5/10/38. Farmhand, age 19, 7th Dist., Rockbridge Co. 1860 census. Enl. Staunton 8/2/61. Ab. 11/12/61-3/6/62. Present 3/7/62-4/30/62. Reenlisted 5/1/62. AWOL 10/19/62-12/31/62. Dropped as a deserter. NFR. Resident of New Hope, Augusta Co. Tradesman, Rapp's Mill, Rockbridge Co. 1910.

BARGER, JOEL J.: Pvt. Co. E. Listed on post war roster. Farmer, Gilmore's Mill, Rockbridge Co. 1897. D. near Fancy Hill 1/8/10. Bur. Covington, Va.

BARGER, JOHN COINER: Pvt., Co. G. B. Augusta Co. 10/2/23. Farmer, age 36, Burkes Mill Dist., Augusta Co. 1860 census. Enl. Staunton 8/2/61. Present 11/61-4/30/62. Reenlisted 5/1/62 as non-conscript over 35 years. Present 5/1/62-8/31/62. AWOL 1 mo. and 17 days on rolls 10/31/62. Present 12/11/62. CM and acquitted 2/12/63. Present 1-12/63. WIA (gunshot wound right thigh and left arm) and capt. Bethesda Ch. 5/30/64. Adm. to Armory Sq. USA Gen. Hosp., Wash., D.C. 6/12/64. Right thigh amputated. DOWs 6/14/64. Age 40 yrs. Bur. Wash., D.C. (incorrectly listed as 2nd Va.). Removed to Lutheran Ch. Cem., Augusta Co. Left widow.

BARGER, WILLIAM A.: Pvt., Co. C. B. Va. 1838? Farmhand, age 22 Burkes Mill Dist., Augusta Co. 1860 census. Pvt., Capt. Koiner's Co., 32nd V.a. Militia 3/62. Enl. Shenandoah Mtn. 4/11/62. Present 5/1/62-7/18/62. Ab. sick 7/19/62-10/31/63. Detailed in hospitals Richmond and Staunton 12/23/63-12/31/63 unfit for field duty — disability in rt. leg and fever. Paroled Staunton 5/13/65, age 26, 5' 8", dk. complexion, dk. hair and dk. eyes. Farmer. D. near Lee Valley, St. Mary's Md. 3/15/94.

BARNETT, BENJAMIN FRANKLIN: Pvt., Co. B (1st). B. Rockbridge Co. 1838? Enl. Fairfield 7/10/61. Present until transf. 9/28/61.

BARTLEY, HENRY A.: Pvt., Co. B (1st). B. Va. 1836? Farmer, age 24. 7th Dist., Rockbridge Co. 1860 census. On postwar roster only.

BARTLEY, WILLIAM J.: Pvt., Co. B. (1st). B. Va. 1828. Collier, age 35, 5th Dist., Rockbridge Co. 1860 census. Enl. Staunton 8/1/61. Present until transf. 9/28/61.

BARRACKS, HENRY W.: Pvt., Co. H. B. Va. 1840? Resident of Mt. Clinton, Rockingham Co. Enl. Staunton 7/23/61. Present 11/61-4/62. Reenlisted 5/1/62. Present 5/1/62-11/22/62. AWOL 11/23/62-12/31/63. Dropped as a des. NFR.

BARTLETT, G.R.: Pvt., Co. B. (2nd). Not on muster rolls. Paid 3rd Qtr. 10/64. NFR.

BASKINS, WILLIAM S.: Pvt., Co. H. B. Augusta Co. 1/18/35. Farmhand, age 26, 1st Dist., Augusta Co. 1860 census. Enl. Staunton 7/23/61. WIA (hip) on picket Alleghany Mtn. 12/13/61. DOWs Camp Alleghany 2/13/62. Age 27 yrs., 5' 9", lt. complexion, lt. eyes, dk. hair. Bur. Shutterlee Cem. on Rt. 728 .2 mi. east of intersection with Rt. 732 at Frank's Mill, Augusta Co.

BATEMAN, ELIJAH: Capt., Co. G. B. Augusta Co. 4/3/25. Limeburner, age 34, 1st. Dist., Augusta Co. 1860 census. Enl. Staunton 8/2/61 as Pvt. Present 11-12/61. Detailed as Commissary Sgt. 1/19/62. Present 1-4/62. Elected Capt. 5/1/62. Ab. sick 8/14-10/18/62. Present 10/19/62-12/31/63. WIA (right arm) Wilderness 5/6/64. Arm amputated near shoulder. Retired 12/2/64. Lumber Merchant, age 45, Mt. Sidney PO, Augusta Co. 1870 census. Railroad employee. Killed by runaway horse Alleghany Co. 5/31/96. Bur. Tinkling Spring Pres. Ch. Cem. near Fishersville.

BATEMAN, WILLIAM F.: Pvt., Co. G. B. Va. 1837? Farmhand, age 23, 1st Dist. Augusta Co. 1860 census. Enl. Staunton 1/2/63. Present 3-5/14/63. Ab. on detached service 5/15/63 and AWOL 5/21/63 to 8/31/63. Fined 3 mo. and 15 days pay. Present 11-12/63. WIA (arm) Wilderness 5/6/64. KIA Bethesda Church 5/30/64.

BAXTON, J.: Pvt., Co. G. Not on muster rolls. Capt. Winchester 9/19/64. Sent to Point Lookout 9/30/64. Arrived Harper's Ferry 10/3/64. NFR.

BAYLOR, ANDREW J.: Pvt., Co. A. B. Va. 1840. Farmhand, age 20, Burkes Mill Dist., Augusta Co. 1860 census. Enl. Staunton 7/16/61. Present 11/61-4/62. Reenl. 5/1/62. Cap. Front Royal 5/30/62. Exchanged 8/5/62. Present 8/31/62-12/31/63. WIA (thigh) Wilderness 5/6/64. Cap. Winchester 9/19/64. Sent to Point Lookout. Exchanged 3/15/65. Paroled Staunton 5/13/65. Age 24, 5' 5", dk. complexion, dk. hair, dk. eyes. Died before 5/03.

BAYLOR, JOHN S.: Pvt. Co. A. B. Augusta Co. 1827? Cook, Churchville, Augusta Co. Enl. Staunton 7/17/61. Present 11/61-4/62.Renl. 5/1/62. WIA (stomach) Port Republic 6/9/62. AWOL 11/23/62-3/5/63. Pay deducted for AWOL. Present 3/5-12/7/63. AWOL 12/8-31/63. Deserted Cumberland, Md. 1/5/64. Took oath and sent north. Age 32, 5' 7", dk. complexion, grey eyes, brown hair. Died Augusta Co. 4/5/08. Bur. Jerusalem Chapel Cem., near Churchville.

BEACH, DABNEY SNEED.: Pvt., Co. C. B. Albermarle Co. 1814? Carpenter, age 46, Burkes Mill Dist., Augusta Co. 1860 census. Enl. Staunton 8/1/61. Ab. detached service 11-12/61. Present 1-2/62. Died of fever in hospital Stribling Springs, Augusta Co. 4/10/62. Buried Thornrose Cem. Staunton.

BEAR, HENRY CLINTON.: Pvt., Co. A. B. Augusta Co. 1/5/39. Laborer, age 21, Burkes Mill Dist., Augusta Co. 1860 census. Enl. Staunton 7/9/61. Present 11/61-4/62. Reenl. 5/1/62. WIA (left leg below the knee) Gaines Mill 6/27/62. Ab. wounded through 12/31/64. Paroled Staunton 5/1/65. Age 25, 5' 10", dk. complexion, dk. hair, grey eyes. Millwright, age 31, Mt. Sidney PO, Augusta Co. 1870 census. Farmer, Lynhurst, 1910 census. Member Stonewall Jackson Camp CV, Staunton. Died Lynhurst 10/11/20. Bur. Riverview Cem., Waynesboro.

BEAR, SAMUEL V.: Pvt., Co. A. B. Va. 1821? Farmhand, age 39, Burkes Mill Dist., Augusta Co. 1860 census. Enl. Staunton 7/15/61. Present 11/61-4/62. Discharged 10/10/62, over age of conscription. Alive 1903.

BEARD, HUGH S.: Pvt., Co. B (1st). B. Rockbridge Co. 1940. Enl. Staunton 8/1/61. Ab. sick when transf. 9/28/61.

BEARD, JOHN D.: 1stSgt., Co. A. B. Va. 1834? Farmhand, age 25, Mint Springs, Augusta Co. 1860 census. Enl. Staunton 7/9/61 age 25. Present 11/61-4/62. Reenl. 5/1/62. Deserted 5/30/62, rank appears as Pvt. NFR. Tanner, Mint Spring 1872. Alive 5/03.

BEARD, WILLIAM BRAMBRIDGE OR BOOKER: Pvt., Co. E. B. Rockbridge Co. 1/7/40. Farmhand, age 18, Colliestown PO, Rockbridge Co. 1860 census. Enl. Staunton 8/1/61. Present 11/61-4/62. Reenl. 5/1/62. WIA (right leg) Gaines Mill 6/27/62. Leg shortened 4 inches. Ab. Wounded through 10/31/62. Ab. on leave 1-4/63. Ab. on detached service with Enrolling Officer, 11th Congressional Dist., Lexington, 4/30-12/31/63. Retired 4/27/64. Farmer, age 28, Buffalo Dist., Rockbridge Co. 1870 census. Died on Black Creek near Collierstown 4/4/95. Bur. Collierstown Pres. Ch. Cem.

BEASLEY, JOHN: Pvt., Co. B. (2nd). Not on muster rolls. Returned to duty from hospital Richmond 1/21/65. Cap. High Bridge 4/6/65. Sent from City Point to Point Lookout 4/15/65. Died Point Lookout 6/8/65 and buried in POW Cem. there.

BEATY, GEORGE.: Pvt., Co. K. B. Rockbridge Co. 1839. Blacksmith's apprentice, age 20, Green Valley PO, Bath Co. 1860 census. Enl. Co. G. 25th Va. Inf. Bath CH 6/1/61. Cap. Rich Mt. 7/12/61. Paroled 7/17/61. Enl. Millwood 11/7/62. AWOL 12/17/62-4/63. Rejoined his original Company — Co. G, 25th Va. Inf. Enl. Co. G, 18th Va. Cav. Williamsville 1/1/63. Blacksmith, age 30, Bath CH PO. Bath Co. 1870 census. Member, General Pegram Camp CV, Randolph Co., W. Va. Died there before 2/15. Bur. there.

BEATY, JOHN: Pvt., Co. K. B. Rockbridge Co. 1831? Blacksmith, age 28, Green Valley PO, Bath Co. 1860 census. Enl. Sheperdstown 9/23/62. AWOL 12/17/62-12/31/63. Dropped as deserter. Enl. Co. G. 18th Va. Cav. Williamsville 1/1/63. Ordered returned to Co. K 52nd Va. Inf. 2/8/64. Paroled Staunton 5/26/65. Age 33, 5' 10", fair complexion, dark hair, blue eyes. Blacksmith, âge 39, Bath CH PO, Bath Co. 1870 census. Died before 9/15/07.

BECKS, GEORGE W.: Pvt., Co. G. B. Rockingham Co. 5/13/44. Farmhand, age 16, Burkes Mill Dist., Augusta Co. 1860 census. Enl. Staunton 8/2/61. Present 11/61-2/62. Ab. sick 3-4/62. Present 4/30-10/30/62. AWOL 24 days, fined $8.88. Present 11/62-4/63. 3 months pay stopped by sentence of CM. Present 4/30-8/31/63. 1 months pay deducted by sentence of CM, $11.00 Present 9-12/63 WIA (right leg) Spotsylvania CH 5/12/64. AB. wounded through 12/31/64. Paroled Staunton 5/20/65. Age 23, 5' 8", light complexion, light hair, grey eyes. Admitted to Lee Camp Old Soldiers Home, Richmond 1/26/17 and discharged 4/13/17.

BELL, MARTIN LUTHER: Pvt., Co. F. B. Va. 1/13/35. Farmhand, age 25, Northern Dist., Augusta Co. 1860 census. Enl. Staunton 7/31/61. Died Augusta Co. of fever 11/10/62. Buried Mossy Creek Old Cem., Augusta Co.

BELLAMY, ABNER H.: Pvt., Co. A. Enl. Staunton 7/15/61. Present 11/61-4/62. Reenl. 5/1/62. In hospital with typhoid fever Winchester 8/8/61. In hospital Staunton until cap. 6/6/64. Sent to Wheeling and Camp Morton. Died Camp Morton of acute diarrhea 3/11/65. Bur. grave #1512 Green Lawn Cem., Indianapolis, Ind.

BERRY, GEORGE E.: Pvt., Co. A. B. Va. 1822? Farmer, Augusta Co. Pvt., Co. I, 160th Va. Militia 2/62 age 40. Not on muster rolls. Sick in hospital Richmond 12/12/63. NFR. Farmer, age 48, Mt. Sidney PO, Augusta Co. 1870 census. Drank ammonia by mistake and died near foot of Blue Ridge Mountains, Rockingham Co. 2/28/92.

BERRY, WILLIAM J.: "Billy". Pvt., Co. C. B. Va. 1843? Student, age 15, Burkes Mill Dist., Augusta Co. 1860 census. Enl. Staunton 1/10/63. Present 1/10-4/30/63. AWOL 6/6/63. Deserted New Creek, W. Va. 7/9/63. Took oath and released. Age 19, 5' 8½", light complexion, grey eyes, light hair, farmer.

BETHEL, JAMES S.: Pvt., Co. K. B. Bath Co. 7/21/44. Farmer, age 16, Cleeks Mill PO, Bath Co. 1860 census. Enl. Shenandoah Mountain 4/9/62. Present 4/9-30/62. WIA McDowell 5/8/62. Died of diphtheria at home in Bath Co. 10/14/62.

BETINELLA, GIOVANNI: Pvt., Co. C. Enl. Bunker Hill 10/10/62 as substitute. Deserted near Port Royal 12/26/62.

BEVERIDGE, WILLIAM E: Pvt. Co. ? Not on muster rolls. Died Western State Hospital Staunton 10/21/17. Res. Monterey, Highland Co. Obit. only record. May have served in 62nd Va. Inf.

BIBY, HENRY GEORGE: Pvt., Co. C. B. Augusta Co. 1828? Enl. Staunton 8/2/61. Present 11/61-4/62. Reenl. 5/1/62. as non-conscript over 35. Present 5/1/62-6/4/63. In hospital with 'Catarrah" 6/15-21/63. Present through 12/31/63. Detailed in Pioneer Corps 1/10-11/64. Present 11-12/64. Surrendered Appomattox 4/9/65. Laborer, age 40, Pastures Dist., Augusta Co. 1870 census. Illiterate. Member, Stonewall Jackson Camp CV, Staunton. Resident of Penrose. Died near Spring Hill, Augusta Co. 11/30/03. Bur. Pleasant View Lutheran Ch. Cem.

BINNS, L. F.: Pvt., Co. K. On postwar roster.

BLACK, ANDREW D.: Pvt., Co. E. B. Rockbridge Co. 1838? Farmhand, age 22, Lexington PO, Rockbridge Co. 1860 census. Enl. Staunton 8/1/61. Ab. on leave 11-12/61. Ab. sick 1-2/62. Present 3-4/62. Reenl. 5/1/62. Ab. on leave 3/30-10/31/62. Present 1/1-3/17/63. Ab. sick 3/18-10/31/63, in hospital Staunton with typhoid fever. Present 11-12/63. WIA and Cap. Spotsylvania 5/19/64. Sent to Point Lookout. Transf. to Elmira 6/3/64. Exchanged 3/3/65. Admitted to hospital Richmond 3/4/65. Cap. in Jackson Hosptial, Richmond 4/3/65. Died there of chronic diarrhea 5/5/65. Bur. grave #100, Hollywood Cem., Richmond.

BLACK, ANDREW H.: Pvt., Co. E. B. Amherst Co. 1837? Farmer, age 22, Collierstown PO, Rockbridge Co. 1860 census. Enl. Staunton 8/1/61. Present 11/61-4/62. Reenl. 5/1/62. KIA Sharpsburg 9/17/62, age 25.

BLACK, GEORGE H.: Pvt., Co. G. Died Albermarle Co. 9/9/11. Obit. only record.

BLACK, JAMES S.: Pvt., Co. E. B. Rockbridge Co. 1839? Tanner, Rockbridge Co. Enl. Staunton 8/1/61. Ab. on leave 11-12/61. Ab. sick 1-4/62. NFR. on muster rolls until KIA Spotsylvania CH 5/12/64.

BLACK, JAMES THOMAS: Pvt., Co. H. B. Va. 7/7/36. Farmer, age 23, 1st Dist., Augusta Co. 1860 census. Enl. Staunton 10/25/64. Present 11-12/64, and 3/23/65. Surrendered Appomattox 4/9/65. Farmer, age 33, Fishersville PO, Augusta Co. 1870 census. Died Stuart's Draft 8/26/20. Bur. Tinkling Spring Pres. Ch. Cem. near Fishersville.

BLACK, JOHN: 2ndCpl., Co. E. B. Rockbridge Co. 1/20/33. Farmhand, age 27, Collierstown PO, Rockbridge Co. 1860 census. Enl. Staunton 8/1/61 as Pvt. Present 11/61-2/62. Ab. sick 3/62-10/31/62. Present 1-2/63. Promoted 2ndCpl. 2/1/63. Present 3-12/63. KIA near Spotsylvania CH 5/19/64 bearing the colors. Bur. Richmond. Brother of William L. Black.

BLACK, WILLIAM L.: SgtMajor, F&S and Co. E. B. Rockbridge Co. 4/5/40. Farmhand, age 19, Collierstown PO, Rockbridge Co. 1860 census. Enl. Staunton 8/1/61 as 2ndSgt. Present 11/61-4/62. Reenl. 5/1/62. WIA (hand) Sharpsburg 9/17/62. Ab. wounded through 10/31/62. Present 1-4/63. Promoted SgtMajor 4/7/63. Present 5-8/31/63. Reduced to Pvt. for desertion 9/16/63. Deserted to the enemy in W. Va. 12/63. Took oath and sent north. Farmer, age 28, Fishersville PO, Augusta Co. 1870 census. Brother of John Black. D. 1870. Bur. Providence, Va.

BODKIN, ABEL: Pvt., Co.D. B. 1837? Farmer Augusta Co. Not on muster rolls. Deserted Tucker Co., W. Va. 5/5/64. Took oath and sent to Wheeling. Age 27, 5' 11", fair complexion, grey eyes, dark hair. Moved to Kansas postwar.

BOON, WALTER: 1stLt. Co. K. B. Bath Co. 1828? Saddler, age 32, Millboro Springs, Bath Co. 1860 census. Enl. Shenandoah Mt. 4/9/62, as 4th Sgt. Present 4/9-30/62. Elected 2ndLt. 9/22/62. Present 10/62 and 1-3/63. Elected 1stLt. 2/23/63. Present 4/30-12/31/63. KIA Bethesda Church 5/30/64. Left widow and four children.

BOOZ, JOHN WESLEY: Drummer, Co. B (2nd). B. Rockingham Co. 3/17/35. Clerk, Waynesboro. Enl. Waynesboro 7/15/61. Ab. sick 11/61-2/62. Present 3-4/62. Requested a discharge as a citizen of Maryland 8/1/62. Age 27, 5' 6½", fair complexion, black eyes, black hair. NFR. Died South River Dist., Augusta Co. 8/15/76.

BOSSERMAN, SAMUEL: (1st). Pvt., Co. B (2nd). Enl. 7/61. Served 3 years. Not on muster roll. Post war roster and pension list. Age 84, West Augusta Precinct Pastures Dist., Augusta Co. 1910 census.

BOSSERMAN, SAMUEL: (2nd). Pvt. Co. D. B. Va. 1825? Farmhand, age 34, Sangersville, Augusta Co. 1860 census. Enl. Staunton 7/16/61. Present 7/16-10/31/61. Ab. sick 11/10/61-4/62. Present 4/30-8/31/62. Deserted 11/23/62, however pay receipt dated 12/31/62 says cap. near Port Royal. NFR. Farmer, age 48, Mt. Sidney PO, Augusta Co. 1870 census. Died Glade, North River Dist., Augusta Co. 10/11/85.

BOSSERMAN, WILLIAM HENRY: Pvt., Co. A. B. Augusta Co. 1/20/26. Farmer, age 33, near Ryan's, Pastures Dist., Augusta Co. 1860 census. Enl. Staunton 9/9/61. Present 11-12/61, acting as wagoner since enlisted. Ab. sick 2/5-4/30/62. Reenl. 5/1/62. Deserted 7/17/62. NFR. Farmer, age 43, Craigsville, Augusta Co. 1870 census. Farmer, Middlebrook 1897. Died Augusta Co. 10/8/10. Bur. Mount Tabor Lutheran Ch. Cem., near Middlebrook.

BOWEN, LEONARD: Pvt., Co. B (2nd). Enl. Waynesboro 7/15/61. Present 11/61-2/62. Ab. on leave 3-4/62. WIA (hip) Sharpsburg 9/17/62. Present 1-12/63. WIA Sommerville Ford '64. WIA Cedar Creek 10/19/64. Ab. wounded through 12/31/64. Paroled Manchester 5/5/65.

BOWERS, DAVID: Pvt., Co. G. B. Augusta Co. 3/9/27. Farmhand, Augusta Co. Enl. Staunton 8/2/61. Present 11-12/61. WIA (leg) Alleghany Mt. 12/12/61. Present 1-4/62 non-conscript. AWOL 1 month 5-8/62 and fined $11.00. Present 9/1/62-7/20/63. AWOL 7/21-8/4/63. Fined $3.30. Present 9-12/63. WIA (leg and shoulder) Bethesda Church 5/30/64. Admitted to hospital Richmond 5/31/64. Furloughed for 40 days 6/28/64. Present 10/31-12/31/64. NFR. Farmhand, age 43, Fishersville PO, Augusta Co. 1870 census. Died Stuart's Draft, Augusta Co. 3/4/78. Bur. Calvary Methodist Ch. Cem.

BOWERS, HENRY: Pvt., Co. H. B. Va. 1829. Farmer, age 31, 1st Dist., Augusta Co. 1860 census. Enl. Staunton 7/23/61, age 33. Present 11/61-4/62. Ab. sick 6/12/62-11/2/63. Returned from desertion 6/2/64. WIA Liberty 6/19/64. Pay deducted for time AWOL (dates not indicated). Present 11-12/64. Capt. Farmville 4/6/65. Sent to Point Lookout. Released 6/19/65. 5' 7¾", florid complexion, red hair, hazel eyes. Farmhand, age 40, Fisherville PO, Augusta Co. 1870 census. Died Churchville, Augusta Co. 6/2/93.

BOWMAN, GEORGE LEWIS: Pvt., Co. H. B. Augusta Co. 1/22/32. Farmer, age 28, Staunton PO, Augusta Co. 1860 census. Enl. Staunton 7/23/61. Entry cancelled on rolls. Died of typhoid fever Arbor Hill, Augusta Co. 9/22/61. Bur. St. John's Reformed Ch. Cem. near Middlebrook. Left wife and three children.

BOWMAN, JOHN: Pvt., Co. B (1st). B. Rockingham Co. 5/4/26. Overseer, age 35, South River Dist., Rockbridge Co. 1860 census. Enl. Staunton 8/1/61. No further details in Staunton when transf. 9/28/61.

BRADSHAW, ALEXANDER K.: Pvt., Co. E. Res. of Fredericksburg, Spotsylvania Co. Enl. Gordonsville 8/1/62. Capt. near Spotsylvania CH 5/19/64. Sent to Point Lookout. Transf. to Elmira. Released 6/27/65. 5' 7", florid complexion, dk. hair, blue eyes.

BRADY, JOHN: Pvt., Co. K. B. Augusta Co. 7/7/07. Farmer, age 53, Burkes Mill Dist., Augusta Co. 1860 census. Conscript sent to Regiment 10/62. NFR. Probably discharged for overage of conscription. Died Augusta Co. 7/3/89. Bur. Tinkling Spring Pres. Ch. Cem. near Fishersville.

BRAGG, BARTHOLOMEW KIDD: Pvt., Co. H. B. Va. 1814? Farmhand, age 45, 1st Dist., Augusta Co. 1860 census. Enl. Staunton 7/23/61, age 47. Present 11/61-2/62. AWOL 2/1-16/62. Present 2/17-4/62. Present 4/30-8/31/62. Discharged 6/10/62 over age of conscription. Died Afton, Nelson Co. of dropsy 11/13/65.

BRANUM, WILLIAM: Pvt., Co. A. Not on muster rolls. WIA (foot) Port Republic 6/9/62. NFR.

BRANE, MICHAEL B.: Pvt., Co. I. B. 1830? Enl. Staunton 7/16/61. Detailed in hospital Staunton with sore eyes 2/4/63 and with fever 10/31/63. Paroled Staunton 5/1/65. Age 35, 5' 5", dk. complexion dk. hair, black eyes.

BRIDGE, ALEXANDER: Pvt., Co. B (2nd). B. Augusta Co. 2/18/63. Farmer, age 25, Sherando, Augusta Co. 1860 census. Enl. Waynesboro 7/15/61. Ab. sick 11/61-2/62. Presence or absence not stated 3-4/62, Marylander non-conscript. WIA date and place unknown. AWOL 7/21-12/62. Discharged from hospital for "necrosis of tibia" 1/16/63. Age 27, 5' 9¾", fair complexion, blue eyes, dark hair, illiterate. Farmhand, age 36, Warm Springs Dist., Bath Co. 1870 census. Farmer, Sherando, Augusta Co. 1897. Died Basic City, Augusta Co. 4/26/15. Bur. Sherando, Augusta Co.

BRIDGE, JEFFERSON: Pvt., Co. B (2nd). B. Augusta Co. 6/17/38. Farmhand, age 21, 1st Dist., Augusta Co. 1860 census. Enl. Waynesboro 7/15/61. Died of disease on Back Creek, Augusta Co. 9/17/61. Fair complexion, grey eyes, dark amber hair.

BRIGHT, JOSIAH F.: Pvt., Co. B. Augusta Co. 1/8/31. Farmer, age 29, Deerfield, Augusta Co. 1860 census. Enl. Staunton 8/21/61. Present 11/61-2/62, detailed as wagoner. AWOL 4/20-8/31/62. Ab. or special duty as teamster 8/31/62-2/28/63. Present 3-4/63. KIA Gettysburg 7/3/63. Left widow and 4 children. Reburied Hollywood Cem., Richmond postwar.

BROOKS, ALEXANDER B.: Pvt., Co. C. B. Va. 1837? Farmer, age 22, Burkes Mill Dist., Augusta Co. 1860 census. Enl. Staunton 7/16/61, age 23. Present 11/61-4/62. Reenl. 5/1/62. WIA (hand) Port Republic 6/9/62. Ab. wounded through 8/31/62. WIA Fredericksburg 12/13/62 (mouth, right jaw factured). Furloughed from hospital Richmond 12/24/62 for 60 days. Deserted to the enemy New Creek, W. Va. 7/9/63. Age 25, 5' 9¾", florid complexion, brown eyes, dark hair.

BROOKS, RICHARD R.: Pvt., Co. G. B. Fishersville, Augusta Co. 2/28/33. Farmhand, age 26, 1st Dist., Augusta Co. 1860 census. Pvt., Capt. Cochran's Co., 93rd Va. Militia, furnished subsitute 3/62. Enl. Staunton 4/17/62. Present 4/30/62-2/63. AWOL 3/26-13/31/63. Dropped as deserter. Farmhand, age 36, 1st Dist., Augusta Co. 1870 census. Died Hickory Hill, Augusta Co. 12/28/83.

BROOKS, ROBERT T.: Pvt., Co. C. B. Powhatan Co. 9/23/24. Laborer, Long Glade, Augusta Co. Enl. Staunton 8/21/61. Present 1-12/61, detailed as wagoner. Ab. sick until discharged 2/7/63, over age of conscription. Age 42, 5' 10", dk. complexion, dk. eyes, dk. hair, illiterate. Resident of Spitler, Augusta Co. Died North River Dist., Augusta Co. 9/23/14. Bur. Mt. Pisgah Meth. Ch. Cem.

BROOMS, A. T.: Pvt., Co. G. Not on muster rolls. Attended reunion Staunton 6/91. Resident of Staunton.

BROWN, JAMES A.: Pvt., Co. D. B. Rockingham Co. 1836? Farmhand, age 23, Northern Dist., Augusta Co. 1860 census. Enl. Staunton 7/16/61. Present 7/16/61-4/62. AWOL 5/3-7/12/62. KIA Cedar Run 8/9/62. Left widow and two children.

BROWN, JAMES W.: (2nd), Pvt., Co. I. B. 6/16/23. Not on muster rolls. Obit. said surrendered Appomattox 4/9/65. Farmer, Augusta Co. Died Augusta Co. 2/19/25. Buried Hebron Pres. Ch. Cem.

BROWN, JAMES W.: (2nd), Pvt., Co. I. B. 6/16/23. Not on muster rolls. Obit. said surrendered Appomattox 4/9/65. Farmer, Augusta Co. Died Augusta Co. 2/19/25. Buried Hebron Pres. Ch. Cem.

BROWN, JOHN W.: Pvt., Co. D. B. Augusta Co. 1842. Carpenter, age 17, Sangersville, Augusta Co. 1860 census. Enl. Staunton 7/16/61. Present 7/16/61-4/62. Reenl. 5/1/62. Wounded accidently 7/4/62. Ab. wounded through 8/31/62. Present in arrest 1-2/63. Present 3-4/63. Fined $34.10 by CM for AWOL 3 months and 3 days and 2 months addition pay. Cap. Fairfield, Pa., 7/5/63 as "deserter". Sent to Ft. Delaware. Requested to take the oath of allegiance 12/64. Rejected for Federal service by Surgeon. Released 5/4/65. 5' 10", dark complexion, dark hair, grey eyes. Farmer, Augusta Co. 1866. In Ohio Soldiers Home 1912. Died Lee Camp Old Soldiers Home, Richmond, 3/10/13. Bur. Hollywood Cem., Richmond.

BROWN, JOSEPH: Pvt., Co. G. B. Va. Resident of Fairfax Co. Enl. Fredericksburg 3/22/63. Present 3/22/63-11/64, on detached service as teamster in Division Train. Paroled Alexandria 6/17/65. 5' 7", light complexion, dark hair, blue eyes.

BROWN, SAMUEL BRADFORD, JR.: Capt., Co. K. B. Augusta Co. 1842. Student, age 17, 1st Dist. of Staunton 1860 census. Att. UVa 59-60. Enl. Staunton 7/16/61 as 3rd Lt. Co. C. Presence or absence not stated 11/61-4/62. Reelected 5/1/62. WIA (foot) Port Republic 6/9/62. Elected Capt. 8/23/62. Elected Capt. Co. K. 9/22/62. Ab. wounded until resigned 2/23/63 for disability. Elected Capt. Co. C, "Lee's Body Guard", 39th Bn. Va. Cav. General R. E. Lee's escort at Appomattox and surrendered 4/9/65. Farmer, Augusta Co. 1878. Died Fort Valley, Ga. 5/10/03 and buried there.

BROWNLEE, JOHN BELL: Pvt., Co. H. B. Greenville, Augusta Co. 7/25/32. Farmer age 27, 1st Dist., Augusta Co. 1860 census. Co. A, 93rd Va. Militia exempt for deafness 3/62. Enl. Staunton 10/26/64. Present 11-12/64. Present Appomattox. Farmer, age 36, Fishersville PO, Augusta Co. 1870 census. Farmer and Stockraiser Greenville 1885. Member, Stonewall Jackson Camp, CV, Staunton. Died near Greenville 7/4/08. Bur. Bethel Pres. Ch. Cem. near Middlebrook.

BROYLES, WALKER: Pvt., Co. A. B. Madison Co. 1833? Enl. Staunton 7/9/61 age 27. Present 11/61-4/62. Reenl. 5/1/62. KIA Gaines Mill 6/27/62.

BRUCE, ROBERT A.: Pvt., Co. B (2nd). Enl. Waynesboro 7/15/61. Present 11/61-4/62. Reenl. 5/1/62. AWOL 6/20-10/26/62. Present 10/27-12/63. KIA Bethesda Church 5/30/64. Buried Old Waynesboro Cem.

BRYAN, JOEL.: Pvt., Co. E. Not on muster rolls. WIA (leg) Port Republic 6/9/62. NFR.

BRYAN, LORENZO S.: Pvt., Co. B (1st). Born 1835? Attended UVa. 1854. Teacher Rockbridge Co. On postwar roster.

BUCHANAN, JAMES MADISON, JR.: Pvt., Co. I. B. Bath Co. 6/7/31. Farmer, age 27, Staunton PO, Augusta Co. 1860 census. Enl. Staunton 7/16/61 age 28. Present 11/61-4/62. Reenl. 5/1/62. WIA (abdomen) McDowell 5/8/62. Ab. wounded through 7/64. Deserted to the enemy Beverly, W. Va. 7/15/64. Took oath and released. Age 32, 5' 11", fair complexion, grey eyes, dark hair. Farmer, age 34, Riverheads Dist., Augusta Co. 1870 census. Died Augusta Co. 4/27/00. Bur. Mt. Hermon Lutheran Ch. Cem. Newport, Augusta Co.

BULL, WILLIAM H.: 2ndSgt., Co. G. B. Timberville, Rockingham Co. 1840? Farmer, age 19, Burkes Mill Dist., Augusta Co. 1860 cenus. Enl. Staunton 8/2/61 as Pvt. Present 11/61-4/62. Reenl. and elected 2ndSgt. 5/1/62. Ab. on leave 5/1-8/31/62. Present 9/62. Furnished substitute and discharged Winchester 9/24/62. Reenl. date unknown and served as teamster. KIA Gettysburg 7/3/63. Age 21, 5' 8", fair complexion, blue eyes, dark hair. Buried on the battlefield.

BUMGARDNER, DAVID: Co. unknown. Not on muster rolls. Attended reunion Carlisle, Pa. 9/28/81.

BUMGARDNER, JAMES JR.: Capt. Co. F. B. Fayette, Howard Co., Mo. 1/18/35. Moved back to Va. 1847. Graduate of Brownsburg Academy. Attended UVA. 52-53. One of founders of Phi Kappa Psi fraternity at UVa. Clerk, School Teacher, Lawyer, Lieutenant "West Augusta Guards" 56-61. Present at Harper's Ferry after John Brown's Raid 59. Lawyer, 1st Dist., of Staunton 1860 census. Enl. Staunton 4/17/61 as 3rdLt., Co. L, 5th Va. Inf. Elected Adjutant 5th Va. 5/61. Present Kernstown 3/23/62. Ab. sick much of time and not reelected 4/12/62. Elected 2ndLt. Co. A, 52nd Va. 5/1/62. Elected Capt. Co. F 8/29/62. Ab. on GCM 8/31/62-2/63. Present 4-7/63. Ab. on special duty Staunton 11-12/63. Present 1-4/64. Commanding Regiment 5/30-7/7/64. Cap. Winchester 9/19/64. Sent to Ft. Delaware. Smuggled letter to his wife in a bar of soap via Captain McFarland who was paroled. Released 6/12/65. 5' 6", sallow complexion, black hair, hazel eyes. Present in over 50 engagements and never wounded. Captain, Virginia Militia 1871. Commonwealth's Attorney of Augusta Co. 1866-83. Delegate to Democratic National Convention in Chicago 1886. Rarrfor U.S. Congress but was defeated. Circuit Court Judge of Augusta Co. Member, Board of Directors Valley RR for 50 years. Member, Stonewall Jackson Camp CV, Staunton. Died Staunton 9/2/17. Bur. Bethel Pres. Ch. Cem. near Middlebrook. "A brave and daring soldier."

BUNCH, DAVID: Pvt., Co. A. B. Va. 1826? Carpenter, age 32, Burkes Mill Dist., Augusta Co. 1860 census. Pvt., Capt. Shoemake's Co. K, 32nd Va. Militia 3/62, exempt because of piles. Enl. Staunton 9/1/62. Ab. sick 11/10-12/31/62. Present 1-4/63. Detailed as nurse in hospital Staunton 5/18/63-12/31/64. WIA Hatcher's Run 2/6/65. WIA (forearm) Petersburg 3/29/65. Cap. in hospital Richmond 4/3/65. Sent to Point Lookout 5/9/65. Released 6/26/65. 5' 9½", dark complexion, black hair, dark eyes, illiterate.

BUNCH, JAMES JR.: Pvt., Co. B (2nd). B. 2/22/45. Tombstone only record. Died Augusta Co. 12/3/26. Bur. Middle River Ch. of Brethren Cem., near New Hope.

BUNCH, JAMES W.: Pvt., Co. B (2nd). B. Fluvanna Co. 1819? Carpenter age 41, Burkes Mill Dist., Augusta Co. 1860 census. Enl. Waynesboro 7/15/61. Present 11/61-4/62, detailed as "Artificial". Ab. on leave 4/30-10/31/62. Discharged 12/16/62, over age of conscription. Age 43, 5' 10", fair complexion, light blue eyes. auburn hair.

BUNCH, JOHN W.: Pvt., Co A. B. Va. 1839? Laborer, age 19, Burkes Mill Dist., Augusta Co. 1860 census. Enl. Staunton 7/9/61 age 21. Present 11/61-4/62. Reenl. 5/1/62. Present 8/31/62-4/63. Capt. Chambersburg, Md. 7/6/63. Sent to Baltimore 8/21/63. Transf. to Hammond General Hospital, Point Lookout 12/63. Died there of chronic diarrhea 12/18/63. Bur. in POW Cem. there.

BUNCH, WILLIAM W.: Pvt., Co. A. Born Va. 1823. Miller, Augusta Co. Pvt. Capt. Western's Co., 32nd Va. Militia 4/62. Exempt as miller. Enl. Martinsburg 6/30/64. Rolls 7/30-10/31/64 illegible. Present 11-12/64. Admitted to hospital Richmond with debilitas 3/30/65. Cap. in hospital there 4/3/65. Escaped from hospital 4/26/65. Farmer, age 44, near Mt. Solon PO, Augusta Co. 1870 census.

BURBY, DAVID.: Pvt., Co. G. On postwar roster.

BURES, THOMAS: Pvt., Co. D. Detailed as teamster 10/62. NFR.

BURGER, JOEL: Pvt., Co. E. B. Botetourt Co. 1840. Farmer, Rockbridge Co. Enl. Staunton 8/1/61. Detailed as teamster 9/10-10/30/61. Present 11/61-4/62. Reenl. 5/1/62. Present 4/30-10/31/62 and 1-12/63. Pay stopped for 1 bayonet lost $5.48. Issued clothing 3/1/64. NFR. (also listed on post war roster of Co. D, 60th Va. Inf. and in Beckner's Co., Burkes Regt.)

BURGER, WILLIAM CRAWFORD: Capt., Co. K. B. Bath Co. 3/17/26. Farmer, age 34, Bath CH PO, Bath Co. 1860 census. Enl. Shenandoah Mt. 4/9/62 as 1stLt. Elected Capt. 6/11/62. Resigned because of ill health, over 35, and to manage a nitre cave on his property in Bath Co. 8/23/62. Reenl. as Pvt. '63. Cap. Gettysburg 7/5/63. Sent to Ft. Delaware. NFR. Farmer, age 44, Williamsville Township, Bath Co. 1870 census. Farmer, Green Valley Bath Co. 1897. Died Bath Co. 1910. Bur. on Cowpasture River near Millboro Springs.

BURNS, AARON WILLIAM: Pvt., Co. K. Born Burnsville, Bath Co. 3/23/45. Farmhand, age 23, Red Hole, Bath Co. 1860 census. Enl. Shenandoah Mt. 4/9/62. Deserted 4/19/62. Farmer, Indianola, Iowa. Died there 10/19/07 and buried there.

BURNS, ABRAHAM W.: Pvt., Co. K. B. Burnsville, Bath Co. 1833? Farmer, age 27, Cleeks Mill PO, Bath Co. 1860 census. On postwar roster. Enl. McClanahan's Va. Battery. Blacksmith, age 37, Williamsville Dist., Bath Co. 1870 census. Died Burnsville 12/31/11. Bur. Burnsville Cem.

BURNS, HUGHART MONROE: Pvt., Co. K. B. Burnsville, Bath Co.7/17/30. Farmer, age 29, Red Hole, Bath Co. 1860 census. Enl. Shenandoah Mt. 4/9/62 age 31. AWOL 4/19-5/19/62. Deserted 6/9/62. Enl. Co. G. 18th Va. Cav. 1/1/63 and WIA (foot) Edenburg '64. Paroled Staunton 5/25/65. 5' 5", dark complexion, black hair, blue eyes. Farmer, age 39, Williamsville Dist., Bath Co. 1870 census. Died Burnsville 10/2/11. Bur. Burnsville Cem.

BURNS, JAMES A.: 1stLt., Co. B (2nd) B. Augusta Co. 1841. Farmhand, age 18, 1st Dist., Augusta Co. 1860 census. Enl. Waynesboro 7/15/61 as 3rd Sgt. Present 11/61-4/62. Elected 2nd Lt. 5/1/62. Promoted 1stLt. 6/8/62. Commanding Co. 8/1/62. WIA (hip) Sharpsburg 9/17/62. Ab. on leave 2/63. Present 4/63. Ab. arresting deserters 5/63. Commanding Co. 9-12/63. Present 1/31/64. Commanding Co. 8/21/64. WIA (stomach) Winchester 9/19/64. DOWs, age 23. Buried Old Waynesboro Cem.

BURNS, JOHN: Pvt., Co. H. B. County Kerry, Ireland 1846? Enl. Staunton 7/23/61, age 25. Discharged by Civil Authority for under age 9/4/61. Enl. Co. G, 11th Va. Cav. WIA '65. Barkeeper, age 23, Staunton 1870 Census. Merchant, Staunton. Died Staunton 1/17/97. Buried Thornrose Cem.

BURNS, JOSEPH F.: Pvt., Co. K. B. Bath Co. 1832? On postwar roster only. Enl. Co. G., 18th Va. Cav. Paroled Staunton 5/24/65. 5' 9", black hair, dark complexion, gray eyes. Died Little Valley, Bath Co., 1900.

BURNS, LEWIS F.: Pvt., Co. K. B. Burnsville, Bath Co. 7/25/34. Laborer, age 26, Green Valley PO, Bath Co. 1860 census. Enl. Shenandoah Mt. 4/9/62. Present 4/9-30/62. AWOL 5/10-9/15/62. Ab. sick 11/1/62-2/11/63. AWOL 2/12-12/19/63. Sentenced by Reg'l CM to forfeit 1 months pay in addition to the time of his absence. WIA (right thigh) and cap. Cedar Creek 10/19/64. NFR. Carpenter, age 37, Williamsville Dist., Bath Co. 1870 census. Farmer, McClung, Bath Co. 1897. Died Bath Co. 11/27/95. Bur. Woodland Ch. Cem., Fairview, Bath Co.

BURNS, METHUSELAH CRODDY GAY: Pvt., Co. K. B. Burnsville, Bath Co. 9/13/39. Enl. Shenandoah Mt. 4/9/62. Deserted to the enemy 4/19/62. NFR. Farmer, Milo, Iowa. Died there 8/13/21 and bur. there.

BURNS, MICHAEL NATHANIEL.: 1stCpl., Co. K. B. Burnsville, Bath Co. 2/25/29. Enl. Shenandoah Mt. 4/9/62, as Pvt. Promoted 1stCpl. WIA McDowell 5/8/62. DOWs in Hospital Staunton 5/12/62. Buried grave #953, Thornrose Cem.

BURNS, WILLIAM H. (1st): Pvt., Co. K. On postwar roster.

BURNS, WILLIAM H. (2nd): 2ndLt., Co. B (2nd). B. Augusta Co. 1839? Farmhand, age 20, 1st Dist., Augusta Co. 1860 census. Enl. Staunton 5/27/61 as Sgt., Capt. R. D. Lilley's Co., 25th Va. Inf. Transf. to Co. B (2nd) 5/1/62 and promoted 3rdSgt. WIA (thigh) McDowell 5/8/62. Promoted 1stSgt. 7/15/62. WIA Sharpsburg 9/17/62. Elected 2ndLt. 10/2/62. Present 11-12/62 and 2-4/63. Ab. 5/63. Presence or absence not stated 6-12/63. KIA Wilderness. 5/6/64, age 24. Bur. Old Waynesboro Cem.

BUSH, CHARLES E.: Pvt., Co. A. B. Henrico Co. 1842? Blacksmith's Apprentice, age 17, Northern District, Augusta Co. 1860 census. Enl. Staunton 7/9/61 age 22. Present 11/61-4/62. Reenl. 5/1/62. Promoted Color Sergeant of the Regt. WIA Sharpsburg 9/17/62 while carrying the colors. DOWs at Shepherdstown 9/18/62.

BUSH, DANIEL L.: Musician, Co. F. B. Va. 1841? Coach Painter, age 19, Northern Dist. Augusta Co. 1860 census. Enl. Sommerville Ford 2/27/64. Present 3-12/64. Surrendered Appomattox 4/9/65.

BUSH, JOHN W.: Pvt., Co. A. B. Page Co. 1839? Blacksmith, Augusta Co. 1860. Not on muster rolls. Cap. and sent to Point Lookout and died of disease there. Bur. POW Cem. there.

BUSH, LIVINGSTON F.: Pvt., Co. B. (2nd). B. Augusta Co. 1827? Laborer, Augusta Co. Enl. Waynesboro 7/15/61. Present 11/61-4/62. WIA (thigh) McDowell 5/8/62. Ab. wounded in hospital Staunton until discharged for disability and chronic rheumatism 2/15/63. Age 36, 5' 9", fair complexion, grey eyes, light hair.

BUSHONG, ISAAC A.: 2ndLt., Co. H. B. Rockingham Co. 2/6/25. Miller, age 36, Burkes Mill Dist., Augusta Co. 1860 census. Enl. Staunton 7/23/61 age 37. Present 11/61-4/62. Not reelected 5/1/62. Died Culpeper Co. 11/29/09. Bur. Bushong-Hawkins Cem., Wayland's Mill Rd., Culpeper Co.

BUZZARD, JOHN M.: Pvt., Co. A. B. Pocahontas Co. 1830? Farmer, Pocohontas Co. Enl. Richmond 7/64. Deserted to the enemy Rowlesburg, W. Va. 9/5/64. Took oath and released. Age 35, 5' 8½", fair complexion, blue eyes, auburn hair.

BYERLY, DAVID MARTIN: Pvt., Co. G. B. Augusta Co. 1/24/42. Enl. Staunton 8/2/61. Present 11/61-1/62. Ab. sick 2/1/62-12/31/64. NFR. Farmer, age 26, Mt. Sidney PO, Augusta Co. 1870 census. Died Mt. Meridian 12/26/78. Bur. St. Paul's Ch. Cem., near Mt. Meridian.

BYERS, JOHN S.: Capt., Co. C. B. Augusta Co. 1819? Farmer, age 40, Burkes Mill Dist., Augusta Co. 1860 census. Enl. Staunton 7/16/61 as 1stLt., Age 41, Presence or absence not stated 11/61-4/62. WIA Port Republic 6/9/62. WIA (arm and groin) 2nd Manassas 8/28/62. Elected Capt. 12/24/62. Ab. wounded through 1/9/63. Detailed as Enrolling Officer Augusta Co. 1/9-10/63. Ab. on leave 30 days 11-12/63. WIA (left foot) Spotsylvania CH 5/12/64 (obit. says 12 members of Company KIA that day.). Ab. wounded until retired 2/23/65. Railroad employee, age 50, Staunton, 1870 census. Thrown from a buggy and killed on South River 7/19/90. Bur. Thornrose Cem., Staunton.

BYERS, ROBERT G.: 3rd Sgt., Co. C. B. Augusta Co. 5/25/40. Farmhand, age 16, Northern Dist., Augusta Co. 1860 census. Enl. Staunton 7/16/61 as Pvt., age 21. Present 11-12/61. Promoted 3rdCpl. 11/26/61. Present 1-4/62. Reenl. 5/1/62. WIA (breast) Port Republic 6/9/62. Ab. wounded through 8/29/62. Present 8/30/62-2/28/62 Promoted 4thSgt 1/1/63. Present 3-10/63. Reduced to Pvt. 9/12/63. Present 11-12/63. Promoted 3rdSgt 64. WIA (right hip) and cap. Cedar Creek 10/19/64, age 22. Sent to Point Lookout 10/28/64 and paroled for exchange same day. WIA (left arm by a shell) and cap. Ft. Stedman 3/25/65. Lower left arm amputated at 3rd Div., 9th Army Corps Hospital, Army of the Potomac. Sent to Lincoln General Hospital, Washington, D.C. Released 6/12/65. 5' 10" light complexion, brown hair, grey eyes. Butcher, Augusta Co. 1872. Living Staunton 1896. Died near Parnassus, Augusta Co. 4/15/00. Bur. Union Pres. Ch. Cem., Augusta Co.

BYERS, SAMUEL J.: Pvt., Co. C. B. Augusta Co. 2/24/32. Farmhand, age 28, Northern Dist., Augusta Co. 1860 census. Pvt., Capt. Shumake's Co., 32nd Va. Militia, furnished substitute 3/15/62, age 25. Enl. Rude's Hill 4/5/63 in 5th Va. Inf. Transf. to this Co. 5/1/63. Ab. sick 8/28/63 until died of fever at home in Augusta Co. 9/5/63. Bur. Union Pres. Ch. Cem., Augusta Co.

BYERS, THOMAS C.: Pvt., Co. E. Not on muster rolls. WIA Cross Keys 6/8/62. NFR.

BYERS, WILLIAM CONRAD: Pvt., Co. E. B. Buchanan, Botetourt Co. 2/14/40. Farmhand, age 20, Natural Bridge Dist., Rockbridge Co. 1860 census. Enl. Staunton 8/1/61. Present 11/61-4/62. Reenl. 5/1/62. WIA Port Republic 6/9/62. Present 1-12/63., detailed Orderly of Regt. 8/4-13/63. Issued clothing 4/29/64 and 6/12/64. AWOL 6/25-12/7/64. Sentenced by Regt'l CM to forfeit pay and clothing allowance for period of AWOL and 1 months pay in addition. Deserted to the Army of the Potomac near Petersburg 3/8/65. Took oath and transportation furnished to Greene Co., Ohio. Laborer, age 30, Natural Bridge Dist., Rockbridge Co. 1870 census. Stockraiser Rockbridge Co. 1888. Died Covington 7/8/19. Buried Cedar Hill Cem. there.

BYRD, JAMES W. E.: Pvt., Co. H. B. Alleghany Co. 1827? Farmer, age 32, Craigsville PO, Augusta Co. 1860 Census. Enl. Staunton 7/21/61 age 34. Present 11/61-2/62. AWOL 4/21/62-1/1/63. Deserted to the enemy Randolph Co. 1/7/63. Sent to Camp Chase. Took oath and released. Age 38, 5' 9", dark eyes, black hair, dark complexion.

BYRD, WILLIAM WALLACE: 2ndLt., Co. K. B. Botetourt Co. 5/8/21. Farmer, age 38, Bath CH PO, Bath Co. 1860 census. Served as Drillmaster 25th N. C. Inf. Enl. Shenandoah Mt. 4/9/62 as 1st Sgt. Elected 2ndLt. same day. Reelected 5/1/62. Present McDowell 5/8/62. Resigned for ill health 7/18/62. Cap. Co. E., 10th Va. Bn. Reserves 4/16/64. Present commanding Co. 4th Bn. Va. Reserves Piedmont 6/5/64. Promoted Major of Bn. Commanded Bn. Va. Reserves to end of war. Farmer, age 49, Warm Springs, Bath Co. 1870 census. Farmer, Mountain Grove PO 1884. Died Bath Co. 5/6/01.

CAFFREY, PHILIP: Pvt., Co. F B. Ireland 1837? Farmer, Augusta Co. Enl. Staunton 7/31/61. Present 1-4/62. Reenl. 5/1/62. KIA Gaines Mill 6/27/62. Age 24, 5' 10" fair complexion, blue eyes, light hair.

CALBRETH, THOMAS.: Pvt., Co. G. B. Va. 1835? On postwar roster. Farmer, age 45, Mt. Sidney PO, Augusta Co. 1870 census. Bur. Tinkling Spring Pres. Ch. Cem., near Fishersville.

CALE, DAVID T.: Pvt., Co. C. B. Augusta Co. 1/26/32. Farmer, age 29, 1st Dist., Augusta Co. 1860 census. Enl. Staunton 9/8/61. Present 11/61-4/62. Reenl. 5/1/62. WIA Port Republic 6/9/62. Ab. wounded through 11/26/62. AWOL 11/27/62-8/31/63. Present in arrest 10/25/63. Escaped from custody of Regt'l guard night of 12/27/63. Ab. confined at Castle Thunder, Richmond by sentence of GCM for 12 months 2/1-12/31/64. NFR. Died Augusta Co. 2/4/12. Bur. Mt. Tabor Lutheran Ch. Cem. near Middlebrook.

CALE, WILLIAM VAN BUREN: Pvt., Co. C. B. Augusta Co. 1836? Wagonmaker, age 22, 1st Dist., Augusta Co. 1860 census. Enl. Staunton 7/16/61 age 24. Present 11/61-2/20/62. AWOL 2/21-4/62. Deserted to the enemy near Deerfield 4/20/62. Took oath and released at Cincinnati, Oh. 9/19/62. Farnhand, age 32, 6th Dist., Augusta Co. 1870 census.

CALHOUN, EPHRIAM: Pvt., Co. E. B. 1839? Farmer, Pendleton Co. Not on muster rolls. Deserted to the enemy Preston Co., W. Va. 3/26/63. Age 24, 6', dark complexion, black eyes, black hair, black whiskers. Sent to Camp Chase. NFR.

CAMDEN, JAMES: Pvt., Co. E. B. Rockbridge Co. 1841? Laborer, age 18, 8th Dist., Rockbridge Co. 1860 census. On postwar roster. Farmer, Rockbridge Co. 1866. Died Rockbridge Co. 4/20/17. Brother of John D. Camden.

CAMDEN, JOHN: Pvt., Co. E. B. 1845? Not on muster rolls. Teamster, age 65, Warm Springs Dist., Bath Co. 1910 census. Resident of Buena Vista, Rockbridge Co. when admitted to Lee Camp Old Soldiers home, Richmond 11/14/32. Died there 2/8/36. Buried Hollywood Cem.

CAMDEN, JOHN D.: Pvt., Co. E. B. Va. 1837? Laborer, age 22, Buena Vista Furnace, Rockbridge Co. 1860 census. Enl. Staunton 8/1/61. Present 11/61-2/62. Ab. sick 3-4/62. Reenl. 5/1/61. Present 5/1-10/31/62. Present in arrest 1-2/63, pay deducted for 38 days AWOL. Ab. sick 6/9-8/31/63. Present 9-12/63. Died of pneumonia at Sommerville Ford, Orange Co. 1/17/64. Age 21. "A good soldier." Brother of James Camden.

CAMDEN, LAYNE OR LANE: Pvt., Co. E. B. Va. 1815? Farmhand, age 44, Buena Vista Furnace, Rockbridge Co. 1860 census. Enl. Staunton 8/1/61. AWOL 11/61-4/62. Reenl. 5/1/62, non-conscript. Present 5/1-10/31/62. Present in arrest 1-2/63. Present 3-4/63, pay deducted for 54 days AWOL. Ab. sick 6/4-12/31/63. Deserted from line of battle near Richmond 6/1/64. NFR. DOD on postwar roster.

CAMPBELL, JOSEPH WILLIAM: Pvt., Co. I. B. Augusta Co. 1841? Farmhand, age 18 1st Dist., Augusta Co. 1860 census. Enl. Staunton 7/16/61. Present 11/61-2/62. Died of fever in hospital Staunton 4/25/62. Bur. Thornrose Cem., grave #946.

CAMPBELL, LAWSON P.: Pvt., Co. H. B. Amherst Co. 1835? Wagonmaker, Craigsville Augusta Co. Enl. Staunton 7/23/61 age 26. Present 11-12/61. AWOL 2/24-3/16/62. Fined $5.00 by CM. Present 3/17-4/62 Ab. sick 6/16/62. AWOL 11/28-12/62. Present 1-2/63, fined 6 months pay besides time AWOL by Brig. CM. Sent to hospital 4/63. Detailed as nurse in hospital Jordan Springs 6/18-7/23/63. Present 9-12/63. Deserted to the enemy Beverly, W. Va. 5/8/64. Took oath and released. Age 29, 6' 3", fair complexion, brown eyes, dark hair. Farmer, Craigsville, 1870.

CAMPBELL, MATHEW BRYAN: 2ndLt. Co. E. B. Rockbridge Co. 8/16/34. Farmer, age 26, Lexington Dist., Rockbridge Co. 1860 census. Enl. Staunton 8/1/61. Present 11/61-4/62. Not reelected 5/1/62. Enl. as Pvt. in 2nd Rockbridge Artillery and promoted Cpl. Surrendered Appomattox 4/9/65. Farmer, age 36, Natural Bridge Dist., Rockbridge Co. 1870 census. Died near Buffalo Mills 5/14/06. Bur. Wesley Chapel Cem.

CAMPBELL, NIMROD MCPHERSON: Pvt., Co. B (1st). B. Va. 6/10/43. Resident, 16, 6th Dist., Rockbridge Co. 1860 census. Enl. Fairfielld 8/10/61. Present until trans. 9/28/61.

CARPENTER, WILLIAM JACOB: Pvt., Co. B (2nd). B. Rockingham Co. 2/1/44. Enl. New Market 5/21/62 as substitute for B. F. Smith. Deserted 6/6/62. Enl. Co. H. 12th Va. Cav. Died Cross Keys, Rockingham Co. 2/4/22. Bur. St. Paul's Ch. Cem., Rockland Mills.

CARPENTER, WILLIAM R. N.: Pvt., Co. K. B. Burnville, Bath Co. 11/8/42. Farmhand, age 17, Cleeks Mill PO, Bath Co. 1860 census. Enl. Shenandoah Mt. 4/9/62. AWOL 4/9-30/62. WIA (lost middle finger right hand) Port Republic 6/9/62. Died of diphtheria and wound at home in Bath Co. 7/13/62, age 20 years and 8 months. Bur. Burnsville Cem.

CARRELL, SAMUEL: 3rdSgt., Co. A. B. Eastern Va. 10/2/30. Farmer, Mt. Solon. Enl. Staunton 7/9/61, as 4thSgt., age 30. Ab. sick 11/17/61-1/8/62. Present 1/9-4/62. Reenl. 5/1/62. Promoted 3rd Sgt. WIA 2nd Manassas 8/28/62 and foot amputated. In Hospital Staunton 5/64. NFR. Builder, contractor, farmer and stockraiser, Mt. Solon 1885. Died Weyer's Cave 4/17/13. Bur. St. Paul's Ch. Cem.

CARSON, JAMES HENRY: Pvt. Co. D? B. 5/6/39. Died on picket near Winchester 12/21/61. Bur. Old Emanuel Ch. Cem. near Mt. Solon. Family bible only record. Believed to have served in Co. D, 5th Va. Inf. Brother of John A. Carson.

CARSON, JOHN ADDISON: 2ndLt., Co. D. B. Augusta Co. 3/6/33. Farmer, age 25, Mt. Solon PO, Augusta Co. 1860 census. Enl. Staunton 7/16/61, as 2ndCpl. Present 7/16/61-4/62. Elected 2ndLt. 5/1/62. KIA McDowell 5/8/62. Age 27, 5' 7", fair complexion, brown eyes, brown hair, left widow. Bur. Old Emanuel Ch. Cem., near Mt. Solon. Brother of James H. Carson.

CARTER, CHARLES B.: Pvt., Co. B (2nd). B. Va. 1824? Shoemaker, age 35, Waynesboro, Augusta Co. 1860 census. Enl. Waynesboro 7/15/61. Cap. on picket Alleghany Mt. 12/13/61. Sent to Camp Chase. Exchanged Vicksburg, Miss. 8/25/62. AWOL since exchanged on rolls 1-2/63. Pres. 3-4/63. Ab. sick 6/13-17/63. Detailed as nurse Jordan Springs hospital 6/18/63-7/25/63. In hospital Staunton with fever 8/14-31/63. Present 9/63-6/13/64 but WIA (breast) Bethesda Ch. 5/30/64. In hospital Richmond 6/14-30/64 debilitas. Transf. to hospital Staunton 7/27/64 "fever in btub" and diarrhea. Present 12/7-31/12/64. In hospital Richmond 2/10/65 "ascites". Capt. in hospital Richmond 4/3/65. Paroled 4/29/65. Des. from hospital 5/11/65. D. Batesville, Albemarle Co. 5/4/71. Bur. Riverview Cem., Waynesboro, Augusta Co.

CARTER, DILLARD: Pvt., Co. B (2nd). WIA McDowell 5/8/62. NFR.

CARTER, THOMAS H.: Pvt. Co. B (2nd). B. Albemarle Co. 1833? Farmhand, age 26, Waynesboro 1860 census. Enl. Waynesboro 7/15/61. Present 11/61-4/62. Reenlisted 5/1/62. AWOL 5/17/62-12/31/62. Dropped as a deserter. Paroled Staunton 5/15/65. Age 31, 6' 1", dark complexion, dark hair, dark eyes. Farmer, Avon, Nelson Co. 1902. Adm. to Lee Camp Old Soldiers Home, Richmond 3/26/19. D. there 4/13/21. Bur. Hollywood Cem., Richmond.

CARVER, VALENTINE: Pvt., Co. B (1st). B. Rockingham Co. 1837? Wagonmaker 8/1/61. Ab. on detail Staunton when transf. 9/28/61.

CARY, PETER MINOR: Pvt., Co. F. B. Nelson Co. 4/22/41. Farmer, Augusta Co. Enl. Richmond 8/6/64. Deserted the same day. Farmer, Augusta Co. 1870.

CASEY, JOHN: Pvt., Co. G. B. Ireland. Enl. 10/16/62 as Sub. Present until des. 11/15/62. NFR.

CASH, BRAXTON D.: Pvt., Co. B (1st). B. near Vesuvius, Rockbridge Co. 1838? Farmhand, age 22, 6th Dist., Rockbridge Co. 1860 census. Enl. Fairfield 7/10/61. Pres. until transf. 9/28/61. Brother of James P. Cash.

CASH, JAMES PATTERSON: Pvt., Co. B (1st). B. Rockbridge Co. 1826? Farmhand, age 24, 6th Dist., Rockbridge Co. 1860 census. Enl. Fairfield 7/10/61. Ab. on detail Staunton when transf. 9/28/61. Brother of Braxton D. Cash.

CASH, JAMES W: Pvt., Co. B. (1st). B. Va. 1824? Wagoner, age 26, 1st Dist., Augusta Co. 1860 census. CM for desertion only record.

CASH, JOHN: Pvt. Co. B (1st). B. Va. 1818. Resident, Rockbridge Co. D. of disease Staunton 1861 on postwar roster.

CASH, JOHN WESLEY: Pvt. Co. B (1st). Enl. Fairfield 8/10/61. Ab. on detail Staunton when transf. 9/28/61.

CASH, JOSEPH BENJAMIN: Pvt., Co. E. B. Augusta Co. 3/4/44. Shoemaker, Rapps Mill, Rockbridge Co. Enl. Staunton 8/1/61. Pres. 11/61-4/62. Reenl. 5/1/62. AWOL 6/6/62-12/31/62. Joined from des. 2/29/64. Sentenced by 2nd. Corps CM to hard labor with ball and chain for 12 months on the public works. Pardoned by Sec. of War 12/6/64. Pres. 12/31/64. Paroled Farmville 4/11-21/65. D. Natural Bridge, Rockbridge Co. 11/19/15. Bur. Natural Bridge Baptist Ch. Cem.

CASH, JOSEPH MARION: Pvt., Co. B (1st). B. Nelson Co. 1836? Enl. Fairfield 7/10/61. Pres. until transf. 9/28/61.

CASH, JOSEPH W.: Pvt., Co. B (1st) B. Nelson Co. 1833? Farmhand, age 27, South River Dist., Rockbridge Co. 1860 census. D. of disease 1861, cause and place unknown on postwar roster.

CASH, WILLIAM: Pvt. Co. B (1st). B. Va. 1842? Farmhand, age 18, 6th Dist., Rockbridge Co. 1860 census. D. of disease in Staunton 1861 on postwar roster.

CASH, WILLIAM HENRY: Pvt., Co. B (1st). B. Rockbridge Co. 9/23/42. Farmhand, age 18, South River Dist., Rockbridge Co. 1860 census. Enl. Staunton 8/1/61. Pres. until transf. 9/28/61.

CASON, JACOB CHRISTIAN.: Pvt., Co. H. B. Augusta Co. 4/22/30. Resident of Middlebrook, Augusta Co. Enl. Camp Alleghany 3/21/62. Des. to the enemy and took oath and released Cincinnati, Oh. 9/19/63. NFR.

CASSIDY, PATRICK: Pvt., Co. D B. 1835? Conscript assigned 6/13/62, at Mt. Meridian. Pres. 6/13/62-8/31/62. Des. 12/13/62. Des. to the enemy Morgan Co., W. Va. 1/5/63. Sent to Camp Chase, Oh. Age 27, 5' 8½", blue eyes, dark hair, sallow complexion, sandy whiskers. NFR.

CAULEY, LEWIS: Pvt., Co. K. B. Bath Co. 5/9/30. Farmhand, age 27, Mtn. Grove PO, Bath Co. 1860 census. Pvt. Capt. Davis' Co., 81st Va. Militia. Enl. Somerville Ford 1/31/64. KIA Bethesda Ch. 5/30/64.

CAULEY, THOMAS JEFFERSON: Pvt., Co. K. B. Bath Co. 5/27/02. Went into Army as Sub. for son so he could stay at home with mother. Enl. Shenandoah Mtn. 4/9/62. Pres. 4/9-30/62. WIA (foot, breast and arm) Port Republic 6/9/62. DOWs in hospital Charlottesville 6/21/62.

CAULEY, WILLIAM BROWN: Pvt. Co. K. B. Bath Co. 9/17/40. Not on muster rolls. Enl. Shenandoah Mts. 4/9/62. Detailed in Nitre Works, Bath Co. due to defective eyesight. Farmer, Ft. Lewis, Bath Co. 1897. D. Bath Co. 2/28/18. Bur. Westminister Chapel Cem., Bath Co.

CAVINDER, MICHAEL C.: Pvt., Co. G. Enl. Staunton 10/6/62 as sub. Des. near Bunker Hill 10/17/62. NFR.

CHANCELLOR, JAMES EDGAR: Surgeon. B. Chancellorsville, Spotsylvania Co. 1/26/16. Att. UVa. Gd. Jefferson Med. Co., Phila '48. Physician, Charlottesville. May have served with the Regt. for a short time.Assigned to General Hosp. Charlottesvill to end of war. Physician, Charlottesville. Instructor in anatomy UVa. Pres. Med. Soc. of Va. Member, State Board Med. Examiners. D. UVa, Charlottesville 9/11/96.

CHANDLER, ANDREW J.: Pvt., Co. C. B. Augusta Co. 1838? Farmhand, age 21, Burkes Mill Dist. Augusta Co. 1860 census. Enl. Staunton 7/16/61 age 23. Pres. 11/61-4/62. Reenl. 5/1/62. KIA 2nd Manassas 8/28/62. Brother of John D. Chandler.

CHANDLER, JAMES H.: Pvt., Co. F. Res. of Mt. Sidney, Augusta Co. Enl. Mt. Meridian 6/15/62. Des. 7/17/62. Capt. Bristoe Station 10/19/63. Took oath Wash., D.C. and sent north. 5' 9", dk. complexkon, black hair, hazel eyes. NFR.

CHANDLER, JOHN DAVID: Pvt., Co. B. (2nd). B. Augusta Co. 6/25/41. Farmhand, age 16, Waynesboro, Augusta Co. 1860 census. Enl. Waynesboro 7/15/61. Pres. 11/61-4/62. Reenl. 5/1/62. AWOL 5/23/62-12/31/63. Pres. 2/27/64. CM 3/7/64. WIA (shoulder) Spotsylvania 5/12/64. Ab in arrest (Div. Guard) 11-12/64. Surr. Appomattox 4/9/65. Farmer, near Waynesboro. D. Waynesboro 11/10/17. Bur. Zion Ch. Cem. Brother of Andrew J. Chandler.

CHANDLER, WILLIAM: Pvt., Co. B (1st). B. Augusta Co. 1834. Shoemaker. On postwar roster.

CHAPLAIN, WILLIAM J.: Pvt., Co. A. Not on muster rolls. Deserted to enemy Clarksburg, W. Va. 9/14/63. Sent north via New Creek. NFR.

CHILDRESS, DAVID DILLARA: Pvt., Co. A. B. Augusta Co. 8/31/42. On postwar roster. Resident, age 67, Waynesboro, Augusta Co. 1910 census. Bur. Riverview Cem., Waynesboro, Augusta Co.

CHILDRESS, FRANCIS M.: Pvt., Co. A. B. Rockingham Co. 1839? Farmer, age 20, Burkes Mill Dist., Augusta Co. 1860 census. Enl. Staunton 7/9/61 age 21. Present 11/61-4/62. Reenl. 5/1/62. Died of typhoid fever Augusta Co. 7/12/62. 5' 11½", dark complexion, blue eyes, light brown hair. Brother of Thaddeus G. Childress.

CHILDRESS, PATRICK HENRY: Pvt., Co. C. B. Augusta Co. 12/12/32. Millwright, Augusta Co. Enl. Staunton 7/16/61 age 20. Present 11/61-4/62. Reenl. 5/1/62. WIA Cedar Run 8/9/62. Ab. wounded until discharged wounds and permanent disease of the spine and "spermatonhuas,"Richmond 10/17/62. Age 27, 5' 10", fair complexion, grey eyes, brown hair. Millwright, age 37, Fishersville PO, Augusta Co. 1870 census. Died South River Dist., Augusta Co. 5/4/14. Buried Thornrose Cem., Staunton.

CHILDRESS, THADDEUS JOHNSON: Pvt., Co. A. B. Rockingham Co. 12/20/42. Farmhand, age 19, Burkes Mill Dist., Augusta Co. 1860 census. Enl. Staunton 7/9/61 age 19, Present 11/61-4/62. Reenl. 5 /1/62. WIA (arm) Gaines Mill 6/27/62. WIA (lost 2 fingers) Fredericksburg 12/13/62. Ab. wounded 12/14/62-12/31/64. Paroled Staunton 5/1/65. Age 23, 5' 10", light complexion, light hair, grey eyes. Shoemaker, New Hope 1866. Living in Illinois 1921. Brother of Francis M. Childress.

2CHILDRESS, WILLIAM M.: Pvt., Co. B (2nd) B. Augusta Co. 12/24/42. Res. Waynesboro. Enl. Waynesboro 7/15/61. Present 11-12/61. Ab. sick 1-2/62. Present 3-4/62. Reenl. 5/1/62. AWOL 6/9/62-4/63. Transf. to Co. B, 39th Bn. Va. Cav. 5/15/63. Carpenter, age 25, Fishersville PO, Augusta Co. 1870 census. Living Basic City, Augusta Co. 1913.

CHRISTIAN, BOLIVAR: Captain and Assistant Commissary of Subsistence, F&S. B. Greenville, Augusta Co. 4/26/25. Grad. Washington Co. '45. Lawyer; Member Va. House of Delegates 1853-58. Member Va. Senate. Capt., Co. A (Staunton), 182nd Va. Militia. Assistant Adjutant General, Harper's Div., Harper's Ferry 4-8/61. Appointed 8/19/61. Reappointed 5/1/62. Present 10/62. Elected Va. State Senate and absent in attendence 2-3/63. Present 4-5/63. Promoted to Lt.Col. and assigned to Special Service. NFR. Member, Va. House of Delegates and State Senator; Attorney Valley RR 1866-78. President Washington College Alumni Association 1870; Member, Board of Trustees, Washington College 1858-91. Died Western State Hospital, Staunton 7/17/00. Bur, Bethel Pres. Ch. Cem. near Middlebrook.

CHISHOLM, H. C.: Pvt., Co. A. On postwar roster. Enl. 1864 and served 6 months.

CHRISTMAN, NOAH LEVI: Pvt., Co.A. B. Augusta Co. 1/31/45. Farmhand, age 15, Burkes Mill Dist., Augusta Co. 1860 census. Enl. at Camp near Hamilton's Crossing 3/10/63. Present 3/10/63-4/30/64. WIA (below the knee) Spotsylvania CH 5/12/64. Returned to duty 7/1/64. Present 7-12/64. Paroled Staunton 5/1/65. Age 20, 5'5", fair complexion, light hair, grey eyes. Farmhand, age 23, Pastures Dist., Augusta Co. 1870 census. Farmer, Augusta Co. Died Staunton 12/26/21. Brother of Thomas F. Christman.

CHRISTMAN, THOMAS F.: Pvt., Co. A. B. Augusta Co. 5/10/34. Farmer age 23, Burkes Mill Dist., Augusta Co. 1860 census. Enl. Staunton 7/16/61. Present 11/61-4/62. Reenl. 5/1/62. Present 8/31-12/31/62. Ab. on leave 2/19-28/63. Present 3/63-12/31/64. Surrendered Appomattox 4/9/65. Age 30, 6', dark complexion, blue eyes, brown hair. Farmer, age 33, Mt. Sidney PO, Augusta Co. 1870 census. Member, Stonewall Jackson Camp CV, Staunton. Died Pastures Dist., Augusta Co. 3/4/14. Bur. Green Hill Cem., Churchville. Brother of Noah L. Christman.

CHRISTOPHER, WILLIAM: Pvt., Co. H. B. Va. 1827? Enl. Staunton 1861 age 34. Marched from Orange CH to Manasses bare footed. WIA (shoulder) 2nd Manassas 8/29/62. Alive 1884.

CHRISTOPHER, WILLIAM F.: Pvt., Co. I. B. Va. 1820? Laborer, age 36, Stuanton, 1860 census. Enl. Staunton 7/23/61 age 40. Present 11/61-8/31/62. WIA Sharpsburg 9/17/62. Ab. wounded 1-12/63. Deserted from hospital 4/23/64. Surrendered Appomattox 4/9/65. Died Riverheads Dist., Augusta Co. 1/2/81.

CLARK, JAMES A.: Pvt., Co. E. B. Rockbridge Co. 1838? Laborer, age 25, Collierstown PO, Rockbridge Co. 1860 census. Enl. Camp of 52nd Va. 2/22/64. Deserted from Camp near Spotsylvania CH 5/19/64. NFR. Laborer, age 32, Buffalo Dist., Rockbridge Co. 1870 census. Resident, age 66, Rapp's Mill, Rockbridge Co. 1900. Borther of Joseph D. and Robert Clark.

CLARK, JAMES MADISON C.: Pvt. Co. K. Val. Valley Mills 4/26/62 as substitute. Present 4/26-30/62 and 5/26/62 over 35. Discharged 2/13/63 over age of conscription. NFR. Bur. Bath Co.

CLARK, JOSEPH DAVID.: Pvt., Co. E. B. Rockbridge Co. 1839? Farmer age 26, Collierstown, Rockbridge Co. 2/64. Enl. Camp of 52nd Va. 2/22/64. WIA (left arm and hip) and cap. near Spotsylvania CH 5/19/64. Sent to Point Lookout. Transf. to Elmira. Released 6/19/65. 5'9" florid complexion, dark hair, grey eyes. Died Collierstown 5/25/02. Bur. Collierstown Pres. Ch. Cem. Brother of James A. and Robert Clark.

CLARK, ROBERT: Pvt., Co. E. B. Rockbridge Co. 1831. Laborer, age 28, Collierstown PO, Rockbridge Co. 1860 census. Enl. Staunton 8/1/61. AWOL 11/61-10/31/62. Dropped as deserter. NFR. Exempt Mar. '62 fistula. Conscripted Collierstown 2/12/64 unit unknown, age 33, fair complexion, blue eyes, light hair, 6' 1", farmer. Laborer, age 39, Buffalo Township, Rockbridge Co. 1870 census. Brother of James A. and Joseph D. Clark.

CLARK, WILLIAM M.: Pvt., Co. E. B. Rockbridge Co. 8/21/37. Farmhand age 23, Collierstown PO, 1860 census. Rockbridge Co. rolls. Listed as POW Elmira in "Staunton Vindicator" 12/1/64. NFR. Died 11/21/15. Bur. Mt. Moriah Ch. Cem. Botetourt Co.

CLARKE, THOMAS JEFFERSON.: Pvt., Co. B (2nd). B. Va. 1832? Carpenter age 27, Waynesboro, Augusta Co. 1860 census. Enl. Waynesboro 7/15/61. Ab. sick 11/61-2/62. Ab. on leave 3-4/62. Reenl. 5/1/62. AWOL 6/28-10/10/62. In hospital Staunton 10/11/62-3/31/63 "chronic nephretics." Detailed as Ward Master 8/26/63-12/31/64. Capt. Ft. Stedman 3/25/65. Sent to Point Lookout. Released 6/19/65. 6' 1", light complexion, dark brown hair, hazel eyes. Carpenter, age 39, Fishersville PO, Augusta Co. 1870 census. Died Western State Hospital Staunton, 6/8/98. Bur. Riverview Cem., Waynesboro. "A Faithful Soldier."

CLARKE, WILLIAM FLACK: Pvt., Co. D. B. Mt. Crawford, Rockingham Co. 9/17/31. Shoemaker, age 27, Churchville, Augusta Co. 1860 census. Enl. Staunton 7/16/61. Present 7/16-12/31/61. Died of fever Alleghany Mt. 1/23/62. Left wife and 3 children. Buried Churchville. (originally served in "Mount Crawford Cavalry" until after Rich Mt.")

CLARKE, WILLIAM NELSON: Pvt., Co. B (2nd). B. Augusta Co. 3/27/45. Farmhand, age 16, New Hope, Augusta Co. 1860 census. Enl. Waynesboro 7/15/61. Ab. sick 11/16-2/62. Present 3-4/62. Reenl. 5/1/62. AWOL 5/9/62-12/31/63. Dropped as deserter. Served in Capt. Avis's Co. Provost Guard, Staunton. Farmer, age 24, Mt. Sidney PO, Augusta Co. 1870 census. Member, Stonewall Jackson Camp, CV, Staunton. Died Augusta Co. 5/13/10. Bur. Laurel Hill Baptist Ch. Cem.

CLAYTOR, GEORGE WASHINGTON.: Pvt., Co. G. B. Augusta Co. 8/9/29. Farmer, New Hope, Augusta Co. Enl. Staunton 8/2/61. Present 11/61-4/62. AWOL 5/15/62-9/63. Deserted to the enemy in W. Va. 9/14/63. Took oath and sent north. Farmer, Augusta Co. Died Red Mills, Augusta Co. 9/10/83.

CLAYTOR, WILLIAM C.: Pvt., Co. G. B. Va. 1815? Farmer, age 44, New Hope, Augusta Co. 1860 census. Enl. Staunton 8/2/61. Present 11/61-4/62. AWOL 5/16/62-9/63. Deserted to the enemy in W. Va. 9/14/63. Took oath and sent north. Father of William H. H. Claytor.

CLAYTOR, WILLIAH H. H., JR.: Pvt., Co. G. B. Augusta Co. 7/15/44? Farmhand, age 15, Burkes Mill Dist., Augusta Co. 1860 census. Pvt., Capt. Koiner's Co. 32nd Va. Militia 3/62 not 18. Not on muster rolls. Deserted to the enemy in W.Va. 9/14/63. Took oath and sent north. Farmer, age 26. Mt. Sidney PO, Augusta Co. 1870 census. Farmer, Greenville, 1897. Son of William C. Claytor.

CLEEK, ADAM GIVEN: Ordance Sgt., F&S. B. on Jackson River, Bath Co. 4/30/26. Sheriff, Bath Co. 1856-60. Farmer, age 34, Warm Springs Dist., Bath Co. 1860 census. Pvt., Capt. T. Hamilton's Co., 81st Va. Militia and Co. G, 11th Va. Cav. Enl. Shenandoah Mt. 4/9/62 as Pvt. Co. K. Present 4/9-30/62. Promoted Ordnance Sgt. of Regt. 5/1/62. Present 5/1-12/31/62. Ab. on leave 2/1/63. Present 3/63-9/64. Elected Sheriff of Bath Co. and discharged 9/64. Sheriff of Bath Co. 1864-1869. Farmer, age 44, Warm Springs Dist., Bath Co. 1870 census. Clerk of Circuit Court 1887-99. Died McClung, Bath Co. 2/16/01. Bur. Woodland Cem., Fairview, Bath Co.

CLEEK, JACOB CRAWFORD.: Pvt., Co. K. B. Bath Co. 5/17/27. Farmer, age 32, Hot Springs PO, Bath Co. 1860 census. Pvt., Capt. Hamilton's Co., 81st Va. Militia. Enl. Shenandoah Mt. 4/9/62. WIA (shoulder) Gaines Mill 6/27/62. Ab. wounded through 12/64. NFR. Died Bath Co. 4/1/90. Bur. Lower Cleek Cem., Bath Co.

CLEEK, JACOB K.: Pvt. Co. K. Enl. Shenandoah Mt. 4/9/62. Present 4/9-30/62. Ab. sick 8/27/62 until died of typhoid fever Warrenton hospital 9/21/62.

CLEMENTS, STEPHEN E.: Pvt., Co. C. Enl. Staunton 5/1/63. AWOL 7/25-11/25/63. Fined $44.00. Present 11-12/63. AWOL 6/8/64. Ab. in arrest 11-12/64, 9 months pay deducted by sentence of GCM 4/2/64. Admitted to hospital Richmond with acute dysentery 12/13/64. Returned to duty 1/17/65. WIA Ft. Stedman 3/25/65. Cap. in hospital Richmond 4/3/65. Escaped 5/4/65. Living in Alexandria 1888.

CLEMMER, JAMES HARVEY: Pvt., Co. B (1st). B. Fairfield, Rockbridge Co. 1841? Cabinetmaker, age 19, South River Dist. Rockbridge Co. 1860 census. On postwar roster. Detailed as wagonmaker, Waynesboro. Brother of William L. Clemmer.

CLEMMER, JAMES W.: Pvt., Co. I. B. Augusta Co. 1845? Farmer, Cap. date and place unknown. Sent to Camp Chase. Died of brain fever there 1/19/65. Bur. there, grave #812. Age 20.

CLEMMER, WILLIAM LEWIS.: 4thSgt., Co. B. (1st) B. Va. 6/3/38. Cabinetmaker, age 22, South River Dist., Rockbridge Co. 1860 census. Enl. Fairfield 7/10/61. Present until transf. 9/28/61. Brother of James H. Clemmer.

CLEVELAND, OLIVER E.: Pvt., Co. G. B. Va. 4/8/23. Farmer, age 37, Burkes Mill Dist., Augusta Co. 1860 census. Enl. Staunton 8/2/61. Ab. detailed as teamster 9/1/61-4/62. NFR. Farmer, age 47, Mt. Sidney PO, Augusta Co. 1870 census. First Postmaster of Annex, Augusta Co. Died Annex 9/21/01. Bur. Belmont Cem., Annex.

CLINE, JOSEPH E.: Capt., Co. F. B. Augusta Co. 5/13/33. Farmer, age 41, 1st Dist., Augusta Co. 1860 census. 2ndLt., Co. D, 182nd Va. Militia. Enl. Staunton 7/61. Ab. sick 11/16-12/31/61. Present 1-4/62. Reelected 5/1/62. Resigned for ill health 8/6/62. Approved 8/11/62. Died Augusta Co. 10/12/63. Bur. Pleasant Valley Cem., Weyers Cave.

CLINE, SAMUEL F.: Pvt., Co. G. B. Rockingham 1838? Farmhand, age 23, Waynesboro Dist., Augusta Co. 1860 census. Enl. Staunton 7/16/61, age 23. Present 11/61-3/62. AWOL 4/22-30/62. Deserted near Orange CH. NFR. May have served in Co. C, 14th Va. Cav. Farmer, age 28, Mt. Sidney PO, Augusta Co. 1870 census.

CLINEBELL, JAMES: Pvt., Co. I. B. Augusta Co. 1840? Farmhand, age 20, Staunton PO, Augusta Co. 1860 census. Enl. Staunton 7/16/61 age 21. Ab. sick 11/61-1/11/62. Present 1/23-4/62. Reenl. 5/1/62. WIA (abdomen) Cross Keys 6/8/62. Died of diphtheria Bunker Hill 10/11/62. Age 22, 5' 6", fair complexion, hazel eyes. black hair.

COCHRAN, ANDREW A.: 3rdSgt., Co. B (1st). B. Augusta Co. 1828? Carpenter, age 32, South River Dist., Rockbridge Co. 1860 census. Enl. Fairfield 7/10/61. Ab. on duty in Rockbridge Co. 1860 census. Enl. Fairfield 7/10/61. Ab. on duty in Rockbridge Co. when transf. 9/28/61.

COCHRAN, GEORGE MOFFETT, JR.: Quartermaster, F&S B. "Elk Meadows" near Stribling Spring, Augusta Co. 2/26/32. Grad. Staunton Academy and UVa. '56. Lawyer, Staunton 1860 census. Served as Ordnance Officer, Harper's Ferry under Gen. T. J. Jackson and at Bull Run under Gen. Joseph E. Johnston. Appointed QM 8/19/61. Reappointed 5/1/62. Present Alleghany Mt., Fredericksburg, Chancellorsville, Gettysburg, Wilderness, Ab. detached service in Shenandoah Valley 7-10/64. Appointed QM of Brigade 9/1/64. Present Petersburg. Surrendered Appomattox as QM of Walker's Brig. Law Partner of Col. John B. Baldwin. Elected to Va. State Legislature 1889. Member, Staunton City Council. Owner of Folly Mills 1897. Died Staunton 4/7/00. "He was an upright, pure and honorable man in all his relationships."

COCHRAN, JOHN LYNN.: 2ndLt., Co. A. B. Augusta Co. 9/30/38. Student, age 20, Staunton 1860 census. Attended UVa. 1860. Doctor, Staunton. Enl. Staunton 7/9/61 age 22. Ab. on detached service Staunton 12/28/61. Died of typhoid fever Alleghany 1/30/62. Bur. Thornrose Cem. "He won for himself the confidence and esteem of all his companions in arms, and where he had distinguished himself by his energy and zeal in the discharge of every duty which devolved upon him as a soldier and an officer."

COFFEY, EDWARD F.: QMSgt., Co. A. On postwar roster. Served 4 years. Member, Stonewall Jackson Camp CV, Staunton. Died Stuart's Draft 1907.

COFFEY, JAMES: Pvt., Co. A. On postwar roster. Enl. 1862 and served 3 years.

COFFEY, MARVEL M.: Pvt., Co. B (1st) B. Nelson Co. 1838? Farmhand, age 22, South River Dist., Rockbridge Co. 1860 census. Enl. Fairfield 7/10/61. Present until transf. 9/28/61. Brother of Peter J., Robert W. and William M. Coffey.

COFFEY, PETER J.: Pvt., Co. B (1st). B. Nelson Co. 1836? Farmhand, age 24, South River Dist., Rockbridge Co. 1860 census. Enl. Fairfield 7/10/61. Present until transf. 9/28/61. Brother of Marvel M., Robert W. and William M. Coffey.

COFFEY, ROBERT W.: Pvt., Co. B (1st). B. Nelson Co. On postwar roster. Brother of Marvel M., Peter J. and William M. Coffey.

COFFEY, WILLIAM MONTABALLE: Pvt., Co. B (1st). B. Nelson Co. 4/15/34. Farmhand, age 26, South River Dist., Rockbridge Co. 1860 census. Enl. Fairfield 7/10/61. Present until transf. 9/28/61. Brother of Marvel M., Peter J. and Robert W. Coffey.

COFFMAN, ALEXANDER STUART "Sandy": 1stSgt., Co. G. B. Augusta Co. 9/8/42. Cabinetmaker's Apprentice, age 17, Burkes Mill Dist., Augusta Co. 1860 census. Enl. Staunton 8/2/61 as Pvt. Present 11/61-4/62, under 18, not reenl. for the war. Present 4/30-12/31/62. Promoted Cpl. 8/62, 3rdSgt. 11-12/62. Present 1-4/63. Promoted 1stSgt. 7/5/63. Ab. on leave 8/21-9/6/63. Present 10/63-12/64. Surrendered Appomattox 4/9/65. 5' 10", fair complexion, blue eyes, light hair, mechanic. Carpenter, age 26, Mt. Sidney PO, Augusta Co. 1870 census. Undertaker and Cabinetmaker. Died Mt. Sidney 7/1/10. Bur. Old Salem Lutheran Ch. Cem., 3 miles west of Mt. Sidney. Brother of Benjamin F. & David.

COFFMAN, BENJAMIN FRANKLIN: Pvt., Co. G. B. Va. 1843? Farmhand, age 17, Burkes Mill Dist., Augusta Co. 1860 census. Enl. Staunton 8/2/61. Ab. sick 12/11/61-4/62. Ab. on leave 4/30-8/31/62. Present 9/1/62-4/63. AWOL 7/25-9/24/63. Present in arrest 9/25/-12/31/63. CM 1/19/64. In hospital Richmond with "cattarrhus" 3/7-9/64 and returned to Castle Thunder. Sentence remitted 7/7/64. Issued clothing 4/21 and 27/64. WIA and cap. near Spotsylvania CH 5/20/64. Sent to Point Lookout. "16 years old and wants to take oath and go to his sister's in Pa." Transf. to Elmira. Applied to take oath 8/64. "Will be 17 on the 10th of October 1864. Attempted to desert 7/23/63 but was captured and kept in prison 9 months". Released 5/29/65, 5' 6½", fair complexion, light hair, blue eyes, resident of Harrisonburg. Bur. in old Coffman Cem., located 1 mile north of Barren Ridge, Augusta Co., on farm of John C. Driver.

COFFMAN, DAVID.: Pvt., Co. G. On postwar roster. Living Des Moines, Iowa 1910. Bur. in old Coffman Cem., located 1 mile north of Barren Ridge, Augusta Co., on John C. Driver farm.

COINER, CYRUS BENTON "BENT": Capt., Co. G. B. Fishersville, Augusta Co. 1/30/42. Attended VMI 60-61. Drillmaster Richmond 4-6/61. Enl. Staunton 8/2/61 as Pvt. Served as Drillmaster. Present 1-4/62. Elected 1stLt. 5/1/62. Commanding Co. 5/1-8/31/62. Present 10-12/62. Ab. on leave 2/22-28/63. Present 4-12/63. Commanding Co. 4/30/64 until WIA (through cheeks) 5/18/64. Near Spotsylvania CH. Returned to duty 7/64 after Monocacy. Commanding Co. 7-12/64. Promoted Capt. 12/2/64. Commanding Regt. 1/28-31/65. Ab. on leave 2/17-25/65. Escaped at Appomattox 4/9/65. Paroled Staunton 5/15/65. Age 23, 6', light complexion, dark hair, hazel eyes. Farmer and Stockraiser, Fishersville. Outlived 4 wives. Member, Stonewall Jackson Camp CV, Staunton. Died Fishersville 4/8/19. Bur. Tinkling Spring Pres. Ch. Cem., near Fishersville. Brother of Joseph S. Coiner.

COINER, DAVID M.,JR.: 2ndSgt. Co. G. B. Augusta Co. 1833? Farm Manager, age 30, Burkes Mill Dist., Augusta Co. 1860 census. Enl. Staunton 8/2/61 as Pvt. Present 11/61-4/62 and 5/1/62-4/63. Appointed 1stCpl. 3/15/63. Ab. sick in hospital Lynchburg 8-28-9/63. Present 10-12/63. WIA near Spotsylvania CH 19/5/64. Promoted 2ndSgt. Present 6-12/64. Cap. Petersburg 4/2/65. Sent to Point Lookout. Released 6/10/65. 5' 10¼", light complexion, dark brown hair, light blue eyes. Traveled to Harper's Ferry by train and walked home. Farmer, age 41, Mt. Sidney PO, Augusta Co. 1870 census. Died Long Meadows, Augusta Co. 9/14/97. Bur. St. James Ch. Cem. Brother of John B. and James W. Coiner.

COINER, DAVID W.: 2ndLt., Co. G. B. Augusta Co. 12/25/17. Farmer, age 42, 1st Dist., Augusta Co. 1860 census. Capt., Tinkling Spring Co., 32nd Va. Militia. Enl. Staunton 8/2/61. Present 11/61-4/62. Reelected 5/1/62. Present 5/1/62. Resigned for ill health 12/20/62. Age 45 years, 10 months. 1stLt., Capt. Hottel's Co., Augusta Co. Reserves 4/64. Farmer, age 51, Fishersville PO, Augusta Co. 1870 census. Died Tinkling Spring 8/13/85. Bur. Bethlehem Ch. of the Brethren Cem., Ladd, Augusta Co.

COINER, JAMES DAVID: 3rdCpl., Co. A. B. Va. 6/11/40. Farmhand, age 20, Burkes Mill Dist., Augusta Co. 1860 census. Attended Roanoke College. Enl. Staunton 7/9/61 as Pvt. age 21. Promoted 4thCpl. 8/17/61 Present 11/61-2/62. Promoted 3rdCpl. Present 3-4/62. Transf. Co. B, 23rd Va. Cav. Farmer, age 30, Mt. Sidney PO, Augusta Co. 1870 census. Farmer, Fishersville 1897. Died near Hermitage, Augusta Co. 8/29/99. Bur. St. James Lutheran Ch. Cem.

COINER, JAMES WILLIAM: Pvt. Co. G. B. Va. 1845? Farmhand, age 15, 1st Dist., Augusta Co. 1860 census. Enl. Alleghany Mt. 3/5/62. Reenl. 5/1/62. Present 5/1/62-12/31/63. Issued clothing 3/1/64 and 4/21/64. WIA (liver) Spotsylvania CH 5/12/64. Ab. wounded through 12/31/64. Paroled Staunton 5/15/65. Age 21, 5' 11½", light complexion, brown hair, blue eyes. Farmer, age 25, Fishersville PO, Augusta Co. 1870 census. Emigrated to California. Alive there in 1916. Died and bur. there date and place unknown. Brother of David M. and John B. Coiner.

COINER, JOHN B.: Pvt., Co. G. B. Va. 1841? Farmhand, age 19, 1st Dist., Augusta Co. 1860 census. Enl. Staunton 8/2/61. Ab. sick 3-4/62. AWOL 5/1-10/31/62. Dropped as deserter. Farmer, age 29, Fishersville PO, Augusta Co. 1870 census. Died Sherando 3/6/16. Bur. Bethlehem Ch. of the Brethren Cem., Ladd. Brother of David M. and James W. Coiner.

COINER, JOHN CALVIN: Pvt., Co. A. B. Augusta Co. 12/18/32. Farmer, Augusta Co. Not on MR. Enl. 6/62. Farmer near Waynesboro 1897. D. 1/7/13. Bur. Trinity Lutheran Ch. Cem.

COINER, JOSEPH SMITH: 1stLt., Co. G. B. Augusta Co. 4/25/35. Farmer, age 23, 1st Dist., Augusta Co. 1860 census. Enl. Staunton 8/2/61 as 3rdLt. Present 1-4/62. Not reelected 5/1/62. Reenlisted as Pvt. same day. WIA (breast) Port Republic 6/9/62. Elected 2ndLt. 10/62. Present 10-11/62. Commanding Co. 12/13/62. Promoted 1stLt. 12/24/62. Commanding Co. 1-3/63. Detailed as Enrolling Officer Augusta Co. 4/63. Commanding Co. 5-12/63. WIA (thigh and main artery cut) Spotsylvania CH 5/12/64. DOWs. Left widow and 3 children. Bur. Tinkling Spring Pres. Ch. Cem. near Fishersville. Brother of Cyrus B. Coiner. "Killed in the memorable charge of Gordon's Division, which checked the successful onset of the enemy . . . As an officer, he possessed the high quality of being a good displinarian — He knew both how to command and how to obey. He was courteous to his superiors, considerate and kind to those under him. He was intelligent, zealous, and determined in the discharge of his duties, he took a just pride in the career which the necessities of his country had called upon him to enter upon. . . ."

COLE, HARRISON: Pvt., Co. F. B. Rockingham Co. 6/6/25. Farmer, age 36, Northern Dist., Augusta Co. 1860 census. Pvt., Capt. Cline's Co. Va. Militia. On postwar roster. Wagoner — detached from company by Col. Harman as Wagonmaster. Served 3½ years. Died Basic City, Augusta Co. 6/16/14. Bur. Cole family Cem., Greenfield, Nelson Co.

COLEY, JOHN WESLEY: Pvt., Co. G. B. Augusta Co 5/9/41. Res of New Hope. Pvt. Co. L, 5th Va. Inf. (disbanded). Enl. Staunton 8/2/61. Present 11/61-1/-12/63. Cap. Fisher's Hill 9/22/64. Sent to Point Lookout. Exchanged 3/17/65. Present Camp Lee 3/19/65. Paroled Staunton 5/1/65. Age 23, 6', fair complexion, dark hair, blue eyes. Wagonmaker, Rolla, Augusta Co. 1897. Died Augusta Co. of injuries received Staunton 10/98. Bur. Verona, Augusta Co.

COLLINS, THOMAS HENRY: Pvt., Co. H. B. Augusta Co. 1840? Stonecutter, age 18, 1st Dist., Augusta Co. 1860 census. Enl. Staunton 7/23/61 age 22. Present 11/61-2/62. Died Camp Alleghany 3/15/62, cause unknown. Bur. at church cem. on Alleghany Mt.

COLLINS, WILLIAM C.: 2ndCpl, Co. G. B. Ireland 1831? Resident, age 29, 1st Dist., Staunton 1860 census. Enl. Staunton 8/2/61 as Pvt. Promoted 2ndCpl. 10/31/61. Ab. sick 13/11-12/31/62. Present 1-2/62. Ab. sick 3-4/62. Died date and place unknown.

CONDON, PATRICK: Pvt., Co. H. B. Ireland. Resident of Staunton. Enl. Bunker Hill 10/3/62. Deserted 10/11/62. NFR.

CONNELL, JAMES H.: 3rdCpl., Co. A. Enl. Staunton 7/16/61 as Pvt. Present 11/61-5/62. WIA (foot) Port Republic 6/9/62. Present 6-12/62. WIA (shoulder) Fredericksburg 12/13/62. Present 1-2/63. Appointed 3rdCpl. 2/28/63. Present 3-8/63. Deserted 9/24/63. NFR. Died before 5/03.

COOKE, JACOB W.: Pvt., Co. D. B. Augusta Co. 1848? Not on muster rolls. Enl. 7/61 and served to the end of war. Farmhand, age 22, Mt. Sidney PO, Augusta Co. 1870 census. Farmer, Valley Mills 1897.

COOKE, JOHN D.: 3rdCpl, Co. D. B. Augusta Co. 1844? Farmhand, age 16, Burkes Mill Dist., Augusta Co. 1860 census. Enl. Staunton 7/16/61 as Pvt. Present 7/16/61-4/62. Reenlisted 5/1/62. Present 5/1-8/31/62. Ab. on leave 2/19-3/5/63. Present 3/6-12/31/63. Promoted 4thCpl. 64. Issued clothing 3/31, 4/21, and 6/16/64. Promoted 3rdCpl. 8/30/64. WIA "gunshot wound back of head, fractured of occipit and ball entered the brain" and cap. Winchester 9/19/64. Sent to USA hospital, West Building, Baltimore, age 22, Transf. to Ft. McHenry and Point Lookout. Exchanged 2/20/65. In hospital Richmond 2/26/65. Furloughed for 30 days 3/6/65, however, readmitted to hospital Richmond 3/9/65. Returned to duty 3/16/65. Retired for disability by Surgeons of Early's Div. 3/24/65. Age 22, 5' 7½", fair complexion, blue eyes, black hair. Paroled Lynchburg 4/15/65. Resident of Ft. Defiance 1888.

COOK, JAMES: Pvt., Co. I. Not on muster rolls. Present 2/10/62. NFR.

COOK, JOHN WILLIAM: Pvt. Co. G. B. Augusta Co. 5/2/43. On postwar roster. Carpenter, age 33, Staunton 1870 census. Farmer, Augusta Co. Died Mt. Sidney 11/6/14. Bur. Mt. Pisgah Meth. Ch. Cem.

COOK, JOHN W.: Pvt., Co. D. B. Nelson Co. 1838? Miller, Staunton. Pvt., Co. A, 160th Va. Militia 3/18/62, exempt, Miller, Staunton Flouring Mills. Not on muster rolls. Farmer, Augusta Co. 1898. Member, Stonewall Jackson Camp CV, Staunton 1902.

COOK, JOHN WILLIAM: Pvt., Co. G. B. Augusta Co. 5/2/43. On postwar roster. Carpenter, age 33, Staunton 1870 census. Farmer, Augusta Co. Died Mt. Sidney 11/6/14. Bur. Mt. Pisgah Meth. Ch. Cem.

COOK, SAMUEL D.: Pvt., Co. D. B. Mt. Sidney, Augusta Co. 5/3/30. Farmhand, age 25, Burkes Mill Dist., Augusta Co. 1860 census. Not on muster rolls. Deserted to the enemy Clarksburg, W Va. 4/2/65. Took oath and released. Farmer, age 38, 5th Dist., Augusta Co. 1870 census. Living Stonewall, Augusta Co. 1900. Died Mt. Sidney 2/12/06. Bur Old Salem Lutheran Ch. Cem., 3 miles west of Mt. Sidney.

COOK, SIMON P.: Pvt., Co. D. B. Augusta Co. 9/15/40. Shoemaker, Augusta Co. Enl. Staunton 7/16/61. Present 7/16/61-4/62. WIA (right arm) 2nd Manassas 8/29/62. Ab. wounded 8/30/62-3/63. Present 4/30-10/31/63. Retired because of wounds 4/27/64. Died Jennings Gap, Augusta Co. 11/18/84.

COOK, THOMAS: Pvt., Co. D. B. Va. 1818. Laborer, age 42, Burkes Mill Dist., Augusta Co. 1860 census. Enl. Staunton 7/16/61. Prsent 7/16/61-2/15/62. AWOL 2/16-28/62. Fined $11.00 by CM. Present 3-4/62. Ab. sick 5/1-6/5/62. AWOL 6/6-8/31/62. Fined $31.16. Discharged at Camp 10/27/62 over age of conscription. NFR.

COX, ALEXANDER HARRISON: Pvt., Co. A. B. Augusta Co. 1/21/41. Farmhand, age 20, Burkes Mill Dist., Augusta Co. 1860 census. Enl. Staunton 7/12/61. Ab. sick 11/3/61-4/63. Discharged from Hospital Staunton for "phihicis palmonalis", bronchitius, rheumatism and debility, 7/22/63. Age 21, 5' 9", dark complexion, hazel eyes, dark brown hair. Farmhand, age 29, Mt. Sidney PO, Augusta Co. 1870 census. Died near New Hope 10/5/15. Bur. Laurel Hill Baptist Ch. Cem.

COX, CHARLES F.: Pvt., Co. A. B. Va. 1843? Carpenter's Apprentice, age 17, Hermitage, Augusta Co. 1860 census. Enl. Staunton 7/9/61. Present 11/61-4/62. Reenl. 5/1/62. Present 5/1-8/31/62. Present 11/23-12/27/62. Present in arrest 12/28/62-2/63. Fined 6 months pay in addition to time AWOL by GCM. Present 4-12/63. Issued clothing 1/28, 2/27, and 4/21/64. Appears on rolls Co. A, Ward's Bn., C.S. Prisoners released from military prison, Lynchburg 7/64 for their participation in the defence of Lynchburg, Cap. Fisher's Hill 9/22/64. Sent to Point Lookout. Joined U.S. service 10/17/64. Alive 1903.

COX, HENRY FRANK.: Pvt., Co. G. B. Augusta Co. 7/12/38. Enl. Staunton 8/2/61. Present 11/61-2/62. Died of pneumonia Monterey 4/2/62. Bur. Eakle family Cem., on Rt. 608 near New Hope.

COX, JACOB: Pvt., Co. I. B. Rockbridge Co. 12/1/40. Farmhand, age 19, Burkes Mill Dist., Augusta Co. 1860 census. Enl. Staunton 7/16/61 age 19. Present 11/61-4/62. Reenl. 5/1/62. AWOL 9/12-10/31/62. Present 1-4/63. WIA (back) Gettysburg 7/3/63. Ab. wounded 7/4-10/63. Present 11-12/63. Issued clothing 3/16, 4/1, 22 and 29/64. Cap. Bethesda Church 5/30/64. Sent to Point Lookout. Transf. to Elmira. Released 6/30/65. Resident of Staunton, 5' 9", fair complexion, light hair, hazel eyes. Farmer, age 29, Fishersville PO, Augusta Co. 1870 census. Died near Old Providence Pres. Ch. 1/2/87 and bur. in cem. there.

COX, JACOB S.: Pvt., Co. A. B. Va. 1841? Enl. Camp Alleghany 3/10/62. Present 3/10-4/62. Cap. date and place unknown (not on Federal POW Rolls). Exchanged. AWOL 11/15-12/62. Present 1-12/63. Issued clothing 1/28 and 30/64. Cap. Waynesboro 9/28/64. Sent to Point Lookout. Exchanged 3/17/65. Present Camp Lee, Richmond 3/19/65. Paroled Staunton 5/1/65. Age 24, 5' 8", fair complexion, brown hair, grey eyes. Saddler and Harness Maker, Steele's Tavern, 1897. Alive 1903.

COX, SAMUEL A.: Pvt., Co. G. B. Augusta Co. 6/6/62. Farmer, age 33, Burkes Mill Dist., Augusta Co. 1860 census. Enl. Staunton 8/2/61. Present 11/61-12/62, on extra duty with Pioneer Corps since 8/28/62. Present 1/63-10/64, detached with Pioneer Corps. Present 11-12/64. Paroled Staunton 5/1/65. Age 32, 5' 6", dark complexion, dark hair, dark eyes. Cabinetmaker, age 38, Mt. Sidney PO, Augusta Co. 1870 census. Died Barren Ridge, Augusta Co. 9/10/94. Bur. Barren Ridge Brethren Ch. Cem.

COX, THOMAS HENRY: 3rdCpl., Co. A. B. Augusta Co. 2/24/39. Farmhand, age 21, Burkes Mill Dist., Augusta Co. 1860 census. Enl. Staunton 7/15/61 as Pvt. Present 11/61-4/62. Reenl. 5/1/62. Present 8/31/62-12/31/63. Promoted 3rdCpl. 64. Issued clothing 4/21/64. Cap. Bethesda Church 5/30/64. Sent to Point Lookout. Transf. to Elmira. Released 6/30/65. Resident of Staunton, fair complexion, dark hair, blue eyes, 5' 8". Farmer, age 31, Mt. Sidney PO, Augusta Co. 1870 census. Died Augusta Co. 12/15/11. Bur. Hildebrand Mennonite Ch. Cem., Madrid, Augusta Co.

COYNER, GEORGE ADAM.: 2ndCpl., Co. B (2nd). B. Augusta Co. 9/20/37. Farmhand, age 22, 1st Dist., Augusta Co. 1860 census Enl. Waynesboro 7/15/61. Cap. on picket Camp Alleghany 13/31/61. Enl. Waynesboro 7/15/61. Cap. on picket Camp Alleghany 12/13/61. Sent to Camp Chase. Took oath and "released by order of Gen. Roscrans 3/28/62 to remain in Illinois until the rebellion close." Wrote President Andrew Johnson 5/20/65 requesting permission to return to Augusta Co., address Nokousi, Montgomery Co., Ill. Carpenter, age 32, Fishersville PO, Augusta Co. 1870 census. Farmer and Fruit Grower, South River Dist., Augusta Co. 1910 census. Died South River Dist., 8/22/14. Bur. Bethlehem Ch. of Brethren Cem., Ladd.

COYNER, SAMUEL BROWN: Pvt., Co. D. B. Augusta Co. 4/11/38. Law Student, age 22, Northern Dist., Augusta Co. 1860 census. Attended Lexington Law School 1860-61. Lawyer, Staunton. Living in Illinois when war began. Served in "West Augusta Guards" militia Co. Enl. Staunton 7/16/61. "Joined with the understanding that he was to be transferred to Cavalry before he was mustered in: this was not done, he was mustered in immediately. About a month later . . .Captain Macon Jordan of Co, D., Ashby's Cavalry offered him a horse, he joined Captain Jordan's Co., and through the aid of General Thomas Jordan was regularly transferred from the 52nd Va. Inf. to Co., D, of Ashby's Cavalry. Elected Captain when Captain Jordan transferred to Gen. Heth's staff. Jordan recommended him for Captain over the other officers because of his gallantry at Romney, W. Va. Fall '61. WIA Culpepper CH 8/20/62. Burned the White House, Columbia and Conrad Store bridges by order of Gen. Jackson during Valley campaign." WIA near Culpeper CH 9/13/63. DOWs Orange CH 9/16/63. Bur. Mossy Creek Old Cem., Augusta Co. "As a soldier he possessed all the attributes that make a hero."

COYNER, WILLIAM PETER: Pvt., Co. G. B. Augusta Co. 1845? On postwar roster. Farmer, age 25, Mt. Sidney PO, Augusta Co. 1870 census. Living Indianapolis, Ind. 1882. Died near Waynesboro, Augusta Co. 8/5/91.

CRAUN, ADAM H.: 2ndLt., Co. F. B. Va. 1/12/40. Farmhand, age 19, 1st. Dist., Augusta Co. 1860 census. Enl. Staunton 7/31/61 as 2ndSgt. Present 11/61-4/62. Reenl. 5/1/62. WIA (arm) Port Republic 6/9/62. Present 8/31-12/31/62 and 1-5/63. Present 11/63-4/64. WIA Wilderness 5/5/64. DOWs 5/6/64. Bur. St. Michael's Reformed Ch. Cem. north of Centerville off Rt. 699. Augusta Co.

CRAUN, FREDERICK S.: Pvt., Co. F. B. Augusta Co. 1/22/30. Farmhand, age 31, Northern Dist., Augusta Co. 1860 census. Pvt., Capt. Rimel's Co., 160th Va. Militia. Enl. Staunton 7/31/61. Pesent 11/61-4/62. Reenl. 5/1/62. In hospital Danville 7/23-30/62 with diarrhea. Unofficial record say discharged '62 for heart disease. NFR. Farmer, age 40, Mt. Sidney PO, Augusta Co. 1870 census. Died Augusta Co. 12/31/04. Bur. Old Salem Lutheran Ch. Cem., 3 miles west of Mt. Sidney.

CRAUN, JACOB HARVEY, SR.: Pvt., Co. F. B. Augusta Co. 11/20/17. Farmer, age 43, Mossy Creek, Augusta Co. 1860 census. Enl. Staunton 7/31/61. Ab. sick 11/5-12/31/61. Present 1-4/62. KIA Port Republic 6/9/62. Age 45, 5' 8", fair complexion, hazel eyes, auburn hair. Left widow and 5 children. Father of Jacob Harvey Craun, Jr.

CRAUN, JACOB HARVEY, JR.: Pvt., Co. F. B. Augusta Co. 9/21/42. Farmhand, age 23, Mt. Sidney, Augusta Co. 1860 census. Enl. Staunton 7/31/61. Present 11/61-4/62. Reenl. 5/1/62. AWOL 11/23/62-2/9/63. Present under arrest 2/10-28/63. Present 4/30/63-4/64. WIA (head) Wilderness 5/6/64. Deserted 9/7/64. NFR. Farmer, age 26, Mt. Sidney PO, Augusta Co. 1870 census. Died near Spring Hill, Augusta Co. 4/29/20. Obit. says crippled by wound. Bur. Old Salem Lutheran Ch. Cem., 3 miles west of Mt. Sidney. Son of Jacob Harvey Craun, Sr.

CRAUN, JUNIUS R.: Pvt., Co. F. B. Va. 1842? Farmhand, age 18, North Dist., Augusta Co. 1860 census. Enl. Staunton 7/31/61. Ab. sick 11/5-12/31/61. Present 1-4/62. Reenl. 5/1/62. WIA (hip) McDowell 5/8/62. Present 8/31-12/31/62. WIA (leg) Fredericksburg 1/9/63. Ab. wounded 1/10-8/31/63. Present 9/63-4/64. KIA Bethesda Church 5/30/64, age 21.

CRAUN, TIMOTHY H.: Pvt., Co. H. B. Va. 1823. Laborer, age 37, Staunton 1860 census. Enl. Staunton 7/23/61. Deserted same day. NFR.

CREEL, MATHEW NATHAN: Pvt., Co. C. B. 1834? Enl. Staunton 7/16/61 age 27. Ab. sick 11-12/61. Present 1-4/62. Reenl. 5/1/62. KIA 2nd Manassas 8/28/62.

CRENSHAW, D.: Pvt., Co. K. Not on muster rolls. Cap. date and place unknown. Sent to Camp Chase. Died Columbus, Oh. date and cause unknown. Bur. Camp Chase Confederate Cem.

CRESS, JOSHUA T.: Pvt., Co. H. B. Va. 1845. Farmhand, age 15, 1st Dist., Augusta Co. 1860 census. Not on muster rolls. Deserted to the enemy Beverly, W. Va. 1/6/63. Sent to Camp Chase. Took oath and released 2/18/63. Resident of Augusta Co. age 18, 5' 8¼", fair complexion, red hair, dark eyes.

CRESS, NICHOLAS: Pvt., Co. H. B. Va. 1834? Farmhand, age 26, 1st Dist., Augusta Co. 1860 census. Enl. Staunton 7/23/61. Ab. sick 11/9/61-2/63. AWOL 3/31-12/31/63. Dropped as deserter. NFR.

CRICKENBERGER, JOSEPH EARMAN: Pvt., Co. G. B. Augsta Co. 5/20/42. Farmhand, age 18, Burkes Mill Dist., Augusta Co. 1860 census. Enl. Staunton 8/2/61. Present 11/61-4/62. Reenl. 5/1/62. Present 5/1-11/13/62. Detailed as wagoner on rolls 11/14/62-12/31/64. NFR. Farmhand, age 28, Mt. Sidney PO, Augusta Co. 1870 census. Farmer, Middle River Dist., Augusta Co. 1910 census. Died Barren Ridge 6/19/18. Bur. Barren Ridge Ch. Cem.

CRISMOND, LEVI: Pvt., Co. A. B. Augusta Co. 1835? Farmer. On postwar roster. Member, Stonewall Jackson Camp, CV, Staunton. Died Augusta Co. 12/26/16.

CRIST, GEORGE M.: 2ndLt., Co. F. B. Spring Hill, Augusta Co. 7/31/43. Farmer, age 17, Northern Dist., Augusta Co. 1860 census. Enl. Staunton 7/31/61 as 4thSgt. Present 11/61-4/62. Reduced to Pvt. and fined $12.16, reason not indicated. Elected 2ndLt. 5/1/62. AWOL 8/30-9/22/62. Ab. under arrest 9/23-12/31/62. Cashiered 10/14/62, resident of Greenfield, Nelson Co. Reenl. Hamilton's Crossing 4/14/63 as Pvt. Present 4/30/63-4/64. WIA (right leg) Bethesda Ch. 5/30/64. In hospital Richmond 6/1-13/64. Present 7-12/64. WIA Petersburg 4/2/65. Cap. in hospital Richmond 5/3/65. Died Loundon, W. Va. 12/24/15.

CRIST, GERALD ENOS: 4thCpl., Co. I. B. Augusta Co. 12/21/35. Farmhand, age 24, 1st Dist., Augusta Co. 1860 census. Enl. Staunton 7/16/61 as Pvt. Present 11/61-4/62. Appointed 4thCpl. 2/5/62. Reenl. 5/1/62. In hospital Charlottesville with "Morb. raril" 8/3-4/62. WIA (side) 2nd Manassas 8/29/62. Present 9-10/62. WIA (right hand and right thigh) Fredericksburg 12/13/62. Ab. wounded through 12/63. WIA (right hand and hip) Bethesda Ch. 5/30/64, age 27. Cap. Nelson Co. 6/11/64. Sent to Camp Chase. Transf. for exchange 3/2/65. Age 28, 5' 10", fair complexion, blue eyes, light hair. NFR. Farmer, age 35, Fishersville PO, Augusta Co. 1870 census. Farmer, Middlebrook 1897-1910. Died Riverheads Dist., Augusta Co. 10/13/15. Bur. New Providence Pres. Ch. Cem., near Brownsburg, Rockbridge Co.

CRIST, JOSEPH W: Pvt., Co. A. B. Rockbridge Co. 2/27/41. Living 1st Dist., Augusta Co. 1860 census. Enl. Staunton 7/9/61 age 19. Present 1-2/62. Ab. sick 3/28-4/62. Reenl. 5/1/62. WIA (thigh) and cap. Sharpsburg 9/17/62. Exchanged 10/62. In hospital Richmond 10/24-11/5/62. Furloughed for 30 days 11/6/62. Deserted from hospital date unknown. Age 20, 6', dark complexion, blue eyes, light hair. NFR. Farmer and Wool Dealer, Sangersville 1897. Died Augusta Co. 10/2/13. Bur. Mossy Creek Old Cem., Augusta Co.

CRITZER, WILLIAM: Pvt., Co. H. B. 1836? Resident, Augusta Co. Enl. Staunton 7/23/61 age 24. Pesent 11/61-4/62. Reenl. 5/1/62. Present 5/1/62-12/63. Deserted Staunton 6/26/64. NFR.

CRITZER, WILLIAM H.: Pvt., Co. B (2nd). B. Nelson Co. 1837? Farmhand, age 23, Burkes Mill Dist., Augusta Co. 1860 census. Enl. Waynesboro 7/15/61. Present 11/61-4/62. Reenl. 5/1/62. Fined $11.00 for 15 days AWOL. Reenl. 5/1/62. WIA Cross Keys 6/8/62. Present 6/62-6/63. AWOL 7/27-8/22/63. Present 9-12/63 and 4/30-6/19/64. Ab. sick 6/20-12/31/64. WIA Ft. Stedman 3/25/65. Admitted to Hospital Richmond 3/27/65. Cap. there 5/3/65, escaped 5/4/65. Living Orange Co. 1902.

CROFT, DANIEL: Pvt., Co. G. B. Augusta Co. 1829? Farmer, age 31 Burkes Mill Dist., Augusta Co. 1860 census. Enl. Staunton 8/2/61. Present 11/61-12/31/62. Discharged 1/12/63 as Dunker by act of Congress passed 10/11/62. Age 38, 5' 10", dark complexion, hazel eyes, auburn hair. Farmer, age 41, Mt. Sidney PO, Augusta Co. 1870 census. Died Staunton 6/11/89. Bur. Croft Family Cem., on Rt. 254, 8/10's mile east of intersection with Rt. 714, Augusta Co.

CROFT, JACOB, JR.: 4thCpl., Co. A. B. Augusta Co. 5/14/43. Farmhand, age 17, Burkes Mill Dist., Augusta Co. 1860 census. Enl. Staunton 7/15/61 as Pvt. Ab. on Leave 12/6/61-5/1/62. Present 1/6-4/62. Reenl. 5/1/62. Present 8/31/62-11/63. Ab. on leave 12/23-31/63. Promoted 4thCpl. '64. Cap. Bethesda Ch. 5/30/64. Sent to Point Lookout. Transf. to Elmira. Released 6/30/65. Resident of Staunton, fair complexion, auburn hair, blue eyes, 5' 9". Farmer and Florist, Annex, Augusta Co. 1897. Member, Stonewall Jackson Camp CV, Staunton. Died on New Hope Road 3 miles from Staunton 8/11/13. Buried Croft family cem., on Rt. 254 8/10's mile east of intersection with Rt. 714, Augusta Co. Brother of Samuel A. Croft.

CROFT, JACOB S.: Pvt., Co. C. B. Augusta Co. 1837? Farmhand, age 23, Burkes Mill Dist., Augusta Co. 1860 census. Enl. Staunton 7/16/61 age 23. Present 11/61-4/62. Reenl. 5/1/62. WIA (neck) Port Republic 6/9/62. AWOL 6/9/62-3/10/63, fined $99.00 Present 4/30-12/31/63. Issued clothing 4/27/64. Cap. Bethesda Ch. 5/30/64. Sent to Point Lookout. Transf. to Elmira. Released 6/30/65. Resident of Staunton, 5' 10", dark complexion, dark hair, dark eyes, illiterate. Farmhand, age 36, Mt. Sidney PO, Augusta Co. 1870 census. Died Beverly Manor Dist., Augusta Co. 2/14/84.

CROFT, JAMES H.: Pvt., Co. C. B. Va. 1841? Farmhand, age 19, Burkes Mill Dist., Augusta Co. 1860 census. Enl. Staunton 7/16/61 age 20. Present 11/61-4/62. Reenl. 5/1/62. WIA 2nd Manassas 8/28/62. AWOL 10/9-11/1/62. Present in arrest 11/7/62-2/63. Fined $18.33. Ab. Sick 3/18-4/63. Present 4-13-12/31/63. WIA (left lung) Spotsylvania CH 5/12/64. DOWs.

CROFT, SAMUEL A.: Pvt., Co. A. B. Augusta Co. 7/16/41. Farmhand, age 18, Burkes Mill Dist., Augusta Co. 1860 census. Enl. Staunton 7/15/61. Present 11/61-4/62. Reenl. 5/1/62. Ab. sick 8/29/62-4/63. Present 4/30-12/31/63. Issued clothing 4/27 and 6/16/64. Cap. near Winchester 7/20/64. Sent to Wheeling. Age 23, 6', fair complexion, blue eyes, light hair. Transf. Camp Chase. Released 6/12/65. Farmer, age 29, Mt. Sidney PO, Augusta Co. 1870 census. Member, Stonewall Jackson Camp CV, Staunton. Died Augusta Co. 11/20/11. Bur. Laurel Hill Baptist Ch. Cem. Brother of Jacob Croft, Jr.

CROSS, GABRIEL: Cpl., Co. A. B. Va. 9/6/13. Enl. Staunton 7/15/61 as Pvt. Present 11/61-4/62. Reenl. 5/1/62. Present 8/31-11/22/62. AWOL 11/23-12/14/62. Present in arrest 1-2/63. Present 3-4/63, fined 4 months pay in addition to time AWOL by GCM. Present 4/30-12/63. Issued clothing 2/29/64. Present 4/30-12/31/64. Promoted Cpl. Cap. Harper's Farm 4/6/65. Sent to Point Lookout. Released 6/8/65. Carpenter, age 56, 6th Dist., Augusta Co. 1870 census. Died Augusta Co. 1/28/83. Buried Green Hill Cem., Churchville.

CROSS, JEREMIAH: Pvt., Co. A. Born 1837? Cooper, Churchville, Augusta Co. Enl. Staunton 7/29/61. Present 1-4/62. Reenl. 5/1/62. Present 8/31/62-4/63. Ab. sick 6/6-10/10/63. AWOL 12/8-31/63., lost 1 knapsack $6.50, 1 haversack .50, 1 canteen $1.25. stoppage for accouterments. Deserted to the enemy in W. Va. 1/5/64. Took oath and sent north. Age 26, 5' 11", dark complexion, black eyes, brown hair. Alive 1903.

CROSS, JOHN A. or D.: Pvt., Co. F. B. Va. 3/17/44. Student, age 15, Ryan's, Augusta Co. 1860 census. Enl. Alleghany Mt. 3/10/62. Deserted 4/10/62. Carpenter, age 65, Buena Vista, Rockbridge Co. 1910 census. Died Rockbridge Co. 12/17/14. Bur. Green Hill Cem., near Buena Vista.

CROSSLEY, SYLVANIUS W.: Pvt., Co. E. Farmhand, age 17, 4th Dist., Rockbridge Co. 1860 census. Enl. Staunton 8/1/61. Ab. on leave 11-12/61. Present 1-4/62. Reenl. 5/1/62. WIA (arm and leg) Gaines Mill 6/27/62. Leg amputated. Paid Lexington 10/25/64. NFR.

CROUSE, GEORGE C.: 5thSgt., Co. B (2nd). B. Va. 1833? Confectioneer, age 27, Burkes Mill Dist., Augusta Co. 1860 census. Enl. Waynesboro 7/16/61, as Pvt. Ab. on leave 11-12/61. Present 1-4/62. Reenl. 5/1/62. Pesent 4/30-10/31/62., promoted 3rdCpl. 7/15/62, promoted 2ndCpl. 10/2/62. Present 11/62-4/63. Promoted 5thSgt. 3/15/3. Present 4/30-8/31/63. Ab. sick in hospital Staunton with ulcers on leg and scabes 9/24/63-10/64. Returned to duty 10/20/64. Present 10/20-1/31/64. Cap. Amelia CH 4/6/65. Sent to Point Lookout. Released 6/14/65. 5' 10", light complexion, brown hair, blue eyes. Merchant, age 35, Mt. Sidney PO, Augusta Co. 1870 census. Died South River Township, Augusta Co. 3/3/71. Bur. Riverview Cem., Waynesboro.

CROUSE, WILLIAM F.: Pvt., Co. H. B. Augusta Co. 1816? Machinist, age 44, Burkes Mill Dist., Augusta Co. 1860 census. Enl. Staunton 7/23/61. NFR. Probably discharged because of occupation or age.

CROUSHORN, DANIEL J.: Pvt., Co. F. B. Augusta Co. 10/7/44. Farmhand, age 18, Burkes Mill Dist., Augusta Co. 1860 census. Enl. Staunton 9/1/62. Present 9/1-12/62. Ab. sick in hospital 1/13/63. AWOL 7/20-10/25/63, and pay deducted. Present 11-12/63. Ab. in hospital 4/30-10/31/64. Deserted 11/1/64. NFR. Farmer, age 28, Mt. Sidney PO, Augusta Co. 1870 census. Died on Naked Creek 9/19/70. Bur. Old Salem Lutheran Ch. Cem., 3 miles west of Mt. Sidney. Brother of William F. Croushorn.

CROUSHORN, DAVID B.: 2ndCpl., Co. G. B. Rockingham Co. 1823? Farmer, age 37, Mt. Sidney PO, Augusta Co. 1860 census. Enl. Staunton 7/31/61. Ab. sick 8/31-12/31/61. Present 1-2/62. Reduced to Pvt. 1/28/62. Present 3-4/62. Deserted 5/30/62. NFR. Alive 1900.

CROUSHORN, WILLIAM F.: Pvt., Co. H. B. Va. 6/10/40. Farmhand, age 20, Burkes Mill Dist., Augusta Co. 1860 census. Enl. Staunton 7/23/61, age 21. Present 11/61-4/62. Reenl. 5/1/62. AWOL 7/20-8/31/62. Present 1-10/63. Ab. on leave 11/63-1/8/64. Detailed in Pioneer Corps 2-9/64. Deserted Brown's Gap 9/28/64. NFR. Farmer, age 30, Mt. Sidney PO, Augusta Co. 1870 census. Died near Stonewall, Augusta Co. 12/9/05. Bur. Old Salem Lutheran Ch. Cem., 3 miles west of Mt. Sidney.

CRUM, JACOB.: Pvt., Co. D. B. Va. 1819? Not on muster rolls. Enl. 7/61. Cap. date and place unknown. Listed as POW Elmira 12/1/64, in "Staunton Vindicator". Served to end of war. Died Augusta Co. 5/2/05. Bur. Green Hill Cem., Churchville.

CRUTCHFIELD, THOMAS J.: Pvt., Co. C. B. Louisa Co. 1831? Farmhand, age 29, Burkes Mill Dist., Augusta Co. 1860 census. Enl. Staunton 7/16/61 age 33. Ab. sick 11/3/61-2/62. Present 3-4/62. AWOL 7/16-8/31/62. Present 9/1/62-2/28/63. AWOL 3/25-7/15/63. Fined $30.00 for being arrested and $70.33 for 110 days AWOL. Present 9-12/63. Issued clothing 4/27 and 29/64. WIA (bruise over sternum) Bethesda Ch. 5/30/64. In hospital Richmond 5/31-6/13/64. Cap. Carter's Farm near Winchester 7/20/64. Sent to Wheeling. Took oath and released 8/27/64. Age 38, 5' 10", dark complexion, dark hair, grey eyes.

CULLEN, COLUMBUS C.: Pvt., Co. G. B. Autusta Co. 1844? Enl. Staunton 8/2/61. Died of pneumonia in hospital Staunton 5/8/62. Age 17, 5' 9", fair complexion, fair hair, blue eyes.

CULLEN, GEORGE BARNHART: Pvt., Co. G. B. Augusta Co. 9/4/19. Carpenter, age 41, Burkes Mill Dist., Augusta Co. 1860 census. Pvt., Capt. Koiner's Co., 32nd Va. Militia, exempt 3/62 for rupture, age 42. Enl. Staunton 8/6/63. Present 8-9/63. Ab. sick 10/2/63. "Irredcible hernia". Ab. detached service in hospital Staunton 11-12/63. Discharged 3/28/64. Age 44, 5' 8", fair complexion, light brown eyes, sandy hair. Carpenter, age 50, Mt. Sidney PO, Augusta Co. 1870 census. Died Augusta Co. 8/21/99. Bur. Pines Chapel Cem.

CULLINAN, THOMAS C.: Musician, Co. B (2nd). B. Clare Co., Ireland, 1830. Stonemason, age 30, Burkes Mill Dist., Augusta Co. 1860 census. Enl. Waynesboro 7/15/61, as Bugler. Present 11-12/61, Chief Musician for Regt. Present 1-2/62. Ab. on leave 3-4/62. Chief Bugler for Regt. Ab. on leave until discharged 2/13/63 for chronic rheumatism, age 30, 5' 5¾'', fair complexion, blue eyes, dark hair.

CULTON, ZACHARIAH JOHNSTON: Pvt., Co. B (1st). B. near Brownsburg, Rockbridge Co. 2/11/22. Enl. Staunton 8/1/61. Ab. on duty in Rockbridge Co. when transf. 9/28/61.

CUPP, FREDERICK W.: 4thSgt., Co. D. B. Rockingham Co. 8/2/42. Farmhand, age 18, Burkes Mill Dist., Augusta Co. 1860 census. Enl. Staunton 7/16/61 as 4thCpl. Present 7/16/61-4/62. Reenl. 5/1/62. WIA (head) McDowell 5/8/62. WIA (mouth) 2nd Manassas 8/29/62. Present 10/1-30/63. Ab. on leave 10/31-11/15/63. Present 11/16-12/63. Promoted 4thSgt. Present 4/30-12/31/64. Surrendered Appomattox 4/9/65. Farmer, age 27, Mt. Sidney PO, Augusta Co. 1870 census. Died near North River Gap, Augusta Co. 11/9/75. Bur. Emanuel Ch. Cem. near Mt. Solon. Brother of John C. Cupp.

CUPP, JOHN C.: 4thCpl., Co. D. B. Va. 1845. Farmhand, age 16, Burkes Mill Dist., Augusta Co. 1860 census. Enl. Fair Oaks 7/1/62, as Pvt. Present 7/1-8/31/62 and 1-12/63. Promoted 4th Cpl. KIA Spotsylvania CH 5/18/64 by an accidental shell in an artillery fight. Brother of Frederick W. Cupp.

CUPP, WILLIAM J.: Pvt. Co. F. B. Va. 1830. Farmhand, age 30, Northern Dist., Augusta Co. 1860 Census. Enl. Staunton 7/31/61. Present 11/61-4/62. Reenl. 5/1/62. WIA (arm) McDowell 5/8/62. WIA (head) Gaines Mill 6/27/62. In hospital Winchester 10/18-12/31/62. Present 1-2/63. AWOL 7/26-8/6/63, pay deducted. Present 8/7-12/31/63 and 3-4/64. KIA Spotsylvania CH 5/12/64. Left widow and 3 children.

CURRIER, JOHN W.: Pvt., Co. A. B. Augusta Co. 1838? Farmhand, age 22, Burkes Mill Dist., Augusta Co. 1860 census. Enl. Staunton 7/9/61. Ab. sick 9/30/61-1/8/62. Present 1/9-4/62. Reenl. 5/1/62. Ab. sick 6/25/62-4/63. Present 4/30-12/31/63. KIA Bethesda Ch. 5/30/64. Brother of Joseph S. and William M. Currier.

CURRIER, JOSEPH S.: Pvt., Co. A. B. Augusta Co. 1839? Farmhand, age 21, Burkes Mill Dist., Augusta Co. 1860 census. Enl. Staunton 7/9/61. Present 11/61-4/62. Reenl. 5/1/62. AWOL 11/23-12/31/62. Present 1-4/63, in arrest. KIA Fredericksburg 5/5/63. Brother of John W. and William M. Currier.

CURRIER, ROBERT H.: Pvt., Co. A. B. Augusta Co. 1842. Not on muster rolls. Enl. Spring 62. WIA (left arm) Gaines Mill 6/27/62. Present Appomattox. Farmhand, Ladd, Augusta Co. 1910 census. Died Augusta Co. 3/25/20. Bur. Springdale Mennonite Ch. Cem., near Ladd.

CURRIER, WILLIAM M.: Pvt., Co.A. B. Augusta Co. 1836? Farmhand, age 24, Burkes Mill Dist., Augusta Co. 1860 census. Enl. Staunton 7/9/61. Ab. sick 12/16/61-3/17/62. Present 3/17-4/62. Reenl. 5/1/62. DOWs received 2nd Manassas 8/29/62. Brother of John W. and Joseph S. Currier.

CURRY, ALEXANDER: Pvt., Co. K. B. Bath Co. 1837? Laborer, age 23, Bath CH PO, Bath Co. 1860 census. Pvt., Capt. Hamilton's Co., 81st Va. Militia. Enl. Shenandoah Mt. 4/9/62. Present 4/9-30/62. AWOL 6/6/62-12/63. Dropped as deserter. Resident of Back Creek, Bath Co. Died 1907. Bur. Mountain Grove Cem., Back Creek, Bath Co.

CURRY, ALEXANDER R.: Pvt., Co. H. B. Va. 1828? Laborer, age 32, Northern Dist., Augusta Co. 1860 census. Enl. Staunton 7/23/61 age 38. Discharged by War Department 8/28/61. Enl. Co. H, 18th Va. Cav. 8/15/63. Paroled Staunton 5/1/65, 5' 10'', blue eyes.

CURRY, EUGENE GRANVILLE: 2ndSgt., Co. C. B. "Glen-Curry" on Naked Creek, Augusta Co. 11/4/39. Attended Loch Willow School under Jedediah Hotchkiss. Attended VMI 59-60. Enl. Staunton 7/6/61 age 21. "served as Drillmaster of company and was a candidate for Captain. A great athlete and splendid specimen. Could 'throw down' every man in Regiment and was said to be the storngest man at VMI — could stand on a half bushel (measure) and shoulder 4 bushels of wheat." Died of typhoid pneumonia at hospital Alleghany 11/5/61. Bur. Augusta Stone Ch. Cem. Brother of George H. A. and William B. Curry.

CURRY, GEORGE HARVEY ANDERSON: Pvt., Co. C. B. Augusta Co. 4/17/44. Farmhand, age 16, Northern Dist., Augusta Co. 1860 census. Enl. Staunton 7/25/61. Ab. on leave 11-12/61. Present 1-4/62. Reenl. 5/1/62. Transf. Co. C, 14th Va. Cav. Died 4/21/63. Brother of Eugene G. and William B. Curry.

CURRY, J. K.: Pvt., Co. B (2nd). Pension application only record. Served 3 years. Living in Roanoke 1902.

CURRY, JAMES H.: 1stSgt., Co. D. B. Augusta Co. 1837? Carpenter, age 23, Burkes Mill Dist., Augusta Co. 1860 census. Enl. Shenandoah Mt. 4/10/62 as Pvt. Present 4/10-30/62. Promoted 1stSgt. 5/30/62. WIA (right leg) Port Republic 6/9/62. Ab. wounded through 8/31/62. Present1-12/63. Reduced to Pvt. by order of Regt'l CM for conduct prejudicial to good order and discipline 12/15/63. Present 4/30-12/31/64. NFR. Admitted to Lee Camp Old Soldiers Home, Richmond from Augusta Co. Died there 3/26/05, age 68. Bur. Hollywood Cem. Brother of William L. Cupp.

CURRY, JOHN M.: Pvt., Co. K. B. Va. 1831? Pvt., Capt. Hamilton's Co., 81st Va Militia. Tombstone only record. Farmer, Sunrise, Bath Co. 1897. Age 79, Warm Springs Dist., Bath Co. 1910 census. Bur. Mountain Grove Cem., Back Creek, Bath Co.

CURRY, JOHN W.: Pvt., Co. A. Not on muster roll. KIA Bethesda Ch. 5/30/64.

CURRY, JOSEPH: Pvt., Co. A. Not on muster roll. KIA Fredericksburg 5/5/63.

CURRY, MARTIN V.: Pvt., Co. K. B. Bath Co. 1845? Farmhand, age 15, Bath CH PO, Bath Co. 1860 census. Enl. Shenandoah Mt. 4/9/62. Present 4/9-30/62. Present 1-12/63. WIA (thigh and neck) and cap. Bethesda Ch. 5/30/64. In 3rd Div., 5th Army Corps, Army of Potomac Hospital DOWs in hospital Richmond, 5/31/64, age 21.

CURRY, PETER L.: 3rdCpl, Co. K. Resident of Back Creek, Bath Co. Enl. Shenandoah Mt. 9/9/62 as Pvt. Present 4/9-30/62. Promoted 3rdCpl. 5/1/62. AWOL 7/21/62-2/29/64. Cap. North Anna 5/22/64. Sent to Point Lookout. Exchanged 11/1/64 Savannah, Ga. Issued clothing 11/20/64. Ab. 11-12/64. NFR. Died before 1917.

CURRY, SAMUEL M.: Pvt., Co. K. B. Bath Co. 1822? Farmer, Highland Co. Enl. Shenandoah Mt. 4/9/62. Present 4/9-30/62 and 5/26/62. WIA (head) Port Republic 6/9/62. Ab. wounded through 4/63. Ab. sick 8/6-31/63. Deserted 9/1/63. Deserted to the enemy Beverly, W. Va. 10/10/64. Age 45, 5' 10'', light complexion, blue eyes, dark hair. Blacksmith, age 48, Williamsville Dist., Bath Co. 1870 census. Died Pocohontas Co., W. Va. 1908.

CURRY, WILLIAM A.: Pvt., Co. E. B. Daugherty Co., Ga. 1838. Planter. Married Augusta Co. 5/1/62. Not on muster rolls. Received commutation for rations Salisbury, N. C. 12/29/64. NFR.

CURRY, WILLIAM BROWN.: Pvt., Co. C. B. Augusta Co. 11/30/46. Student, age 13, Northern Dist., Augusta Co. 1860 census. Not on muster rolls. WIA (leg) Gaines Mill 6/27/62. DOWs Richmond 7/29/62, age about 17. Brother of Eugene G. and George H. A. Curry.

CURRY, WILLIAM L.: Pvt., Co. D. B. Augusta Co. 1842? Farmhand, age 18, Burkes Mill Dist., Augusta Co. 1860 census. Enl. Staunton 7/16/61. Present 7/16/61-4/62. AWOL 6/4-20/62. Fined $5.86. Wounded accidentally 7/9/62. Ab. wounded 7/9-8/31/62. Present 1-12/63. WIA (side) Spotsylvania CH 5/19/64. DOWs. Brother of James H. Curry.

DABNEY, EDWARD MOON: Capt., Co. C. B. Memphis, Tenn. 3/18/39. (Mother died soon after birth and he was brought to Va. and raised by his uncle Major William S. Dabney of Albemarle Co). VMI '62. Drillmaster, Richmond 4-6/61. Enl. Staunton 7/16/61 age 22. Present 11/61-4/62. Reelected 5/1/62. WIA (right arm, bone shattered from wrist to elbow) McDowell 5/8/62. Ab. wounded through 8/31/62. Commanding Regt. 10/30/62. WIA (grape shot through both hips while dismounting his horse) Fredericksburg 12/13/62, while acting Major of Regt. DOWs 12/23/62. Over 6' 4". "He was one of the best officers in the service and prospects of promotion were bright."

DAGGY, FRANKLIN ALEXANDER: Pvt., Co. D. B. Va. 1847? Farmhand, age 13, Deerfield PO, Augusta Co. 1860 census. Not on muster rolls. Deserted to the enemy Oakland, Md. 9/16/62. Sent to Cincinnati, Oh. Took oath and released 9/19/62. Age 16, 5' 6", fair complexion, dark hair, dark eyes. Brother of William H. Daggy.

DAGGY, JOHN HENRY: Pvt., Co. K. B. Augusta Co. 2/21/39. Wagonmaker, age 21, Cleeks Mill PO, Bath Co. 1860 census. Enl. Shenandoah Mt. 4/9/62. Present 4/9-30/62. Ab. sick 1/30-9/30/63. AWOL 10/1/63-5/24/64, pay deducted. Ab. sick 6/20/12/31/64. NFR. Carpenter, age 31, Williamsville Dist., Bath Co. 1870 census. Wheelwright, Burnsville, Bath Co. 1897. Died Burnsville 2/23/13. Bur. Burnsville Cem.

DAGGY, JOHN W.: Pvt., Co. D. B. Va. 1842? Farmhand, age 18, Northern Dist., Augusta Co. 1860 census. Enl. Staunton 7/16/61. Present 7/16/61-4/62. Reenl. 5/1/62. AWOL 7/18-8/6/62., fined $6.60. Present 8/7-31/62. Deserted 12/16/62. Deserted to the enemy Tucker Co., W. Va. 5/5/64. Took oath and released. Age 22, 6' 4", fair complexion, blue eyes, dark hair, resident of Mossy Creek, Augusta Co. Blacksmith, Middlebrook 1890. and Blacksmith, Stover, Augusta Co. 1897.

DAGGY, JONAS FRANKLIN.: Pvt., Co. D. B. Long Glade, Augusta Co. 8/29/40. Enl. Staunton 7/16/61. Present 7/16/61-4/62. AWOL 7/18-9/20/62., fined $11.00. WIA (leg) 2nd. Manassas 8/29/62. In hospital Charlottesville with debilitis 5/30/63. Deserted from hospital 7/1/63. WIA (left leg) Mine Run 12/1/63. In hospital Richmond 12/3-9/63. WIA (head) Spotsylvania CH 5/12/64. Ab. wounded in hospital Richmond through 12/31/64. Cap. Amelia CH 4/6/65. Sent to Point Lookout. Released 6/12/65. 5' 9½", dark complexion, black hair, dark hazel eyes. Carpenter, age 30, Mt. Sidney PO, Augusta Co. 1860 census. Undertaker, Dorcas, Augusta Co. 1897. Member, Stonewall Jackson Camp CV, Staunton. Died near Mt. Solon, Augusta Co. 2/11/26. Bur. Mt. Olivet Cem.

DAGGY, WILLIAM H.: Pvt., Co. C. B. Va. 1836? Farmhand, age 24, Northern Dist., Augusta Co. 1860 census. Enl. Staunton 7/16/61 age 24. Present 11-12/61. AWOL 2/21-28/62. Present 3/19-4/62. Deserted near Deerfield 4/20/62. Deserted to the enemy Oakland, Md. 9/16/62. Took oath and released Cincinnati, Oh. Age 26, 5' 10", dark complexion, blue eyes, dark hair, resident of Deerfield. Reported detailed to Nitre Bureau, Bath Co. 8/4/63. Applied for pension in Oklahoma 1915. Brother of Franklin A. Daggy.

DALE, SAMUEL K.: Pvt., Co.E. B. Rockbridge Co. 1844? Weaver, age 16, Kerr's Creek, Rockbridge Co. 1860 census. Enl. Bunker Hill 10/11/62. Ab. on leave 10/11-31/62. AWOL 11/1/62-12/63. Ab. in arrest under charges for desertion 4/30-12/21/64. NFR. Living Buena Vista 1890.

DANDRIDGE, THOMAS: Lt., Co. ?. Not on muster rolls. WIA Cross Keys 6/8/62 and discharged. Enl. in Cav. Resident Charlestown, W. Va.

DANNER, GEORGE W.: Pvt., Co. F. B. Va. 1843. Shoemaker, age 17, Burkes Mill Dist., Augusta Co. 1860 census. Pvt., Capt. Shuemake's Co., 32nd Va. Militia, 3/62, exempt for scrofula. Enl. Staunton 10/4/62. Ab. sick 6/5/63. In hospital Richmond with rheumatism 7/5-24/63. Returned to duty. Deserted NFR. Enl. Co. E, 18th Va. Cav. 12/3/63.

DAVIS, JOHN HENRY: Pvt., Co. I. B. Augusta Co. 2/5/42. Shoemaker age 18, Newport, Augusta Co. 1860 census. Enl. Staunton 7/16/61 age 19. Present 11/61-4/62. Reenl. 5/1/62. AWOL 7/19-8/13/62, fined $30.00. AWOL 9/15/62-1/19/63, fined $45.47. Present 1/20-12/31/63 and 4/30-12/7/64. AWOL 12/8/64. Deserted to the enemy Beverly, W. VA. 12/28/64. Took oath and released. Age 23, 5' 8", fair complexion, dark eyes, brown hair. Married Augusta Co. 11/30/65.

DAVIS, JOSEPH H.: Pvt., Co. A. B. Louisa Co. 1836? Farmhand, age 24, Staunton PO, Augusta Co. 1860 census. Enl. Staunton 7/16/61. Present 11/61-4/62. Reenl. 5/1/62. WIA (side) Port Republic 6/9/62. WIA (head) 2nd Manassas 8/28/62. Ab. wounded through 12/31/62. Present 1-4/63. WIA (bruise on crest of illum from shell) Fredericksburg 5/5/63. AWOL 7/6-8/31/63. Present 9-12/63. AWOL 12/8-20/63. Present in arrest 2/21-31/63. Ab. sick in hospital 7/1-5/64. Deserted to the enemy Beverly, W. VA. 10/18/64. Took oath and released. Age 27, 5' 9", dark complexion, blue eyes, light hair. Laborer, age 30, 1st Dist., Augusta Co. 1870 census. Living Clifton Forge 1903.

DAVIS, ROBERT COLEMAN: Capt., Co. A. B. Augusta Co. 1840? Enl. Staunton 8/9/61 as 2ndCpl., age 23. Present 11/61-4/62. Elected 2ndLt. 1/16/62. Elected 1stLt. 5/1/62. WIA (arm) Port Republic 6/9/62. Elected Capt. 8/28/62. Present 8/31/62-12/31/63. WIA (head) Wilderness 5/6/64. Cap. Cedar Creek 10/19/64. Sent to Ft. Delaware. Released 6/17/65. 5' 6", dark complexion, dark hair, dark eyes. Coachmaker, age 30, Mt. Sidney PO, Augusta Co. 1870 census. Alive 1903.

DAVIS, ROBERT HAMILTON: Pvt., Co. A. B. Augusta Co. 12/19/45. Farmhand, age 18, Burkes Mill Dist., Augusta Co. 1860 census. Enl. Rapidan 10/1/63 age 17. Present 10/1-12/7/63. AWOL 5/15/64. Deserted to the enemy at Beverly, W. Va. 10/18/64. Took oath and released. Age 20, 5' 9", light complexion, grey eyes. light hair. Admitted to Lee Camp Old Soldiers Home, Richmond 7/7/04. Died there 8/3/04. Buried Clifton Forge.

DEAKINS, ANDREW J.: 4thSgt., Co. H. B. Va. 1822? Enl. Staunton 7/23/61. Discharged by Civil Authority as a citizen of Md. 12/61. Hotel Clerk, age 48, Mt. Sidney PO, Augusta Co. 1870 census.

DEARING, CALVIN E.: Pvt., Co. B (2nd). Enl. Hamilton's Crossing 3/13/63. Detailed as Wagon Master in Division Commisarry Train 3/13/63-12/31/64. Surrendered Appomattox 4/9/65.

DECKER, HOWARD WESLEY: Pvt., Co. B (1st). B. Va. 12/19/42. Enl. Staunton 8/1/61. Present until transf. 9/28/61.

DEDRICK, HENRY H.: Pvt., Co. B (2nd). B. Rockingham Co. 1836? Farmer, Rockingham Co. Enl. Waynesboro 7/15/61. Present 11/61-4/62. Reenl. 5/1/62. WIA Cross Keys 6/8/62. WIA Gaines Mill 6/27/62. AWOL 7/18/62-4/19/63. Fined all pay 7/18/62-8/1/63. Present 7/3-27/63. AWOL 7/27-10/63. Deserted to the enemy Clarksburg, W. Va. 10/24/63. Took oath and sent north. Age 24, 5' 8" dark complexion, blue eyes, dark hair. Farmer, age 34, Fishersville PO, Augusta Co. 1870 census. Farmer, age 74, South River Dist., Augusta Co. 1910 census. Died Sherando, Augusta Co. 11/10/21. Bur. Sherando Methodist Ch. Cem.

DEDRICK, JAMES M.: Pvt., Co. B. (2nd). B. Rockingham Co. 1838? Farmhand, age 22, 1st Dist., Augusta Co. 1860 census. Pvt., Capt. Hall's Co., 32nd Va. Militia 3/62 exempt for fistula, age 24. Enl. Bunker Hill 10/3/62. Present 10/3/62-2/63. AWOL 3/4-10/63. Deserted to the enemy Clarksburg, W. Va. 10/24/63. Took oath and sent north. Age 26, 5' 9", dark complexion, black eyes, dark hair, resident of Rockingham Co. Farmer, age 33, Fishersville PO, Augusta Co. 1870 census. Farmer, Age 72, South River Dist., Augusta Co. 1910 Census.

DEFENBAUGH, BENJAMIN ARNI: Pvt., Co. F. B. Augusta Co. 7/17/42. Farmhand, age 18, Burkes Mill Dist., Augusta Co. 1860 census. Pvt., Co. C, 160th Va. Militia exempt 3/62 defective eyesight, age 19. Enl. Staunton 9/15/62. Sent back to wagons 11/24/62 and not heard from since. AWOL 1-12/63. Dropped as deserter. Farmer, age 28, Mt. Sidney PO, Augusta Co. 1870 census. Farmer, Long Glade 1897. Died Springhill, Augusta Co. 12/18/17. Bur. Parnassus Methodist Ch. Cem.

DEFENBAUGH, HENRY: Pvt., Co. C. B. Va. 6/12/20. On postwar roster. Served 1 year. Farmer, Laurel Hill 1897. Died Augusta Co. 1/18/03. Bur. Laurel Hill Baptist Ch. Cem.

DEMASTIS, ROBERT WILLIAM: Pvt., Co. C. B. Va. 1843? Farmhand, age 17, 1st Dist., Augusta Co. 1860 census. Enl. Staunton 7/16/61 age 17. Present 11-12/61, fined $11.00 for desertion. Present 1/1-4/20/62. WIA (thigh) Port Republic 6/9/62. In hospital Staunton 11/21/62. DOWs.

DEPRIEST, JOHN W.: 4thCpl., Co. C. B. Va. 1826. Plasterer, age 34, Burkes Mill Dist., Augusta Co. 1860 census. Enl. Staunton 7/16/61 age 35, as Pvt. Present 11/61-4/62. Reenl. 5/1/62. AWOL 5/15-8/62. Present 8/31/62-4/63. Appointed 4thCpl. 4/1/63. Deserted in W. Va. 6/6/63. Took oath Wheeling and released. Age 37, 5' 11", florid complexion, blue eyes, black hair.

DICKERSON, C. J.: Pvt., Co. E. Pension application only record. Boot and Shoemaker, New Hope, 1872. Living in Augusta Co. 1888.

DICKINSON, JOHN CARTER: 2ndLt., Co. B (1st). B. Va. 1827? Farmer, age 33, Walker's Creek Dist., Rockbridge Co. 1860 census. Enl. Fairfield 7/10/61 as Pvt. Appointed 2ndLt. 7/27/61. Present until transf. 9/28/61.

DIDDLE, WILLIAM: Pvt., Co. B. (2nd). B. Va. 1837? Farmhand, age 23, Deerfield, Augusta Co. 1860 census. Enl. Camp Alleghany 3/15/62. AWOL 4/15/62-12/31/62. Dropped as deserter. NFR.

DINKEL, CHARLES A.: 4thCpl, Co. A. B. Rockingham Co. 4/30/32. Farmer, Augusta Co. Enl. Stuanton 7/17/61 as Pvt. Present 11/61-4/62. Promoted 4thCpl. 1/18/62. Reenl. 5/1/62. KIA Port Republic 6/9/62. Bur. Melanchthon Chapel Cem., near Weyer's Cave.

DINKLE, GEORGE WILLIAM: Pvt., Co. C. B. Augusta Co. 4/21/40. Enl. Staunton 7/16/61 age 21. Died of typhoid fever at his home on Mossy Creek, Augusta Co. 11/12/61. Bur. Parnassus Methodist Ch. Cem.

DIXON, CHARLES S.: Pvt., Co. A. B. Va. 1842? Laborer, age 18, Burkes Mill Dist., Augusta Co. 1860 census. Enl. Staunton 7/9/61 age 18. Present 11/61-4/62. Reenl. 5/1/62. Deserted from hospital date unknown. On rolls 8/31/62-2/63. NFR. Brother of Francis M. Dixon.

DIXON, FRANCIS MARION: Pvt., Co. A. B. Augusta Co. 8/18/43. Carpenter's Apprentice, age 18, Burkes Mill Dist., Augusta Co. 1860 census. Enl. Staunton 7/15/61. Ab. sick 11/20/61-2/7/62. Present sick 2/8-28/62. Died of typhoid fever Camp Alleghany 3/12/62. Bur. Augusta Stone Ch. Cem. Brother of Charles S. Dixon.

DIXON, WILLIAM A.: Pvt., Co: E. B. Rockbridge Co. 1813. Carpenter, Rockbridge Co., 1860 census. Enl. Staunton 8/1/61. Present 11/61-2/62. Ab. sick 3-4/62. Ab. on leave 5/1-10/31/62. Discharged 11/1/63 over age of conscription. Laborer, age 54, Natural Bridge Dist., Rockbridge Co. 1870 census.

DOBBS, JOHN A.: Pvt., Co. B. (2nd). B. Albemarle Co. 1829? Laborer, age 31, 1st Dist., Augusta Co. 1860 census. Enl. Waynesboro 7/15/61. Present 11/61-4/62. Died of pneumonia Sherando, Augusta Co. 6/13/62. Age 33, left widow.

DOLD, JAMES ADDISON: Capt., Co. H. B. Bath Co. 1838? Farmhand, age 22, 1st Dist., Augusta Co. 1860 census. Enl. Staunton 7/23/61 as 2ndSgt., age 23. Present 11/61-4/62. Elected 3rdLt. 5/1/62. Present 5/1-8/31/62. Elected 2ndLt. 9/17/62. Present 10/62. Commanding Co. 12/1/62. Elected 1stLt. Ab. on leave 7 days 2/63. Commanding Co. 3-8/63. Ab. on leave 10/30-11/14/63. Present 11/15-12./63. Elected Capt. 12/19/63. KIA Bethesda Ch. 5/30/64, age 27. Left widow. Bur. Tinkling Spring Pres. Ch. Cem. near Fishersville. "He was a brave and gallant officer and his gentlemanly deportment and a (illegible) qualities endeared to all his friends and acquaintenses." Brother of William F. Dold.

DOLD, WILLIAM F.: 1stLt., Co. H. B. Bath Co. 1836? Teacher, age 24, 1st Dist., Augusta Co. 1860 census. Enl. Staunton 7/23/61 as 3rd Sgt., age 24. Present 11/61. Elected 1stLt. 11/23/61. Present 12/61-4/62. Reelected 1stLt. 5/1/62. Present 5/1-8/31/62. KIA Sharpsburg 9/17/62. Age 25, 5'11", dark complexion, black eyes, black hair. Bur. Old Reformed Ch. Cem., Shepherdstown, W. Va. Brother of James A. Dold.

DONAGHE, BRISCOE BALDWIN: "Bruce". Surgeon, F&S. B. Staunton 3/23/24. Attended UVa. and Jefferson Medical College, Phil., Pa. Doctor, age 36, 1st Dist., Staunton, 1860 census. Not on muster rolls. Medical Director, Gen. Harper's Div., Va. Militia at Harper's Ferry 4/29/61. Assistant Surgeon Augusta Raid Guard 12/5/63. NFR. Doctor, age 46, Staunton 1870 census. Died Staunton 9/10/79. Bur. Thornrose Cem.

DONAHOO,M JOHN W.: Pvt., Co. I. B. Va. 1846? Student, age 14, 1stDist., Augusta Co. 1860 census. Not on muster rolls. Paroled Staunton 5/17/65. Age 19, 5' 8", dark complexion, light hair, brown eyes. Farmhand, age 68, Middlebrook Precinct, Greenville Dist., Augusta Co. 1910 census.

DONALD, WILLIAM KEYS, JR.: 1st Sgt. Co. B (1st). B. Va. 5/20/28. Wash. Col. 47-48. Farmer, age 30, South River Dist. Rockbridge Co. 1860 census. Enl. Fairfield 7/10/61. Present until transf. 9/28/61.

DONAVANT, CORNELIUS: Pvt., Co. H. B. Ireland. Enl. Bunker Hill 9/24/62 as substitute. AWOL 10/11/62-12/31/63. Dropped as deserter. NFR. Described by Lt.Col. Lilley as a professional deserter, making "a business of enlisting, drawing bounty, and then skip."

DORMAN, ANANIAS: Pvt., Co. B (2nd). Resident of Rockingham Co. Date of Enl. unknown. Issued clothing 2/19 and 3/1/64. Deserted to the Army of the Potomac at Sommerville Ford 3/23/64. Took oath and transportation furnished to Philadelphia, Pa. 5' 11", light complexion, black hair, hazel eyes.

DOYLE, ELIAS H.: Pvt., Co. C. B. Southampton Co. 1830. Miner, age 30, 1st Dist., Augusta Co. 1860 census. Enl. Staunton 7/16/61 age 30. AWOL 11-12/61, fined $11.00. Present 1-3/62. Ab. sick 4/4-19/62. Ab. on detached service at Ivy Furnace as a machinist 4/20/62-12/31/64. NFR. Died near Midvale, Rockbridge Co. 1/7/88.

DOYLE, JOHN FLETCHER: Pvt., Co. B (1st). B. Augusta 1841? Farmhand, age 17, South River Dist., Rockbridge Co. 1860 census. Enl. Fairfield 7/19/61. Present until transf. 9/28/61.

DRAIN, DAVID C. C.: Pvt., Co. E. B. Rockbridge Co. 1836? Farmer Rockbridge Co. Not on muster rolls. Died of erysipelas, typhoid fever, Staunton 11/18/61 age 24. Bur. Thornrose Cem. grave #785.

DRAIN, LEWIS C.: Pvt., Co. E. B. Rockbridge Co. 1832? Blacksmith, age 28, Lexington Dist., Rockbridge Co. 1860 census. Served in Militia. Enl. Staunton 8/1/61. Present 11/61-2/62. Transf. 2nd Rockbridge Artillery by order Gen. Johnson 3/20/62 to take effect 1/1/62.

DRAIN, RILAND: Pvt., Co. G. Born Rockbridge Co. 1839? Blacksmith, age 19, Lexington Dist., Rockbridge Co. 1860 census. Enl. 9/17/62. Present 9/17-10/62. WIA (right leg) Fredericksburg 12/13/62. Leg amputated. DOWs in hospital Richmond 1/13/63, age 23.

DRAKE, B. C.: Pvt., Co. A. Not on muster rolls. Died in hospital Staunton 5/19/62, cause unknown. Bur. grave #960, Thornrose Cem.

DUDLEY, EDWARD AUGUSTUS: Pvt., Co. H. B. Jennings Gap, Augusta Co. 2/6/29. Tanner, age 30, Burkes Mill Dist., Augusta Co. 1860 census. Pvt., Co. D. 160th Va. Militia 3/25/62, detailed as tanner. Enl. Staunton 10/25/64. Present 11-12/64. Present Appomattox. Tanner, Cattle Dealer, Augusta Co. Tax Collector, Clifton Forge, Alleghany Co. Died Clifton Forge 5/23/01.

DULL, AUGUSTUS B.: Pvt., Co I. B. Augusta Co. 11/7/38? Laborer, age 23, Staunton 1860 census. Enl. Staunton 7/16/61 age 24. Ab. sick 11/61-1/9/62. Present 1/10-4/62. Reenl. 5/1/62. AWOL 7/18-8/13/62. Fined $9.16. Present 1-12/63 and 4/30 until WIA (nose) Bethesda Ch. 5/30/64. Present 6/1-12/31/64. Paroled Staunton 5/1/65. Age 21, 5' 8'', fair complexion, brown hair, blue eyes. Laborer, age 27, 1st Dist., Augusta Co. 1870 census. Brother of George L., Jacob H. and William H. Dull, Sr.

DULL, GEORGE LEWIS: 2ndSgt., Co. I. B. Augusta Co. 6/11/43. Farmhand, age 16, Staunton 1860 census. Enl. Staunton 7/16/61 age 18, as Pvt. Present 11/61-4/62. Reenl. 5/1/62. Present 8/30-10/23/62. Ab. sick 10/24/62-4/63. Present 4/30-12/31/63. Promoted 2ndSgt. 11-12/63. Cap. near Winchester 9/14/64. Sent to Washington, D.C. Took oath and transportation furnished to Frederick, Md. Farmer, age 27, Fishersville PO, Augusta Co. 1870 census. Farmer and Stock Raiser, Middlebrook, 1885. Merchant and Wool Dealer, Spottswood, 1898. Brother of Augustus B., Jacob H. and William H. Dull, Sr.

DULL, JACOB H.: Pvt., Co. A. B. Va. 3/15/33. Laborer, age 27, 1st. Dist., Augusta Co. 1860 census. Pvt., Co. E. 93rd Va. Militia 3/25/62, exempt as ironworker, age 25. Enl. Camp Valley Mills 4/31/62. Reenl. 5/1/62. In hospital with diarrhea Mt. Jackson 6/1/62. WIA (leg) 2nd Manassas 8/28/62. Leg amputated below the knee. Paid 8/31/63. NFR. Laborer, age 37, 1st Dist., Augusta Co. 1870 census. Farmer, Arbor Hill, Augusta Co. 1910 census. Died Augusta Co. 2/24/11. Bur. Mt. Tabor Lutheran Ch. Cem., near Middlebrook. Brother of Augustus B., George L., and William H. Dull, Sr.

DULL, WILLIAM HENRY, SR.: Pvt., Co. I. B. Augusta Co. 9/19/34. Enl. Staunton 7/16/61 age 26. Not received. Name cancelled on rolls. Died of fever near Middlebrook, Augusta Co. 7/20/62. Bur. Mt. Tabor Lutheran Ch. Cem. Brother of George L., Augustus B. and Jacob H. Dull.

DULL, WILLIAM HENRY, JR.: Pvt., Co. I. B. Augusta Co. 10/6/42. Resident of Staunton, age 13, 1860 census. Enl. Staunton 7/16/61 age 17. Ab. sick 11/61-1/5/62. Present 6/1/62 until he died of pneumonia Camp Alleghany 1/25/62. Age 17, 5' 3'', fair complexion, blue eyes, fair hair, farmer. Effects: 1 blanket, 1 quilt, sent with his body to his father. 1 overcoat, 1 pr. pants, 1 shirt, 1 pr. socks, 1 port monaia containing $1, money due $29.60. Bur. Mt. Tabor Lutheran Ch. Cem.

DUNLAP, JOHN J.: Pvt., Co. H. B. Augusta Co. 1827. Farmer, age 30, 1st Dist., Augusta Co. 1860 census. Enl. Staunton 7/23/61. Present 11/61-2/62. AWOL 3/5-18/62. Present 3/19-4/62. WIA (shoulder) Port Republic 6/9/62. Ab. wounded through 4/63. Detailed as nurse in hospital Staunton 4/63-1/31/64. Present 4/1-25/64. Ab. sick 4/26-12/31/64. NFR.

DUNLAP, ROBERT B.: 1stSgt., Co. C. B. Augusta Co. 1834. Farmer, age 25, 1st Dist., Augusta Co. 1860 census. Capt., Cav. Co., 93rd Va. Militia. Enl. Staunton 7/16/61 age 25. Present 11-12/61. Discharged by order of Secretary of War 1/28/62. Iron Master, Forrer's Furnace, Mossy Creek, Augusta Co. Enl. "Churchville Cavalry", 14th Va. Cav. Enl. Co. G. 18th Va. Cav. 2/29/64. Paroled Staunton 5/13/65. 5' 11'', light complexion, light hair, hazel eyes. Farmer, Churchville 1885. Died near Zetta, Augusta Co. 10/23/05. Bur. Green Hill Cem., near Churchville.

DUNLAP, ROBERT BAILEY: Pvt., Co. A. B. Augusta Co. 9/3/39. Farmer, age 25, Burkes Mill Dist., Augusta Co. 1860 census. Enl. Staunton 7/16/61. Present 11/61-4/62. Reenl. 5/1/62. Ab. sick 11/1/62-2/28/63. Ab. wounded through 8/31/63. WIA (hip) Port Republic 6/9/62. WIA (middle of clavicle) and cap. Bethesda Ch. 5/30/64. Admitted to 1st Div., 5th Army Corps, Army of Potomac Hospital near Hanover Junction. Sent to Old Capitol Prison, Washington. Transf. to Elmira. Exchanged Savannah, Ga. 11/15/64. Paroled Staunton 5/16/65. Age 32, 6', dark complexion, black hair, black eyes. Dr. Hunter McGuire removed minie ball from bladder 1874. Farmer, North River Dist., Augusta Co. 1885. Died near Lone Fountain, Augusta Co. 12/19/34. Bur. Green Hill Cem.

EAKIN, JAMES MOORE: Pvt., Co. B. (1st). Enl. Fairfield 7/10/61. Present until transf. 9/28/61.

EAST, CHARLES H., JR: Pvt., Co. C. B. Augusta Co. 1838? Laborer, age 22, 1st Dist., Augusta Co. 1860 census. Enl. Staunton 7/16/61, age 23. Present 11/61-4/62. Reenl. 5/1/62. Present 5/1-8/31/62. Ab. sick 11/1/62-2/28/63. Present 3-4/63. AWOL 6/6-12/31/63. Dropped as deserter. NFR.

EAST, WILLIAM T.: Pvt., Co. C. B. Rockbridge Co. 1832? Carpenter, age 28, 1st Dist., Augusta Co. 1860 census. Enl. Staunton 7/16/61 age 29. Present 11-12/61. Ab. on leave with train of sick 2/26/62. Present 3-4/62. Reenl. 5/1/62. Ab. sick with hernia 8/27/62-4/63. Detailed as nurse and guard in hospital Richmond and Staunton 4/30/63-12/31/64. Paroled Staunton 5/10/65. Age 38, 5' 10'' fair complexion, sandy hair, grey eyes. Laborer, age 38, 1st Dist., Augusta Co. 1870 census.

ECHARD, SILAS C.: Pvt., Co. F. B. Augusta Co. 12/13/38. Pvt., Capt. Rimel's Co. 160th Va. Militia 3/17/62 exempt for breast complaint age 23. Enl. Richmond 8/6/64. WIA (right thigh) and cap. Winchester 9/19/64. Ball removed 10/15/64. Transf. to U.S. hospital, West Building, Baltimore 12/20/64. Sent to Ft. McHenry. Released 6/9/65. Farmer, age 32, Mt. Sidney PO, Augusta Co. 1870 census. Farmer, Greenville 1885. Died while visiting his daughter near Mt. Sidney 6/90. Bur. Mt. Pisgah Meth. Ch. Cem.

ECHARD, WILLIAM: Pvt., Co. E. B. Rockbridge Co. 1842? Farmer, age 18, South River Dist., Rockbridge Co. 1860 census. Enl. Bunker Hill 9/26/62. Present 9/26/62-4/63. Returned to Co. G, 58th Va. Inf. where he had enlisted prior to joining this Co. KIA Winchester 9/5/64.

EDDY, JOSEPH: Pvt., Co. F. B. Va. 1845? Laborer, age 15, Burkes Mill Dist., Augusta Co. 1860 census. Enl. Staunton 7/31/61. Present 11/61-4/62. Reenl. 5/1/62. Present 8/31/62-2/63. Issued clothing 4/6 and 5/19/63. Cap. Gettysburg 7/3/63. Sent to Ft. McHenry. Transf. Ft. Delaware. Took oath and joined Captain Ahl's U.S. Battery 7/27/63.

EDMONDSON, WILLIAM H.: Assistant Surgeon, F&S. B. N. C. 1840? Doctor. Enl. Co. G. 12th N. C. Inf. as Pvt. 5/16/61, age 23, at Halifax, N. C. Present or accounted for until promotion Assistant Surgeon, 52nd Va. Inf. 2/1/63. Present 5/63 until KIA near Spotsylvania CH 5/10/64.

ELLINGER, ALEXANDER: Pvt., Co. H. Not on muster rolls. Enl. Staunton 1861 age 40. Discharged 10/16/62 over age of conscription.

ELLINGER, HENRY: Pvt., Co. H. B. Augusta Co. 1814? Laborer, Middlebrook, Augusta Co. Enl. Staunton 7/23/61 age 47 as substitute. Present 11-12/61, fined $11.00 by CM. AWOL 1/5-16/62. Present 1/17-4/62. AWOL 4/21/62-8/9/63. Present 8/10-10/63. Ab. sick 11/8-12/63. Age 50 12/26/63. Requested discharge 1864 claiming to have never reenl. Age 47, 5' 1'', fair complexion, blue eyes, dark hair. NFR.

ELLIOTT, STEPHEN D.: Cpl., Co. F. B. Augusta Co. 1829? Laborer, age 31, Burkes Mill Dist., Augusta Co. 1860 census. Enl. Staunton 7/31/61. Present 11/61-1/62. Ab. sick 2/22-28/62. Present 3-4/62. Reenl. 5/1/62. Present 8/31/62-2/63. Sent to hospital 6/3/63. Ab. sick through 8/31/63. Present 9-12/63. Promoted Cpl. 63. Present 3-4/64. WIA (hip) near Spotsylvania CH 5/16/64. Cap. Augusta Co. 6/10/64. Age 36, 5' 10", dark complexion, dark hair, grey eyes. Sent to Wheeling. Transf. Camp Chase. Released 5/16/65. Age 38, 5' 10½", florid complexion, dark hair, blue eyes. Admitted to Jackson, Richmond, 5/22/65. Died there of chronic diarrhea 5/27/65. Bur. Hollywood Cem., lot #140. Left widow and 1 child.

ENGLEMAN, FRANCIS M.: Pvt., Co. I. B. Va. 1844? Farmhand, age 16, 1st Dist., Augusta Co. 1860 census. Pvt., Co. E. 93rd Va. Militia 3/62. Enl. Staunton 11/15/64. Present 11/15-12/31/64. Paroled Staunton 5/1/65. Age 22, 5' 5", dark complexion, black hair, dark eyes. Farmer, age 27, 1st Dist., Augusta Co. 1870 census. Died 5 miles south of Staunton 8/30/04. Bur. Mt. Tabor Lutheran Ch. Cem. "A brave and faithful soldier."

ENGLEMAN, JOHN J.: Pvt., Co. I. B. Va. 1839? Farmhand, age 21, Staunton PO, 1860 census. Enl. Staunton 7/16/61 age 23. Ab. detached service 11/61-1/4/62. Present 1/5-4/62. Reenl. 5/1/62. WIA (arm) Port Republic 6/9/62. Ab. wounded through 4/63. Detailed as nurse in hospital Richmond 5/18-25/63. Died of diphtheria at his father's house in Augusta Co. 7/22/63. Bur. St. John's Ch. Cem., near Middlebrook.

ENTSMINGER, JOHN AYERS: Pvt., Co. E. B. Rockbridge Co. 2/16/42. Enl. Staunton 8/1/61. Present 11/61-4/62. Reenl. 5/1/62. AWOL 2 months and 20 days 4/30-10/31/62. Present in arrest 1-2/63. Present 3-12/63. Deserted on the march from camp near Somersville Ford to the Wilderness 5/5/64. NFR. Farmer, age 28, Buffalo Township, Rockbridge Co. 1870 census. Died Alpin, on upper Colliers Creek, Rockbridge Co. 1/7/20. Bur. Collierstown Pres. Ch. Cem.

EPPERLY, A.: Pvt., Co. A. Pension application only record. Living in Montgomery Co. 1888.

ESKRIDGE, JOHN B.: 2ndLt., Co. A. B. Va. 10/3/30. Doctor, Staunton. Enl. Staunton 7/9/61 age 30. Died of consumption Staunton 8/14/61. Bur. Thornrose Cem.

EVANS, WILLIAM H.: Pvt., Co. B. (2nd). B. Shenandoah Co. 3/14/28. Laborer, age 27, Waynesboro, Augusta Co. 1860 census. Enl. Waynesboro 7/15/61. Present 11/61-4/62. Reenl. 5/1/62. AWOL 5/9/62-12/31/63. Dropped as deserter. NFR. Painter, age 38, 6th Dist., Augusta Co. 1870 census. Died Riverheads Dist., Augusta Co. 12/7/88.

EWING, WILLIAM DAVIS.: Assistant Surgeon, F&S. B. Rockbridge Co. 10/30/28. Attended Washington College 49-50, UVa. and Jefferson Medical College, Philadelphia, Pa. Doctor, age 29, Burkes Mill Dist., Augusta Co. 1860 census. Enl. Staunton 8/2/61 as Pvt., Co. G. Present 11/61-2/62, and 3-4/62, detailed as Hospital Steward. Ab. on extra duty 4/30-8/31/62. Ab. on extra duty as Assistant Surgeon 9-12/62. Acting Hospital Steward of Regt. 1/63-1/64. Appointed Hospital Steward by Sec. of War 1/19/64. 5/12/64. Cap. Winchester 9/19/64 rank indicated as Assistant Surgeon. Sent to West Building Hospital, Baltimore. Exchanged 2/2/65. NFR. Doctor, age 35, Mt. Sidney PO, Augusta Co. 1870 census. Doctor and Farmer, Weyer's Cave 1897. Died Weyer's Cave 1/22/02. Bur. Mt. Horeb Pres. Ch. Cem.

FADLEY, DAVID: Pvt., Co. D. B. Va. 1827? Farmhand, age 33, Burkes Mill Dist., Augusta Co. 1860 census. Enl. Staunton 7/16/61. Present 7/16-12/61, detailed as wagoner. Present 3-4/62. AWOL 5/15-6/12/62, fined $10.26. Detailed as wagoner in Div. Train 6/17/62-3/15/64. AWOL 3/16-12/31/64. Dropped as deserter. NFR. Farmhand, age 41, Mt. Sidney PO, Augusta Co. 1870 census.

FAIRBURN, ANDREW JACKSON: Pvt., Co. F. B. Rockingham Co. 11/29/38. Farmhand age 21, Northern Dist., Augusta Co. 1860 census. Enl. Staunton 7/31/61. Ab. sick 11/5-12/61. Present 1-4/62. Reenl. 5/1/62. WIA Cedar Run 8/9/62. Ab. wounded through 4/64. Present 4/30-12/31/64. WIA (right shoulder) Hatcher's Run 2/6/65. Admitted to hospital Richmond 2/10/65. Furloughed for 60 days 2/24/65. NFR. Farmer, age 31, Mt. Sidney PO, Augusta Co. 1870 census. Died Parnassus 5/15/24. Bur. Parnassus Methodist Ch. Cem.

FAUBER, ANDREW S.: Pvt., Co. C. B. Va. 1838? Resident of Union Ch., Augusta Co. Enl. Staunton 7/16/61 age 23. Present 11-12/61. AWOL 2/26-28/62. Present 3-4/62. Reenl. 5/1/62. Fined $11.00 for AWOL plus $6.96 for days absent. Deserted near Gordonsville 7/21/62. NFR.

FAUBER, JOHN WESLEY: Pvt., Co. C. B. Va. 1840? Farmer, age 20, Northern Dist., Augusta Co. 1860 census. Enl. Staunton 7/16/61 age 21. Present 11/61-4/62. Reenl. 5/1/62. WIA (arm) McDowell 5/8/62. Ab. wounded through 8/30/62. Present 8/31/62-4/63. AWOL 6/6/63. Deserted to the enemy New Creek, W. Va. 7/9/63. Took oath and released. Age 23, 6' 3¾", dark complexion, dark eyes, dark hair. Farmer, age 33, Mt Sidney PO, Augusta Co. 1870 census. Farmer, Vesuvius, Rockbridge Co. 1886. D. Staunton 6/16/15. Bur. Thornrose Cem.

FAUVER, JOHN ALEXANDER: 1stLt., Co. F. B. Parnassus, Augusta Co. 9/16/40. Farmhand, age 20, Northern Dist., Augusta Co. 1860 census. Enl. Staunton 7/31/61 as 1stCpl. Present 11/61-4/62. Elected 2ndLt. 5/1/62. Present 5/1-12/31/62. Commanding Co. 12/31/62, and 2/28/63. Elected 1stLt. 8/11/62. Present 6-12/63 and 3-6/64. WIA Carter's Farm near Winchester 7/20/64. Ball was in left shoulder blade for 4 years before being removed. Returned to duty 1/28/65. Commanding Regt. 2/28/65 as Brevet Captain. Present Ft. Stedman, Sailor's Creek, surrendered Appomattox 4/9/65. Farmer, Augusta Co. 4 years. RR Builder in Va. and W. Va. until 1875. Grocer and Orchardist, Staunton, 1910 census. Member, Stonewall Jackson Camp CV, Staunton. Died Staunton 6/1/20. Bur. Thornrose Cem., Brother of Joseph M. Fauver.

FAUVER, JOSEPH MICHAEL: 2ndSgt., Co. F. B. Augusta Co. 9/24/43. Student, age 16, 1st Dist., Augusta Co. 1860 census. Enl. Staunton 7/31/61 as Pvt. Ab. sick 11/5/61-4/62. Reenl. 5/1/62. Present 8/31-11/25/62. Sent to hospital Middletown 11/24/62. Present 1/1-8/31/63. Promoted 2ndSgt. 4/30/63. In hospital Richmond with dysentery 5/3-22/63. Ab. sick 9/10-12/31/63. Present 3-5/64. Ab. sick 6/20-10/31/64. Deserted 12/1/64. NFR. Carpenter, age 33, Mt. Sidney PO, Augusta Co. 1870 census. Farmer, near Staunton 1897. Died Staunton 7/5/31. Bur. Thornrose Cem. Brother of John A. Fauver.

FELLERS, GIDEON: Pvt., Co. A. B. Augusta Co. 1838? Carpenter, New Hope, Augusta Co. Enl. Staunton 7/9/61 age 23. Ab. sick 9/10/61-2/25/62. Present 2/26-4/62. Deserted 6/9/62. Present in arrest 4/30-8/31/63, pay and allowances forfeited for desertion. Deserted 9/24/63. NFR. Died before 5/03.

FENDALL, THOMAS E.: Pvt., Co. E. Not on muster rolls or Federal POW rolls. Cap. date and place unknown. Sent to Point Lookout. Died there date unknown. Bur. in Confederate POW Cem. there.

FIFER, GEORGE W.: 1stCpl., Co. D. B. Sangersville, Augusta Co. 2/24/40. Laborer, age 20, Burkes Mill Dist., Augusta Co. 1860 census. Enl. Staunton 7/16/61 as Pvt. Present 7/16/61-4/62. Reenl. 5/1/62. AWOL 7/18-8/16/62, fined $7.50. Present 8/7-31/62 and 1-4/63. Ab. sick in hospital Lynchburg 8/27-9/22/63. Present 9/23-12/63, and 4/30-12/31/64. Cap. Burkesville 4/6/65. Sent to Point Lookout. Released 6/12/65. 5' 8½", light complexion, brown hair, grey eyes. Farmer, Sangersville, 1877. Died Augusta Co. 12/16/13. Bur. Old Emanuel Ch. Cem., 3 miles west of Mt. Sidney.

FIGGATT, W. R.: Pvt., Co. E. Not on muster rolls. Member, William Watts Camp CV, Lynchburg.

FIREBAUGH, JOHN: Pvt., Co. H. B. Rockbridge Co. 6/5/21. Farmer, age 39, Burkes Mill Dist., Augusta Co. 1860 census. Pvt., Capt. T. Koiner's Co, 32nd Va. Militia 7/28/62 exempt, furnished William Murray as substitute, age 41. Served in Co. F. 4th Va. Bn. Reserves. Enl. Staunton 10/28/64. Present 11-12/64. WIA (right thumb) Hatcher's Run, 2/8/65. Paroled Staunton 5/1/65. Age 44, 5' 11½", dark complexion, dark hair, black eyes. Farmer, age 49, Mt. Sidney PO, Augusta Co. 1870 census. Died New Hope 5/5/89. Bur. Middle River Ch. of the Brethren Cem., near New Hope.

FISHER, ADDISON K.: Pvt., Co. C. B. Rockingham Co. 1822? Miller, age 38, 1st Dist., Staunton 1860 census. 3rdSgt., Co. A, 160th Va. Militia, Staunton 1861. Enl. Staunton 7/16/61 age 38. Ab. on detached service 11/61-4/62. Ab. detailed in QM Dept. Staunton 5/1/62-12/63. Died Staunton before 3/27/64.

FISHER, HENRY: Pvt., Co. B. (2nd). B. Rockingham Co. 1823? Farmhand age 37, Burkes Mill Dist., Augusta Co. 1860 census. Enl. Waynesboro 7/15/61. Present 11-12/61. AWOL 1-2/62. Present 3-4/62. Reenl. 5/1/62. fined $12.00. WIA (hand and thigh) McDowell 5/8/62. AWOL 7/18-9/22/62. Present 9/23-10/31/62. Ab. sick 12/11-31/62. Discharged 2/4/63. Age 42, 5' 10½", fair complexion, blue eyes, dark hair. Farmhand, age 48, Mt. Sidney PO, Augusta Co. 1870 census. Died near Dooms, Augusta Co. 1/30/90.

FISHER, J.: Pvt., Co. K. B. Va. 1819? Cabinetmaker, age 41, Northern Dist., Augusta Co. 1860 census. Not on muster rolls. Died Richmond 4/20/65. Bur. Hollywood Cem. as "J. Fishard."

FISHER, JAMES S.: Pvt., Co. E. B. Va. 1840? Resident, age 20, Rapp's Mill, Rockbridge Co. 1860 census. Enl. Staunton 8/1/61. Ab. on leave 1-12/61. Ab. sick 1-2/62. Present 3-4/62. Reenl. 5/1/62. Cap. Frederick, Md. 9/14/62. Exchange 9/30/62. Ab. on leave 10/1-31/62. Present 1-12/63. Pay stopped for 1 cartridge box $6, bayonet $5.48, shoulder strap with canvas $1.50, bayonet scabbard $2.57 and cap pouch $2.23. WIA (side) near Spotsylvania CH 5/19/64. DOWs.

FISHER, JOHN ALEXANDER: 1stSgt. Co. E. B. Rockbridge Co. 1843? Farmhand age 17, Lexington PO, Rockbridge Co. 1860 census. Served in militia. Enl. Staunton 8/1/61 as 4thSgt. Present 11/61-4/62. Reenl. and promoted 1stSgt. 5/1/62. WIA (right thigh) Cedar Run 8/9/62. Leg amputated below the knee. Discharged from Hospital Charlottesville 10/6/62. Age 20, 5' 7", light complexion, blue eyes, auburn hair. Still in hospital Charlottesville 12/31/64. School teacher, Craigsville, Augusta Co. 1886.

FITZPATRICK, WILLIAM: Pvt., Co. K. Enl. Cedarville 11/8/62. Deserted 11/19/62. NFR.

FIX, WILLIAM J.: Pvt., Co. C. B. Va. 1826? Wagonmaker, age 34, Buffalo Gap, Augusta Co. 1860 census. Enl. Staunton 7/16/61 age 28. Present 11-12/61. AWOL 2/21-28/62. Fined by CM. Present 3-4/62. Reenl. 5/1/62. Present 5/1/62-7/63. Detailed as blacksmith with Ordnance Train 7/1-12/31/63. Cap. Bethesda Ch. 5/30/64. Sent to Point Lookout. Transf. to Elmira. Requested to take oath 3/15/65. "Volunteered 7/16/61 for 12 months. Was conscripted at expiration of enlistment. His wife and five small children are at Buffalo Gap, Augusta Co., Va. and are in destitute circumstances. He desires to go to Frederick City, Md., where he has friends residing, there to make some arrangements to remove his family if practicable, to a place of safety where they have the benefit of his labor and support. Born in Va., 37 years old, was always opposed to the course of the South, wished to take the Oath and go to Pittsburg, Penn." Released 5/13/65. Resident of Staunton, fair complexion, black hair, grey eyes, 5' 8½". Wagonmaker, age 41, Augusta Springs, Augusta Co. 1870 census. Died near Pond Gap. Augusta Co. 1/17/01.

FLEMING, HAY: Rank and Co. not indicated. Not on muster rolls. Paroled Richmond 4/24/65.

FLICK, BENJAMIN F.: Pvt., Co. D. B. Rockingham Co. 6/13/26. Laborer, age 32, Sangersville, Augusta Co. 1860 census. Enl. Staunton 7/16/61. Present 7/16/61-4/62. AWOL 6/7-8/11/62, fined $23.46. Present 8/11-31/62. Deserted 10/29/62. NFR. Farmer, Linville Dist., Rockingham Co. 1885. Died Rockingham Co. 11/4/89. Bur. Linville Cem.

FLICK, WILLIAM M.: Pvt., Co. D. B. Edam, Rockingham Co. 1846? Resident, Sangersville, Augusta Co. Enl. Staunton 7/16/61. Present 7/16-12/31/61. AWOL 2/16-28/62. Left company on a scout near Buffalo Gap 4/23/62 and never returned. Deserted 4/24/62. NFR. Butcher and Farmer, Sangersville. Alive 1911.

FOLEY, EDWARD.: Pvt., Co. F. Pension application only record. Alive 1900.

FOLEY, SAMUEL H.: Pvt., Co. F. B. Rockingham Co. 12/21/36. Carpenter age 23, Burkes Mill Dist., Augusta Co. 1860 census. Enl. Staunton 7/31/61. Present 11-12/61, wagoner. Present 1-4/62. Reenl. 5/1/62. WIA Sharpsburg 9/17/62. Detailed as teamster in Div. Train (ordnance) 11/23/62-12/31/64. Paroled Staunton 4/30/65. Age 29, 5' 1", fair complexion, brown hair, blue eyes. Carpenter, age 30, Mt. Sidney PO, Augusta Co. 1870 census. Died Middle River Dist., Augusta Co. 8/21/13. Bur. Old Salem Lutheran Ch. Cem., 3 miles west of Mt. Sidney.

FORD, JOHN THOMAS: Pvt., Co. B (1st) b. Va. 1839? Farmer, age 21, 6th Dist., Rockbridge Co., 1860 census. Enl. Fairfield 7/10/61. Present until transf. 9/28/61.

FORD, WILLIAM A.: Pvt., Co. E. B. Amherst Co. 6/25/35. Laborer, age 27, Collierstown, Rockbridge Co. 1860 census. Enl. Staunton 8/1/61. Ab. on leave 11-12/61. Ab. sick 1-2/62. Present 3-4/62. Reenl. 5/1/62. AWOL 8 months, 16 days 5/1/62-2/63, and pay deducted. Present 3-5/63. Ab. sick 6/4-10/31/63. Present 11-12/63. Pay stopped for 1 cartridge box $6.00, waist belt canvas $1.50, cap pouch $2.23, lost while AWOL. WIA near Spotsylvania CH 5/19/64. Ab. wounded through 10/31/64. AWOL 11-12/64. NFR. Laborer, age 37, Buffalo Dist., Rockbridge Co. 1870 census. Died Collierstown 11/22/24. Bur. Collierstown Pres. Ch. Cem.

FORD, WILLIAM ALEXANDER: Pvt., Co. B. (1st). Enl. Fairfield 7/10/61. Present until transf. 9/28/61.

FORRER: Pvt. Co. H. Resident, Augusta Co. On postwar roster. Enl. New Market '64. Present Appomattox. Alive 1884.

FORRER, JOHN K.: Pvt., Co. C. B. Pa. 1832? Farmhand, age 28, Northern Dist., Augusta Co. 1860 census. Served in 160th Va. Militia. Enl. Staunton 7/16/61 age 30. Ab. on detail 11-12/61. Discharged by order Sec. of War 1/28/62. Paroled Staunton 5/20/65. Age 34, 5' 11", dark complexion, dark hair, grey eyes. Iron Master, Mossy Creek or Forrer Furnace and Miller. Farmer, Stuart's Draft, 1897. Died Red Hill, Albermarle Co. 7/10/10. Brother of Samuel Forrer.

FORRER, SAMUEL: Pvt., Co. C. B. Page Co. 7/9/38. Iron Worker, Mossy Creek, Augusta Co. Served in 160th Va. Militia. Enl. Staunton 7/16/61 age 23. Ab. on detail 11-12/61. Discharged by order Sec. of War 1/28/62. Paroled Staunton 5/19/65. Age 26, 5' 5", dark complexion, dark hair, dark eyes. Manager of Furnace, Pastures Dist., Augusta Co. 1870 census. Farmer and Miller, Mossy Creek, 1885 and 1897. Director, Rockingham Co. National Bank. Died Mossy Creek 3/1/16. Bur. Mossy Creek Ch. Cem. Brother of John K. Forrer.

FORRER, WILLIAM: Pvt., Co. H. B. Va. 12/20/43. Resident, age 16, 1st Dist., Augusta Co. 1860 census. Enl. Roanoke Co. 10/25/64. Present 11-12/64. Cap. Ft. Stedman 3/25/65. Sent to Point Lookout. Released 6/12/65. 5' 6½", fair complexion, light brown hair, greyish eyes. Supervisor in Iron Works, Mossy Creek, Augusta Co. 1870 census. Died Augusta Co. 9/1/11. Bur. Mt. Vernon Ch. of the Brethren near Ladd, Augusta Co.

FOSTER, DAVID C.: Pvt., Co. K. B. Va. 1835? Laborer, age 25, Bath CH PO, Bath Co., 1860 census. Enl. Shenandoah Mt. 4/9/62. Present 4/9-30/62. WIA (hand) Gaines mill 6/27/62. Ab. wounded through 4/63. AWOL 5/1-12/31/63. Issued clothing 4/21/64. KIA Spotsylvania CH 5/12/64.

FOX, MATHIAS A.: Pvt., Co. D. B. Highland Co. 1835? Resident of Sangersville, Augusta Co. Enl. Staunton 7/16/61. Present 7/16-12/61. Ab. sick in hospital Staunton 1/2-2/62. Present 3-4/62. Reenl. 5/1/62. WIA (finger and left leg) Gaines Mill 6/27/62. Ab. wounded through 8/31/62. Present 1-12/63. WIA (arm) Wilderness 5/6/64. Present 6-10/64. AWOL 11/2-12/31/64. Deserted to the enemy Clarksburg, W. Va. 4/2/65. Took oath and released. Illiterate. Laborer, Sangersville. Died Rockingham Co. 11/25/12. Bur. Otterbine EUB Ch. Cem., on Rt. #613, Rockingham Co.

FRASIER, SANDY: Pvt., Co. F. Enl. Staunton 7/31/61. Ab. sick 11/5-12/31/61. Present 1-4/62. Reenl. 5/1/62. WIA Cedar Run 8/9/62. AWOL 9/62-1/7/63. Present in arrest 1/8-2/63. Present 3/1-7/27/63. AWOL 7/28-8/10/63, pay deducted. Present 9/63-4/64. WIA Spotsylvania CH 5/12/64. WIA (face) Bethesda Ch. 5/30/64. Ab. wounded through 12/31/64. Deserted to the enemy New Creek, W. Va. 3/10/65. Took oath and sent to Ohio. Died before 1900.

FRAZIER, L. F.: Pvt., Co. F. Not on muster rolls. Listed as deserter 2/63. Resident of Shenandoah, Augusta Co. NFR.

FREED, GIDEON DORSEY: Cpl., Co. G. B. Va. 7/28/41. Laborer, Augusta Co. 1843? Enl. Staunton 7/16/61 as Pvt. Present 11/61-4/62. Reenl. and promoted Cpl. 5/1/62. KIA 2nd Manassas 8/29/62, age 19. Bur. Zion Lutheran Ch. Cem. near Fishersville.

FREEMAN, JAMES T.: Pvt., Co. I. B. Augusta Co. 1838? Farmhand, age 22, Glenwood Dist., Rockbridge Co. 1860 census. Enl. Staunton 7/16/61 age 29. Present 11/61-1/62. AWOL 2/26-28/62. Fined $11.00. Present 3-4/62. Reenl. 5/1/62. Deserted 5/9/62. NFR. Farmer, age 33, Fishersville PO, Augusta Co. 1870 census.

FRETWELL, JAMES FRANKLIN: Pvt., Co. B. (2nd). Resident of Waynesboro. Enl. Waynesboro 7/15/61. Present 11/61-2/62. AWOL 3-4/62, fined $9.00. AWOL 6/9/62-12/31/63. Cap. Port Republic 9/26/64. Sent to Point Lookout. Joined U.S. Army 10/18/64. Merchant, New Hope. D. there 2/16/89.

FREY, JOSEPH: 2ndLt., Co. B. Va. 7/30/37. Laborer, Augusta Co. Enl. Staunton 8/2/61 as 4thSgt. Present 11/61-4/62. Reenl. 5/1/62. Present 5/1-10/31/62, and 11-12/62, rank shown as 1stSgt. Present 1-5/63. Elected 2ndLt. 3/15/63. On detached service hunting deserters 6-8/63. Presence or absence not stated 9-12/63. WIA (left arm) Spotsylvania CH 5/12/64. WIA Winchester 9/19/64. DOWs 9/23/64, age 27. Bur. Old Salem Lutheran Ch. Cem., 3 miles west of Mt. Sidney.

FREY, WILLIAM HENRY: 4thCpl., Co. B. (2nd). B. Waynesboro, Augusta Co. 1/11/43. Student, age 17, Northern Dist., Augusta Co. 1860 census. Attended Roanoke College. Enl. Staunton 7/16/61 age 18 as Pvt. Present 11/61-4/62. Promoted 4thCpl. 3/25/62. Reenl. 5/1/62. WIA (breast) McDowell 5/8/62. DOWs 5/12/62, age 19, 5'9", fair complexion, dark eyes, dark hair, farmer. Bur. Old Salem Lutheran Ch. Cem., 3 miles west of Mt. Sidney. "A young man of rare intellectual abilities, fine social qualities, excellent and moral character, a brilliant student, and a good and faithful soldier."

FRY, NEWTON ASBERRY: Pvt., Co. B (2nd) B. Augusta Co. 3/25/44. Student, age 16, Burkes Mill Dist., Augusta Co. 1860 census. Enl. West View 5/1/62. WIA (breast and arm) McDowell 5/8/62. Arm amputated. Retired to Invalid Corps 4/16/64. Merchant, age 26, Mt. Sidney PO, Augusta Co. 1870 census. Merchant, Waynesboro. Died there 6/19/85. Bur. Waynesboro Old Cem.

FRY, WILLIAM HARVEY: 3rdCpl., Co. F. B. Augusta Co. 3/5/30. Cooper, age 30, Northern Dist., Augusta Co. 1860 census. Enl. Staunton 7/31/61. Present 11/61-4/62. Reenl. 5/1/62. WIA (breast and thigh) McDowell 5/8/62. DOWs 5/13/62. Bur. St. Michael's Ch. Cem., Augusta Co.

FULTZ, GEORGE W.: Pvt., Co. C. B. Augusta Co. 1838? Laborer, age 22, Burkes Mill Dist., Augusta Co. 1860 census. Enl. Staunton 7/16/61 age 23. Present 11/61-1/62. AWOL 2/26-28/62. Ab. sick 4/21-30/62. Reenl. 5/1/62. KIA Port Republic 6/9/62.

FULWIDER, JAMES: Pvt., Co. I. B. Va. 6/18/37. Farmhand, age 23, 1st Dist., Augusta Co. 1860 census. Enl. Staunton 7/16/61 age 22. Present 11/61-4/62. Reenl. 5/1/62. WIA (thigh) McDowell 5/8/62. Ab. wounded through 4/63. Detailed in American (Hotel) Hospital Staunton unfit for field service 4/30-10/30/63. WIA (left leg) Mine Run 11/27/63. Returned to duty 12/9/63. Detailed Regt'l Teamster 12/26-31/63. Detailed as Brigade Teamster 1/27-12/31/64. Transf. to Supply Train 6/27/64. Paroled Staunton 5/16/65. Age 27, 5'6", dark complexion, dark hair, dark eyes. Farmhand, age 33, Walker's Ceek Dist., Rockbridge Co. 1870 census. Farmer, Brownsburg, Rockbridge Co. 1884. Died 5/19/08. Bur. Bethel Pres. Ch. Cem., near Middlebrook.

FURR, EDWARD M.: 4thSgt. Co. F. B. Augusta Co. 3/7/41. Farmhand, age 19, Burkes Mill Dist., Augusta Co. 1860 census. Enl. Staunton 7/31/61. Present 11/61-4/62. Reenl. 5/1/62. Present 8/31/62-2/63. Ab. on leave 8/21-9/6/63, rank shown as 2ndCpl. Present 9/7-12/63. Present 3-4/64. WIA (thigh) Bethesda Ch. 5/30/64. Promoted 4thSgt. 9/1/64. Sent to hospital 9/30/64. Ab. sick through 10/31/64. Present 11-12/64. Paroled Staunton 5/11/65. Age 24, 5'8", fair complexion, dark hair, dark eyes. Farmhand, age 30, Mt. Sidney PO, Augusta Co. 1870 census. Farmer, Rolla, 1897. Died Rolla Mills, Augusta Co. 1/3/20. Bur. Laurel Hill Baptist Ch. Cem.

GAINS, HENRY H.: Pvt., Co. C. B. Va. 1829? Laborer, age 31, Waynesboro, Augusta Co. 1860 census. Enl. Staunton 7/16/61 age 33. Present 11/61-4/62. Reenl. 5/1/62. AWOL 7/21-12/25/62. Ab. in arrest 12/26/62-2/28/63. Ab. in arrest by sentence of CM 3/7-4/63 and fined $47.60. Present 4/30-12/31/63. WIA (right knee) Spotsylvania 5/12/64. DOWs in hospital Richmond 6/15/64. Possessions 1 shirt value $2.50 and 1 pr. socks value 50¢. Bur. Hollywood Cem.

GALT, WILLIAM "Willie": Adjutant, F&S. B. Va. 1843? Resident, "Glenarnon", Columbia, Va. Attended VMI 60-61. Drillmaster, Richmond 4-6/61. Enl. Staunton 7/9/61 as Pvt., Co. A. Elected 2ndLt. 9/19/61. Resigned 1/14/62, because of ill health. Served as Adjutant of Post Staunton under Col. Baldwin. 11-12/61. Reentered VMI and present with them during McDowell campaign. Reelected 2ndLt. 5/62. WIA (wrist) Port Republic 6/9/62. Returned to duty and promoted 1stLt. 10/2/62. Promoted Adjutant 12/6/62. Present 2-12/63. Ab. sick 4/64. Present 5/2/64 until WIA (leg) Winchester 9/19/64. DOWs at Winchester 9/29/64, age 23. "He was worth to the Army a hundred men."

GARBER, DAVID W.: Pvt., Co. G. B. Augusta Co. 8/10/41. Farmhand, age 19, Burkes Mill Dist., Augusta Co. 1860 census. Enl. Staunton 8/2/61. Present 11/61-4/62. Furnished substitute and discharged Winchester 5/27/62, age 22, 5'11", auburn hair, blue eyes, fair complexion, NFR. Farmer, New Hope, 1866.

GARBER, EDWARD VALENTINE: Capt., Co. A. B. Augusta Co. 5/26/36. Mechanic. Resident of Mo. Returned when Va. seceded. Enl. Staunton 7/9/61 as 1stLt., age 25. Present 11-12/61, acting Adj. of Regt. since 12/26/61. Present 1-2/62, relieved as acting Adj. 1/20/62. Commanding Co. 3-4/62. Elected Capt. 5/1/62. Present Harrisonburg 6/6/62. KIA 2nd Manassas 8/29/62, age 25. "While leading his company in a charge against the enemy...in all relations of life he uniformly bore himself with rare and undeviating faultlessness. He aproved himself to his comrades brave, generous, self-sacrificing. He encountered all the vicissitudes of the trying service in which he was engaged with unwavering fortitude. He discharged all the duties of this office with signal ability and fidelity: he met all the difficulties with unshaken equanimity." Bur. Thornrose Cem.

GARDNER, FRANKLIN: Pvt., Co. B (2nd). B. 1840? Cooper, Waynesboro. Enl. Waynesboro 7/15/61. Present 11/61-4/62, detailed as teamster. AWOL 5/20/62-12/31/63. Dropped as deserter. Deserted to the enemy Charleston, W. Va. 3/13/65. Took oath and sent north. Age 25, 5'6", fair complexion, grey eyes, fair hair.

GARDNER, JOHN B.: 3rdSgt., Co. B. (2nd). Laborer, age 35, 1st Dist., Augusta Co. 1860 census. Mexican War Veteran. Enl. Waynesboro 7/15/61 as 3rdCpl. Present 11/61-4/62. Reenl. 5/1/62. WIA Cross Keys 6/8/62. Present 6-10/62., promoted 5thSgt. 7/15/62, and 4th Sgt. 10/2/62. Present 11/62-12/63. Promoted 3rd Sgt. 3/63. WIA (shoulder) Spotsylvania CH 5/12/64. Returned to duty 5/25/64. WIA Winchester 9/19/64. Returned to duty 10/25/64. Present 11-12/64. Cap. Ft. Stedman 3/25/65. Sent to Point Lookout. Released 6/27/65. 5' 7", fair complexion, auburn hair, blue eyes. Lived in Albemarle Co. 1920.

GARRISON, JOHN W.: Pvt., Co. K. B. Augusta Co. 1831? Shoemaker, Green Valley, Bath Co. Enl. Shenandoah Mt. 4/9/62. AWOL 4/22/62-2/63. Present in arrest 3-4/63. Cap. Gettysburg 7/3/63. Sent to Ft. Delaware. Died of dysentery there 10/9/63. Bur. Finn's Point National Cem., N. J. Also listed Co. G. 18th Va. Cav.

GAULDING, JAMES HENRY: Pvt., Co. B (2nd). B. Albemarle Co. 1839? Laborer. Enl. Waynesboro 7/25/61. Ab. sick 12/11/61. Died of Fever at home in Albemarle Co. 1/10/62. Age 23, 5' 10½", dark complexion, blue eyes, dark hair.

GAY, D. H.: Pvt., Co. B (2nd). Not on muster rolls. Listed as deserter 2/63.

GAYLOR, JOHN R.: Pvt., Co. H. B. 1837? Enl. Staunton 7/23/61 age 24. Present 11/61-3/62. Ab. sick 4/4/-30/62. Reenl. 5/1/62. AWOL 6/17-10/6/62. Ab. sick 11/23/62-12/31/63. Deserted Highland Co. 5/29/64. Farmer, age 31, Craigsville PO. Augusta Co. 1870 census. Alive 1885.

GAYLOR, WILLIAM: Pvt., Co. B (1st). Farmhand, age 20, 5th Dist., Rockbridge Co., 1860 census. Enl. Fairfield 7/16/61. Present until transf. 9/28/61.

GENTRY, JAMES H.: Pvt., Co. B (2nd). B. Nelson Co. 7/4/26. Cabinetmaker, age 32, Burkes Mill Dist., Augusta Co. 1860 census. Enl. Waynesboro 7/15/61. Ab. sick 11/61-2/62. Reenl. 5/1/62. Present 3-4/62. Present 11/62-10/64. detailed Regt'l QM Dept., as wagonmaster since 11/1/62. Present 11-12/64. Surrendered Appomattox 4/9/65. Cabinetmaker and Lumberman, Mt. Sidney PO, Augusta Co. 1870 census. Farmer and Cabinetmaker. Died near Cave Station, Augusta Co. 2/5/02. Bur. Melanchthon Chapel Cem., near Weyer's Cave.

GISINGER, JOHN TIMOTHY DWIGHT: Pvt., Co. B (1st). B. Berkley or Rockingham Co. 1844? Attended Brownsburg Academy. Resident of Brownsburg, Rockbridge Co. Enl. Farifield 7/10/61. Present until transf. 9/28/61.

GIBSON, ABRAHAM: Pvt., Co. A. Not on muster rolls. Enl. Staunton 1861. WIA (hip and arm) Winchester 7/19/64. Living Charlottesville 1908. Died there 9/6/10.

GIBSON, BURWELL: Pvt., Co. H. B. Louisa Co. 1808? Laborer, age 52, Burkes Mill Dist., Augusta Co. 1860 census. Enl Staunton 7/23/61 age 55. Present 11/61-2/62. AWOL 2/9-18/62. Discharged for overage 12/9/62. Age 57, 6', fair complexion, grey eyes, light hair, illiterate. Farmhand, age 63, Mt. Sidney PO, Augusta Co. 1870 census. Died Staunton 5/30/91. Bur. Thornrose Cem.

GIBSON, JOHN ST. P.: Surgeon, F&S. B. Culpeper Co. 5/11/32. Moved to Md. 1842. Graduate UMd. in medicine. Doctor, Grafton and Preston Co. until 7/62. Entered Confederate service as Surgeon in hospitals Aldie and Winchester, 1862. Appointed Assistant Surgeon 52nd Va. 12/4/62. Present 2-4/63. Promoted Surgeon 4/18/63. Present 5-12/63, and 1-10/64. Promoted Surgeon Pegram's Brig. Fall 1864. Surrendered Appomattox 4/9/65. Doctor, Waynesboro 1865-77. Doctor, Staunton 1878-98. Surgeon, Stonewall Jackson Camp CV, Staunton. Died Staunton 10/30/98. Bur. Thornrose Cem.

GIBSON, JOHN W.: Pvt., Co. C. B. Orange Co. 10/22/42. Farmhand, age 18, 1st Dist., Augusta Co. 1860 census. Enl. Staunton 7/16/61 age 19. Present 11/61-4/62. Reenl. 5/1/62. Present 5/1-8/31/62. WIA (right hip) Fredericksburg 12/13/62. Ab. wounded through 10/63. Present 11-12/63. Detailed for conscript duty 1/7-10/31/64. Present 11-12/64. WIA (eye) Ft. Stedman 3/25/65. Surrendered Appomattox 4/9/65. Illiterate. Farmhand, age 27, Mt. Sidney PO, Augusta Co. 1870 census. Member, Stonewall Jackson Camp CV, Staunton. Carpenter, age 68, Staunton, 1910 census. Died Staunton 10/4/12. Bur. Thornrose Cem.

GILBERT, ANDREW F.: Pvt., Co. E. B. Va. 1842? Farmhand, age 18, Lexington PO, Rockbridge Co. 1860 census. Enl. Staunton 8/1/61. Present 11/61-4/62. Reenl. 5/1/62. WIA (hand) Gaines Mill 6/27/62. Ab. on leave 7/1-10/31/62. Present 1-4/63. Ab. sick 7/20/63-10/21/64. Present 11-12/64, forfeit and clothing allowance for 3/1-12/15/64 while AWOL and 1 month pay in addition, sentence of Regt'l CM. Deserted to the Army of the Potomac near Petersburg 3/8/65. Took oath and transportation furnished to Reading, Pa. Transportation furnished to Rockbridge Co. 7/15/65. Laborer, age 28, Natural Bridge Dist., Rockbridge Co. 1870 census. Saw Mill Engineer, age 66, Buffalo Dist., Rockbridge Co. 1910 census. Brother of James M. Gilbert.

GILBERT, JAMES: Pvt., Co. K. B. Va. 1833. On postwar roster. Died near Starr Chapel, Bath Co. 4/1/24.

GILBERT, JAMES M.: Pvt., Co. E. B. Va. 1844? Farmhand, age 16, Lexington PO, Rockbridge Co. 1860 census. Enl. Staunton 8/1/61. Present 11/61-4/62. Present 5/1-8/7/62. Ab. sick 8/8/62-2/63. Present 3-4/63. WIA Gettysburg 7/3/63. DOWs 7/5/63, age 18. "A good soldier." Brother of Andrew F. Gilbert.

GILKESON, JOHN A.: Pvt., Co. H. B. Augusta Co. 5/6/24. Farmhand, age 36, 1st Dist., Augusta Co. 1860 census. Pvt., Co. B. 32nd Va. Militia requesting discharge 3/15/62. Enl. Staunton 10/25/64. Present 11-12/64. Paroled Staunton 5/15/65. Age 41, 5' 10", dark complexion, black hair, blue eyes. Farmer, Barterbrook, Augusta Co. 1897. Died there 12/30/91. Bur. Tinkling Spring Pres. Ch. Cem. near Fishersville.

GILLESPIE, JOHN W.: Pvt., Co. K. B. Bath Co. 1844? Farmhand, age 16, Bath CH PO, Bath Co. 1860 census. Enl. Valley Mills 4/25/62. Present 4/25-30/62. Deserted 5/17/62. Enl. Co. G. 11th Va. Cav. 5/26/62 Millboro Springs. In arrest 1-2/63. NFR. Resident Green Valley, Bath Co.

GILLETT, ANDREW W.: Pvt., Co. K. B. Pocahontas Co. 1839? Farmhand, age 21, Cleeks Mill PO, Bath Co. 1860 census. Pvt., Capt. Hamilton's Co. 81st Va. Militia. Enl. Shenandoah Mt. 4/9/62. Present 4/9-30/62, and 1-4/63. WIA (shell wound left foot; lost 2 toes) Fredericksburg 5/5/63. Ab. wounded through 12/31/64. NFR. Farmer, age 31, Warm Springs Dist., Bath Co. 1870 census. Died Flood, Bath Co. 3/18/21.

GILLETT, JAMES: 1stSgt., Co. K. B. Bath Co. 1/4/33. Laborer, age 27, Bath CH PO, Bath Co. 1860 census. Enl. Shenandoah Mt. 4/9/62, as Pvt., Present 1-12/62, rank shown as 1stSgt. Cap. Winchester 9/19/64. Sent to Point Lookout. Exchange 3/15/65. NFR. Carpenter, age 37, Warm Srings Dist., Bath Co. 1870 census. Carpetner, Cleeks Mill, 1897. Died Warm Springs Dist., 4/9/24. Bur. Starr Chapel or Lower Cleek Cem., Bath Co. Brother of John W. and William R. Gillett.

GILLETT, JOHN WESLEY: Pvt., Co. K. B. Bath Co. 8/30/36. Pvt., Capt. Hamilton's Co., 81st Va. Militia. Enl. Shenandoah Mt. 4/9/62. Present 4/9-30/62. Detailed work in nitre caves, Bath Co. 8/9/62-12/64. WIA (spine) Ft. Stedman 3/25/65. Cap. in hospital Richmond 4/3/65. Sent to Newport News. Released 6/15/65. 5' 10", fair complexion, light hair, grey eyes. Farmer, age 33, Warm Springs Dist., Bath Co. 1870 census. Died Hively, Bath Co. 1/22/89. Bur. Starr Chapel or Lower Cleek Cem., Bath Co. Brother of James and William R. Gillett.

GILLETT, WILLIAM ROLFE: Capt., Co. K. B. Bath Co. 6/10/31. Enl. Shenandoah Mt. 4/9/62 as Pvt. Present 4/9-30/62. Cap. date and place unknown. Exchanged 9/62. Elected 1stLt. 9/22/62. Present 10-12/62. Commanding Co. 1-12/63. Elected Capt. 2/23/63. WIA (right hand) Spotsylvania CH 5/12/64. WIA and cap. Winchester 9/19/64. Sent to Ft. Delaware. Released 6/15/65. 5' 10", light complexion, light hair, grey eyes. Farmer, age 39, Warm Springs Dist., Bath Co. 1870 census. Farmer, Cleeks Mill, 1897. Died Warm Springs Dist. 2/2/17. Bur. Starr Chapel or Lower Cleeks Cem. Brother of James and John W. Gillett. LtCol. Watkins sworn testimony before a board of officers considering him for promotion to Captain. "I consider him one of the best administrative officers in the Regt...he manages his company as well as any other officer...He is very faithful and attentive to all duty and his company is as orderly and under as good discipline as any...other....I have observed him in battle on several occassions and he always behaves well himself and manages his company well. He has always been prompt in obeying orders and in enforcing their obedience in his subordinates...Lt. Gillett has a tolerable good knowledge of tactics..."

GILLILAND, JAMES M.: Pvt., Co. K. B. Alleghany Co. 1835? Laborer, age 25, Green Valley Dist., Bath Co. 1860 census. Enl. Sheperdstown 9/25/62. Present 1-4/63. Ab. sick 7/25/63. Died Edinburg 7/30/63. Left widow and 2 children.

GILMER, JAMES.: Pvt., Co. A. B. Rockingham Co. 1829? Laborer, age 31, Northern Dist., Augusta Co. 1860 census. Enl. Capt. S. M. Crawford's Co. L, 5th Va. Inf. 4/19/61, Disbanded 6/61. Enl. Staunton 7/16/61. Ab. detached service Staunton 12/28/61-1/6/62. Present 1/7-4/62. Reenl. 5/1/62. WIA (breast and arm) Port Republic 6/9/62. DOWs in Albemarle Co. 6/16 or 21/62. Left widow and 1 child in Mt. Sidney.

GILMORE, ANDREW JACKSON: Pvt., Co. E. B. Va. 1838. Mechanic, age 30, Saunders Store Dist., Rockbridge Co. 1860 census. Enl. Staunton 8/1/61. Present 11/61-4/62. Reenl. 5/1/62. WIA Gaines Mill 6/27/62. WIA (foot) Sharpsburg 9/17/62. Ab. wounded through 2/63. Present 3-4/63. Trans. Co. C, 1st Va. Cav. 5/15/63. Laborer, age 41, Natural Bridge Dist., Rockbridge Co. 1870 census. Died near Broad Creek Ch., Rockbridge Co. 12/29/15. Bur. Natural Bridge Baptist Ch. Cem.

GILMORE, SAMUEL DAVIDSON: 1stSgt., Co. E. B. Kerr's Creek, Rockbridge Co. 3/1/36. Farmhand, age 24, 5th Dist., Rockbridge Co. 1860 census. Enl. Staunton 8/1/61 as Pvt. Present 11/61-4/62. Reenl. and promoted 4thSgt. 5/1/62. WIA (shoulder) Gaines Mill 6/27/62. Present 7-12/62, and 1-2/63, promoted 1stSgt. 1/1/63. Present 3-7/63. Ab. on detail in Rockbridge Co. arresting deserters 7/28-11/18/63. Listed as Farmer, age 28, Kerr's Creek, 2/64., fair complexion, blue eyes, light hair, 6'. WIA Winchester 9/19/64. Ab. wounded through 12/31/64. Surrendered Appomattox 4/9/65. Farmer, age 35, South River Dist., Rockbrige Co. 1870 census. Teacher, Merchant, Accountant. Died Staunton 3/31/02. Bur. Stonewall Jackson Cem., Lexington.

GIPSON, ABRAM: Pvt., Co. A. Enl. Winchester 9/12/64. Present 9/12-10/31/64. AWOL 12/8-31/64. Deserters notice dated 2/20/65 to be found in Albemarle Co. near Charlottesville. NFR.

GLADWELL, ANDREW J.: Pvt., Co. C. B. Augusta Co. 1827? Laborer, age 33, 1st Dist., Augusta Co. 1860 census. Enl. Staunton 7/16/61 age 30. Present 11/61-4/62. AWOL 14 days fined $4.03. Reenl. 5/1/62. Ab. sick 6/6/62-2/28/63. Present 3-4/63. Ab. sick 6/4-10/63. Deserted to the enemy Clarksburg, W. Va. 10/13/63. Took oath and sent north. Died Augusta Co. 11/30/72. Bur. Hebron Pres. Ch. Cem., near New Hope.

GLASS, DAVID W.: Pvt., Co. H. B. Augusta Co. 1838? Farmhand, age 18, 1st Dist., Augusta Co. 1860 census. Enl. Staunton 7/23/61 age 19. Died in hospital Staunton 12/25/61, cause unknown. Bur. Thornrose Cem.

GLENN, JAMES MADISON: Pvt., Co. E. B. Va. 1832? Farmer, age 28, Lexington Dist., Rockbridge Co. 1860 census. Enl. Staunton 8/1/61. Present 11/61-2/62. Ab. sick 2/25-4/62. Reenl. 5/1/62. WIA (arm) McDowell 5/8/62. Ab. wounded through 12/31/64. NFR. Member Stonewall Jackson Camp CV, Staunton. Killed by train 3 miles east of Staunton 4/11/94. Bur. Croft's Cem., Augusta Co.

GLENN, JOSEPH E.: Pvt., Co. E. B. Rockbridge Co. 6/31. Wagonmaker, Rockbridge Co. Enl. Staunton 8/1/61. Present 11/61-4/62. Reenl. 5/1/62. Present 5/1-9/62. WIA (hand) Sharpsburg 9/17/62. Present 10/1-31/62, and 1-12/63. WIA and cap. Carter's Farm near Winchester 7/21/64. Sent to Point Lookout. Died there of typhoid fever 7/64. Bur. in Confederate POW Cem. there.

GLENN, ROBERT J.: Pvt., Co. E. Not on muster rolls. Tombstone only record. Bur. Pine River Cem., Bayfield, Colo.

GOCHENOUR, MARTIN J.: Pvt., Co. B. Va. 1/20/40. Farmhand, age 20, Burkes Mill Dist., Augusta Co. 1860 census. Enl. Staunton 7/16/61 age 21. Ab. sick 12/15/61-4/62. Reenl. 5/1/62. Present 5/1/62-2/28/63. AWOL 3/25-12/31/63. Dropped as deserter. NFR. Farmer, age 28, Mt. Sidney PO, Augusta Co. 1870 census. Farmer, age 70, Walker's Creek Dist., Rockbridge Co. 1910 census. Died Rockbridge Co. 4/13/16. Bur. Goshen Baptist Ch. Cem.

GOINS, JOSEPH D.: Pvt., Co. F. B. Va. 1844? Laborer, age 16, 1st Dist., Augusta Co. 1860 census. Enl. Staunton 7/31/61. Present 11/61-3/62. Ab. sick 4/4-30/62. Discharged 10/16/62 under age. NFR.

GOLLADAY, ABRAHAM K.: Pvt., Co. A. B. Augusta Co. 1823? Laborer age 37, 1st Dist., Augusta Co. 1860 census. Enl. Staunton 7/9/61 age 36. Present 11/61-4/62. Fined $7.00 by sentence of CM for AWOL. Reenl. 5/1/62. Ab. sick 6/18-12/31/62. Present 1-6/63. AWOL 7/6-12/31/63. Dropped as deserter. NFR. Died Augusta Co. 12/24/67. Brother of Jacob F. Golladay.

GOLLADAY, JACOB F.: Pvt., Co. I. B. Augusta Co. 1839. Laborer, age 23, Greenville, Augusta Co. 1860 census. Enl. Staunton 7/16/61 age 23. Present 11/61-4/62. Reenl. 5/1/62. WIA (knee) 2nd Manassas 8/29/62. Leg amputated. Ab. wounded through 12/31/64. NFR. Lumberman, Buckingham Co. postwar. Brother of Abraham K. Golladay.

GONGUER, JAMES P.: Pvt., Co. A. B. Va. 1838. Farmhand, age 18, Burkes Mill Dist., Augusta Co. 1860 census. Enl. Staunton 7/9/61 age 18. Present 11/61-4/62. Reenl. 5/1/62. Present 8/31/62-12/63, and 4/30-5/64. WIA (shell wound below the knee) Spotsylvania 5/12/64. Present 6-12/64. Paroled Staunton 5/1/65. Age 22, 5' 11", light complexion, light hair, grey eyes. Alive 1903.

GOODE, PETER: Pvt., Co. D. B. Augusta Co. 1825? Laborer, age 35, Burkes Mill Dist., Augusta Co. 1860 census. Enl. Staunton 7/16/61. Present 7/16/61-4/62. WIA (foot) 2nd Manassas 8/27/62. DOWs Middletown 9/27/62. Left widow and 3 children.

GOODNIGHT, JOHN B.: Pvt., Co. I. B. Augusta Co. 1827? Farmer, age 33, 1st Dist., Augusta Co. 1860 census. Pvt., Co. E, 93rd Va. Militia. Enl. Staunton 7/16/61 age 33. Present 11/61-4/62. Reenl. 5/1/62. WIA (arm) Gaines Mill 6/27/62. DOWs Greenwood hospital 7/9/62. Age 35, 5' 8", dark complexion, black hair, hazel eyes. Bur. St. Johns Ch. Cem., near Middlebrook.

GOOLSBY, JAMES A.: Pvt., Co. B (1st). B. Rockbridge Co. 1828? Carpenter, age 32, South River Dist., Rockbridge Co. 1860 census. Enl. Fairfield 7/10/61. Ab. sick in hospital when transf. 9/28/61.

GOOLSBY, WILLIAM CYRUS: Pvt., Co. B (1st). B. Amherst Co. 1841? Carpenter, age 19, South River Dist., Rockbridge Co. 1860 census. Enl. Fairfield 7/10/61. Present until transf. 9/28/61.

GORDON, JAMES W.: 3rdSgt., Co. K. B. Rockbridge Co. 1836? Farmer, age 24, Walker's Creek Dist., Rockbridge Co. 1860 census. Enl. Shenandoah Mt. 4/9/62 as Pvt. Present 4/9-30/62, and 1-4/63. Promoted 3rdSgt. Ab. on leave 8/23-31/63. Present 9-12/63. KIA Bethesda Ch. 5/30/64.

GORDON, ROBERT: Pvt., Co. F. B. Rockbridge Co. 1830? Laborer, age 30, Burkes Mill Dist., Augusta Co, 1860 census. Enl. Staunton 7/31/61. Married 9/17/61. Died 12/27/61 Alleghany Mt., cause not stated. Left widow.

GORDON, THOMAS: Pvt., Co. B (1st). Died of disease in 1861 on post war roster.

GORMAN, JOHN: Pvt., Co. E. Enl. Co. K, 11th Va. 7/24/61. WIA 5/5/62 Williamsburg and disabled. NFR. Listed in Co. E, 52nd Va. 2/26/65. NFR.

GRABILL, BENJAMIN H.: Pvt., Co. C. B. 1842? Farmer, Augusta Co. Enl. Staunton 7/16/61 age 19. Present 11/61-1/62. Ab. sick 2/5-28/62. Present 3-4/62. Reenl. 5/1/62. Present 5/1/62-4/63. AWOL 6/6-7/63. Deserted to the enemy New Creek, W. Va. 7/20/63. Sent to Camp Chase, age 20, 5' 9½", dark complexion, dark eyes, black hair. Enl. U.S. Navy 7/20/64.

GRABILL, CHRISTIAN: Pvt., Co. D. B. Rockingham Co. 1826? Mechanic. Enl. Staunton 7/16/61. Present 7/16/61-1/62. Died of fever Camp Alleghany 2/6/62. Age 35, 5' 11", fair complexion, dark hair, dark eyes. Left widow and 1 child.

GRAHAM, WILLIAM: Pvt., Co A. B. Baltimore, Md. 1840? Shoemaker. Enl. Staunton 4/63. Deserted to the enemy Beverly, W. Va. 10/18/64. Took oath and released. Age 24, 5' 8", light complexion, blue eyes, light hair.

GRADSTAFF, ISAAC M.: Pvt., Co. G. B. 1843? Enl. Fredericksburg 3/22/63. Detailed as teamster with Div. Train 3/22/63-12/31/64. Paroled Edinburg 5/4/65. Age 22, 6', fair complexion, light hair, grey eyes.

GRANT, JEREMIAH C. "Jerry".: Pvt., Co. F. B. Augusta Co. 1833. Laborer, age 27, Burkes Mill Dist., Augusta Co. 1860 census. Enl. Staunton 7/31/61. Present 11/61-4/62. Reenl. 5/1/62. AWOL 5/9-9/62. Deserted to the enemy Oakland, Md. 9/7/62. Took oath and released. Age 27, 5' 8", dark complexion, grey eyes, dark hair, resident of Staunton. Farmer, Augusta Co. 1867. Died Riverheads Dist., Augusta Co. 12/24/14. Bur. Alms House Cem. Augusta Co.

GRANT, WILLIAM H.: Pvt., Co. E. Not on muster rolls. Enl. 1861. Served 4 years. NFR.

GRASS, WILLIAM E.: Pvt., Co. B. (2nd). B. Augusta Co. 1825. Blacksmith, age 35, 1st Dist.,Augusta Co. 1860 census. Enl. Waynesboro 7/15/61. Present 11/61-4/62, detailed as Brigade Blacksmith. Ab. sick 8/10-10/62. Discharged 10/21/62 over age of conscription. Age 42, 5' 7", fair complexion, dark hair, dark eyes. Died Sherando, Augusta Co. 9/12/89.

GRAY, DAVID H.: Pvt., Co. B (2nd). Enl. Waynesboro 7/15/61. Ab. sick 11/61-2/62. Present 3-4/62. AWOL 5/17-6/25/62 and 7/18-10/16/62. AWOL 11/16/62-12/31/63. Dropped as deserter. NFR.

GREAVER, JACOB S.: 1stSgt., Co. H. B. Augusta Co. 1840? Farmhand, age 20, 1st Dist., Augusta Co. 1860 census. Pvt., Co. E. 93rd Va. Militia. Enl. Staunton 7/23/61 age 21 as Pvt. Present 11-12/61, promoted 3rdSgt. 11/23/61. Present 1-4/62. Reenl. and promoted 1stSgt. 5/1/62. KIA Port Republic 6/9/62. Age 22, 5' 11", dark complexion, black eyes, black hair. Bur. St. John's Ch. Cem., near Middlebrook.

GREGORY, THOMAS A.: Pvt., Co. I. B. Va. 1833? Shoemaker, age 27, Walker's Creek Dist., Rockbridge Co. 1860 census. Detailed from militia as shoemaker, Staunton Shoe Factory 3/62. Enl. Staunton 11/18/64. AWOL 11/19-12/64. Deserted to the enemy Buckhannon, W. Va. 12/10/64. Took oath and released.

GREINER, WILLIAM H.: Pvt., Co H. Va. 1830? Carpenter, age 30, 1st Dist., Augusta Co. 1860 census. Pvt., Co. F, 160th Va. Militia 3/62 exempt for sore leg, age 33. Enl. Staunton 10/29/64. Present 11-12/64. Cap. Ft. Stedman 3/25/65. Sent to Washington, D.C. Transf. to Point Lookout. Released 5/12/65. Farmer, age 41, 6th Dist., Augusta Co. 1870 census..

GREGER, JOSEPHUS: Pvt., Co. G. B. 1841? Enl. Staunton 8/2/61. Present 11/61-12/31/63. WIA (left foot) Spotsylvania CH 5/12/64. WIA (foot) Bethesda Ch. 5/30/64. Ab. wounded through 12/31/64. Paroled Staunton 5/1/65. Age 24, 5' 6", fair complexion, fair hair, blue eyes.

GRIFFIN, LEWIS ANDREW: Pvt., Co. B (1st). B. Rockbridge Co. 1/6/23. Farmhand, age 35, South River Dist., Rockbridge Co. 1860 census. Enl. Fairfield 7/10/61. Present until transf. 9/28/61.

GRIFFITH, WILLIAM H.: Pvt., Co. B. 1822? Laborer, age 38, 7th Dist., Rockbridge Co. 1860 census. Enl. Staunton 8/1/61. Present 11/61-4/62. Reenl. 5/1/62. AWOL 5/1-10/31/62. Present 1-4/63. Ab. sick 6/4-8/31/63. Present 9-12/63. Cap. Keezletown 9/26/64. Sent to Point Lookout. Exchanged 3/17/65. Present Camp Lee, Richmond 3/19/65. NFR. Admitted Lee Camp Soldiers Home, Richmond 3/1/94 from Rockbridge Co. Died there 6/11/08. Bur. Hollywood Cem.

GRIM, ELIJAH PIPES: Pvt., Co. I. B. Augusta Co. 6/18/42. Potter, age 21, Staunton, 1860 census. Enl. Staunton 7/16/61 age 23. Ab. sick 11/61-2/62. Present 3-4/62. Reenl. 5/1/62. AWOL 7/19/62-8/31/63. Dropped as deserter. "Supposed to be in 'Churchville Cavalry'", 14th Va. Cav. Farmer, Augusta Co. 1866. Died Augusta Co. 12/4/08. Bur. McKinley Baptist Ch. Cem., Augusta Co.

GROOMS, ARTHUR T.: Pvt., Co. G. B. Mt. Sidney, Augusta Co. 1837? Laborer, age 23, Burkes Mill Dist., Augusta Co. 1860 census. Enl. Staunton 8/1/61. Ab. on leave 4/30-8/31/62. Present 9/1-11/23/62. AWOL 11/24-12/18/62. Present in arrest 12/19/62-2/63. Fined 2 months and 24 days pay by sentence of CM. Present 3-4/63. WIA (left arm) Gettysburg 7/3/63. Arm amputated above the elbow. Cap. 7/5/63. Sent to Decamp Gen. Hospital, David's Island, N.Y. Paroled for exchange 8/24/63. NFR. Tollgate Keeper and Farmer, Mt. Sidney 1877. Member, Stonewall Jackson Camp CV, Staunton. Admitted to Lee Camp Soldiers Home, Richmond 12/22/09. Died there 5/4/11. Bur. Hollywood Cem. Brother of John S. Grooms.

GROOMS, JOHN S.: Pvt., Co. G. B. Augusta Co. 1840? Laborer, age 20, Burkes Mill Dist., Augusta Co. 1860 census. Enl. Mt. Meridian 6/16/62. AWOL 6/16-8/31/62. WIA (right hip) and cap. Sharpsburg 9/17/62. Ab. wounded until discharged for disability 6/20/63. Age 24, 5' 10", dark complexion, dark hair, resident of Mt. Meridan. Died Mt. Sidney 10/14/90. Brother of Arthur T. Grooms.

GROSS, WILLIAM A.: Pvt. Co. K. Enl. Shenandoah Mt. 4/9/62. Present 4/9-30/62, and 5/26/62 over 35. NFR. Died before 8/17.

GROVE, ABRAHAM J.: 2ndSgt., Co. A. B. Va. 1844? Farmhand, age 16, 1st. Dist., Augusta Co. 1860 census. Enl. Staunton 7/9/61 age 22 as Pvt. Present 11/61-4/62. Reenl. 5/1/62. WIA (breast) Gaines Mill 6/27/62. Present 8/31-12/31/62, promoted 2ndSgt. 10/12/62. Present 1/1-4/16/63. Ab. sick 4/17-5/63. Present 6-12/63. KIA Bethesda Ch. 5/30/64, age 26. "This gallant Soldier fell in bloom of early youth and at the post of duty. . . .Having participated in all it's battles and marches-his Soldierly career was an honorable as his death-zeal and energy in the discharge of his duty and courage in battle distinguished him as a Soldier. His cheerful spirit was missed on the weary march and in the circle around the campfire."

GROVE, JOHN H.: Pvt., Co. A. B. Va. 5/6/24. Farmhand, age 27, Burkes Mill Dist., Augusta Co. 1860 census. Pvt., Capt. J. K. Koiner's Co. 32 Va. Militia exempt 3/62 furnished substitute, age 29. Enl. Staunton 10/29/64. Present 11-12/64. Paroled Staunton 5/1/65. Age 31, 5' 8", fair complexion, light hair, blue eyes. Farmer, age 38, Mt. Sidney PO, Augusta Co. 1870 census. Living Choctaw Nation, Indian Territory 1874 but moving to San Francisco, Calif. Died Hermitage, Augusta Co. 2/27/94. Bur. Hildebrand Mennonite Ch. Cem.

GROVES, JOHN, JR.: Pvt., Co. K. B. Bath Co. 1826. Wagoner, age 35, Bath CH PO, Bath Co. 1860 census. Enl. Shenandoah Mt. 4/9/62. Present 4/9-30/62. Cap. Ft. Royal 5/30/62. Exchanged 6/62. Present 6/18-7/18/62. AWOL 7/19-11/11/62. Present 1/1-9/15/63. Ab. sick 9/16-12/29/63, nephritus. KIA Bethesda Ch. 5/30/64, age 41. Left widow.

GUINN, GEORGE HAMILTON: Pvt., Co. A. B. Va. 11/12/36. Enl. Staunton 7/16/61. Present 11/61-4/19/62. AWOL 4/20-30/62. Detailed in Nitre Bureau, Bath Co. 7/23/62-12/31/64. Paroled Staunton 5/19/65. Age 27, 5' 7", dark complexion, dark hair, grey eyes. Cattle Dealer and Magistrate, Bath Alum Springs, Bath Co. 1897. Farmer, age 73, Walker's Creek Dist., Rockbridge Co. 1910 census. Died Goshen 2/16/29. Bur. Goshen Baptist Ch. Cem. (Reported to have crossed the Rockies with Kit Carson and 40 other miners during the Gold Rush.)

GUINN, WILLIAM C.: Pvt., Co. A. Highland Co. 1825? Resident of Deerfield, Augusta Co. Enl. Staunton 7/16/61. Present 11/61-4/17/62. Ab. sick 4/18-30/62. Deserted from hospital date unknown. Enl. Co., C, 14th Va. Cav. Died near Deerfield 6/14/94. Bur. Starr Chapel or Lower Cleek Cem., Bath Co.

GUTSHALL, JACOB: Pvt., Co. C. B. Va. 1833? Blacksmith, age 27, Burkes Mill Dist., Augusta Co. 1860 census. Enl.Staunton 7/10/61 age 29. Discharged 9/23/61 "appendix". Detailed in Recruiting Service, Staunton 4-11/62. 2ndLt. Provost Guard, Staunton 9/1-10/31/63. Serving as Scout 2/64. Commanding Provost Guard Staunton 4/64. NFR.

HALE, JOHN WILLIAM: Pvt., Co. F. B. Rockingham Co. 12/31/18. Farmer age 42, Northern Dist., Augusta Co. 1860 census. Not on muster rolls. Enl. Valley Mills 4/62. WIA Gaines Mill 6/27/62. Furnished substitute and was discharged 10/62. Paroled Staunton 5/1/65. Age 47, 6', dark complexion, black hair, brown eyes. Farmer, age 51, Mt. Sidney PO, Augusta Co. 1870 census. Died Long Glade, Augusta Co. 12/9/89. Bur. Old Salem Lutheran Ch. Cem., 3 miles west of Mt. Sidney. Father of Samuel H. Hale.

HALE, SAMUEL HENRY: Pvt., Co. F. B. Va. 3/9/44. Farmhand, age 16, Northern Dist., Augusta Co. 1860 census. Enl. Valley Mills 4/20/62 as substitute for father. Present 8/31/62-4/64. WIA (contusion in middle of back) Bethesda Ch. 5/30/64. Returned to duty 6/15/64. Deserted to the enemy New Creek, W. Va. 3/20/65. Took oath and sent to Ohio. Farmer, age 26, Mt. Sidney PO, Augusta Co. 1870 census. Died Middle River Dist., Augusta Co. 9/14/14. Bur. Old Salem Lutheran Ch. Cem., 3 miles west of Mt. Sidney. Son of John W. Hale.

HALL, E. A.: Pvt., Co. E. Not on muster rolls. Enl. 1862 or 64 Present Ft. Stedman 3/25/65. Cap. Five Forks 4/1/65. (Not on Federal POW rolls). Resident of Louisa Co. Living Loudon Co. 1902. Pension application only record.

HALL, H. LEWIS: Pvt., Co. E. B. Va. 1824? Listed on postwar roster. Laborer, age 36, Lexington, Rockbridge Co. 1870 census.

HALL, JEREMIAH WESLEY: Pvt., Co. D. Va. 9/0/39. Enl. Staunton 7/16/61. Present 7/16/61-4/62. AWOL 6/9-20/62, fined $4.03. Present 6/21-8/31/62, and 1/1-8/31/63. Transf. Co., D, 7th Va. Cav. 8/21/63. Carriage Maker and Postmaster Mt. Solon, 1880. Died Augusta Co. 4/30/11. Bur. St. Paul's Lutheran Ch. Cem., near New Hope.

HALL, JOHN S.: Pvt. Co. G. B. Augusta Co. 1846? Resident, age 14, Northern Dist., Augusta Co. 1860 census. Pvt., Long Meadow Co., Militia, Augusta Co. exempt 3/62, underage. Enl. Staunton 11/15/62. Present 11/15/62-2/15/63. Ab. sick 2/16-4/63. Present 4/30-9/13/63. AWOL 9/14-12/31/63. Dropped as deserter. NFR.

HALL, JOHN W.: Pvt., Co. F. B. 1834. Not on muster rolls. Enl. 4/62. WIA (right leg and right arm) Bethesda Ch. 5/30/64. NFR. Resident, Meyerhoffer's Store, Rockingham Co. 1898. Died Rockingham Co. 1914. Bur. Freidan's Lutheran Ch. Cem. near Mt. Crawford.

HALL, JOSEPH D.: Pvt., Co. E. B. Rockbridge Co. 1838? Laborer, age 22, Fancy Hill, Rockbridge Co. 1860 census. Enl. Staunton 8/1/61. Deserted the day he was mustered into service. NFR. Carpenter, Rockbridge Co. 1868.

HALL, REUBEN WYMAN: Pvt., Co. B (2nd). B. Va. 1844? Laborer, age 16, Burkes Mill Dist., Augusta Co. 1860 census. Enl. Hamilton's Crossing 3/26/63. Present 3/26-7/31/63. AWOL 8/1-14/63. Present 8/15-12/31/63. WIA (hand) near Spotsylvania CH 5/19/64. Returned to duty 11/10/64. Present 11/10-12/64. NFR.

HALL, SAMUEL H.: Pvt., Co. B (2nd). Enl. Waynesboro 7/15/61. Present 11/61-2/62. AWOL 3-4/62. Present 5/1-7/19/62. AWOL 7/20-9/30/62. Present 11/1-7/26/63. AWOL 7/27-8/7/63. Present 9-12/63. KIA Wilderness 5/6/64. Could be Samuel Hall (Va) bur. Fredericksburg Confederate Cem.

HALL, WINFORD J.: Pvt., Co. A. B. Va. 1840? Enl. Staunton 7/9/61 age 21. Present 11/61-4/62. Fined $11.00 by sentence of CM for AWOL. WIA (hip) 2nd Manassas 8/29/62. Present 8/31/62-7/26/63. AWOL 7/27-8/21/63, fined $9.16. Present 8/22-12/63. WIA (right arm) Bethesda Ch. 5/30/64. Ab. wounded through 10/31/64. Present 11-12/64. Surrendered Appomattox 4/9/65. Alive 1903.

HALL, WILLIAM L. (1st): Pvt., Co. E. B. Va. 1838? Farmer, age 22, 7th Dist., Rockbridge Co. 1860 census. Enl. Staunton 8/1/61. AWOL 11-12/61. Present 1-4/62. Reenl. 5/1/62. WIA (thigh) McDowell 5/8/62. Ab. wounded and on leave to 4/63. Present 4/30-8/27/63. Ab. sick 8/28-31/63. Present 9-12/63. WIA (left thigh) Spotsylvania CH 5/12/64. Ab. wounded through 10/31/64. Present 11-12/64. WIA Petersburg 4/1/65. Admitted to hospital Richmond 4/2/65. Cap. there 4/3/65. Sent to Newport News. Released 7/1/65. 5' 10", dark complexion dark hair, dark eyes. Alive in Rockbridge Co. 1913.

HALL, WILLIAM L. (2nd): Pvt., Co. A. On postwar roster. Enl. 1861. Detailed NFR.

HALL, WILLIAM LEWIS: Pvt., Co. I. B. near Fishersville, Augusta Co. 8/29/41. Shoemaker, age 18, 1st Dist., Augusta Co. 1860 census. Pvt., Co. E, 93rd Va. Militia. Enl. Staunton 7/16/61 age 20. Present 11/61-4/62, and 4/30-7/17/62. AWOL 7/18-8/4/62, fined $30.00, and $5.26 deducted. Present 8/5/62 until WIA (knee) Sharpsburg 9/17/62. Ab. wounded through 4/63. Detailed as shoemaker, Richmond 5/25/63-12/31/64. Paroled Manchester 4/20/65. Home: Manchester. Shoemaker, Van Lear, Augusta Co. Died Stuart's Draft, Augusta Co. 5/9/15. Bur. White Hill Ch. of the Brethren Cem. near Stuart's Draft.

HALL, WILLIAM T.: Pvt., Co. E. B. Botetourt Co. 10/26/41. Farmhand age 17, Lexington Dist., Rockbridge Co. 1860 census. Enl. Staunton 8/1/61. Died of diarrhea or typhoid fever in hospital Staunton 12/28/61. Bur. grave #112, Thornrose Cem.

HAM, JOHN F.: Pvt.,Co. C. B. Penn. 1/27/18. Machinist. Enl. Staunton 7/16/61 age 44. AWOL 11-12/61, fined $11.00 by CM. Present 1-2/62. Detailed as Machinist in Clifton Forge foundry for Navy Dept. 3/19/62-12/31/64. 6' 3", light complexion, dark hair, blue eyes. Miner and Moulder, age 52, Pastures Dist., Augusta Co. 1870 census. Admitted to Lee Camp Soldiers Home, Richmond 8/19/98. Left 4/14/99. Died 8/6/02. Bur. Cedar Hill Cem, Covington.

HAMILTON, ALEXANDER C.: Pvt., Co. E. B. Rockbidge Co. 1823? Collierstown, Rockbridge Co. Enl. Staunton 8/1/61. Ab. on leave 11-12/61. Ab. sick 1-2/62. Present 3-10/62, and 1-12/63. Detailed in Pioneers Corps 1/10-11/30/64. Ab. on leave 12/25-31/64. NFR.

HAMILTON, CHARLES ALEXANDER: Pvt., Co. K. B. Bath Co. 1840? Laborer, age 20, Bath CH PO, Bath Co. 1860 census. Pvt., Capt. Hamilton's Co. 81st Va. Militia. Enl. Shenandoah Mt. 4/9/62. Present 4/9-30/62, 1-12/63, and 4/30-12/31/64. NFR. Farmer, Bath Co. Died Bath Co. 10/23/66, age 30.

HAMILTON, CHARLES B. (1st) Pvt., Co. K. B. Bath Co. 1842? Farmhand age 18, Bath CH PO, Bath Co. 1860 census. Enl. Shenandoah Mt. 4/9/62. Present 4/9-30/62. Cap. Front Royal 5/30/62. Exchanged 8/5/62. Ab. sick 8/62-4/63. AWOL 4/30-12/31/63. Dropped as deserter. NFR.

HAMILTON, CHARLES B. (2nd): Pvt., Co. K. B. Bath Co. 1837? Laborer, age 23, Bath CH PO, Bath Co. 1860 census. On postwar roster. Miller, age 35, Warm Springs PO, Bath Co. 1870 census.

HAMILTON, HARVEY: Pvt., Co. B. (1st). B. Donaldsburg, Rockbridge Co. 1832? Farmer, Rockbridge Co. 1860. Enl. Fairfield 7/16/61. Ab. sick when transf. 9/28/61.

HAMILTON, JAMES HARVEY: Pvt., Co. C. B. Va. 6/19/39. Farm Manager, age 20, Burkes Mill Dist., Augusta Co. 1860 census. Pvt., Capt. J. K. Koiner's Co. 32nd Va. Militia, exempt 3/62 heart disease, age 22. Enl. Staunton 11/23/64. Ab. not reported on rolls 11-12/64. NFR. Farmer and Stockraiser, Fishersville 1885 and 1897. Died Christian's Creek, Augusta Co. 1/24/17. Bur. Tinking Spring Pres. Ch. Cem., near Fishersville.

HAMILTON, JAMES WILLIAM: 2ndSgt., Co. G. B. Augusta Co. 1833? Farmer age 27, 1st Dist., Augusta Co. 1860 census. Enl. Staunton 8/2/61. Ab. detailed as Commissary Sgt. 9/11-12/11/61. Ab. sick 12/12/61-2/62. Present 3-4/62. Reenl. and reduced to Pvt. 5/1/62. KIA 2nd Manassas 8/28/62, age 29. Left widow near Fishersville. Bur. Tinkling Spring Pres. Ch. Cem.

HAMILTON, JOHN EDWARD: 1stLt., Co. G. B. Brands, Augusta Co. 1/22/38. Farmhand, age 22, Burkes Mill Dist., Augusta Co. 1860 census. Enl. Staunton 8/2/61 as 1stSgt. Present 11/61-4/62. Reenl. and elected 3rdLt. 5/1/62. Ab. on leave 5/1-8/31/62. Ab. sick 9-10/62. Present 11-12/62, promoted 2ndLt. Present 1/1-2/27/63. Ab. sick 2/28-4/63. Present 4/30/63-5/64. WIA (shoulder) Bethesda Ch. 5/30/64. Ab. sick with dysentery 6/12-30/64. Presence or absence not stated 7-10/64. Present 11-12/64, promoted 1stLt. 12/2/64. Ab. on leave 1/19-28/65. Paroled Staunton 5/1/65. Age 27, 5' 11", fair complexion, light hair, blue eyes. Unofficial report indicates he commanded Bn. Sharpshooter last 2 years of war and was present Appomattox but escaped. Farmer, age 32, Mt. Sidney PO, Augusta Co. 1870 census. Stockraiser, Land assessor, and County Treasurer, Augusta Co. Died Brands 7/22/89. Bur. Tinkling Spring Pres. Ch. Cem.

HAMILTON, JOHN F.: Pvt., Co. B (1st). B. Va. 1837? Farmhand, age 23, South River Dist., Rockbridge Co. 1860 census. Enl. Fairfied 7/10/61. Present until transf. 9/28/61.

HAMILTON, WILLIAM LEWIS: Pvt., Co. B. (1st). B. 1831? Farmhand, age 19, South River Dist., Rockbridge Co. 1860 census. Enl. Fairfield 7/10/61. Present until transf. 9/28/61.

HAMNER, SAMUEL AUSTIN: 1stCpl., Co. B (2nd). B. Albemarle Co. 1840? Mechanic. Enl. Waynesboro 7/15/61, as Pvt. Present 11/61-4/62. Reenl. 5/1/62 Promoted 4thCpl. 10/2/162. Present 11/62-4/63, promoted 2ndCpl. Present 4/30-8/31/63, promoted 1stCpl. Present 9-12/63. WIA (through right lung) Bethesda Ch. 5/30/64. Returned to duty 10/29/64. Ab. recommended for light duty by Medical Examining Board and sent to Richmond 11-12/64. Retired for disability 2/15/65. Age 25, 5' 10½", dark complexion, dark eyes, dark hair. Resident, age 30, Mt. Sidney PO, Augusta Co. 1870 census.

HAMNER, WILLIAM H.: 4thCpl., Co. B (2nd). Enl. Camp Shenandoah 7/18/62. Present 4/18/62-12/63, promoted 4thCpl. Present 9-10/63. KIA Winchester 9/19/64.

HAMPSHIRE, J. H.: 2ndSgt., Co. B (2nd). On postwar roster. Enl. 1861. Served 4 years.

HANGER, GEORGE ROBERTSON: Pvt., Co. G. B. Near Shutterlee's Mill, Augusta Co. 1830. Farmer, Augusta Co. Enl. Co. L, 5th Va. Inf. (disbanded). Enl. Staunton 8/2/61. Present 11/61-4/62. detailed asst. Commissary Sgt. 12/11/61. Reenl. 5/1/62. Present 5/1/62-12/63. Cap. Spotsylvania CH 5/12/64. Sent to Point Lookout. Released 6/27/65. 5' 8", fair complexion, light brown hair, blue eyes. Farmer, Augusta Co. 1868. Jeweler and Clock Repairer, Staunton. Member, Stonewall Jackson Camp CV, Staunton. Died Augusta Co. 9/5/03. Bur. Hanger family Cem., near Shutterlee's Mill.

HANGER, JAMES J.: Pvt., Co. I. B. near Mt. Solon, Augusta Co. 9/36? Farmer, age 22, Staunton PO, 1860 census. Enl. Staunton 7/16/61 age 23. Died of fever Camp Alleghany 12/27/61. Effects: 1 overcoat, 1 pr. pants, 2 shirts, 2 pr. socks, 1 pr. blankets, delivered to his brother.

HANNA, JOHN NEWTON: 2ndLt., Co. D. B. near Mt. Solon, Augusta Co. 1839? Farmhand, age 21, Northern Dist., Augusta Co. 1860 census. Enl. Staunton 7/16/61 as 1stCpl. Present 7/16/61-4/62. Reenl. and elected 2ndLt. 5/1/62. WIA (thumb) Port Republic 6/9/62. Presence or absence not stated on rolls 6-8/62. Present 1-5/63. Presence or absence not stated on rolls 6-12/63. WIA (side) Bethesda Ch. 5/30/64. Cap. Winchester 9/19/64. Sent to Ft. Delaware. Released 6/17/65. 5' 10", dark complexion, dark hair, grey eyes. School Teacher and Farmer Augusta Co. Died near Mt. Crawford 1/29/07. Bur. Mt. Solon.

HARDWICK, JAMES W.: Pvt., Co. G. B. Augusta Co. 1807? Carpenter, age 53, Burkes Mill Dist., Augusta Co. 1860 census. Enl. 10/24/62 as a substitute. Present 10/24-12/62. Ab. sick 1/30/63-12/31/64. NFR. Died Weyer's Cave 4/1/76. Father of James W. Hardwick, Jr.

HARDWICK, JAMES W., JR.: Pvt., Co. A. B. Va. 1843? Carpenter's Apprentice, age 17, Burkes Mill Dist., Augusta Co. 1860 census. Enl. Staunton 7/9/61 age 18. Ab. sick 11/16/61-1/18/62. Present 1/19-4/62. Reenl. 5/1/62. Present 8/31/62-11/63. AWOL 12/8/63. Lost knapsack $6.50, 1 haversack 50¢, 1 canteen $1.25, stoppage for assessments. Deserted to the enemy in W. Va. 12/25/63. Took oath and sent north. Age 20, 6' ¾", light complexion, blue eyes, light hair. Alive 1903. Son of James W. Hardwick, Sr.

HARDWICK, JOHN W.: Pvt., Co.A. B. 1837? Farmer, Augusta Co. Enl. Staunton 7/9/61 age 24. Ab. on detached duty Staunton 12/28/61-1/6/62. Present 1/7-4/62. Reenl. 5/1/62. Present 8/31/62-11/63. AWOL 12/8/63. Lost 1 knapsack $6.50, 1 haversack 50¢, 1 canteen $1.25, stoppage for assessments. Deserted to the enemy in W. Va. 12/25/63. Took oath and sent north. Age 26, 6' 1", light complexion, blue eyes, light hair. Alive 1903.

HARLAN, JAMES F.: Pvt., Co. B (2nd). Not on muster rolls until 10/23/64. Enl. Staunton 1861 and served until 4/62 when he was disabled by a fractured limb and assigned to light duty until Fall '64. Enl. Staunton 10/23/64. Present 10/23-12/31/64. Cap. Amelia CH 4/7/65. Sent to Point Lookout. Released 6/13/65. Resident of Augusta Co., 5' 9¼", fair complexion, brown hair, blue eyes. Living in Charlottesville 1920. Brother of John, Patrick and William H. Harlan.

HARLAN, JOHN: Pvt., Co. F. B. Ireland 1823? Laborer, age 27, Beverly Manor Dist., Augusta Co. 1850 census. Post war roster only. Enl. as substitute. Brother of James F., Patrick and William H. Harlan.

HARLAN, PATRICK.: Pvt., Co. B (2nd). Born Ireland 1822? Laborer on railroad, age 38, 1st. Dist., Augusta Co. 1860 census. Illiterate. On postwar roster. Brother of James F., John and William H. Harlan.

HARLAN, WILLIAM H.: Pvt., Co. B (2nd). Enl. Hamilton's Crossing 3/13/63, Present, detailed as Wagon Master in Div. Commissary Train 4/30/63-12/31/64. Surrendered Appomattox 4/9/65. Living in Botetourt Co. 1902. Brother of James F., John and Patrick Harlan.

HARLOW, WILLIAM JOHN: Pvt., Co. B (2nd). B. King William Co. 1838? Wagonmaker, age 22, Burkes Mill Dist., Augusta Co. 1860 census. Enl. Waynesboro 7/15/61. Present 11/61-4/62. KIA 2nd Manassas 8/29/62. Age 23, hazel eyes, dark hair, fair complexion, 5' 8½".

HARMAN, GEORGE WASHINGTON: 4thCpl., Co. D. B. Augusta Co. 4/14/39. Cooper, age 20 Mt. Solon PO, Augusta Co. 1860 census. Enl. Staunton 7/16/61 as Pvt. Present 7/16/61-4/62, promoted 4thCpl. 1/13/63. Reenl. 5/1/62. WIA (thigh) McDowell 5/8/62. Ab. wounded through 8/31/62. Ab. sick 2/16-4/63. Present 4/30-12/31/63. Ab. sick 5/27-6/16/64. Cap. Winchester 9/19/64. Sent to Point Lookout. Exchanged 3/15/65. NFR. Farmer, Mossy Creek, 1897. Died Augusta Co. 4/10/04. Bur. Mossy Creek Old Cem.

HARMAN, JOHN I.: Pvt., Co. D. B. Augusta Co. 1839? Cooper, Augusta Co. Enl. Staunton 7/16/61. Present 7/16/61-4/62. KIA McDowell 5/8/62. Effects $47.75, 1 knapsack containing his clothing, 1 blanket. Age 23, 6' 4", dark complexion, dark hair, dark eyes. Bur. Mossy Creek Old Cem.

HARMAN, LEWIS: 1stLt., Co. C. Born Staunton 12/41/45. VMI 60-61. Drillmaster, Richmond 4-6/61. Enl. Staunton 7/16/61 as Pvt. Elected 1stLt. Present 11/61-6/9/62. WIA (arm) Port Republic 6/9/62. Resigned 8/30/62. Enl. Co. I, 12th Va. Cav. Served as Adj. 2nd Manassas and mortally wounded Col. Broadhead, 1st. Mich. Cav. Promoted Capt. 3/6/64, by Gen. Rosser for "distinguished valor and skill". WIA and cap. Verdiersville 5/5/64. Sent to Ft. Delaware. One of the "Immortal 600" at Morris Island, S. C. and Ft. Pulaski, Ga. Released 5/10/65. Farmer and Stockraiser, Mt. Solon, Augusta Co. Railroad Contractor, Clifton Forge. Steward, Western State Hospital, Staunton. Deputy U.S. Revenue Collector and employee State Treasurer's office, Richmond. Served on monument committee to erect same over Confederate soldiers graves, Thornrose Cem. Member, Stonewall Jackson Camp, CV, Staunton. Died Richmond 3/22/20. Bur. Thornrose Cem. Son of Col. Michael G. Harman.

HARMAN, MICHAEL GARBER "Mike": Col., F&S. B. Staunton 8/22/23. Proprietor, age 37, Staunton 1860 census. Hotel and Stage Line owner and operator, Staunton. Appointed Major and QM State of Va. 4/61. QM Staunton 4/61-1/62. Also elected Capt. Co., H. 34th Va. Inf. Elected LtCol. 52nd Va. 8/19/61, and served in dual capacity until resigning his commission as QM 1/8/62. Acting Chief QM Gen. T. J. Jackson 2/62. Elected Col. 5/1/62. WIA (right arm) McDowell while commanding the Regt. Returned to duty and present Sharpsburg, 9/17/62. Ab. 10/62-2/63. Present 4-5/63. Resigned 6/6/63 for disability. Served as QM Staunton to end of war. Part owner Natural Bridge Hotel 1864. Hotel owner, Stage Line Operator, President, Valley Railroad; Mail Contractor; President, Central Livestock Co., Staunton; VMI Board of Visitors 65-66. Died Richmond 12/17/77. Bur. Thornrose Cem. Father of Lewis Harman.

HARNSBARGER, ROBERT S.: Pvt., Co. H. B. near Port Republic, Augusta Co. 8/23/23. Miller, Mt. Meridan Mill, Augusta Co. Name cancelled on muster roll. Exempt as Miller. Served in Augusta Co. Reserves and was present Piedmont 6/5/64. Farmer and Stockraiser, Augusta Co. Died Augusta Co. 7/26/83. Bur. Augusta Stone Ch. Cem.

HARRIS, ALEXANDER: Pvt., Co. E. Resident, Thompson's Landing, Rockbridge Co. Enl. Staunton 8/1/61. Present 11/61-4/62. Reenl. 5/1/62. Present 5/1-10/31/62, AWOL 1 month, 15 days. AWOL 11/23/62-1/23/63. Sentenced by GCM to 12 months hard labor and forfeiture of pay for same period, and $23.00 for musket lost, $2.50 for cartridge box, $5.00 for bayonet, $1.00 for bayonet scabbard and $1.00 for belt. Present in arrest 1/23-2/63. Ab. in arrest 3/1-7/30/63. Pardoned and sent to duty by the Presidential Proclamation of 8/1/63. Present 8-12/63. AWOL 5/22-27/64. Pardoned by Sec. of War. Present 9/8-12/31/64, stoppage for 1 gun and accoutriments lost $45.75. Deserted to the Army of the Potomac near Petersburg 3/8/65. Sent to Washington, D.C. Took oath and transportation furnished.

HARRIS, JAMES FRANKLIN: Pvt., Co. I. B. Augusta Co. 5/1/41. Farmhand, age 19, 1st Dist., Augusta Co. 1860 census. Enl. Staunton 7/16/61 age 20. Present 11/61-4/62. Reenl. 5/1/62. WIA (right arm) Cedar Run 8/9/62. Ab. wounded through 10/31/62. Present 1/63-12/31/63. Cap. 1865. Escaped from Libby USA Prison, Richmond 4/10/65. Farmhand, age 29, Fishersville PO, Augusta Co. 1870 census. Watchman for Mines, Riverheads Dist., Augusta Co. 1910 census. Died Lorton, Augusta Co. 6/1/19. Bur. Pines Chapel Cem.

HARRIS, JESSIE L.: 2ndSgt., Co. G. B. Va. 1814? Carpenter, age 46, Northern Dist., Augusta Co. 1860 census. Enl. Staunton 8/1/61 as Drummer. Present 11/61-4/62. Reenl. 5/1/62. Present 5/1-12/31/62. Present 1/1-8/31/63, promoted 2ndSgt. for good conduct on the field at Gettysburg. Present 9-12/63. KIA Spotsylvania CH 5/12/64.

HARRIS, JOHN A.: Pvt., Co. H. B. Va. 3/30/43. Blacksmith, age 32, 1st Dist., Augusta Co. 1860 census. Name cancelled on roll. Die New Hope 4/17/11. Bur. Middle River Ch. of the Brethren near New Hope.

HARRIS, JOHN EDWARD: Pvt., Co. I.B. Augusta Co. 1829? Pvt., Co. D. 93rd Va. Militia. Enl. Camp Alleghany 3/30/62. Present 4/62. Reenl. 5/1/62. Ab. sick 6/6-12/62. WIA (groin and out hip) Fredericksburg 12/13/62. Ab. wounded 1-12/63. Cap. 6/19/64. Exchanged 2/65. Admitted to hospital Richmond with scurvy as paroled POW 3/6/65. Furloughed for 60 days 3/16/65. Illiterate. Farmer, age 41, Fishersville PO, Augusta Co. 1870 census. Died near Greenville 10/09. Bur. Pines Chapel Cem.

HARRIS, JOHN L.: Pvt., Co. E. B. Va. 1835? Farmhand, age 25, Lexington Dist., Rockbridge Co. 1860 census. Enl. Staunton 8/1/61. Present 11/61-4/62. Reenl. 5/1/62. Ab. on leave 5/1-10/31/62. Present 1-11/63. AWOL 12/6-31/63. Died Gordonsville 4/8/64, cause unknown.

HARRIS, JOSEPH W.: Pvt., Co. H. B. Rockbridge Co. 1834? Laborer, age 28, 5th Dist., Rockbridge Co. 1860 census. Enl. Staunton 7/23/61 age 25. Died of measles Camp Alleghany 11/8/61. Effects 1 overcoat, sold for $13.00 and sent to legal represenative. Left widow. Bur. Alleghany Mt.

HARRIS, PETER P.: Pvt., Co. I. B. Va. 2/7/30. Farmhand, age 27, 1st Dist., Augusta Co. 1860 census. Enl. Camp Alleghany 3/30/62. Present 4/62, and 8/30-10/31/62. AWOL 12/7/62-1/9/63. Ab. sick 1/15-12/63. Present 4/30-10/31/64. AWOL 12/8-31/64. Paroled Staunton 5/15/65. Age 35, 5' 11", dark complexion, dark hair, dark eyes. Farmer, age 39, Fishersville PO, Augusta Co. 1870 census. Died Augusta Co. 8/19/07. Bur. Mt.Vernon Ch. of the Brethren Cem., near Ladd.

HARRIS, SAMUEL F.: Pvt., Co. I. B. Augusta Co. 4/19/36. Farmer, age 23, 1st Dist., Augusta Co. 1860 census. Enl. Staunton 7/16/61 age 24. Present 11/61-4/62. KIA Port Republic 6/9/62. Effects 1 watch and $1.25. Age 24, 5' 9", fair complexion, hazel eyes, dark hair. Left widow.

HARRIS, SAMUEL JACKSON: Pvt., Co. I.B. Augusta Co. 10/10/43. Farmhand, age 17, 1st Dist., Augusta Co. 1860 census. Enl. Staunton 7/16/61 age 18. Present 11/61-4/62. Reenl. 5/1/62. Present 8/30-10/31/62, and 1-12/63. Cap. Spotsylvania CH 5/12/64. Sent ot Ft. Delaware. Released 6/20/65. 5' 5", light complexion, red hair, grey eyes. Farmer, age 26, Fishersville PO, Augusta Co. 1870 census. Farmer, near Lovingston, Nelson Co., 1910.

HARRIS, WILLIAM HARRISON.: Pvt., Co. I. B. Augusta Co. 2/3/46. Student, age 13, Staunton 1860 census. Enl. Sommerville Ford 1/24/64. Present 4/30-12/31/64. WIA Ft. Stedman 3/25/65. Admitted to hospital Petersburg 3/28/65. Transf. 4/2/65. Paroled Staunton 5/1/65. Age 19, 5' 6", fair complexion, brown hair, grey eyes. Farmer, Stuart's Draft until 1890. Moved to Middlebrook. Elder and Trustee, St. John's Ch. for 20 years. Died near Middlebrook 3/18/24. Bur. St. John's Ch. Cem.

HARUFF, ANDREW J.: Pvt., Co. A. B. Augusta Co. 11/8/35. Enl. Staunton 7/17/61. Pesent 11/61-4/62. Reenl. 5/1/62. Present 8/31/62-10/9/62. Ab. sick 10/11/62-10/63. Present 11-12/63. Cap. Bethesda Ch. 5/30/64, age 34. Sent to Point Lookout. Transf. Elmira. Exchanged 10/29/64. Paroled Staunton 5/1/65. Age 30, 5' 8", dark complexion, black hair, grey eyes. Carpenter and Millwright, Augusta Co. Died Augusta Co. 5/9/97. Bur. Hebron Pres. Ch. Cem. near Weyer's Cave.

HARUFF, GEORGE W.: Pvt., Co. A. B. Augusta Co. 1833? Farmer, Augusta Co. Resident of Churchville. Enl. Staunton 7/17/61. Present 11/61-4/62. Reenl. 5/1/62. WIA (neck) Port Republic 6/9/62. AWOL 8/31-12/27/62. Present in arrest 12/28/62-2/63. Sentenced by GCM to forfeit 6 months pay in addition to pay while AWOL. Present 3-12/63. WIA (compound fracture of arm) Bethesda Ch. 5/30/64. Furloughed for 60 days 7/20/64. Ab. wounded through 12/64. NFR. Died before 5/03.

HARUFF, JACOB A.: Pvt., Co. K. B. Va. 1836? Farmer, age 24, Cleeks Mill PO, Bath Co. 1860 census. Enl. Shenandoah Mt. 4/9/62. Ab. sick 4/9-30/62. Deserted to the enemy 5/1/62. Enl. 53rd Ill. Inf. Kendallville, Ind. for 1 year 12/8/64 as substitute. Discharged Indianapolis, Ind. 5/8/65. Age 27, 5' 7", brown hair, hazel eyes. Resident Back Creek, Bath Co.

HARSHBARGER, JOHN: 4thCpl., Co. F. B. Augusta Co. 12/42. Enl. Staunton 7/31/61 as Pvt. Present 11-12/61, detailed as wagoner. Present 1-4/62. Reenl. 5/1/62. Detailed as teamster 8/31/62-2/63. Present 4/30-12/63. Promoted 4thCpl. '64. WIA and cap. Bethesda Ch. 5/30/64. Sent to Point Lookout. Transf. Elmira. Released 5/30/65. Resident of Staunton, 5' 7", dark complexion, dark hair, blue eyes. Farmer, age 28, Mt. Sidney PO, Augusta Co. 1870 census. Living in Ga. 1900. Died 2/22/22.

HARSHER, JOHNATHAN: Pvt., Co. I. B. Augusta Co. 1840? Farmer, Middlebrook, Augusta Co. Enl. Staunton 9/19/62. Present 10/1-30/62. AWOL 11/22/62-4/63. Discharged for disability 6/8/63, "assitus." Age 23, 5' 8", dark complexion, blue eyes, black hair.

HART, RICHARD M.: Pvt., Co. I. B. Augusta Co. 1839? Farmer, age 21, 1st Dist., Augusta Co. 1860 census. Enl. Staunton 7/16/61 age 22. Present 11/61-4/62. Reenl. 5/1/62. Ab., reason not stated, 6/5-10/31/62. Present 1/1-10/3/63. Ab. sick 10/4-31/63. Present 11-12/63. WIA (knee) Bethesda Ch. 5/30/64. Furloughed for 40 days 6/28/64. Returned to duty 11/14/64. Present 11/14-12/31/64. WIA (head) Ft. Stedman 3/25/65. Admitted to hospital Richmond 4/1/65. Cap. in hospital Richmond 4/3/65. Paroled 5/11/65. Farmer, age 32, Fishersville PO, Augusta Co. 1870 census. Died Riversheads Dist., Augusta Co. 3/4/89.

HAWKINS, JOHN: Pvt., Co. G. B. Rockingham Co. 1834? Farmer, age 26, Northern Dist., Augusta Co. 1860 census. Enl. Staunton 8/2/61. Ab. sick 11/12/61-3/8/62. Present 3/9-4/62. AWOL 5/62-12/63. Paid Staunton 12/22/62. NFR. Farmer, age 38, Mt. Sidney PO, Augusta Co. 1870 census. Living Crimora, Augusta Co. 1896.

HAWPE, GEORGE CROBARGER: Pvt. Co. D. B. Greenville, Augusta Co. 4/15/38. On postwar roster. Served also "Augusta Lee Rifles" 25th Va. Inf. Farmer and sawmill operator on South River, Riverheads Dist., Augusta Co. 1910 census. Died 2/1/16, near Greenville. Bur. Geenville Baptist Ch. Cem.

HAWPE, HENRY THOMAS: 3rdSgt., Co. I. B. Augusta Co. 2/20/43. Farmer, age 17, 1st Dist., Augusta Co. 1860 census. Enl. Staunton 7/16/61 as Pvt., age 18. Present 11/61-4/62. Reenl. 5/1/62. Present 8/30-10/31/62. Present 1-4/63. Appointed 4thCpl. 4/1/63. WIA hand and capt. Gettysburg 7/3/63. Sent to Ft. McHenry. Exchanged 7/31/63. Ab. wounded through 10/63. Present 11-12/63. Appointed 3rdSgt. 1/11/64. WIA nose Cold Harbor 6/2/64 DOWs in hosp. Richmond 6/6/64. Bur. Bethel Pres. Ch. Cem. near Middlebrook. Brother of James W. Hawpe.

HAWPE, JAMES WILLIAM: Pvt. Co. I. B. Augusta Co. 4/14/44. Resident, age 14, 1st. Dist., Augusta Co. 1860 census. Enl. Greenville 11/27/63. Present 12/63. Capt. Bethesda Ch. 5/30/64. Sent to Pt. Lookout. Transf. to Elmira. Released 6/30/65. 5' 6", florid complexion, dk. hair, blue eyes. Farmer, age 25, Walker's Creek Dist., Rockbridge Co. 1870 census. D. near Spottswood Augusta Co. 2/18/29. Bur. New Providence Pres. Ch. Cem. near Brownsburg, Rockbridge Co. Brother of Henry T. Hawpe.

HAYSLETT, ANDREW J.: Pvt., Co. B (1st). B. Va. 1821? Farmer, age 39, Kerr's Creek Dist. Rockbridge Co. 1860 census. On postwar roster.

HAYSELTT, ANDREW JACKSON: Pvt. or Company Surgeon, Co B (1st). B. on upper Kerr's Creek, Rockbridge Co. 11/25/29. Grad. Washington College 1856, UVa. Medical School and Jefferson Medical College, Philadelphia. Doctor, age 41, Kerr's Creek Dist., Rockbridge Co. 1860 census. Enl. Staunton 8/1/61 as Pvt. (Company Surgeon on postwar roster). Present until transf. 9/28/61.

HAYSLETT, BENJAMIN FRANKLIN: Pvt., Co. E. B. Rockbridge Co. 1838? Farmhand, age 22, Collierstown PO, Rockbridge Co. 1860 census. Enl. Staunton 8/1/61. Present 11/61-4/62. Reenl. 5/1/62. Present 5/1-10/31/62. Present 1-12/63. Ab. on leave 12/24-31/63. Present 4/30-6/64. Des. while on the march from Lynchburg to Staunton near Lexington 6/25/64. NFR. Living in Mo. 1881. Brother of Richard M. and William Hayslett.

HAYSLETT, BRADLEY: Pvt., Co. E. B. Rockbridge Co. 1846? Farmhand, age 14, Collierstown PO, Rockbridge Co. 1860 census. Enl. Rockbridge Co. 8/1/63. Present 8-12/63, and 4/30-6/64. Des. on the march from Lynchburg to Staunton near Lexington 6/25/64. Des. to the enemy Charleston, W. Va. 5/10/65. Age 20, 6', fair complexion, blue eyes, black hair. Took oath and released. Farmer, age 50, Rockbridge Co. 1898.

HAYSLETT, RICHARD M.: Pvt., Co. E. B. Rockbridge Co. 1837? Farmhand, age 23, Collierstown PO, Rockbridge Co. 1860 census. Enl. Staunton 8/1/61. Present 11/61-2/62. Ab. sick 3-4/62. WIA McDowell 5/8/62. DOWs in hosp. Staunton 5/23/62. Age 27, 5', dk. complexion, hazel eyes, dark hair. Bur. Thornrose Cem. grave #970. Brother of Benjamin F. and William Hayslett.

HAYSLETT, SARGENT MCDONALD: Pvt., Co. E. B. Rockbridge Co. 1842? Farmhand, age 18, Collierstown PO, 1860 census. Enl. Staunton 8/1/61. Present 11/61-4/62. Reenl. 5/1/62. AWOL 5-10/62. Present 1-12/63, and 4/30-12/31/64. NFR. Illiterate. Farmer, Rockbridge Co. 1865.

HAYSLETT, WILLIAM, SR.: Pvt., Co. E. B. Collierstown, Rockbridge Co. 11/18/26. Farmer, age 30, Collierstown PO, Rockbridge Co. 1860 census. Enl. Staunton 8/1/61. Present 1-2/62. Ab. sick 3-4/62. Reenl. 5/1/62. WIA (arm) Port Republic 6/9/62. Absent wounded through 8/31/62. AWOL 9/1/62-12/31/63. Dropped as a deserter. Farmer, age 37, Buffalo Dist., Rockbridge Co. 1870 census, illiterate. Died Collierstown 11/18/13. Bur. Laurel Hill Cem., Rockbridge Co. Brother of Benjamin F. and Richard M. Hayslett.

HEATON, C.: Pvt., Co. H. Not on muster rolls. WIA (hip) Cross Keys 6/8/62. NFR.

HEATON, ELIJAH H.: Pvt., Co. G. B. Augusta Co. 1841? Farmhand, age 19, Burkes Mill Dist., Augusta Co. 1860 census. Enl. Staunton 8/2/61. Present 11/61-4/62. Reenl. 5/1/62. Ab. on leave 5-8/62. Ab. sick 9/29-12/62. Present 1-4/63. WIA (compound fracture left thigh) and capt. Gettysburg 7/3/63. DOWs Letterman Hosp., Gettysburg 8/9/63. Bur. Gettysburg Cem., 4th sect. Gen. Hosp., grave #11. Removed to Hollywood Cem., Richmond 6/20/72. Brother of Jacob H. and James H. Heaton.

HEATON, JACOB HARVEY: Pvt., Co. G. B. Augusta Co. 1845? Farmhand, age 15, Burkes Mill Dist., Augusta Co. 1860 census. Enl. Brandy Station 10/31/63 as substitute. Present 10-12/63. Deserted to the enemy Mt. Crawford 10/1/64. Sent to New Creek, W. Va. Took oath and released. Age 20, 5' 6", lt. complexion, grey eyes, dk. hair. Farmer, age 26, Mount Sidney PO, Augusta Co. 1870 census. Married Augusta Co. 9/29/77. Brother of Elijah H. and James H. Heaton.

HEATON, JAMES H.: Cpl., Co. G. B. Rockingham Co. 1839? Farmhand, age 21, Burkes Mill Dist., Augusta Co. 1860 census. Enl. Staunton 8/2/61 in 33rd Va. Inf. Present 7/25-10/31/62, joined from 33rd Va. Inf. on 7/27/62. Ab. on detached service with Commissary Wagon 11/62-2/63. Present 3-4/63, 3 mos. pay stopped by sentence of CM. Present 4/30-8/31/63. Fined 1 mo. pay by sentence of CM for AWOL. Present 9-12/63. Promoted Cpl. KIA near Spotsylvania CH 5/19/64. Brother of Elijah H. and Jacob H. Heaton.

HEATWOLE, MANASSAS: Pvt.,Co. F. B. Rockingham Co. 8/7/43. Enl. Valley Mills 4/20/62. WIA date and place unknown. AWOL 5/8-9/26/62. Present 9/27/62-7/27/63. AWOL 7/28-10/63. Des. to the enemy Clarksburg, W. Va. 10/13/63. "Went to Ohio until war ended." NFR. Farmer, Dayton, Rockingham Co. D. 11/25/90. Bur. Weavers Mennonite Ch. Cem. on Rt. #33. west of Harrisonburg.

HEFLIN, JOHN MARSHALL: Pvt., Co. B. (2nd). B. Strasburg, Shenandoah Co. 11/2/28. Mexican War Veteran. Enl. Bunker Hill 9/24/62. Present 9/24-1/63. Detailed in hosps. Richmond and Staunton 2/19/63-12/31/64. Surrendered Appomattox 4/9/65. Gardener, age 39, Staunton 1870 census. D. Augusta Co. 2/24/03. Bur. Thornrose Cem., Staunton.

HEIZER, WILLIAM J.: Pvt., Co. K. B. Augusta Co. 1839. Farmer, age 21, Cleeks Mill PO, Bath Co. 1860 census. Enl. Shenandoah Mtn. 4/9/62. Detailed in Brigade Pioneer Corps 12/28/62-12/63. Ab. sick 4/30-12/31/64. Des. to the enemy Clarksburg, W. Va. 4/8/65. Took oath and released. D. 1903.

HELLER, JONAS: Pvt., Co. A. B. Germany or Austria 1819? Enl. Staunton 7/16/61. Ab. sick 9/10/61-1/28/62. Present 1/29-4/62. AWOL 10/1-12/31/62. Discharged for chronic rheumatism 3/2/63. Age 39, 5' 2", dk. complexion, blue eyes, bk. hair, pedler. Took oath Richmond 4/15/65. Age 46, merchant, residence between 3rd and 4th Sts., Richmond. D. before 5/03.

HELMS, BENJAMIN: Pvt., Co. C. B. 1840? Resident of Miss. Enl. Staunton 7/16/61, age 20. Des. 8/16/61. NFR.

HEMP, JOHN, JR.: Co. unknown. B. 9/12/17. Not on muster rolls. D. in hosp. Richmond cause unknown. 8/12/62. Bur. St. John Ch. Cem. near Middlebrook.

HEMP, WILLIAM C.: Pvt., Co. C. B. Augusta Co. 3/30/34. Farmer, age 24, 1st Dist., Augusta Co. 1860 census. Member Long Meadow Co. Va. Militia, age 27, exempt 3/62 but volunteered. Enl. Shenandoah Mtn. 5/1/62. Present 4/1-30/62. WIA (leg) Malvern Hill 7/1/62. Ab. wounded through 8/31/62. Present 9/1/62-2/28/63. Detailed as wagoner 3-4/63. Ab. sick in hosp. Gordonsville with chronic diarrhea 6/4-8/31/63. Present 9-12/63. Ab. in hosp. Richmond with rheumatism 12/13-31/63. Present 4/30-9/4/64. AWOL 9/5-10/31/64. Present 11-12/64. Capt. Petersburg 5/2/65. Sent to Pt. Lookout. Released 6/14/65. 5' 7", florid complexion, dk. hair, light blue eyes. Illiterate. D. of typhoid fever, contracted in prison, at his home near Annex, Augusta Co. 6/24/65. Left widow. Bur. Mt. Tabor Lutheran Ch. Cem. near Middlebrook.

HENKEL, SAMUEL: Pvt., Co. H. B. Rockingham Co. 1814? Laborer, Rockingham Co. Enl. Staunton 7/23/61, age 47. Present 11/61-3/62. Ab. sick 4/21-30/62. Ab. sick 6/18-8/31/62. Disch. overage of conscription, 11/25/62. Age 47, 5' 5", fair complexion, blue eyes, dark hair. D. before 1890.

HENKLEY, HENRY: Pvt., Co. C. On postwar roster. Served 3 yrs. Living Pulaski Co. 1902.

HENNE, JOHN G.: Pvt., Co. G. B. 2/15/15. Res of Long Meadows, Augusta Co. Enl. Staunton 8/2/61. Ab. on leave 11/24-12/31/61. Present 1-8/62. WIA (head) Cross Keys 6/8/62. AWOL 10/17/62-12/63. Dropped from rolls as deserter. Shoemaker, Fishersville. D. Fishersville 5/8/06. Bur. Hildebrand Mennonite Ch. Cem., Madrid, Augusta Co.

HEPLER, JACOB A.: Pvt., Co. K. On postwar roster.

HEPLER, JOSEPH P.: Pvt., Co. K. B. Va. 1833? Farmer, age 27, 5th. Dist., Rockbridge Co. 1860 census. Enl. Valley Mills 4/26/62. Present 4/9-5/16/62. Ab. sick 5/17-12/7/62. D. of smallpox in hosp. Richmond 12/8/62.

HERNDON, WILLIAM H.: Pvt., Co. B. (2nd). B. Va. 1843? Carpenter's Apprentice, age 17, Waynesboro, Augusta Co. 1860 census. Enl. Waynesboro 7/15/61. Present 11/61-4/62, fined $11.00 for 9 days AWOL. Reenl. 5/1/62. AWOL 6/20/62-12/63. Dropped as deserter. NFR.

HESLEP, JOSEPH SPRIGGS: 2nd.Cpl., Co. B. (1st). B. Rockbridge Co. 1838? Farmhand, age 21, South River Dist., Rockbridge Co. 1860 census. Enl. Fairfield 7/10/61. Ab. on detail Staunton when transf. 9/28/61.

HEVNER, ZEBULON: Pvt., Co. K. B. Pendleton Co. 1815? Farmhand, age 45, Millboro Springs PO, Bath Co. 1860 census. Enl. Shenandoah Mtn. 4/9/62. AWOL 4/9-30/62. Dropped as des. (may have been disch. overage of conscription.) Farmhand, age 52, Warm Springs PO, Bath Co. 1870 census.

HEW(S), W. F.: Pvt., Co. A. Not on muster rolls. D. 12/28/61. Bur. Thornrose Cem., Staunton, grave #860.

HICKS, WILLIAM H. JR.: Pvt., Co. E. B. Rockbridge Co. 1838? Laborer age 22, Lexington Dist., Rockbridge Co. 1860 census. Served in Militia. Enl. Staunton 8/1/61. Ab. on leave 11-12/61. AWOL 1-2/62. Fined 2 mos. pay by CM. Present 3-4/62. Reenl. 5/1/62. WIA (ankle) McDowell 5/8/62. DOWs in hosp. Staunton from gangrene. Bur. Thornrose Cem., Staunton, grave #996. "A good soldier."

HILDEBRAND, BENJAMIN FRANKLIN: 2ndLt., Co. A. B. Augusta Co. 2/22/43. Farmhand, age 17, Burkes Mill Dist., Augusta Co. 1860 census. Enl. Staunton 7/15/61 as Pvt. Present 11/61-4/62. Reenl. 5/1/62. WIA Gaines Mill 6/27/62. Present 8/31/62-8/31/63. Elected 2ndLt. 9/28/63. Presence or absence not stated 9-12/63. WIA Cedar Creek 10/19/64. Ab. wounded through 11/29/64. Commanding Co. 12/31/64. Commanding 3 companies when surrendered Appomattox 4/9/65. Farmer, age 27, Mt. Sidney PO, Augusta Co. 1870 census. Died near Fishersville 8/17/07. Bur. Hildebrand Mennonite Ch. Cem., Madrid, Augusta Co.

HILDEBRAND, GABRIEL: Pvt., Co. A. B. Augusta Co. 11/23/23. Farmer, age 36, Burkes Mill Dist., Augusta Co. 1860 census. Pvt., Capt. J. K. Koiner's Co. Long Meadow Co., 32nd Va. Militia exempt 3/62 "applexy", age 39. Enl. Staunton 10/25/62. Present 11-12/64. Paroled Staunton 5/12/65. Age 44, 5' 9", light complexion, dark hair, hazel eyes. Farmer, age 46, Mt. Sidney PO, Augusta Co. 1870 census. Died Barren Ridge 1/11/93. Bur. Hildebrand Mennonite Ch. Cem., Madrid, Augusta Co.

HILL, C. P.: Pvt., Co. E. On postwar roster. Enl. 1861. Served 4 years.

HILL, DAVID: Pvt., Co. C. B. Va. 1832? Laborer, age 28, Burkes Mill Dist., Augusta Co. 1860 census. Enl. Staunton 7/16/61 age 25. Present 11/61-4/62. Reenl. 5/1/62. Deserted near Grodonsville 7/22/62. NFR.

HILL, JOHN A.: Pvt., Co. C. B. Nelson Co. 7/5/29. Farmer, age 31, 1st Dist. Augusta Co. 1860 census. Enl. Staunton 7/16/61 age 27. Ab. sick 7/16/61-7/20/62. Discharged for pulmonary disease and general debility 7/20/62. Age 31, 5' 10", light complexion, blue eyes, light hair. Farmhand, age 49, Riverheads Dist., Augusta Co. 1870 census. Died Riverheads Dist., 11/18/93. Bur. Mt. Tabor Lutheran Chem. Cem., near Middlebrook.

HILL, JOHN LYON: Pvt., Co. C. B. Augusta Co. 1838? On postwar roster. B. Augusta Co. 1837? Enl. Staunton 8/62. WIA (foot) by piece of shell Waynesboro 3/65. Married Jennings Gap, Augusta Co. 4/3/66. Farmhand, Dayton, Rockingham Co. 1905.

HINDMAN, JACOB.: Pvt., Co. H. B. 1844? Farmer, Albemarle Co. Not on muster rolls. Deserted to the enemy Cumberland, Md. 4/22/65. Took oath and sent north. Age 21, 5' 4", dark complexion, blue eyes, dark hair.

HINKLE, WILLIAM H.: Pvt., Co. E. B. 5/31/44. Enl. Pulaski Co. 12/9/64. Present 12/9-31/64. Deserted to the Army of the Potomac near Petersburg 3/8/65. Sent to Washington, D. C. Took oath and transportation furnished to Greene Co., Ohio. Died Rockingham Co. 4/8/05. Bur. Millcreek Ch. of the Brethren Cem., near Port Republic.

HINTLEY, WILLIAM HENRY: Pvt., Co. B. (1st) B. Bath Co. 1842. Carpenter, age 27, South River Dist., Rockbridge Co. 1860 census. Enl. Staunton 8/1/61. Present until transf. 9/28/61.

HISE, EMANUEL: Pvt., Co. D. B. Nurtemburg, Germany 5/18/23. Enl. Staunton 7/16/61. Ab. on leave 7/31-10/31/61. Present 11/61-8/31/62. Ab. on leave 2/19-3/5/63. Present 3-6/63. Ab. sick 6/4-12/31/63. Ab. detailed in hospital Staunton 4/26-12/31/64. Discharged for disability just before the surrender. Weaver, age 46, Mt. Sidney PO, Augusta Co. 1870 census. Died North River, Augusta Co. 5/22/96. Bur. St. Paul's Lutheran Ch. Cem., near Weyer's Cave.

HITE, BENJAMIN LEWIS.: Pvt., Co. I. B. Augusta Co. 6/6/18. Farmer, Augusta Co. Enl. Staunton 7/16/61 age 42. Present 11/61-4/62. Reenl. 5/1/62. Present 8/30-12/31/62, and 4/30-12/63. Ab. detailed as teamster 1/20-12/31/64. NFR. Farmer, age 51, Fishersville PO, Augusta Co. 1870 census. Died near Greenville 10/8/96. Bur. Haines Chapel Cem., near Vesuvius, Rockbridge Co.

HITE, JOHN MARTIN: PVT., Co. B. (2st) B. Rockbridge Co. 4/29/39. Enl. Fairfield 7/10/61. Present until transf. 9/28/61.

HITE, NATHANIEL WILSON: Pvt., Co. B. (1st). Enl. Fairfield 7/10/61. Present until transf. 9/28/61.

HITE, SAMUEL.: Pvt., Co. B. (1st) B. Va. 1824? Farmhand, age 36, 6th. Dist., Rockbridge Co. 1860 census. Enl. Fairfield 7/0/61. Ab. detailed in Staunton when transf. 9/28/61.

HITE, WILLIAM PAUL: Pvt., Co. B. (1st). B. Augusta Co. 6/19/45. Farmhand, age 15, South River Dist., Rockbridge Co. 1860 census. Enl. Fairfield 7/12/61. Present until transf. 9/28/61.

HIVELY, GEORGE W.: 4th Sgt., Co. K. B. Va. 1833? Chairmaker, age 27, Hot Springs PO, Bath Co. 1860 census. Enl. Shenandoah Mt. 4/9/62, as Pvt. Present 4/9-30/62, and 1-4/63. Appointed 4th Sgt. 4/15/63. WIA (left thigh) and cap. Gettysburg 7/3/63. Sent to DeCamp General Hospital, David's Island, N. Y. Exchanged 9/16/63. Ab. wounded through 10/31/64. Present 11-12/64. Cap. Ft. Stedman 3/25/65. Sent to Washington, D. C. Transf. Point Lookout. Released 5/13/65. Died Bath Co. before 8/17. Bur. below Hot Springs at Crowdertown RR Crossing on Rt. 615.

HODGES, THOMAS S.: Pvt., Co. H. B. 1831? Resident, Augusta Co. On postwar roster. Enl. Staunton 1861, age 30. Exempt Goshen Bridge Feb. 64, age 37, 5' 7", fair complexion, gray eyes, lt. hair. Farmer.

HODGES, WHITEHEAD: Pvt., Co. H. B. Goochland Co. 1815? Farmer, age 45, 1st Dist., Augusta Co. 1860 census. Enl. Staunton 7/23/61 age 47 as substitute. Ab. sick 11/14/61-4/62. Discharged for general bad health, dropsical effusion in chest and abdomen. Age 48, 5' 10", dark complexion, dark eyes, dark hair. Died before 1885.

HOGSHEAD, NEWTON HENDERSON: Pvt., Co. F. B. Va. 1843? Farmhand, age 17, Burkes Mill Dist., Augusta Co. 1860 census. Enl. Staunton 7/31/61. Died of fever in Augusta Co. 10/28/61.

HOGUE, SAMUEL E.: Pvt., Co. E. B. Staunton, Augusta Co. 1838? On postwar roster. Farmer, Churchville 1877. Died Staunton 9/15/95. Bur. Thornrose Cem.

HOGUE, WILLIAM MADISON: Pvt., Co. E. B. Rockbridge Co. 4/21/33. Farmer, Rockbridge Co. Enl. Staunton 8/1/61. Present 11/61-2/62. Ab. sick 3-4/62. Reenl. 5/1/62. Present 5/1-10/31/62, and 1-12/63. WIA (right arm and left lung) Spotsylvania CH 5/12/64. Ab. wounded through 12/31/64. NFR. Farmer, age 41, Natural Bridge Dist., Rockbridge Co. 1870 census. Died Roaring Run near Natural Bridge 8/5/04. Bur. High Bridge Pres. Ch. Cem., near Natural Bridge.

HOLBERT, EDWARD A.: Pvt., Co. I. B. Albemarle Co. 1846? Resident age 14, 1st Dist., Augusta Co. 1860 census. Enl. Greenville 8/8/63. Present 8-12/63. and 4/30-7/20/64. AWOL 7/21-12/31/64. NFR. Collier, age 25, 6th Dist., Augusta Co. 1870 census. Living Augusta Springs 1904. Laborer, age 65, Pond Gap. Augusta Co. 1910 census. Died 1920. Brother of William T. Holbert.

HOLBERT, WILLIAM THOMAS: Pvt., Co. I. B. Albemarle Co. 1846. Farmhand, age 16, 1st. Dist., Augusta Co. 1860 census. Enl. Staunton 7/19/61 age 19. Present 11/61-4/62. Reenl. 5/1/61. AWOL 7/20-11/7/62, and fined $39.27. WIA (over right temple) Fredericksburg 12/13/62. Present 1-12/63. WIA (through right breast and out shoulder and in arm) Wilderness 5/6/64. Ab. wounded through 12/31/64. Surrendered Appomattox 4/9/65. Resident, age 67, Staunton 1910 census. Died 1910. Bur. Asbury Methodist Ch. Cem., Augusta Co. Brother of Edward A. Holbert.

HOLLER, JACOB B.: Pvt., Co. B. (1st). B. Shenandoah Co. 2/7/39. Farmer, age 23, South River Dist., Rockbridge Co. 1860 census. Enl. Fairfield 7/10/61. Ab. sick when transf. 9/28/61.

HOLLIS, HARRISON B. "Henry": 2ndSgt. Co. H. B. 1844? Resident, Winchester. Enl. Staunton 7/23/61, age 17, as Cpl. Present 11/61-2/62. Promoted 4th Sgt. 1/1/62. Present 3-4/62. Reenl. 5/1/62. Present 5/1-8/31/62. Promoted 3rdSgt. 9/1/64. Present 1-12/63. Promoted 2ndSgt. '64. Capt. Bethesda Ch. 5/30/64. Sent to Point Lookout. Transf. to Elmira. Exchanged Savannah, Ga. 11/15/64. D. there 12/26/64. Bur. Grove Cem., Savannah, Ga. "Was a splendid soldier."

HOLT, CYRUS.: 2ndCpl., Co. G. B. Va. 1828? Farmer, age 32, 1st Dist., Augusta Co. 1860 census. Enl. Staunton 8/2/61 as Pvt. Promoted 2ndCpl. 10/31/61. Present 11/61-2/62. Ab. sick 3-4/62. Renenl. 5/1/62. Ab. on leave 5/1-6/5/62. Detailed in hosp. Staunton as Hosp. Steward 6/6/62-10/31/64. NFR. Steward, Va. Deaf, Dumb and Blind Inst., Staunton. D. Corinth, Ky. 10/12/75.

HOOVER, JACOB A.: Pvt., Co. I. B. Bath Co. 1823? Farmer, age 37, Cleeks Mill Dist., Bath Co. 1860 census. Pvt., Capt. Davis's Co., 81st Va. Militia. Enl. Shenandoah Mtn. 4/9/62. Present 4/9-30/62. Discharged overage of conscription. Enl. Co. G., 18th Va. Cav. 12/3/62. Paroled Staunton 5/25/65. Age 41, 5' 10½", dark hair, grey eyes. D. Western Dist., Bath Co. 10/6/88.

HOOVER, JOHN "Lum": Pvt., Co. G. B. Augusta Co. 1836? Plasterer, Augusta Co. Enl. 2/23/62 for the war. Present 3-4/62. KIA Gaines Mill 6/27/62. Age 26, 5' 6", fair complexion, blue eyes.

HOOVER, JOHN A.: Pvt., Co. K. B. Bath Co. 1823? Farmer, age 37, Hot Springs Dist., Bath Co. 1860 census. Pvt., Capt. Davis's Co., 81st Va. Militia. Enl. Shenandoah Mtn. 4/9/62. Present 4/9-30/62. WIA (hip and heel) McDowell 5/8/62. NFR. May have served in Co. G, 18th Va. Cav. Farmer, age 47, Bath CH PO, Bath Co. 1870 census. D. Western Dist., Bath Co. 12/12/82. Bur. Jones Cem., Bath Co.

HOOVER, JOHN J.: Pvt. Co. I. Not on muster rolls. WIA 2nd Manassas 8/28/62. NFR.

HOOVER, SAMUEL: 2ndCpl., Co. K. B. Va. 1830? Farmer, age 30, Hot Springs Dist., Bath Co. 1860 census. Pvt., Capt. Davis's Co., 81st Va. Militia. Enl. Shenandoah Mtn. 4/9/62 as Pvt. Present 4/9-30/62. Ab. sick in hosp. Chalottesville "deb. Aprlites" 2/25-4/8/63. Present 4-12/63. WIA Liberty 6/19/64. DOWs Gen. Hosp., Liberty 8/5/64. Left $10.75 and sundries. Left widow and 3 children.

HOOVER, WILLIAM A.: 2ndSgt., Co. K. B. Cedar Creek, Bath Co. 9/1/35. Farmer, age 22, Millboro Springs, Bath Co. 1860 census. Pvt., Capt. Davis's Co., 81st Va. Militia. Enl. Shenandoah Mtn. 4/9/62 as Pvt. Present 4/9-30/62. WIA McDowell 5/8/62. AWOL 9/13-10/25/62, tried by CM and acquitted. Present 1-4/63, promoted 2ndSgt. 4/1/63. Present 5-12/63, WIA (shoulder) Bethesda Ch. 5/30/64. Furloughed for 40 days 6/28/64. Ab. wounded through 10/31/64. Present 11-12/64. Capt. Ft. Stedman 3/25/65. Sent to Washington, D. C. Transf. to Pt. Lookout. Released 5/13/65. Farmhand, age 34, Bath CH PO, Bath Co. 1870 census. Farmer, Hot Springs 1897. D. Cedar Creek Dist., Bath Co. 6/6/13. Bur. Jones Cem., Bath Co.

HOSTETTER, HENRY, Pvt., Co. E. B. Va. 1820? Laborer, age 40, 5th Dist., Rockbridge Co. 1860 census. Enl. Staunton 8/1/61. Present 11/61-4/62. Reenl. 5/1/62. Present 5/1-10/31/62. Present 4/30/63-12/31/64. Detailed as wagoner, Ordnance train. NFR. Pension application indicates WIA eye date and place unknown. Laborer, age 45, Buffalo Dist., Rockbridge Co. 1870 census. D. Collierstown 4/2/98 "from wounds, old age and debility." Bur. House Mountain Cem.

HOSTETTER, LEWIS J. M.: Pvt., E. B. Va. 1838? Farmhand, age 22, Collierstown PO, Rockbridge Co. 1860 census. Enl. Staunton 8/1/61. Present 11/61-4/62. Reenl. 5/1/62. Present 5/1-6/30/62. Ab. sick 7/1/62-12/63. Deserted from his home in Rockbridge Co. where he had been permitted to go sick by Surgeon of hospital. Reported by Capt. Rogers, Enrolling Officer of Rockbridge Co. as having gone to the enemy, date unknown. No Federal Records. Laborer, age 29, Buffalo Dist., Rockbridge Co. 1870 census. Died Collierstown 12/8/80.

HOSTETTER, MORGAN: Pvt., Co. E. Not on muster rolls. Reported to be present by Lt. Joseph Knick 9/2/61. NFR.

HOTTEL, JOSEPH FRANKLIN: Capt., Co. D. B. Augusta Co. 10/21/16. Constable, age 43, Northern Dist., Augusta Co. 1860 census. Enl. Staunton 7/16/61. Present 7/16/61-4/62. Not reelected 5/1/62. Capt., 3rd Bn. Va. Reserves 4/64. Present battle of Piedmont 6/5/64. Hotelkeeper, age 54, Mt. Solon 1870 census. Merchant. Died Mt. Solon 5/1/89. Bur. Sangersville Ch. of the Brethren Cem.

HOULTZ, WILLIAM H.: Sgt., Co. G. B. 1842. Enl. Staunton 8/2/61 as Pvt. Ab. sick 9/10-12/31/61. Present 1-4/62. Reenl. 5/1/62. Promoted Sgt. WIA 2nd Manassas 8/29/62. WIA (elbow) Chantilly 9/1/62. Ab. wounded 9-10/62. WIA (leg) Fredericksburg 12/13/62. Leg amputated. DOWs 12/24/62, age 21. Bur. Rocky Springs Cem., Deerfield, Augusta Co.

HOUSER, JOHN L.: Pvt., Co. I. B. Augusta Co. 1842? Blacksmith, age 18, Staunton PO, Augusta Co. 1860 census. Enl. Staunton 7/16/61 age 20. Present 11/61-4/62. Reenl. 5/1/62. Present 8/31-10/31/62. Ab. detailed as Brigade Butcher 1/19-12/31/63. WIA (ball passing through right groin into abdomen cavity) Wilderness 5/6/64. DOWs from "peretonitis" at General Hospital, Charlottesville 5/14/64.

HOWARD, THOMAS A.: Pvt., Co. H. B. Va. 1842. Farmhand, age 18, Burkes Mill Dist., Augusta Co. 1860 census. Enl. Staunton 7/23/61 age 18. Ab. sick 12/15/61-2/62. Present 3-4/62. Reenl. 5/1/62. WIA (thigh) Port Republic 6/9/62. Leg shortened. Ab. wounded through 12/64. Retired to Invalid Corps 12/29/64. Applied for pension because of wound 7/13.

HOWDYSHELL, LORENZA, JR.: Pvt., Co. D. B. Rockingham Co. 8/16/14. Farmer, Rockingham Co. Enl. Staunton 7/16/61. Present 7/16/61-4/62. AWOL 4/24-5/3/62. Present 5/4-7/20/62. AWOL 7/21-8/31/62, fined $14.66. Discharged 12/5/62 over age of conscription. Age 42, 5' 3", dark complexion, dark hair, dark eyes. Farmhand, age 51, Mt. Sidney PO, Augusta Co. 1870 census. Died Thorny Branch, North River Dist., Augusta Co. 6/9/89. Bur. Emanuel Ch. Cem., near Mt. Solon.

HUFF, DANIEL.: Pvt., Co. C. B. Augusta Co. 10/23/33. Farmhand, age 27, Burkes Mill Dist., Augusta Co. 1860 census. Enl. Staunton 7/16/61 age 27. Present 11/61-4/62. Reenl. 5/1/62. Cap. Port Republic 6/15/62. Exchanged 8/5/62. WIA (arm and shoulder) 2nd Manassas 8/28/62. Present 8/31/62-7/63. Cap. Gettysburg 7/4/63. Sent to Ft. Delaware, Rejected for enlistment in the U. S. Army 8/30/63. Released 5/65. 5' 8", light complexion, grey hair, blue eyes, illiterate. Farmer, age 35, Mt. Sidney PO, Augusta Co. 1870 census. Hit by automobile and died of injuries Staunton 11/19/09. Bur. Thornrose Cem.

HUFF, DAVID E.: Pvt., Co. C. B. Augusta Co. 1837? Miller, Augusta Co. Enl. Staunton 7/16/61 age 24. Present 11/61-1/62. AWOL 2/26-28/62. Fined $11.00, and forfeit $6.96. Reenl. 5/1/62. Cap. Port Republic 6/15/62. Exchanged 8/5/62. WIA 2nd Manassas 8/28/62. Present 8/31/62-6/63. AWOL 12/63. Deserted to the enemy Hampshire Co. 8/10/63. Sent to Camp Chase. Took oath and released. 10/2/63. Age 26, 5' 9¾", florid complexion, grey eyes, dark hair.

HUFF, JOHN HENRY, JR.: Pvt., Co. G. B. Augusta Co. 12/25/37. Farmhand, age 22, Burkes Mill Dist., Augusta Co. 1860 census. Enl. Staunton 8/2/61. Ab. on leave 5/1-8/31/62. Present 11/62-12/63. WIA 7/15/64. Ab. wounded through 10/64. Deserted to the enemy 10/10/64. NFR. Farmer, Augusta Co. 1870. Died Augusta Co. 3/29/02. Bur. Pleasant Valley Ch. of the Brethren, near Weyer's Cave. Brother of William Huff.

HUFF, NOAH W.: 1stSgt., Co. G. B. Va. 1836? Enl. Staunton 8/2/61 as 1stCpl. Present 11/61-4/62. Reenl. 5/1/62 Cap. Cross Keys 6/9/62. Exchanged 8/62. Age 26, 5' 7". Present 8/62-2/63, rank shown as 2ndSgt. Present 3-4/63, promoted 1stSgt. 3/15/63. Deserted to the enemy Gettysburg 7/5/63. Sent to Ft. Delaware. Took oath and joined 3rd Md. Cav. U.S. Army 9/22/63.

HUFF, WILLIAM: Pvt., Co. G. B. Augusta Co. 1840? Farmer, Augusta Co. Enl. Westview 4/16/62. KIA Cross Keys 6/8/62. Age 22, 5' 6", fair complexion, fair hair, blue eyes. Bur. Middle River Cem. Brother of John H. Huff, Jr.

HUFFER, ABRAHAM: Pvt., Co. D. B. Augusta Co. 4/7/24. Farmhand, age 37, Burkes Mill Dist., Augusta Co. 1860 census. Enl. Staunton 7/16/61. Ab. on leave 7/16-10/31/61. Present 11/61-3/62. Sent to hospital Staunton from Camp Alleghany 4/1/62. Present 4/30-7/28/62. AWOL 7/29-8/31/62, fined $12.10. Present 1-3/63. Detailed as wagoner in Div. Train 3/13/63-12/31/64. NFR. Died Augusta Co. 12/8/84. Bur. Emanuel Ch. Cem., near Mt. Solon.

HUFFER, SAMUEL: Pvt., Co. D. B. Va. 11/17/27. Farmer, age 35, Burkes Mill Dist., Augusta Co. 1860 census. Enl. Staunton 7/16/61. AWOL 7/16-10/31/61. Present 11/61-3/62. AWOL 4/24-5/2/62. Present 5/3-7/11/62. AWOL 7/12-8/31/62, fined $18.33. Present 1-3/63. AWOL 7/15-8/7/63, fined $8.06. Present 9-12/63. Cap. Bethesda Ch. 5/30/64. Sent to Point Lookout. Transf. to Elmira. Released 6/30/65. 5' 9", florid complexion, auburn hair, blue eyes. Farmer, age 43, Mt. Sidney PO, Augusta Co. 1870 census. Died North River, near Mt. Solon 4/20/90. Bur. Emanuel Ch. Cem.

HUFFMAN, ALBERT HENRY: 2ndCpl., Co. G. B. Va. 1832? Farmhand, age 18, Burkes Mill Dist., Augusta Co. 1860 census. Enl. Staunton 8/2/61 as Pvt. Present 11/61-4/62. Reenl. 5/1/62. WIA (hip) Cross Keys 6/8/62. Present 5/1/62-12/63, promoted 3rdCpl. 11-12/63, 2ndCpl. '64. WIA Winchester 9/19/64. DOWs 9/29/64, age 20.

HUFFMAN, JOHN R.: Pvt., Co.H. B. Va. 1843? Farmhand, age 17, Burkes Mill Dist., Augusta Co. 1860 census. Enl. Staunton 7/23/61 age 17. Present 11/61-10/31/62. Discharged at Camp 10/16/62 under age. Farmer, age 25, Mt. Sidney PO, Augusta Co. 1870 census. Alive 1885.

HUFFMAN, JOHN S.: Pvt., Co. F. B. Rockingham Co. 8/7/46. Student, age 13, Burkes Mill Dist., Augusta Co. 1860 census. Enl. Winchester 6/1/62. Deserted 6/2/62. NFR. Resident of Shenandoah, Augusta Co. Farmer, age 25, Mt. Sidney PO, Augusta Co. 1870 census. Died Augusta Co. 7/10/19. Bur. St. Paul's Ch. Cem., near Weyer's Cave.

HUFFMAN, SYLVESTER F.: Pvt., Co. F. B. Rockingham Co. 1835? Stone mason, age 25, Burkes Mill Dist., Augusta Co. 1860 census. Enl. Mt. Meridian 6/15/62. Present 8/31/62-2/63. AWOL 7/28/63. Deserted to the enemy Clarksburg, W. Va. 10/13/63. Took oath and sent north. Farmer, Augusta Co. Died Mt. Solon 10/15/88. Brother of William D. Huffman.

HUFFMAN, WILLIAM DANIEL: Pvt., Co. A. B. Rockingham Co. 6/24/44. Painter, age 23, Burkes Mill Dist., Augusta Co. 1860 census. Pvt. Capt. Shoemake's Co. 32 Va. Militia exempt 3/62 "varoiascele", age 25. Enl. Rappahannock 8/22/62. Ab. sick 8/26/62-1/63. Discharged for vancocele and chronic rheumatism from hospital Staunton 1/29/63. Age 25, 6', fair complexion, grey eyes, light hair. Deserted to the enemy Clarksburg, W. Va. 10/14/63. Took oath and sent north. Farmer, Augusta Co. Died Mt. Sidney 10/7/18. Bur. Frieden's Ch. Cem., near Mt. Crawford. Brother of Sylvester F. Huffman.

HUGHES, DAVID E.: Pvt., Co. E. B. Rockbridge Co. 1845? Farmhand, age 20, Collierstown PO, Rockbridge Co. 1860 census. Enl. Staunton 8/1/61. Present 11/61-4/62. Reenl. 5/1/62. Present 5/1-10/31/62, AWOL 3 months. Present in arrest 1-2/63. Present 3-6/63. WIA Gettysburg 7/3/63. Present 7-10/63, and 4/30-5/21/64. Deserted while joining line of battle near Hanover Junction 5/22/64. Present 11-12/64. NFR. Farmer, age 25, near Lexington, Rockbridge Co. 1870 census. Farmer, Kerr's Creek Rockbridge Co. 1884. Brother of Johnathan F., Lewis and Robert H. Hughes.

HUGHES, ELIJAH M.: Pvt., Co. B (1st). B. near Fairfield, Rockbridge Co. 11/1/30. Farmer, age 27, South River Dist., Rockbridge Co. 1860 census. Enl. Fairfield 7/12/61. Present until transf. 9/28/61. Brother of Joseph P. Hughes.

HUGHES, JOHNATHAN FRANKLIN: Pvt., Co. E. B. Rockbidge Co. 1835? Farmhand, age 25, Collierstown PO, Rockbridge Co. 1860 census. Enl. Staunton 8/1/61. Present 11/61-4/62. Reenl. 5/1/62. Present 5/1-10/31/62, and 1-2/63, detailed as wagoner. Present 3-6/63. AWOL 7/26-9/28/63, returned under arrest and left in charge of Provost Guard, Orange CH 10/8/63. Present in arrest 11-12/63. Deserted from Provost Guard while ab. in arrest, date and place unknown 4/30-10/31/64. Appears on rolls of Co. A, Ward's Bn, C. S. Prisoners recently released from prison, Lynchburg 7/64 for their participation in the defense of Lynchburg. Pardoned by the President. NFR. Farmhand, age 35, Buffalo Township, Rockbridge Co. 1870 census. Died Kerr's Creek 8/3/08. Bur. Collierstown Pres. Ch. Cem. Brother of David E., Lewis and Robert H. Hughes.

HUGHES, JOSEPH PRESTON: Pvt., Co. B (1st). B. Rockbridge Co. 1822? Farmer, age 38, South River Dist., Augusta Co. 1860 census. Enl. Fairfield 7/12/61. Ab. on detail Staunton when transf. 9/28/61. Brother of Elijah M. Hughes.

HUGHES, LEWIS: Pvt., Co. E. B. Rockbridge Co. 1824? Blacksmith, age 36, Colliestown PO, Rockbridge Co. 1860 census. Enl. Staunton 8/1/61. Present 11/61-4/62. Reenl. 5/1/62. Present 5/1-10/31/62, AWOL 1 month, fined $15.00 for arrest as deserter. Present 1-5/63. WIA Fredericksburg 5/5/63. DOWs and typhoid fever at home in Rockbridge Co. 6/16/63. Bur. Collierstown Pres. Ch. Cem. Brother of David E., Johnathan F. and Robert H. Hughes.

HUGHES, ROBERT H.: Pvt., Co. E. B. Rockbridge Co. 1823? Farmer, age 37, Collierstown PO, Rockbridge Co. 1860 census. Enl. Camp 52nd Va. Inf. 2/22/64. Present 2/22-5/21/64. Deserted while in line of battle near Hanover Junction 5/22/64. NFR. Shoemaker, age 47, Buffalo Township, Rockbridge Co. 1870 census. Died on Toad Run, Rockbridge Co. 2/8/85. Bur. Collierstown Pres. Ch. Cem. Brother of David E., Johnathan F. and Lewis Hughes.

HUGHART, CHARLES HARVEY: Pvt., Co. K. B. Bath Co. 10/25/21. Laborer, age 38, Green Valley Dist., Bath Co. 1860 census. Enl. Shenandoah Mt. 4/9/62. Present 4/9-30/62. Present McDowell and other battles. Ab. sick 5/26/62. Discharged 7/25/62. Elected Justice of the Peace, Williamsville Dist. Bath Co. Enl. Co. E, 10th Bn. Va. Res. 4/16/64, age 46, 6', dark complexion, dark hair, dark eyes, whiskers. Cabinetmaker, age 48, Bath CH. PO, Bath Co. 1870 census. Justice of the Peace Williamsville Dist., 16 years and Millboro 8 years. Living Valley Mills, Augusta Co. 1900. Died Westview, Augusta Co. 1/31/01. Bur Windy Cove Pres. Ch. Cem., Millboro Springs, Bath Co.

HULTZ, DAVID W.: Pvt., Co. G. B. Augusta Co. 1830? Farmer, Augusta Co. Enl. Staunton 2/4/63. Never joined Co. Discharged for "cronic articular rheumatism" 2/19/63. Age 32, 6', dark complexion, dark hair, grey eyes. Farmer, age 40, Fishersville PO, Augusta Co. 1870 census.

HUMPHREY, JAMES W.: Pvt., Co. C. B. Augusta Co. 12/1844. Farmhand, age 15, 1st Dist., Augusta Co. 1860 census. Enl. Staunton 8/16/61. Present 11-12/61. Died of fever in hospital Camp Alleghany 1/14/62, age 17.

HUMPHREY, JESSE.: Pvt., Co. C. B. Augusta Co. 1835? Laborer, age 25, Staunton 1860 census. Enl. Staunton 7/15/61 age 24. Present 11/61-3/62. AWOL 4/22-30/62. Reenl. 5/1/62. Deserted Port Republic 6/9/62. NFR. Resident of Summerdean, Augusta Co.

HUMPHREY, THOMAS H.: Pvt., Co. C. B. Augusta Co. 1830? Laborer, Augusta Co. Enl. Staunton 7/16/61 age 30. AWOL 11-12/61, fine $11.00 by CM. Present 1-3/62. AWOL 4/22-30/62. KIA 2nd Manassas 8/28/62, age 26. Left widow.

HUMPHREYS, HENRY H: Pvt., Co. H. Name cancelled on rolls. NFR. Resident, Alleghany Co. 1922.

HUMPHREYS, JAMES G.: Pvt., Co. B (1st). B. Eastern Va. 1841? Died of typhoid fever 9/1/61 age 20 on post war roster.

HUMPHREYS, JOHN MOORE.: Capt., Co. I. B. near Bethel Ch. Augusta Co. 2/29/20. Farmer, age 40, 1st Dist., Augusta Co. 1860 census. Enl. Staunton 7/16/61 as 2ndLt., age 41. Present 11/61-4/62. Elected Capt. 11/21/61. Reelected Capt. 5/1/62. Present 5/1-8/62. WIA (mouth and lost several teeth) McDowell 5/8/62. Returned to duty 7/29/62. WIA (arm) 2nd Manassas 8/29/62. Commanding Co. Fredericksburg 12/13/62. Present 1-12/63. WIA (left arm) Wilderness 5/6/64. WIA (foot) Bethesda Ch. 5/30/64, acting Major of Regt. WIA (left foot) Cold Harbor 6/3/64. Returned to duty 7/4/64. Commanding Regt. 8/31, 9/30 and 10/19/64. WIA and cap. Cedar Creek 10/19/64. Sent to Ft. Delaware. Released 6/12/65. 6' 2" light complexion, brown hair, grey eyes. Farmer, age 50, South River Dist., Augusta Co. 1870 census. Died Spottswood, Augusta Co. 3/1/02. Bur. Bethel Pres. Ch. Cem. "A brave Confederate soldier."

HUMPHRIES, JAMES GREEN: Pvt., Co. G. B. Nelson Co. 5/19/29. Teacher, Augusta Co. Enl. Staunton 8/2/61. Ab. sick 11/10/61-4/62. Discharged for injury of lower part of back and palsy of the right leg 1/30/63. Age 36, 5' 10", fair complexion, blue eyes, light hair. Shoemaker, Augusta Co. 1867. Died Augusta Co. 7/3/96. Bur. Mt. Bethel Methodist Ch. Cem., near Crimora.

HUMPHRIES, JAMES H.: Pvt., Co. B (1st). B. Va. 1841. Farmhand, age 19, South River Dist., Rockbridge Co. 1860 census. Enl. Fairfield 7/10/61. Present until transf. 9/28/61.

HUNTER, CYRUS W.: 1stLt., Co. F. B. Augusta 1836? 1stLt., Co. G, 182nd Va. Militia. Enl. Staunton 7/31/61. Ab. sick 11/12/61-4/62. Not reelected 5/1/62. Farmer, age 34, Mt. Sidney PO, Augusta Co. 1870 census. Book Seller and Stationer, Staunton. Moved to Richmond 1887. Wagonmaker, Stonewall, 1897. Died near Roman, Augusta Co. 8/98.

HUNTER, DAVID S.: Pvt., Co. A. B. Va. 2/7/12. Farmer, Mt. Sidney, Augusta Co. Enl. Staunton 7/15/61. Ab. sick 11/16/61-4/27/62. Present 4/28-30/62. Reenl. 5/1/62. Deserted 7/3/62. Farmer, age 59, Mt. Sidney PO, Augusta Co. 1870 census. Died Augusta Co. 3/28/09. Bur. Old Salem Lutheran Ch. Cem., 3 miles west of Mt. Sidney.

HUNTER, JOHN T.: Pvt., Co. A. B. Augusta Co. 2/27/40. Farmhand, Augusta Co. Enl. Staunton 7/15/61. Ab. sick 11/7/61-2/25/62. Present 2/26-4/62. Reenl. 5/1/62. WIA (leg) Port Republic 6/9/62. Deserted from hospital date unknown 9-10/62. Enl. Capt. McNeill's Co., 18th Va. Cav. Farmhand, age 30, Mt. Sidney PO, Augusta Co. 1870 census. Farmer and Merchant, Spring Hill, Augusta Co. Died 8/18/93. Bur. Old Salem Lutheran Ch. Cem., 3 miles west of Mt. Sidney.

HUPMAN, GEORGE T.: Pvt., Co. A. B. Augusta Co. 1844? Enl. "on the march" 12/1/62. Present 12/62-12/63. Ab. in hospital Petersburg with impetigo 2/19-4/2/64. Present 4/30-10/21/64. Ab. sick 10/22-12/31/64. Returned to duty 4/2/65. Paroled Staunton 5/1/65. Age 21, 5' 5", fair complexion, light hair, blue eyes. Member, Stonewall Jackson Camp CV, Staunton. Farmer, near Kansas City, Mo. 1900. Alive 7/18/11. Bur. Confederate Cem, Higginsville, Mo.

HUPMAN, JOHN A. (1st): Pvt., Co. K. B. Augusta Co. 1841? Resident of Highland Co. Not on muster rolls. Listed as a deserter 2/63. Cap. date and place unknown. Possibly _____ Huppman, killed Camp Chase, Oh. 12/20/64. age 23, resident of Back Creek, Bath Co. Bur. Camp Chase National Cem.

HUPMAN, JOHN ALEXANDER (2nd): Pvt., Co. A. B. Augusta Co. 10/15/45. Enl. Staunton 7/18/61. Present 11/61-4/62. Reenl. 5/1/62. Present 8/31-12/31/62. Ab. sick 2/15-4/63. Issued clothing 5/19 and 28/63. Cap. Gettysburg 7/4/63. Sent to U. S. Army Hospital, Baltimore. Exchanged 9/27/63. In hospital Richmond 9/28-10/11/63. Furloughed for 60 days. Ab. through 12/63. Issued clothing 4/21 and 29/64. WIA (left hand) Bethesda Ch. 5/30/64. Cap. Carter's Farm, near Winchester 7/20/64. Sent to Wheeling. Age 22, 5' 9", dark complexion, dark hair, dark eyes, Farmer, Augusta Co. Sent to Camp Chase. Exchanged 3/10-12/65. Paroled Staunton 5/1/65. Age 23, 5' 7", fair complexion, fair hair, hazel eyes. Farmer, age 27, Mt. Sidney PO, Augusta Co. 1870 census. Deputy Tax Collector, Beverly Manor Dist.; Farmer and Stockraiser, Churchville 1884. Deputy Sheriff, Staunton 1910. Member, Stonewall Jackson Camp CV, Staunton. Died Staunton 8/1/19.

INGRAM, ALEXANDER: Pvt., Co. H. B. Rockbridge Co. 1842? Laborer, age 18, Goshen Bridge, Rockbridge Co. 1860 census. Enl. Staunton 7/23/61 age 20. Present 11/61-4/62. WIA (shoulder) Cross Keys 6/8/62. DOWs "tetnus seguelas" in hospital Staunton 6/28/62. Bur. Thornrose Cem., grave #1079.

INGRAM, HUGH: Pvt., Co. H. B. Rockbridge Co. 1830? Farmer, age 30, 1st Dist., Augusta Co. 1860 census. Enl. Staunton 7/23/61 age 38. Present 11-12/61, fined $11.00 for AWOL by CM 10/27-30/61. Present 1-2/62. AWOL 3/5/62-2/8/63. Present in arrest 2/9-4/63. Sentenced by CM for desertion to 2 years confinement in penitentiary and pay forfeited to 8/1/63. Present 9-12/63. Ab. 4/30-10/31/64 sentenced to hard labor for 12 months for desertion by Corps CM. Charged with loss of bayonet scabbard, haversack, knapsack, canteen, gun sling, 40 rounds of ammunition. Present 11-12/64. NFR. "Left Reg. after surrendered at Appomattox and came Home," on pension application. Farmhand, age 46, Walker's Creek Dist., Rockbridge Co. 1870 census. Died Buffalo Gap, Augusta Co. 5/16/05.

INGRAM, JOSEPH.: Pvt., Co. H. B. 1841? Farmer, Goshen Bridge, Rockbridge Co. Enl. Staunton 7/23/61 age 20. Present 11/61-4/62. AWOL 5/9/62-12/63. Deserted to the enemy Beverly, W. Va. 2/13/64. Took oath and released. Age 22, 5' 6½", fair complexion, black hair, blue eyes.

IRWIN, GEORGE: 4thCpl., Co. E. Enl. Staunton 8/1/61 as Pvt. Present 11/61-4/62. Reenl. 5/1/62. Appointed 4thCpl. WIA (thigh) Cedar Run 8/9/62. In hospital 8/11-10/11/62 when deserted. Ab. wounded through 2/63. Present 3-12/63. Issued clothing 3/1 and 4/21/64. Deserted from line of battle near Hanover Junction 5/22/64. NFR.

JACK, JOHN H.: Pvt., Co. K. B. Va. 1844? Farmhand, age 16, Cleeks Mill PO, Bath Co. 1860 census. Enl. Shenandoah Mt. 4/9/62. Present 4/9-30/62. WIA (leg) McDowell 5/8/62. WIA (leg) Port Republic 6/9/62. Present 1-2/63. Ab. sick 3/18-5/21/63. Cap. Gettysburg 7/3/63. Sent to Ft. Delaware. Transf. to Point Lookout. In hospital there 12/9-23/63. Died Point Lookout 3/27/64, cause unknown. Resident of Millboro Springs, Bath Co. Bur. in POW Cem. there.

JACK, WILLIAM Z. B.: 3rdCpl., Co. K. B. Va. 1844? Farmhand age 16, Bath CH PO, Bath Co. 1860 census. Enl. Shenandoah Mt. 4/9/62. Sick in hospital Winchester with "rubeola" 9/10/62. Present 1-12/62, appointed 3rdCpl. 9/1/63. Cap. Bethesda Ch. 5/30/64. Sent to Point Lookout. Transf. to Elmira. Released 5/19/65. 5' 9½", florid complexion, red hair, blue eyes, resident of Millboro, Bath Co. Died Before 8/17.

JACKSON, ADDISON: Pvt., Co. C. B. Nelson Co. 4/4/35. Carpenter age 25, 1st Dist., Staunton 1860 census. Enl. Bunker Hill 10/10/62. Present 10/10/62-3/24/63. AWOL 3/25-4/63. Cap. Waynesboro, Pa. 7/8/63. Sent to Ft. Delaware. Took oath and joined 3rd Md. Cav. (U.S.) 9/22/63. Died Augusta Co. 6/15/72. Bur. Old Cem., Waynesboro.

JACKSON, ANDREW GEORGE: Pvt., Co. K. B. Alleghany Co. 1822? Farmhand, age 38, Millboro Springs PO, Bath Co. 1860 census. Pvt., Capt. Davis' Co., 81st Va. Militia. Enl. Bath Co. 3/3/64. WIA (hip) near Spotsylvania CH 5/19/64. DOWs 6/4/64, age 41. Left Widow. Also reported to have Enl. Co. G, 18th Va. Cav. 1/2/63.

JACKSON, JACOB H.: Pvt., Co. H. B. Augusta Co. 1827? Farmer, age 33, Pond Gap, Augusta Co. 1860 census. Enl. Staunton 7/23/61. AWOL 7/23/61-12/63. Dropped as deserter. NFR. Farmer, age 43, Walker's Creek Dist., Rockbridge Co. 1870 census.

JAMISON, WILLIAM T.: Pvt., Co. A. B. Va. 1841? Farmhand; age 19, Burkes Mill Dist., Augusta Co. 1860 census. Enl. Staunton 7/9/61 age 18. Present 11/61-4/62. Reenl. 5/1/62. WIA (wrist) Sharpsburg 9/17/62. Ab. wounded through 4/63. Present 4/30-12/62. Cap. Bethesda Ch. 5/30/64. Sent to Point Lookout. Transf. Elmira. Released 6/19/65. 5' 9". dark complexion, auburn hair, grey eyes, illiterate, resident of Staunton. Died before 5/03.

JARVIS, JAMES EVERETT: 2ndCpl., Co B (1st). B. Va. 4/22/43. Enl. Fairfield 7/10/61. Present until transf. 9/28/61.

JARVIS, JOHN: Pvt., Co. B. (1st). On postwar roster.

JEFFRIES, JAMES W.: Pvt., Co. A. Enl. Martinsburg 7/30/64. Cap. Winchester 9/19/64. Sent to Point Lookout. Took oath and released by order of the President dated 2/18/65 on 2/21/65.

JENKINS, JOHN A. J.: Pvt., Co. E. B. near Natural Bridge, Rockbridge Co. 12/4/26. Farmer, age 32, Natural Bridge Dist., Rockbridge Co. 1860 census. Enl. Staunton 8/1/61. Present 11/61-4/62. Reenl. 5/1/62. WIA (hip) McDowell 5/8/62. WIA Cross Keys 6/8/62. Present 6-10/62. Present 1-12/63, pay stopped for 1 cartridge box $6.00, canvas shoulder strap $1.50, canvas waist belt $1.50, cap pouch, $2.23, lost. WIA Middletown 7/15/64. Ab. wounded through 10/31/64. Present 11-12/64. Injured by fall from High Bridge near Farmville 4/7/65. Surrendered Appomattox 4/9/65. Farmer, Fancy Hill, Rockbridge Co. Killed by falling tree 2/27/97. Bur. Natural Bridge Baptist Ch. Cem.

JOHNS, SAMUEL S.: Pvt., Co. A. B. Va. 1842. Laborer, age 18, Burkes Mill Dist., Augusta Co. 1860 census. Enl. Staunton 7/16/61. Present 11/61-4/62. Reenl. 5/1/62. Present 8/31-11/22/62. AWOL 11/23-12/14/62. Present in arrest 12/15/62-2/63. Sentenced to forefeit 6 months pay in addition to pay while AWOL. Present 4/30-12/63. Issued clothing 4/3 and 20/64. WIA (left arm) Spotsylvania CH 5/12/64. WIA Winchester 9/19/64. Ab. wounded through 12/31/64. NFR. Carpenter, age 25, Churchville, Augusta Co. 1870 census.

JOHNSON, BENJAMIN F.: Pvt., Co. B. (2nd). B. Louisa Co. 1842? Carpenter, age 18, 1st Dist., Augusta Co. 1860 census. Enl. Waynesboro 7/15/61. Present 11/61-4/62. Reenl. 5/1/62. WIA (shoulder) McDowell 5/8/62. Ab. from wounds until discharged for loss of use of arm and wound still discharging 4/27/63. Age 24, 6', fair complexion, grey eyes, light hair. Paroled Gordonsville 5/20/65.

JOHNSON, ROBERT W.: 4thCpl., Co. B (1st). B. 1833? Enl. Fairfield 7/10/61. Present until transf. 9/28/61.

JOHNSON, SAMUEL M.: Pvt., Co. A. B. Augusta Co. 1838? Farmhand, age 22, Burkes Mill Dist., Augusta Co. 1860 census. Enl. Staunton 7/9/61 age 24. Died of typhoid pneumonia Camp Alleghany 11/9/61.

JOHNSON, WILLIAM F.: Sgt. Co. B. (1st). Farmer, 18, Oakdale PO, Rockbridge Co. 1860 census. Not on muster roll. Enl. 1861.

JOHNSON, WILLIAM M: Pvt., Co. A. B. Va. 1840? Farmhand, age 20, Lexington Dist., Rockbridge Co. 1860 census. Unofficial souce indicates enl. 4/15/62. NFR.

JOHNSTON, JOHN BLACK: Pvt., Co. A. B. Augusta Co. 1843? Farmer, Augusta Co. Enl. Staunton 7/19/61 age 18. Present 11/61-4/62. Reenl. 5/1/62. WIA Gaines Mill 6/26/62. DOWs 6/28/62. Age 20, 6' ½", fair complexion, black eyes, black hair.

JOHNSTON, OSBORNE S.: Pvt., Co. I. B. Rockbridge Co. 1846? Farmhand, age 14, Staunton PO, 1860 census. Enl. Valley Mills 4/30/62. Present 8/30-1/31/62 and 1/63-11/1/64. AWOL 12/8-31/64. Des. to the enemy Beverly, W. Va. 12/28/64. Took oath and transportation furnished to Parkersburg. Age 18, 5' 7", fair complexion, hazel eyes, black hair, farmer.

JONES, CALVIN H.: Pvt., Co. B. (2nd). B. Va. 1837? Farmhand, age 23, Burkes Mill Dist., Augusta Co. 1860 census. Enl. Waynesboro 7/15/61. Present 11/61-4/62. Reenl. 5/1/62. WIA McDowell 5/8/62. Transf. Co. H. 5th Va. Inf. 5/14/63. WIA arm and through both legs. Illiterate. Farmer, Mechum's River, Albemarle Co. 1897. Died 1914.

JONES, JOHN G.: Pvt., Co. B (2nd). B. Va. 1831? Farmhand, age 29, Burkes Mill Dist., Augusta Co. 1860 census. Not on muster rolls. WIA (both thighs) McDowell 5/8/62. NFR.

JONES, OLIVER: Pvt., Co. E. B. Va. 1838? Farmhand, age 22, Lexington Dist., Rockbridge Co. 1860 census. Served in militia. Enl. Staunton 8/1/61. Present 11/61-2/62. Died of pneumonia Camp Alleghany 3/6/62.

JONES, RICHARD S.: Pvt., Co. D. B. Va. 1820? Joiner, age 40, Sangersville, Augusta Co. 1860 census. Enl. Staunton 7/16/61. Present 7/16-12/61. Present 1-2/62, detailed to mould candles for commissary 2/13/62. Present 3-4/62, left Co. at Ryan's 4/19 and returned 5/2/62. AWOL 5/19-8/6/62, fined $28.60. Present in arrest 1-2/63. Discharged over age of conscription 3/27/63. Age 40, 5' 11½", fair complexion, blue eyes. Indebted to government for loss of 1 Hall's rifle with bayonet $30.00, cartridge box $2.50, cap pouch $1.00, waist belt 75¢, bayonet scabbard $1.00.

JONES, THOMAS FRANKLIN: Pvt., Co. K. B. Rockingham Co. 8/10/27. Tailor, Back Creek, Bath Co. Enl. Bunker Hill 10/16/62. Deserted 11/29/62. NFR. Died Naked Creek, Augusta Co. 10/71.

JONES, WILLIAM: Pvt. Co. B (2nd). Substitute. Not on muster rolls. Listed as deserter 2/63. NFR. Possibly William H. Jones B. 1/15/38. Died 11/9/61. Bur. Old Cem., Waynesboro.

JONES, WILLIAM H.: Pvt., Co. C. B. Augusta Co. 1837? Resident of Deerfield, Augusta Co. Enl. Staunton 7/16/61 age 24, Present 11/61-2/62, fined by CM $5.00 and 6 days pay deducted. Present 3-4/62. Reenl. 5/1/62. Cap. Port Republic 6/9/62. Sent to Ft. Delaware, age 23, 5' 11". Exchanged 8/5/62. Also shown as taking the oath Ft. Delaware 8/10/62, illiterate. Died Welch, W. Va. 10/17. Bur. Thornrose Cem., Staunton.

JORDAN, JAMES H.: Pvt., Co. D. B. Pendleton Co. 1827? Laborer, age 33, Burkes Mill Dist., Augusta Co. 1860 census. Enl. Staunton 7/16/61. Present 7/61-11/61. Died in hospital at Yagar's, Alleghany Mt. of fever 12/30/61. Resident of Mt. Solon. Left widow and 5 children.

KASHER, JOHN: Pvt., Co. I. On postwar roster.

KEATON, JAMES W.: Pvt., Co. G. B. 1/10/43. Pvt., Capt. Western's Co. 32nd Va. Militia. Enl. 4/17/62, age 20. Ab. sick 4/30-10/31/62. Deserted to the enemy Clarksburg, W. Va. 10/14/63. Took oath and sent north. Died Rockingham Co. 9/15/26. Bur. Summit Ch. of the Brethren Cem., on Rt. 690 west of U. S. 11, Rockingham Co.

KEATON, JOHN THOMAS: Pvt., Co. G. B. Brown's Cove, Albemarle Co. 1837? Farmer, Rockingham Co. Enl. Staunton 8/2/61. Present 11/61-4/62. Reenl. 5/1/62. AWOL 5/1-8/31/62. Ab. sick 9/29/62-10/63. Present 11-12/63. Issued clothing 3/1 and 4/1/64. WIA (hip) Spotsylvania CH 5/12/64. Cap. Harrisonburg 10/2/64. Sent to Point Lookout. Released 5/13/65. Died near Bridgewater, Rockingham Co. 1/27/13. Bur. Beaver Creek Ch. of the Brethren Cem., north east of Spring Creek on Rt. 752.

KEATON, JOSEPH: Pvt., Co. H. On postwar roster. Enl. Staunton 6/62. NFR.

KEBLINGER, WILLIAM J.: Pvt., Co. H. B. Albemarle Co. 1844? Resident, Charlottesville. Enl. Staunton 7/23/61 age 17. Transf. "Albemarle Rifles", 19th Va. Inf. 11/27/61. WIA (through left ankle) Gettysburg 7/3/63. Living Charlottesville 1888.

KELLEY, J. H.: Pvt., Co. B. (1st). On postwar roster.

KENNEDY, ALBERT: Pvt., Co. C. Not on muster rolls. Cap. Petersburg 4/2/65. Sent to Point Lookout. Released 6/10/65. Resident of Augusta Co., 5' 8", florid complexion, red hair, grey eyes.

KENNEDY, ANDREW JACKSON: Pvt., Co. C. B. Baltimore, MD. 1843. Apprentice Distiller, age 16, Staunton PO, 1860 census. Enl. Staunton 7/16/61 age 18. Present 11/61-4/62. WIA Cedar Run 8/9/62. Returned to duty 9/15/62. Present 9/16/62-12/63. Issued clothing 4/21/64. Present 4/30-7/19/64. Ab. sick in hospital Staunton 7/20-12/31/64. NFR. Teamster, Fairfax Co. 1902. Died Herndon, Va. 11/20/12. Bur. Chestnut Grove Cem., Herndon.

KENNEDY, DAVID: Pvt., Co. H. B. Augusta Co. 10/15/28. Enl. Bunker Hill 10/3/62. AWOL 11/22/62-12/63. Dropped as deserter. NFR. Farmer, age 30, Fishersville PO, Augusta Co. 1870 census. Died Augusta Co, 1/23/02. Bur. Hildebrand Mennonite Ch. Cem., Madrid.

KENNEDY, GEORGE WASHINGTON: Pvt., Co. G. B. Augusta Co. 5/44. Farmhand, age 16, 1st Dist. Augusta Co. 1860 census. Enl. Staunton 8/2/61. Present 11/61-4/62. WIA (through cheek under left eye and lost sight in eye) Port Republic 6/9/62. Present 8/31-10/31/62, AWOL 15 days fined $5.60. NFR. Farmer, age 25, Fishersville PO, Augusta Co. 1870 census. Farmer, Crimora Station, 1897. Farmer, near Scottsville, Albemarle Co. 1906-13 Member, Stonewall Jackson Camp, CV, Staunton. Died Alleghany Co. 1926. Bur. Crown Hill Cem, Clifton Forge.

KENNEDY, JOHN H.: Cpl., Co. C. B. Ireland, 10/30/25. Carpenter, age 34, Burkes Mill Dist., Augusta Co. 1860 census. Enl. Staunton 7/16/61 as 2ndCpl. Present 11/61-4/62. Reenl. 5/1/62. WIA (shoulder) Port Republic 6/9/62. Ab. wounded through 4/63. Present 4/30-12/63. WIA (right knee) Bethesda Ch. 5/30/64. Returned to duty 6/13/64. WIA and cap. Winchester 9/19/64. Sent to Point Lookout. Exchanged 3/18/65. NFR. Farmer, Hermitage, Augusta Co. 1872. Died Gilmor, Ill, 3/28/91. Bur. Wenger Cem., there.

KENNEDY, WILLIAM A.: Pvt., Co. B (2nd). B. Augusta Co. 1837? Painter, age 23, Burkes Mill Dist., Augusta Co. 1860 census. Enl. Waynesboro 7/15/61. Present 11/61-4/62. Reenl. 5/1/62. AWOL 5/23-8/14/62. Present 8/15-10/31/62 and 11/62-2/63. Detailed as teamster in Regt'l Train 3-4/63. Discharged for epileptic attacks 5/31/63. Age 25, 5' 10", fair complexion, blue eyes, light hair.

KERR, DAVID S.: 3rdSgt., Co. C. B. Va. 1837? Machinist. Enl. Staunton 7/16/61 age 24. Ab. sick 11-12/61. Present 1-2/62. Detailed as machinist in Richmond 3/10/62-12/64, and reduced to Pvt. Issued clothing 11/30 and 12/17/64. NFR. Farmer, age 33, Mt. Sidney PO, Augusta Co. 1870 census.

KERR, LORENZO DOW: Pvt., Co B (1st). Enl. Fairfield 7/10/61. Present until transf. 9/28/61.

KERR, SAMUEL A.: 2ndSgt., Co. I. B. Augusta Co. 1816. Carpenter, age 44, 1st Dist., Augusta Co. 1860 census. Enl. Staunton 7/16/61 age 45. Ab. on detail in Staunton working on wagons 1/17/62 until discharged 2/10/63. Overage of conscription. Age 45, 5' 11½", fair complexion, blue eyes, fair hair, mechanic. Farmer, age 55, Fishersville PO, Augusta Co. 1870 census. Died Valley Mill, Augusta Co. 6/5/81. Father of William S. Kerr.

KERR, WILLIAM D.: Pvt., Co. B (1st). Enl. Fairfield 7/10/61. Present until transf. 9/28/61.

KERR, WILLIAM S.: 1stCpl., Co. C. B. Va. 5/23/41. Farmhand, age 19, Burkes Mill Dist., Augusta Co. 1860 census. Enl. Staunton 7/16/61 age 20. Ab. sick 11-12/61. Present 1-4/62. Reenl. 5/1/62. WIA (hip) Gaines Mill 6/27/62. Ab. wounded through 2/63. Present 3-4/63. Transf. Co. C. 39th Bn. Va. Cav. 5/1/63. Farmer, age 25, Fishersville PO, Augusta Co. 1870 census. Farmer, Summerdean. Died there 4/16/75. Bur. Sherariah Ch. Cem., near Middlebrook. Son of Samuel A. Kerr.

KERSH, ADAM W.: Pvt., Co. F. B. Va. 11/16/28. Cabinetmaker, age 31, Burkes Mill Dist., Augusta Co. 1860 census. Enl. Staunton 7/31/61. Present 11/61-4/62. Reenl. 5/1/62. Present 8/31/62-12/63, and 3-4/64. WIA Bethesda Ch. 5/30/64. AWOL 7/1-12/31/64. Paroled Staunton 5/11/65. Age 35, 6', dark complexion, dark hair, blue eyes. Cabinetmaker and Fiddle Player, Centerville, Augusta Co. Died there 3/22/05. Bur. St. Michael's Ch. Cem.

KERSHNER, JAMES ADDISON: Pvt., Co. C. B. Augusta Co. 8/26/41. Resident of Deerfield, Augusta Co. Enl. Staunton 8/21/61. Ab. sick 11/61-4/62. Deserted to the enemy near Deerfield 4/20/62. Joined Ohio Heavy Artillery. NFR. Moved to Ohio, Illinois. Michigan, Texas 1876, Mo. and Indian Territory. D. Iberia, Mo., 7/21. Bur. Union Cem. there.

KERSHNER, JOHN R.: Pvt., Co. C. B. Augusta Co. 1/3/39. Farmhand, age 21, 1st Dist., Augusta Co. 1860 census. Enl. Shenandoah Mt. 4/16/62. Present 4/16-30/62. WIA (thigh) Port Republic 6/9/62. Ab. wounded through 12/64. NFR. Married Lebanon Springs, Augusta Co. 11/7/72.

KESTERSON, JACOB MYERS: Pvt., Co. I. B. Augusta Co. 10/3/39. Stonemason, age 20, 1st. Dist., Augusta Co. 1860 census. Enl. Staunton 7/15/61 age 22. Present 11/61-4/62. Reenl. 5/1/62. Present 8/30-10/31/62 and 1-4/63. Appointed Drummer 3/25/63. Ab. on leave 8/22-31/63. Present 9-12/63. Issued clothing 3/16 and 4/21 and 29/64. WIA Winchester 9/19/64. Ab. wounded through 12/31/64. Surrendered Appomattox 4/9/65. Farmer, age 30, Fishersville PO. Augusta Co. 1870 census. Died Augusta Co. 2/6/87. Bur. Hanger Cem. on Rt. 670 between Greenville and Middlebrook. Brother of Samuel B. and William G. Kesterson.

KESTERSON, JOHN MARSHALL: Pvt., Co. K. B. Augusta Co. 12/10/33. Carpenter, Back Creek, Bath Co. Enl. Shenandoah Mt. 4/9/62. Present 4/9-30/62. Present McDowell, Front Royal, Harrisonburg. AWOL since 6/9/62 on rolls to 10/63. Present 11/12/63, sent to wagons sick 6/8/62. He absented himself from the wagons and went home. He says he reported himself to Col. Harman in Staunton 11/62 and was placed in hospital guard without any lawful detail and received all his monthly pay due him. He was sent to the Regt. on the 3rd instant with a recommendation for hospital duty which was refused him. Issued clothing 4/1/64. WIA Charlestown 9/22/64. Ab. wounded through 12/31/64. Deserted to the enemy Charleston, W. Va. 2/2/65. Took oath and sent north. Age 28, 5' 9", light complexion, brown eyes, light hair. Moved to Mercer Co., W. Va. and then to Radford. Member, G. C. Wharton Camp CV, Radford. Died Radford 8/1/17.

KESTERSON, SAMUEL B.: Pvt., Co. I. B. Va. 1838? Stonemason, age 22, 1st Dist., Augusta Co. 1860 census. Enl. Staunton 7/16/61 age 23. Present 11/61-4/62. Reenl. 5/1/62. Present 8/30-10/31/62. AWQL 11/23/62-3/7/63, fined $37.50. Present 3/8-12/63. Issued clothing 6/12/64. WIA (ball entered upper lobe right lung, exited near spinal column) and cap. Cedar Creek 10/19/64, age 26. Sent to hospital Baltimore. Transf. Point Lookout. Exchanged 2/14-15/65. In hospital Richmond 2/15/65. Present Camp Lee, Richmond 2/19/65. Furloughed for 60 days 2/20/65. Paroled Staunton 5/16/65. Age 27, 5' 11", dark complexion, dark hair, grey eyes, illiterate, resident of Middlebrook. Stonemason, age 30, Fishersville PO, Augusta Co. 1870 census. Brother of Jacob M. and William G. Kesterson.

KESTERSON, WILLIAM GRASS: Pvt., Co. I. B. Va. 1843? Stonemason, age 17, 1st Dist., Augusta Co. 1860 census. Enl. Staunton 7/16/61 age 18. Present 11/61-4/62. Reenl. 5/1/62. WIA (hand) 2nd Manassas 8/29/62. Ab. on leave 2/19-28/63. Present 3-12/63. Issued clothing 3/16 and 4/1/64. Present 4/30-10/31/64. AWOL 12/8-31/64. WIA (shoulder) Hatcher's Run 2/8/65. Surrendered Appomattox 4/9/65. Paroled Staunton 5/23/65. Age 22, 5' 11", dark complexion, dark hair, grey eyes. Farmhand, age 27, Fishersville PO, Augusta Co. 1870 census. Died near Raphine, Rockbridge Co. 4/11/97. Bur. Fairfield Cem., Rockbridge Co. Brother of Jacob M. and Samuel B. Kesterson.

KEY, NELSON: Pvt., Co. G. B. Augusta Co. 1803? Carpenter, age 57, 1st Dist., Augusta Co. 1860 census. Enl. Staunton 8/2/61. Ab. sick 11/61 until died of disease in Staunton Hospital 11/30/61.

KEYSER, HARRISON G.: Pvt., Co. K. B. Bath Co. 2/6/31. Pvt., Capt. Davis' Co., 81st Va. Militia. Enl. Shenandoah Mt. 4/9/62. Present 4/9-30/62. Ab. sick 5/17-12/62. WIA (lost index finger) Fredericksburg 5/5/63. Present 11-12/63, and 4/30-10/31/64. Surrendered Appomattox 4/9/65. Farmer, age 38, Bath CH PO, Bath Co. 1870 census. Farmer, Healing Springs 1897. Died Cedar Creek Dist., Bath Co. 10/10/12. Bur. Healing Springs. Cem.

KEYSER, MARSHALL D.: Pvt., Co. K. B. Bath Co. 9/12/34. Pvt., Capt. Davis' Co., 81st Va. Militia. Enl. Shenandoah Mt. 4/9/62. Present 4/9-30/62. Ab. sick 10/12/62-4/63. AWOL 5/1-12/63. Cap. Hanover Junction 5/22/64. Sent to Point Lookout. Transf. Elmira. Exchange 11/15/64. Present Camp Lee, Richmond 11/22/64. NFR. Farmer, age 36, Bath CH PO, Bath Co. 1870 census. Farmer, Healing Springs, 1897. Died Cedar Creek Dist., Bath Co. 8/31/13. Bur. Healing Springs Cem.

KIDD, JOHN P.: Pvt., Co. E. B. Va. 5/11/26. Miller, age 32, Glenwood Dist., Rockbridge Co. 1860 census. Enl. Staunton 8/1/61. Ab. on leave 11-12/61. Present 1-4/62. Reenl. 5/1/62. Ab. on leave 5/1-10/31/62. Present 1-12/63. WIA (arm, bone shattered just below elbow) Bethesda Ch. 5/31/64. Ab. wounded through 12/31/64. Illiterate. NFR. Laborer, age 41, Natural Bridge Dist., Rockbidge Co. 1870 census. Died Gilmore's Mill, Rockbridge Co. 10/9/02. Bur. Natural Bridge Baptist Ch. Cem.

KINCAID, THOMAS M. D.: Pvt., Co. K. B. Alleghany Co. 1821? Farmer, age 39, Cleeks Mill PO, Bath Co. 1860 census. Pvt., Capt. Davis' Co. 81st Va. Militia. Enl. Shenandoah Mt. 4/9/62 age 42. Present 4/9-30/62. Ab. sick 7/8/62-8/31/63. Deserted 10/1/63. Enl. 34th Bn. Va. Cav. (later 19th Va. Cav.) 12/1/63. Carpenter, age 47, Warm Springs Dist., Bath Co. 1870 census. Died near Healing Springs before 8/17.

KING, CLINTON M.: 1stLt., Co. B. (2nd) B. Missouri 9/2/34. Merchant, age 25, Waynesboro, Augusta Co. 1860 census. Enl. Waynesboro 7/15/61, as 3rdLt. Ab. sick 11/16-1/16/62. Present 1/17-2/62, rank shown as 2ndLt. Ab. on leave 3-4/62. Reelected 5/1/62. Promoted 1stLt. 5/12/62. KIA Cross Keys 6/8/62, age 26. Left widow and 3 children. Bur. Riverview Cem., Waynesboro.

KINNEY, ALEXANDER FISHER: 1stLt. & Adjutant, F&S. B. Staunton 5/12/36. Clerk of Circuit Court, age 24, Staunton 1860 census. Orderly Sgt., Co. C, 160th Va. Militia. Enl. Staunton Artillery 1861 as Lt. Served as Clerk in office of Surgeon S. S. Lewis, Surgeon in Charge, Staunton Hospital 11/62-12/63. Acting Commissary Officer of 52nd 3/64. Promoted 1stLt. and Adj. of Regt. 12/21/64. NFR. Collector of Customs and Disbursing Agent of the Treasury for the Dist. of Staunton 4th Quarter 1864. Teller, Valley National Bank, Staunton 1903. Died Staunton 5/12/04. Bur. Thornrose Cem.

KINNEY, RICHARD STEVENSON: 2ndLt., Co. A. B. "Walnut Grove", near Staunton 2/28/41. VMI '60 (6 months) West Point '61, resigned 4/22/61. Originally Cadet 2ndLt. Co. A. WIA (leg) Fredericksburg 5/5/63. Present 5/31-12/63. WIA (face) Spotsylvania CH 5/12/64. Ab. detached as Inspector, Gen. Dearing's Cav. Brig. 9/15-12/31/64. Dearing's request for his assignment included "...Lt. Kinney's military education at West Point peculiarly fits for organizing, drilling and disciplining the dismounted Cavalry. He has served earnestly and efficiently during this war and behaved with marked gallantry having been twice wounded...." Promoted Capt., Confederate States Army 11-12/64. Commanded dismounted cavalry of Dearing's Brig. to end of war. Teacher; Clerk in QM Dep., Ft. Sam Houston, Texas. Died Staunton 1/31/23. Bur. San Antonio, Tex.

KIRACOFF, JAMES MCCUTCHEN "Cutch": Pvt., Co. D. B. Augusta Co. 4/6/32. Farmhand, age 27, Northern Dist., Augusta Co. 1860 census. Enl. Staunton 7/16/61. Present 7/16-12/61. AWOL 1/6-16/62. Present 1/17-5/62. Reenl. 5/1/62. Present 5/1-7/18/62. AWOL 7/19-8/6/62, fined $6.00. Present 8/7-31/62. Transf. Co. I, 5th Va. Inf. 1/31/63. Farmer, Age 37, Mt. Sidney PO, Augusta Co. 1870 census. Died near Sangersville 4/4/91. Bur. Methodist Episcopal Ch. Cem., Sangersville.

KIRACOFE, NELSON BITTLE: Pvt., Co. D. B. near Sangersville, Augusta Co. 1839. Resident of Sangersville. Enl. Staunton 7/16/61. Deserted 8/5/61. Wrote denial in "Rockingham Recorder" 12/13/61. "Came home sick with the measles, now sick with fever. Can be found at residence of Lucy Coakley near Mole Mill, Rockingham Co." Died near Spring Creek, Rockingham Co. before 6/23/71. Bur. Bank Mennonite Cem., Rockingham Co.

KIRACOFE, ROBERT G.: Pvt., Co. F. B. Long Glade, Augusta Co. 9/17/42. Blacksmith, age 17, Northern Dist., Augusta Co. 1860 census. Enl. Staunton 7/31/61. Present 11/61-4/62. Reenl. 5/1/62. Died of fever at his father's house, Augusta Co. 5/24/62. Age 20, 5' 8", fair complexion, blue eyes, light hair. Bur. Lytton family Cem., on Rt. 613, 9/10 mile north of intersection with Rt. 607 at Springhill.

KIRBY, OBEDIAH: Pvt., Co. B (2nd). B. Augusta Co. 1824? Farmhand, age 36, Burkes Mill Dist., Augusta Co. 1860 census. Enl. Waynesboro 7/15/61. Ab. sick 11/61-2/62. Present 3-4/62. WIA (side and arm) McDowell 5/8/62. Ab. wounded until discharged 12/10/62. Age 43, 5' 11½", fair complexion, dark hair. Farmhand, age 50, Mt. Sidney PO, Augusta Co. 1870 census.

KIRKPATRICK, WILLIAM R.: Pvt., Co. K. B. Bath Co. 1826? Enl. Shenandoah Mt. 4/9/62. Present 4/9-30/62. Ab. sick 6/12 until he died of "phthisis pulmonalis" in hospital Staunton 8/13/62. Bur. grave #1169, Thornose Cem., age 34.

KNICK, JOSEPH C.: Pvt., Co. E. B. Rockbridge Co. 1839? Farmhand, age 21, Collierstown PO, Rockbridge Co. 1860 census. Enl. Staunton 8/1/61. Present 11/61-4/62. Reenl. 5/1/62. WIA (leg) Port Republic 6/9/62. Ab. wounded through 12/64. NFR. Died Rockbridge Co. 3/20/70 as results of wounds. Bur. Collierstown Pres. Ch. Cem. Brother of William V. Knick.

KNICK, WILLIAM V. "Billie": 2ndLt., Co. E. B. Rockbridge Co. 6/20/33. Farmer, age 23, Collierstown PO, Rockbridge Co. 1860 census. Enl. Staunton 8/1/61. Present 11/61-4/62. Reelected 5/1/62. WIA (hand) Gaines Mill 6/27/62. WIA Cedar Run 8/9/62. WIA Fredericksburg 5/5/63. DOWs at home in Rockbridge Co. 9/8/63. Left widow. Bur. Collierstown Pres. Ch. Cem. Brother of Joseph C. Knick. These two brothers and six more were known as the "Big Knick" brothers, because of their size and strength. Each weighed about 250 lbs. All eight brothers served in the Confederate Army.

KNIGHT, JOHN GARLAND: Pvt., Co. C. B. Va. 1843? Enl. Staunton 7/16/61 age 18. Present 11/61-4/62. Reenl. 5/1/62. WIA (leg) Port Republic 6/9/62. Ab. wounded through 8/31/62. AWOL 12/6/62-2/28/63, fined $44.00. Ab. 3-4/63. Ab. 3-4/63. AWOL 5/17-12/63. Dropped as deserter. NFR.

KNOPP, ABRAHAM LYTTON: Pvt., Co. F. B. 8/22/49. Resident of Shenandoah, Augusta Co. Enl. Staunton 7/31/61. Deserted 11/17/61. NFR. Died Staunton 6/3/20. Bur. Green Hill Cem., Churchville.

KNOTT, CORNELIUS STALEY: Pvt., Co. D. B. Lighterburg, Washington Co., Md. 8/2/38. Farmhand, age 21, Mt. Solon, Augsuta Co. 1860 census. Pvt., Co. K., 160th Va. Militia exempt 2/62 for rupture, age 24. Enl. Martinsburg 9/22/62. Present 1-12/63. Issued clothing 3/31 and 4/29/64. Cap. Cedar Creek 10/19/64. Sent to Point Lookout. Exchanged 3/28/65. NFR. Farmer, age 33, Mt. Sidney PO, Augusta Co. 1870 census. Member, Stonewall Jackson Camp CV, Staunton. Died as result of a fall, Stokesville, Augusta Co. 3/15/12. Bur. Mt. Zion Ch. Cem.

KNOTT, N.: Pvt., Co. F. Not on muster rolls. WIA (chest) Gettysburg 7/3/63. DOWs in Federal hospital 7/6/63.

KOINER, GEORGE MARION K.: 5thSgt., Co. B (2nd). B. Waynesboro, Augusta Co. 12/31/41. Farmhand, age 19, 1st Dist., Augusta Co. 1860 census. Enl. Waynesboro 7/15/61 as 4th Cpl. Present 11/61-4/62. WIA (arm) McDowell 5/8/62. Present 6-7/62., promoted 1stCpl. 6/15/62. Present 8-10/62, promoted 5thSgt. Discharged 12/12/62. Enl. Co. C, 39th Va. Bn. Cav. Farmer, Stockraiser, Clerk of School Board South River District, Augusta Co. 25 years. Died near Waynesboro 9/11/12. Bur. Trinity Lutheran Ch. Cem. Brother of Martin P. Koiner.

KOINER, GEORGE MICHAEL: Pvt., Co. B (2nd). B. Augusta Co. 6/29/43. Enl. Waynesboro 7/15/61. Ab. sick 11/61-3/13/62. Present 3/14-4/62. Reenl. 5/1/62. WIA (left hand) Port Republic 6/9/62. Ab. wounded until discharged for loss of use of left hand 1/8/63. Age 20, 5' 9", fair complexion, blue eyes, dark hair, farmer. Enl, Co. C. 1st Va. Cav. Farmer and Miller, Waynesboro 1885.

KOINER, HARMER D.: Pvt., Co. B (2nd). B. 1844? Not on muster rolls. Cap. Clear Springs, Md. 7/9/63. Sent to Camp Chase. Age 19, 6' 2½", dark complexion, dark eyes, black hair, farmer, Augusta Co. NFR.

KOINER, HIRAM CARTER, : Pvt., Co. B (2nd). B. 1835? Farmhand, age 25, 1st Dist., Augusta Co. 1860 census. Enl. Waynesboro 7/15/61. Present 11/61-4/62. Reenl. 5/1/62. AWOL 5/17-6/5/62. WIA Cross Keys 6/8/62. AWOL 7/18-9/27/62. Present 9/28/62-2/63. AWOL 3/4-4/19/63, fined 4 months pay by CM. Present 4/20-12/63. Cap. Bethesda Ch. 5/30/64. Sent to Point Lookout. Transf. Elmira. Released 5/29/65. Resident of Waynesboro, 5' 8", dark complexion, dark hair, hazel eyes. Furnished transportation from Washington, D. C. to Waynesboro, 5/31/65. Farmer, age 39, Fishersville PO, Augusta Co. 1870 census. Moved to Golden City, Mo. 1893.

KOINER, JOHN CALDWELL: Pvt., Co. B (2nd). B. Va. 4/21/41. Farmhand, age 19, 1st Dist., Augusta Co. 1860 census. Enl. Waynesboro 7/15/61. Ab. sick 11/61-2/62. Detailed as nurse in hospital 3-4/62. Reenl. 5/1/62. AWOL 5/1-10/26/62. Present 10/27/62-3/64. Ab. sick 4/6-30/63. Present 5/1-10/63. Ab. sick in hospital with chronic dysentary 11/27/63 until he died in hospital Staunton 5/8/64. Bur. Zion Lutheran Ch. Cem., near Waynesboro.

KOINER, JOSEPH: SgtMajor, F&S. Not on muster rolls. WIA (arm) Port Republic 6/9/62. NFR.

KOINER, MARCUS: Cpl. Co. B (2nd). Not on muster rolls. Enl. 61-62. Promoted Cpl. WIA McDowell 5/8/62. Discharged Fall '62. Enl. Co. C. 39th Bn. Va. Cav. 2/64. Alive 1913.

KOINER, MARTIN DILLER: Pvt., Co. B (2nd). B. Augusta Co. 11/26/43. Farmhand, age 17, 1st Dist., Augusta Co. 1860 census. Enl. Camp Alleghany 3/13/62. Present 3-4/62. WIA McDowell 5/8/62. Present 5-10/62. AWOL 10/27-12/62. Detailed in Div. Ordnance Train as teamster 1-4/63. Cap. Williamsport, Md. 7/8/63. Sent to Camp Chase. Transf. Ft. Delaware. Age 19, 6' 2½", dark complexion, dark eyes, black hair. Transf. Point Lookout. Died in Hammond General Hospital there of chronic diarrhea 10/29/63. Bur. Zion Lutheran Ch. Cem., near Waynesboro. Brother of George M. K. Koiner.

KRAFT, JOHN F.: Pvt., Co. B (2nd). Enl. Waynesboro 7/15/61. Present 11/61-4/62. Reenl. 5/1/62. Deserted 7/3/63. NFR. Resident of Waynesboro.

LACKEY, WILLIAM HARVEY: Ensign, F&S. B. Wabash Co., Ind. 1842 Enl. Staunton 8/1/61 as Pvt., Co. E. Present 11/61-4/62. Reenl. 5/1/62. Present 5/1-10/31/62. Promoted 5thSgt. 1/10/63. Ab. on leave 2/19-28/63. Present 3-12/63. Recommended for Ensign of Regt. by Col. Skinner, and endorsed by Col. Hoffman, Commanding the Brig., Gen. Early, Gen. Ewell and Gen. R.E. Lee. "I recommend, to his Excellency the President, Sgt. William H. Lackey, of Co. E. of this Rgt., a resident of Rockbridge Co. and a citizen of Va. for appointment to the office of "Ensign" of the Rgt. Sgt. Lackey is a gallant and otherwise meritorious youth, who has borne the colors of the Rgt. for upwards of a year past, having volunteered to do so at a time when it being a "post of danger" alone made it a "post of honor". The rank now attached to the office is therefore clearly his own." Appointed Ensign 4/12/64. Present 4-10/64. Ab. sick 12/25/12-2/26/65. NFR. Died Roanoke 5/29/92. Obit states "...present entire war, never missed a roll call, bore Regt'l colors in 38 battles and 13 other engagements without receiving a scratch, not withstanding the fact that the flag was shot from his hands time and time again and his clothing was preferated with bullets."

LAM, JOSEPH: Pvt., Co. E.B. Va. 3/12/32. Laborer, age 29, South River Dist., Rockbridge Co. 1860 census. Enl. Bunker Hill 10/11/62. Present 10/11-31/62. Ab. on leave 12/11/62-2/63. Present 3-4/63. AWOL 2 months, 15 days and pay deducted. Present 4/30-12/63. WIA (left thigh) Spotslyvania CH 5/12/64. Ab. wounded through 12/31/64. NFR. Died near Lexington 5/29/80. Bur. New Monmouth Pres. Ch. Cem.

LAMB, DAVID W.: Pvt., Co. C.B. Augusta Co. 1844? Farmhand, age 16, Northern Dist., Augusta Co. 1860 census. Enl. Staunton 7/16/61 age 21. Present 11/61-4/62. Reenl. 5/1/62. KIA Port Republic 6/9/62. Left widow.

LAMB, JAMES M.: Pvt., Co. A.B. Va. 7/8/42. Farmer, age 24, 5th Dist., Rockbridge Co. 1860 census. Enl. Staunton 7/15/61. Present 11/61-4/62. Reenl. 5/1/62. Desert 9/13/62. NFR. Died near Crimora, Augusta Co. 11/29/28. Bur. East Point Cem.

LAMBERT, ELBINE.: Pvt., Co E. B. 1841? Farmer, Pendleton Co. Not on muster rolls. Cap. Preston Co. 3/26/63. Sent to Camp Chase..Took oath and released. Age 21, 5' 10", florid complexion, dark eyes, dark hair.

LAMBERT, JACOB N.: Pvt., Co. D. B. Va. 1832? Enl. Staunton 7/16/61. Present 7/16/61-3/62. AWOL 4/24-5/1/62. Reenl. 5/1/62. Present 5/2-6/1/62. AWOL 6/2-2062, fined $21.26. Present 6/21-8/31/62 and 1-5/63. Deserted 9/6/63. NFR.

LAMBERT, JOHN MOORE.: 2nd Lt., Co. I.B. near Bethel Ch, Augusta Co. 7/3/39. Carpenter, age 20, Greenville., Augusta Co. 1860 census. Enl. Staunton 7/16/61 age 22. Ab. sick 12/15/61-2/9/62. Present 2/10-4/62. Reelected 5/1/62. WIA (hip) 2nd Manassas 8/28/62. Ab. wounded through 12/62. Commanding Co. 1-2/63. Present 3-12/63. WIA (left thigh) Spotsylvania CH 5/12/64. WIA and Cap. Bethesda Ch. 5/30/64. Sent to Ft. Delaware. Transf. Ft. Pulaski, Ga. and back to Ft. Delaware. Released 6/16/65. Resident Greenville, 6', fair complexion, auburn hair, blue eyes. Carpenter, age 30, 1st Dist., Augusta Co. 1870 census. Building contractor, Waynesboro. Died there 5/21/05. Bur. Riverview Cem.

LAMBERT, PETER, JR.: Pvt., Co. F. B. Augusta Co. 1837 Farmhand, age 23, Northern Dist., Augusta Co. 1860 census. Enl. Staunton 7/31/61. Present 11/61-4/62. Ab. sick 4/22-30/62. Reenl. 5/1/62. Ab. in hospital Winchester 10/16/62. Present 1-12/63 and 3-12/64. KIA Ft. Stedman 3/25/65.

LAMBERT, SAMUEL A.: Capt., Co. I. B. Augusta Co. 3/31/35. Farmer, age 23, 1st Dist., Augusta Co. 1860 census. Enl. Staunton 7/16/61 age 24. Died of typhoid fever Camp Alleghany 10/31/61. Left widow. Bur. Pilson Cem., on Rt. 694 half way between US 11 and RT 340.

LAMBERT, SAMUEL HARVEY: Pvt., Co. H. B. Rockingham Co. 1827? Farmhand, age 33, Northern Dist., Augusta Co. 1860 census. Enl. Camp Alleghany 2/18/62. Deserted 2/21/62 resident Pendleton Co. NFR. Farmer, age 43, Mt. Sidney PO, Augusta Co. 1870 census. Died 2 miles north of Staunton 4/25/09. Bur. Pleasant View Lutheran Ch. Cem. near Springhill.

LAMBERT, THOMPSON WILSON: Pvt., Co. I. B. Augusta Co. 3/30/45. Resident age 14, 1st Dist., Augusta Co. 1860 census. Enl. Greenville 11/27/63. Present 12/63. WIA (left thigh) Spotsylvania CH 5/12/64. Furloughed from hospital for 60 days 5/25/64. In hospital Charlottesville 9/26-27/64. Transf. to hospital Lynchburg. WIA Cedar Creek 10/19/64. Returned to duty 1/7/65. Cap. Ft. Stedman 3/25/65. Sent to Point Lookout. Released 6/14/65. 5'8½", dark complexion, black hair, hazel eyes. Carpenter, age 25, Fisherville PO, Augusta Co. 1870 census. Manufacturer, Lambert Organ, Waynesboro, 1880. Farmer and Manufacturer of patent neck yoke and wagon, Avis, Augusta Co., 1885. Sawmill Operator and Chair Manufacturer, Mint Spring, 1897.

LANCE, M.H.: Pvt., Co. C. On Postwar roster. Lived Kansas City, Mo. postwar.

LANDES, DANIEL B. (1st): Pvt. Co. H. B. Augusta Co. 2/2/43. Enl. Staunton 11/7/64. Present 11-12/64. Present Hatchers Run 2/8/65. "Threw down his gun and ran, was caught and brought in but wouldn't stay." Paroled Staunton 5/1/65. Died Augusta Co. 7/30/04. Bur. Pleasant Valley Ch. of the Brethren near Weyer's Cave.

LANDES, DANIEL B. (2nd): Pvt., Co. F.B. Augusta Co. 1838? Enl. Staunton 9/17/62. Present through 12/31/62. Ab. 1-2/63, arrested by order 1/2/63 and sent to his Co. Paroled Staunton 5/1/65. Age 30, 5'9", fair complexion, brown hair, blue eyes. Farmer, age 34, Mt. Sidney PO, Augusta Co. 1870 census. Sawyer, Tresher, and Farmer, Augusta Co. Died Augusta Co. 10/9/01. Bur. Old Salem Lutheran Ch. Cem., 3 miles west of Mt. Sidney. Brother of David, George H. and William R. Landes. (May have served in Co. B., 23rd Va. Cav.).

LANDES, DAVID: Pvt., Co. F.B. Augusta Co. 1839? Farmhand, age 21, Northern Dist., Augusta Co. 1860 census. Enl. Staunton 7/31/61. Ab. sick 11/3/61-4/62. Deserted 7/29/62. Enl. Co. C, 14th VaCav. and died soon after in 1862. Brother of Daniel B., George H. and William R. Landes.

LANDES, GEORGE HENRY: Pvt., Co. G.B. Augusta Co. 3/25/45. Farmhand, age 16, Northern Dist., Augusta Co. 1860 census. Enl. Staunton 8/2/61. Present 11/61-4/62. Reenl. 5/1/62. Ab. on leave 5/1-8/31/62. WIA (thigh and knee) and cap. Sharpsburg 9/19/62. Ab. wounded through 10/63. AWOL 11-12/63. Dropped as deserter. Enl. Co. B, 23rdVaCav. Farmer, Augusta Co. Died 9/20/14. Bur. Old Salem Lutheran Ch. Cem., 3 miles west of Mt. Sidney. Brother of Daniel B., David and William R. Landes.

LANDES, WILLIAM R.: Pvt., Co. G.B. Augusta Co. 1835? Farmhand, age 25, Northern Dist., Augusta Co. 1860 census. Enl. Staunton 8/2/61. Ab. sick 10/2-12/31/61. Present 1-3/62. Died of fever in hospital 4/5/62. Brother of Daniel B., David, and George H. Landes.

LANDON, JOHN W.: Pvt., Co. H.B. Sullivan Co., Tenn. Farmer, Augusta Co. 1834? Enl. Staunton 7/23/61 age 27. Present 11/61-2/62. Ab. sick 3/62-8/31/62. In hospital Lynchburg 12/62. Present 1-6/63. WIA and cap Gettysburg 7/3/63. Sent to Ft. Delaware. Transf. Point Lookout. Took oath and released 12/64. Resident of Augusta Co. 5'5", fair complexion, light hair, light eyes.

LANDRUM, WILLIAM F.: 1stSgt., Co. C.B. 1830? Enl. Staunton 7/16/61 age 31 as Pvt. Present 11/61-2/62, promoted 4thCpl 2/20/62. Present 3-4/62, promoted 4thSgt 3/25/62. Reenl. 5/1/62. Present 5/1-8/31/62. Ab. on leave 9/1/62-2/28/63. Appointed 1stSgt. 12/13/62. Present 3-12/63. KIA Bethesda Ch. 5/30/64.

LANG, ROBERT A.: Pvt., Co. E.B. Va. 1845? Enl. Mt. Meridian 6/15/62. Ab. sick 6/18/62-12/63. Dropped as deserter. NFR Laborer, age 25, Kerr's Creek Township, Rockbridge Co. 1870 census.

LANGE, WILLIAM F.: Pvt., Co. A.B. Augusta Co. 7/41. Farmhand age 20, 1st Dist., Augusta Co. 1860 census. Enl. Staunton 7/16/61. Present 11/61-3/62. AWOL 4/20-30/62. Reenl. 5/1/62. Present 8/31/62-12/63., and 4/30-10/31/64. AWOL 12/8-30/64. Surrendered Appomattox 4/9/65. Farmer, age 65, Deerfield, Augusta Co. 1910 census.

LAWHORN, JOSEPH K.: Pvt., Co. E.B. Rockbridge Co. 1838? Enl. Staunton 8/1/61. Present 11/61-4/62. Reenl. 5/1/62. Present 5/1-10/31/62, and 1-12/63. WIA (jaw) Wilderness 5/6/64. Ab. wounded through 12/31/64. NFR. Laborer, age 32, Natural Bridge Dist., Rockbridge Co. 1870 census. Died on Roaring River 6/10/10. Bur. High Bridge Pres. Ch. Cem., Rockbridge Co.

LAWHORN, PRESTON T.: Pvt., Co. B(1st). Enl. Fairfield 7/10/61. Present until transf. 9/28/61.

LAWHORN, SAMUEL KIRKPATRICK: Pvt., Co. E.B. Rockbridge Co. 10/4/42. Enl. Staunton 8/1/61. Ab. on leave 11-2/61. Present 1-2/62, fined $11.00 by Regt'l CM. Present 3-4/62. Reenl. 5/1/62. WIA Gaines Mill 6/27/62. Ab. wounded through 10/31/62. Present 1-4/63. Ab. sick 7/20-8/31/63. Present 9-12/63. Deserted on the march from Lynchburg to Staunton near Lexington 6/25/64. NFR. Laborer, age 28, Natural Bridge Dist., Rockbridge Co. 1870 census. Died Roanoke 10/25/16.

LAWHORNE, WILLIAM: Pvt., Co. B(1st). Enl. Fairfield 7/10/61. Present until transf. 9/28/61.

LAWRENCE, W.B.: Pvt., Co. I. On postwar roster. Served 4 years.

LAYNE, JOSEPH: Pvt., Co. E. B. Amherst Co. 1/14/48. Pension application only record. Farmer, Rockbridge Co. Died Rockbridge Baths, 1889. Bur. Bethesda Pres. Ch. Cem.

LEE, JAMES H.: 4th Sgt., Co. H. B. Rockingham Co. 1833? Carpenter, age 27, Northern Dist., Augusta Co. 1860 census. Enl. Capt. S. M. Crawford's Co. L, 5th Va. Inf., 4/61, disbanded. Enl. Staunton 7/23/61 age 28 as Pvt. Present 11/61-4/62. Reenl. 5/1/62. Present 5/1-8/31/62, appointed 1st Cpl. 6/1/62. Present 1-2/63, promoted 4th Sgt. Present 3-4/63. WIA (leg) Gettysburg 7/3/63. Ab. wounded through 8/31/63. Ab. in arrest and under charges for conduct prejudicial to good order and military discipline since 10/1/63 through 12/63. Cap. Bethesda Ch. 5/30/64, rank shown as Pvt. Sent to Point Lookout. Transf. Elmira. Released 6/30/65. Resident of Staunton, 5'9", fair complexion, light hair, blue eyes. Alive 1890.

LEE, JEREMIAH M.: Cpl., Co. H. B. Rockingham Co. 4/1/29. Enl. Staunton 7/23/61 age 32. Present 11-12/61, rank shown as Pvt. Ab. sick 2/26-28/62. Present 3-4/62. Reenl. 5/1/62. Ab. sick 6/8-8/31/62. Deserted to the enemy on unofficial roster. Died Middle River Dist., Augusta Co. 7/6/95.

LEE, JOHN M.: Cpl., Co. H.B. Rockingham Co. 1827? Laborer. Enl. Staunton 7/23/61 age 34. Present 11/61-4/62. WIA (leg) Gaines Mill 6/27/62. Ab. wounded until discharged over age of conscription 11/20/62. Age 41, 5'7", light complexion, blue eyes, dark hair. Died before 1885.

LEECH, JAMES GRANVILLE: Pvt., Co. B. (1st). Res., age 15, 5th Dist. Rockbridge Co., 1860 census. Enl. Staunton 8/1/61. Ab. sick when transf. 9/28/61.

LEECH, LUCIAN T.: Pvt., Co. B. (1st). B. Rockbridge Co. 1843? Farmhand, age 18, 5th Dist., Rockbridge Co., 1960 census. Enl. Staunton 7/16/61. Present until transfer. 9/28/61.

LEWIS, JAMES M.: 1st. Cpl., Co. B. (2nd). B. Va. 11/23/24. Farmer, age 35, 1st Dist., Augusta Co. 1860 census. Enl. Waynesboro 7/15/61. Ab. on leave 1-2/62. Present 4-6/62. WIA Cross Keys 6/8/62. Present 7-10/62, promoted 3rd Cpl. 10/2/62. Present 11/62-4/63, rank shown as 1st Cpl. Present 4/30-7/63. Present detailed as Teamster in Regt'l Train and reduced to Pvt. 7/20/63-12/31/64. Surrendered Appomattox 4/9/65. Lumberman, age 45, Fishersville PO, Augusta Co. 1870 census. Merchant, Greenville. Died there 8/6/01. Bur. Thornrose Cem., Staunton.

LEWIS, JOHN: Surgeon, F&S. B. 1833. Attended UVA 52-54 MD. Doctor, Charlottesville, Appointed Assistant Surgeon 8/19/61. Reappointed 5/1/62. WIA Cross Keys 6/8/62. Appointed Surgeon 9/26/62. Present 9-12/62. Present 2/63 until transf. hospital duty 4/3/63. Sick in hospital Charlottesville 7/26-8/19/63. NFR. Doctor, Charlottesville, 1872.

LEWIS, JOHN WILLIAM: Adjutant, F&S. B. Ft. Lewis, Bath Co. 10/8/37. Grad. VMI '59. Attended UVA '60. Assistant Professor of Math and Tactics, VMI '59-60. Appointed Adj. 8/19/61. Reelected 5/1/62. WIA Port Republic 6/9/62. Promoted Capt., CS Army 9/16/62 and ordered to Trans-Miss. Dept. Served as AAG on staffs of Gen. Holmes, Price and Magruder. Price's Div. and ordered to Headquarters, Dist. of Ark. 3/15/65. NFR. Member of faculty VMI 1869. Commandant and Professor of Math, Ark. Military Academy, St. John's, Ark: Cotton Planter and Railroad Builder. Died Homan or Little Rock, Ark. 2/12/82.

LIGGETT, JOSEPH LEMUEL: Pvt., Co. B(2nd). B. Moorfield, Hardy Co. Va. 8/7/38. Enl. Waynesboro 7/15/61. Present 11/61-4/62. Present 5/1-10/31/62, detailed as wagoner since 8/8/62. Present 11/62-12/63, as teamster in Reg'l Train. AWOL 6/29-12/31/64. Dropped as deserter. Paroled Winchester 4/25/65. Age 26, 5'7", dark complexion, dark hair, black eyes. Farmhand, age 29, Fishersville PO, Augusta Co. 1870 census. Farmer, Stuart's Draft 25 years. Moved to Navarel, Kan. 1884. Died Herrington, Kan. 6/26/20. Bur. Sunset Hill Cem., there.

LIGHTNER, GEORGE PILSON: QMSgt., F&S. B. Greenville, Augusta Co. 4/8/39. Farmhand, age 21, 1st Dist., Augusta Co. 1860 census. Enl. Staunton 7/16/61 age 23 as Pvt., Co. I. Present 11/61-4/62. Ab. sick 2/28-4/62. Present 8/30-10/31/62. Present 1-2/63, acting QMSgt. since 12/7/63, rank shown as 4th Sgt. Present 3-12/63, appointed QMSgt 4/14/63. Present 4/30-10/31/64. Surrendered Appomattox 4/9/65. Farmer, age 31, Fishersville PO, Augusta Co. 1870 census. Farmer and School Teacher, Raphine, Rockbridge Co. 1885. Died near Spottswood, Augusta Co. 1/31/25. Bur. New Providence Pres. Ch. Cem., near Brownsburg, Rockbridge Co.

LIGHTNER, THOMAS ARMSTRONG: Pvt., Co. I. B. Augusta Co. 1/10/22. Merchant, age 37, 1st Dist., Augusta Co. 1860 census. Pvt., Co. C. 93rd Va. Militia 3/62 furnished substitute, age 41. Enl. Staunton 10/28/64. Present 11-12/64. Detailed 3/2/65. NFR. Farmer, age 48, Fishersville PO, Augusta Co. 1870 census. Blacksmith, Greenville 1872. Died near Greenville 12/27/97. Bur. Bethel Pres. Ch. Cem.

LIGHTNER, THOMAS RAY: Pvt., Co. I. B. Va. 1842? Farmer, age 18, 1st Dist., Augusta Co. 1860 census. Pvt., Co. B. 93rd Va. Militia 2/62 exempt deformed arm, age 18. Enl. Sommerville Ford 4/15/64. Cap. Bethesda Ch. 5/30/64. Sent to Point Lookout. Transf. Elmira. Died there of chronic diarrhea 12/5/64. Bur. grave 1032, Woodlawn Nation Cem.

LILLEY, JAMES CAMPBELL: 4th Sgt., Co. I. B. Augusta Co. 1844? Surveyor, age 16, 1st Dist., Augusta Co. 1860 census. Enl. Staunton 7/16/61 age 17 as 4th Sgt. Ab. sick 11/61-1/3/62. Present 1/4-4/62. Present 8/30-10/31/63, detailed as Courier, Gen. Jackson, 7/23/62, resigned as Sgt. Ab. detailed as Courier 4/30-12/63. Present 4/30-7/25/64. Ab. with Sharpshooters 7/26-10/31/64. Present 11-12/64, appointed 5th Sgt. 10/31/64. Paroled Staunton 5/23/65. Age 21, 5'7", fair complexion, light hair, blue eyes. Studied engineering under Jed Hotchkiss and at Washington College 71-72. Civil Engineer: City Engineer of Staunton; Railroad Engineer C&O and Mexican Central RR's. Died Mexico City, Mexico 10/26/01. Bur. there. Brother of John D. and Gen. Robert D. Lilley.

LILLEY, JOHN DOAK: Lt. Col., F&S. B. Greenville, Augusta Co. 9/5/41. Surveyor, age 18, 1st Dist., Staunton 1860 census. VMI '64. Drillmaster, Richmond 4-6/61. Enl. Staunton 7/23/61 age 20 as 1st Lt., Co H. Promoted Capt. 11/23/61. Present 11/61-4/62. Reelected 5/1/62. Present 5/1 until WIA (left thigh and arm) 2nd Mannassas 9/29/62. Assisted in the burial of Gen. Jackson. Present 11-12/63. Elected Major 12/19/63. Member GCM 1-5/64. Present 5/64 until WIA (left arm and leg, right hand lost 2 fingers) near Spottsylvania CH 5/19/64. Detailed on GCM until 10/64. Recommended for promotion to Lt. Col. by Col. Hoffman, Commanding Pegram's Brig. 10/27/64. "...(He) is the next senior officer in the regiment, and entirely competent to the position. I therefore recommend that he be appointed LtCol. of the Rgt...." Promoted LtCol 11/64. Present 10/19 until hand and arm crushed while building fortifications 10/28/64. Commanding the Brigade 10/27-30/64 and 11/30/64. WIA (arm) Hatcher's Run 2/8/65. Granted 20 days leave 2/24/65. Present Waynesboro with Gen. Early 3/5/65 and escaped. Started for Lynchburg 4/65 with arm still in sling and met soldiers from Lee's Army who had surrendered. Paroled Staunton 5/23/65. Age 23, 5'9", light complexion, dark hair, blue eyes. Colonel, Virginia Militia 1876. Farmer, County Surveyor, and School Board Member, Augusta Co. Member, Stonewall Jackson Camp CV, Staunton. Died "Buffalo Hill" near Greenville 6/13/13. Bur. Bethel Pres. Ch. Cem. Brother of James C. and Gen. Robert D. Lilley.

LINSAY, I.C.: Pvt., Co. K. On postwar roster.

LINDSAY, JOHN ALEXANDER SR. 3rdLt., Co. K. B. Bath Co. 3/19/29. Farmhand, age 30, Bath CH PO, Bath Co. 1860 census. Enl. Shennandoah Mt. 4/9/62. Present 4/9-30/62. Ab. sick 7/19/62-4/63. WIA (shell fragment heel and right arm) Fredericksburg 5/5/63. Ab. wounded through 8/31/63. Ab. on leave 10/29-12/28/63. Resigned because of wounds 3/10/64. Farmer, Cleeks Mill Dist, 1897. Died Hively, Bath Co. 3/24/06. Bur. Jacob Cleek Cem. near Starr Chapel, Bath Co. Brother of Paul McN. Lindsay.

LINDSAY, PAUL MCNEEL: Pvt., Co. G. B. Pocahontas Co. 4/17/45. Not on muster rolls. Enl. Co. G. 18th VaCav. 3/25/64. Paroled Staunton 6/1/65 5'10", fair complexion, light hair, blue eyes. Farmer, age 65, Lexington Dist., Rockbridge Co. 1910 census. Died East Lexington 4/9/17. Bur. Stonewall Jackson Cem. Brother of John A. Lindsay.

LINKSWILER, JAMES: Pvt., Co. K. B. Bath Co. 1830? Laborer, Millboro Springs, Bath Co. 1860 census. Enl. Valley Mills 4/26 62. Present 4/26-30/62. AWOL 5/26/62 over 351 claimed present 1-12/63. WIA (right ankle) and cap. Bethesda Ch. 5/30/64. Foot amputated. DOWs Lincoln General Hospital, Washington, D.C. from gangrene and exhaustion 8/12/64, age 34. Bur. "Soldiers Burial Ground." Left widow.

LINKSWILER, JOSEPH: Pvt., Co. K. B. Bath Co. 1825. Laborer, age 35, Millboro Springs PO, Bath Co. 1860 census. Enl. Valley Mills 4/26/62. Present 4/26-30/62. Deserted 5/16/62. Killed by police at home 8/26/62.

LIPSCOMB JOHN SINCLAIR: 2ndLt., Co. B(2nd). B. Madison Co. 10/19/36. Farmer, age 20 1st Dist., Augusta Co. 1860 Census. Enl. Staunton 4/17/61 in Co. L, 5thValnf. Joined by transf. 3/8/63 as Pvt. Present 4/30-8/31/63, promoted 1stSgt 6/1/63. Present 9-12/63. Issued clothing 1/28, 4/21 and 28/64. WIA (through left side and exited near spine) near Spotsylvania CH 5/19/64, while commanding the Co. Returned to duty 8/20/64. Present 8/20-12/31/64, promoted 2ndLt 11/25/64, and commanding Co. WIA (through both legs) Hatcher's Run 2/6/65, while commanding Co. Paroled Staunton 5/22/65. Age 21, dark complexion, brown hair, grey eyes. Superintendent of gas works and Merchant, Staunton. Life Insurance Agent, Charlottesville, 1907. Died Bristol 8/2/07. Bur. Thornrose Cem., Staunton. Brother of Oscar C. and Robert M. Lipscomb.

LIPSCOMB, OSCAR C.: 1stLt., Co. A. B. Madison Co. 1840? Farmer age 20, 1st Dist., Augusta Co. 1860 census. Enl. Staunton 7/9/61 age 21, as 3rdSgt. Present 11/61-4/62. Reenl. 5/1/62. Present 8/31-12/31/62, elected 2ndLt. 10/2/62 and 1stLt 12/21/62. Present 1-6/63. Ab. on detached service hunting deserters 7/31-8/31/63. Commanding Co. 12/63. WIA (face) Bethesda Ch. 5/24/64. WIA and cap. Winchester 9/19/64. Sent to Ft. Delaware. Released 6/6/65. 5'9" light complexion, light hair, grey eyes, resident of Waynesboro. Farmer and Distiller Crozet and Mechum's River, Albermarle Co. 1897. Had eye removed from effect of wounds 1904. Deputy Treasurer of Albermarle Co. 29 years. Died Staunton 12/30/04. Bur. Hillsboro Ch. Cem., near Crozet. Brother of John S. and Robert M. Lipscomb. "A good and brave soldier."

LIPSCOMB, ROBERT M.: 1stSgt., Co. B(2nd). B. Madison Co. 1844? Farmhand, age 16, 1st Dist., Augusta Co. 1860 census. Enl. Waynesboro 7/15/61 as 5thSgt. Present 11/61-4/62. WIA (foot) 2nd Manassas 8/29/62. Ab. wounded through 4/31/63, promoted 1stSgt 10/2/62. Reduced to Pvt. at own request 6/1/63. Ab. detailed as clerk in hospital Staunton 10/21/63-1/64. WIA (right leg) Spotsylvania CH 5/12/64. Present 6-12/64, Forage Master with Major Snodgrass, Div. Train. Paid 1/31/65. Surrendered Appomattox 4/9/65. Brother of John S. and Oscar C. Lipscomb.

LIPTRAP, DAVID D.: Pvt., Co. K. B. Mill Creek, Bath Co. 8/8/35. Farmhand, age 25, Cleeks Mill PO, Bath Co. 1860 census. Enl. Shenandoah Mt. 4/9/62, age 26. Present 4/9-30/62. WIA (shoulder) Port Republic 6/9/62. WIA (leg) Gaines Mill 6/27/62. Present 1-2/63. Ab. sick 3/18-4/63. Present 4/30/63 until WIA (shoulder) Gettysburg 7/3/63. Present 7-12/63. WIA (leg) Wilderness 5/6/64. WIA Spotsylvania CH 5/12/64. Ab. wounded through 12/31/64. NFR. Farmer, age 27, Williamsville Dist., Bath Co. 1870 census. Died Mill Creek, Bath Co. 6/23/78.

LIVELY, JOHN S: Pvt., Co. B(2nd). Enl. Waynesboro 7/15/61. Present 11/61-4/62. Reenl. 5/1/62. AWOL 7/18-8/13/62. WIA Sharpsburg 9/17/62. In hosp. Charlottesville 11/18/62-6/13/63 when he deserted. Returned 1/11/63. Present in confinement 1-2/63. Ab. in arrest 3-4/63, fined 8 mos. pay by CM Ab. sick 6/30-12/31/63. Dropped as a deserter. NFR.

LIVICK, ALEXANDER J.: Pvt., Co. F.B. Augusta Co. 3/6/37. Farmhand, age 22, Burkes Mill Dist., Augusta Co. 1860 census Pvt. Co. E, 160th Va. Militia exempt for varoicocile 3/62, age 24. Enl. Staunton 10/6/62. Present 10/62-10/63. Ab. sick in hosp. Gordonsville 11/27-12/31/63. Present 3-4/64. Ab. sick 8/25-10/31/64. Present 11-12/64. Paroled Staunton 5/1/65. Age 26, 5'9", fair complexion, brown hair, hazel eyes. Farmer, age 30, 1st Dist., Augusta Co. 1870 census. Carpenter and Window Shade Manufacturer, Staunton 1897. Died near Hebron Ch., Augusta Co. 5/10/19. Bur. Thornrose Cem., Staunton.

LIVICK, JAMES HENRY: Pvt., Co. H.B. Augusta Co. 3/25/45. Not on muster rolls. Listed as deserter, 2/63. Resident near Staunton. Laborer, age 26, Pastures Dist., Augusta Co. 1870 census. Farmer, age 66, Middle River Dist., Augusta Co. 1910 census. D. Ft. Defiance, Augusta Co. 10/8/23. Bur. Pleasant View Lutheran Ch. Cem.

LLOYD GRANVILLE: Pvt., Co. C. Enl. Staunton 4/30/62. WIA (shoulder) McDowell 5/8/62, Ab. wounded through 2/63. D. of pneumonia in hosp. Richmond 4/9/63. Effects, 1 pocketbook containing in stamps 65 cents, 1 knapsack of clothing, 1 finger ring, delivered to Capt. Morfit AQM 4/20/63.

LOAN, SAMUEL: Pvt., Co. K. B. Bath Co. 1830? Farmer, age 30, Cleeks Mill PO, Bath Co. 1860 census. Enl. Valley Mills 4/26/62. Present 4/26-30/62. Charged with disloyalty 6/62. Discharged 2/4/63, over age of conscription. Enl. Co. E. 10thVa. Bn. Res. 4/64. Deserted to the enemy Clarksburg, W.Va. 4/8/65. Took oath and released. D. before 8/17.

LOAN, THOMAS: Pvt., Co. K. B. Va. 1818. On postwar roster. Also served Co. E., 10th Va. Bn. Res.

LOEWNER, EMANUEL: Cp., Co. A.B. Austria 1839? Painter, age 28, Burkers Mill Dist., Augusta Co. 1860 census. Enl. Staunton 7/9/61 age 22. Ab. sick 12/15/61-4/62. Discharged 7/24/62 for chronic infection of the left knee joint, of more than two months standing. Age 23, 5'5", dark complexion, blue eyes, dark brown hair. D. Harrisonburg 2/2/94.

LONG, JAMES M: Pvt., Co. E. B. Botetourt Co. 1842? Farmhand, age 18, Collierstown PO, Rockbridge Co. 1860 census. Enl. Staunton 8/1/61. Present 11/61-62. Reenlisted 5/1/62. WIA Bristoe Sta. 8/26/62 and 2nd Manassas 8/28/62. Ab. wounded through 3/7/63. AWOL 3/8-12/63. Dropped as deserter. NFR. Pension record states captured date and place unknown. Released 6/15/65. Moved to Mo. and Ohio. Carpenter, age 65, Staunton 1910 census. D. Staunton 1/27/14. Bur. Thornrose Cem.

LONG, WILLIAM: Capt. Co. B(2nd). B. Va. 1830? Saddler, age 30, Burkes Mill Dist., Augusta Co. 1860 census. Mexican War Veteran. 2nd Lt., "Waynesboro Greys", Va. Militia. Enl. Waynesboro 7/15/61. Present 11/61-4/62. Reelected 5/1/62. WIA (head) McDowell 5/8/62. DOWs 5/12/62. Bur. Riverview Cem., Waynesboro. "A brave and faithful officer."

LONG, WILLIAM P. "BILL":, Pvt., Co. E. B. Rockbridge Co. 10/19/44. Farmhand, age 20, Collierstown PO, Rockbridge Co. 1860 census. Enl. Staunton 8/1/61. Ab. sick 11/61-2/62. Present 3-4/62. Reenlisted 5/1/62. Ab. on leave 5/1-10/31/62. Present 1-4/63. AWOL 7/13-12/63. Dropped as deserter. Obit says missed only one battle. Farmhand, age 29, Buffalo Dist., Rockbridge Co. 1870 census. D. on Black's Creek, 4/9/09. Bur. Collierstown Pres. Ch. Cem.

LOVING, DAVID H.: Pvt., Co. A. B. Va. 1842? Laborer, age 18, Burkes Mill Dist., Augusta Co. 1860 census. Enl. Staunton 7/9/61, age 19. Ab. sick 11/25/61-2/12/62. Present 2/13-4/62. Reenlisted 5/1/62. Captured Front Royal 5/30/62. Exchanged 8/5/62. AWOL 11/23/62-1/3/63. Present in arrest 1-2/63, pay deducted for AWOL. Present 3-12/63. Issued clothing 2/5 and 4/27/64. Ab. sick 5/28-6/17/64. Deserted near Waynesboro 9/30/64. Resident of Christian's Creek. NFR. Carpenter, age 26, Mount Sidney PO, Augusta Co. 1870 census. D. Staunton 3/21/14. Bur. Thornrose Cem.

LOVING, ROBERT H.: Pvt., Co. C. B. Augusta Co. 1833? Farmhand, age 27, Burkes Mill Dist., Augusta Co. 1860 census. Enl. Staunton 7/16/61, age 29. Ab. sick 11-12/61. Present 1-4/62. Reenlisted 5/1/62. AWOL 6/11-8/30/62. Present 8/31/62-2/63. AWOL 3/25-12/63. Dropped as deserter. Farmer, age 38, Mount Sidney PO, Augusta Co. 1870 census.

LOWE, GEORGE W.: Pvt., Co. C. B. Botetourt Co. 1835? Farmhand, Augusta Co. 1860. Enl. Staunton 5/6/62. Ab. sick 5/10-8/31/62. AWOL 10/8/62-9/24/63. Present in arrest 9/25-12/63. Confined Castle Thunder, Richmond for 14 mos. by sentence of GCM, 2/1/64. Present 11-12/64. Paroled Staunton 5/12/65. Age 30, 6', light complexion, brown hair, gray eyes. Living in Page Co. 1902.

LOYD, PATRICK: Pvt., Co. F. B. Va. 1845? Farmhand, age 15, Northern Dist., Augusta Co. 1860 census. Enl. Staunton 7/31/61. Present 11/61-4/62. Reenl. 5/1/62. WIA (leg) McDowell 5/8/62. Ab. sick 8/25-12/31/62. Present 1-7/63. Detailed to care for wounded left at Gettysburg 7/4/63. Captured and sent to DeCamp Richard's Ireland, NY. Exchanged 9/8/63. Present Camp Lee, Richmond 9/14/63. Present 3-4/64. WIA (head) Bethesda Ch. 5/30/64. Furloughed from hosp. for 60 days 7/9/64. WIA (right foot, fractured metatareal bone) and captured Cedar Creek 10/19/64. Sent to West Building Hosp., Baltimore, Md., age 20. Transf. to Ft. McHenry and Point Lookout. Released 6/26/65. 5'10", fair complexion, light hair, blue eyes, illiterate. Farmer age 66, Milneville, Augusta Co. 1910 census. Member, Stonewall Jackson Camp Confederate Veterans, Staunton. D. Western State Hosp., Staunton 7/20/17.

LUCAS, DANIEL J.: Pvt., Co. I. B. Augusta Co. 1845. Farmhand, age 18, 1st Dist., Augusta Co. 1860 census. Enl. Staunton 7/16/61 age 16. Ab. sick 11/61-2/62. Present 3-4/62. Reenl. 5/1/62. Ab. sick 7/25-10/31/62 and 11/2-12/63. Present 3-12/63. Detailedas Teamster for Brigade 4/30-12/31/64. Paroled Staunton 5/20/65. Age 22, 5'11", dark complexion, dark hair, grey eyes. Farmer, age 25, Kerr's Creek Dist., Rockbridge Co. 1870 census.

LUCAS, PETER, JR.: Pvt., Co. I. B. Va. 4/27/25. Farmer, age 35 6th Dist., Rockbridge Co. 1860 census. Obit only record. Enl. 4/62. Listed 1/20/65 as a Farmer, Moffett's Creek, age 39, fair complexion, gray eyes, dark hair, 6'. Farmer, Zack, Rockbridge Co. 1897. D. Moffett's Creek, Rockbridge Co. 11/20/07. Bur. Mount Hermon Luthern Ch. Cem., Newport, Augusta Co.

LUCK, L.T.: Pvt., Co. B(1st). On postwar roster.

LUSK, JOHN ANDREW MONTGOMERY 2ndLt., Co. B(1st). B. Rockbridge Co. 6/12/27. Farmer, age 33, South River Dist., Rockbridge Co. 1860 census. Enl. Fairfield 6/15/61. Present until transf. 9/28/61.

LYLE, WILLIAM A.: Pvt., Co. K. B. Bath Co. 1843? Farmhand, age 17, Bath CH PO 1860 Census. Enl. Shenandoah Mt. 4/9/62. Present 4/9-30/62. Ab. sick 1/30/63 until died of smallpox in hospital Lexington 3/4/63.

LYNN, JAMES DALLAS: 4th Sgt., Co. C.B. Augusta Co. 9/11/44. Enl. Staunton 7/16/61 as Pvt., age 17. Present 11/61-4/62. Reenl. 5/1/62. Present 5/1-8/31/62. Ab. on leave 9/1/62-2/28/63. Present 3-4/63. Appointed 4thSgt. from 3rdCpl 4/1/63. Present 4/30-12/63. Issued clothing 4/29/64. Capt. Cedar Creek 10/19/64. Sent to Point Lookout. Exchanged 3/28/65. Paroled Staunton 5/27/65. Age 20, 5'6", dark complexion, dark hair, grey eyes. Plasterer, Staunton, 1875.

LYNN, JOHN CALVIN: Pvt., Co. B(1st). B. Rockbridge Co. 5/22/17. Carpenter, age 43, Fairfield, Rockbridge Co. 1860 census. Enl. Fairfield 7/16/61. Present until transf. 9/28/61.

LYNN, ROBERT R.: Pvt., Co. G. B. Augusta Co. 3/20/30. Blacksmith, Augusta Co. Enl. Staunton 8/2/61. Present 11/61-10/31/62. AWOL 1 month fined $11.00. Present 1-12/63, on extra duty as blacksmith since 11/18/62. Ab. sick 6/26-10/64. Deserted to the enemy 10/27/64. NFR. Died Aqua, Rockbridge Co. 4/12/13. Bur. Ludwick Cem., near Fairfield.

LYNN, SAMUEL: Pvt., Co. B(2nd). Pension request only record. Enl. 7/61. Served to end of war. Living in Augusta Co. 1888.

MACKEY, JOHN MCBRIDE: Co. B.(1st) B. 8/26/22. On postwar roster. "A brave and accomplished officer."

MAHONE, JAMES C.: Pvt., Co. E.B. Va. 1845? Enl. Richmond 9/15/64. Cap. Fisher's Hill 9/22/64. Sent to Point Lookout. Took oath and enl. U.S. Army 10/15/64. Cooper, age 25, 1st. Dist., Augusta Co. 1870 census.

MALOY, PATRICK: QMSgt., F&S. B. Ireland 1815. Attended Washington Col. 48-49. School Teacher. Enl. Staunton 8/2/61 as Pvt., Co. G. Present 11/61-2/62, detailed as QMSgt. Present 3-4/62, QMSgt., non-conscript over 45 and not enlisted but kept against his will. Present 4/30-10/31/62. Discharged for over age of conscription 11/18/62. Age 46, 5'6", fair complexion, blue eyes, fair hair. Died Highland Co. 8/29/93. Bur. McDowell Cem., Highland Co.

MANLEY, MILES K.: Pvt., Co. B(2nd). B. 1822? Farmer, age 38, 1st. Dist., Augusta Co. 1860 census. Enl. Waynesboro 7/15/61. Present 11-12/61. Ab. on leave 1-10/62. Discharged for over age of conscription 12/10/62.

MANN, JOHN A.: Pvt., Co. B(1st). B. Va. 1824? Farmer, age 36, 5th Dist., Rockbridge Co. 1860 census. Enl. Fairfield 7/10/61. Present until transf. 9/28/61.

MANN, WILLIAM H.: Pvt., Co. A. B. Va. 1842? Farmhand, age 18, 1st Dist., Augusta Co. 1860 census. Enl. Staunton 7/9/61 age 21. Present 11/61-4/62. Reenl. 5/1/62. Ab. sick 10/18-12/31/62. Present 1-12/63. WIA (right hand) Spotsylvania CH 5/12/64. KIA Liberty 6/20/64. Illiterate.

MARION, THOMAS: Pvt., Co. B(2nd). Enl. Waynesboro 7/15/61. Present 11/61-4/62. Reenl. 5/1/62. AWOL 5/7-11/8/62. Present 11/9/62-2/63, confined to Regt'l Guard House. $20.00 paid for his arrest. Ab. 3-4/63. AWOL 6/1-8/18/63. Present 8/19-12/63. Issued clothing 3/16 and 4/24/64. KIA Bethesda Ch. 5/30/64.

MARSHALL, JAMES L.: Pvt., Co. G. Resident of Staunton. Enl Staunton 8/2/61. Present 11/61-4/62, non-conscript under 18 not reenl. AWOL 8/12/62-10/63. Dropped as deserter. NFR.

MARSHALL, JAMES M.: Pvt., Co. B(2nd). B. Va. 1817? Farmer, age 43, Burkes Mill Dist., Augusta Co. 1860 census. Enl. Staunton 8/2/61. Present 11/61-4/62, non-conscript over 45 not reenlisted. AWOL 4/30-8/31/62, fined $22.00. Present 9/1-12/62, accepted as substitute for Pvt. Koiner 12/12/62. Present 1-4/63. AWOL 7/31-8/13/63. Present 8/13-12/63. Issued clothing 1/26, 3/31 and 4/27/64. Present 4/30-5/22/64. Ab. sick in hospital 5/23-7/20/64. Present 8-12/64. Surrendered Appomattox 4/9/65. Farmer, age 53, Mt. Sidney PO, Augusta Co. 1870 census.

MARSHALL, JAMES T.: Pvt., A.B. Va. 1828? Farmhand, age 32, Burkes Mill Dist. Augusta Co. 1860 census. Enl. Staunton 7/16/61. Detailed as Wagoner 9/6-11/24/61. Present 11/25/61-4/62. Reenl. 5/1/62. WIA (thigh) 2nd Manassas 8/29/62. Present 8/31/62-2/18/63. Ab. on leave 2/19-28/63, fined $9.16. Present 8/22-12/7/63. AWOL 12/8/63, lost 1 knapsack $6.50, 1 canteen $1.25, 1 haversack 50c stoppage for accouterments. Deserted to the enemy in W. Va. 12/25/63. Took oath and sent north. Age 34, 5'11½", dark complexion, dark eyes, dark hair, illiterate. Living in Augusta Co. 1903.

MARSHALL, JAMES WILLIAM: "Cyclone Jim". 4thSgt., Co. D.B. Augusta Co. 3/31/44. Farmhand, age 16, Burkes Mill Dist., Augusta Co. 1860 census. Enl. Staunton 7/16/61 as Pvt. Present 7/16-10/31/61. Ab. sick 11/61-2/62. Present 3-4/62. Reenl. and promoted 4thSgt 5/1/62. WIA McDowell 5/8/62. WIA (hip) 2nd Manassas 8/28/62. Present 1-12/63. Issued clothing 3/31 and 4/21/64. WIA (left thigh) Spotsylvania CH 5/12/64. Ab. wounded through 12/31/64. Paroled Staunton 5/24/65. Age 21, 5'11", light complexion, dark hair, gray eyes. Attended Roanoke College 1870. Studied law. Lawyer and Commonwealth's Attorney Craig Co. 1870-75. Served 4 years in State Senate; Member House of Delegates 1882-83, served 4 or more years as Commonwealth's Attorney, Craig Co.; State Senator 1891-92., Member U.S. Congress 1893-95. Died 1911.

MARSHALL, MOTE: Pvt., C. G. On postwar roster.

MARSHALL, R.: Pvt., C. D. Pension request only record. Living Russell Co. 1888.

MARTIN, JOHN J.: Pvt., Co. B(2nd). B. Va. 5/1/23. Carpenter, age 37, 1st Dist., Augusta Co. 1860 census. Exempt as Ironworker 3/24/62. Enl. Albermarle Co. 3/2/64. Issued clothing 4/1/64. Listed as AWOL 4/30-10/31-64, however issued clothing 4/27 and 8/11/64 and shown on rolls Co. A, Ward's Bn. C.S. Prisoners 7/64 at Lynchburg pardoned by the President for their participation in the defense of Lynchburg. In hospital Harrisonburg 9/20/64 and Charlottesville 11/8/64-1/14-65. Listed as deserter 2/4/65, can be found at or near Covesville, Albermarle Co. WIA (through left thigh) while charging an enemy battery Ft. Stedman 3/25/65. NFR. Carpenter, age 46, 1st Dist., Augusta Co. 1870 census. Illiterate. Living Dutch Creek, Nelson Co. 1/84. Died 12/15/04. Bur. Bethlehelm Methodist Ch. Cem., Ladd, Augusta Co.

MARTIN, WILLIAM W.: Pvt., Co. B(2nd). B. Va. 1840? Baker, age 20, 1st Dist., Staunton, 1860 census. Enl. Waynesboro 7/15/61. Present 11/61-4/62. Reenl. 5/1/62. AWOL 6/20-9/25/62, fined $33.27. Present 9/26/62-2/63. WIA (twice and shin bone split) Gettysburg 7/3/63. Present 8-12/63. Issued clothing 3/16/64. AWOL 4/18-28/64. Listed on rolls Co. A, Ward's Bn., C.S. Prisoners paroled by the President 7/64 for their participation in the defense of Lynchburg. Issued clothing 8/5/64. In hospital Charlottesville with "cephalalgra" 8/17-9/7/64. Present 11-12/64, lost 1 cartridge box and thirty-five rounds of ammunition. Stoppage $2.50 for cartridge box and 25 cents per round of ammunition. Surrendered Appomattox 4/9/65. Resident, Willow Bank, Nelson Co. 1905.

MASINCUPP, HENRY J.: Pvt., Co. A.B. Nelson Co. 1842. Farmer, age 33, Churchville, Augusta Co. 1860 census. Enl. Staunton 7/16/61. Present 11/61-2/62. Died of typhoid pneumonia Camp Alleghany 3/4/62. Age 30, 5'11½", dark complexion, grey eyes, dark brown hair. Left widow. Bur. St. James Methodist Ch. Cem., Churchville.

MASINCUPP, JOHN H.: Pvt., Co. A.B. Augusta Co. 3/9/41. Farmhand, age 29, Churchville, Augusta Co. 1860 census. Enl. Staunton 7/18/61. Ab. sick 9/10-12/31/61. Present 1-4/62. Reenl. 5/1/62. Deserted 6/18/62. NFR. Farmer, age 33, Mt. Sidney PO, Augusta Co. 1870 census. Blacksmith and Wheelwright, Staunton, 1897. Died there 4/11/22. Bur. Union Ch. Cem., near Churchville.

MASON, CLAIBORNE RICE. QM, F&S. B. near Troy, N.Y. 1832. Construction Engineer. Surveyed Midlothain RR 1829 (first RR in Va). Construction Engineer, Petersburg & Weldon RR 1831-32; Baltimore & Ohio RR and its Baltimore and Washington branch 1833-35; Richmond, Fredericksburg and Potomac RR 1835-36; Louisa RR 1836-49 (forerunner of Virginia Central RR which became Chesapeake & Ohio RR), President of Louisa RR Company; Built tunnels through Rockfish Gap 1851-57; Orange & Alexandria RR between Charlottesville and Lynchburg 1855-60. Moved to Staunton 1855 and then to Swoope, Augusta Co. Formed Mason Syndicate (still in business today). Was worth $1 million dollars 1861. Described as "...5'9", broad of shoulder and with long sinewy, powerful arms." Enl. Staunton 7/23/61 age 57, as Capt. Co. H. Appointed QM of Rgt 9/4/61. Detailed by Gen. Henry R. Jackson to build and maintain roads between Staunton and Greenbrier River and Warms Springs and Huntersville. Transf. to Engineers 10/61 and promoted LtCol. (but preferred to be called Captain). Appointed Engineer Officer on "Stonewall" Jackson's staff 5/62. Built boats to ferry ambulances across flooded North River at Mt. Crawford 5/62. Bridged the Shennandoah River at Port Republic 6/62 by driving army wagons into the river, fastening them together and laying planks across them. When sent forth by Jackson to bridge the Chicahominy, Mason told the courier: "Tell the General I'll be there directly, as soon as I finish cooking and eating my greens and bacon." Two minutes later Major William Allen and 2 other members of Jackson's staff arrived. "Come along," said Major Allen, "you've kept the General waiting too long already." "Tell the General he will have to wait only eight more minutes, I'll be there as soon as I finish eating these greens and bacon," replied Mason. He arrived at Jackson's headquarters in exactly eight minutes. "How long will it take you to build a bridge across the river?" Jackson asked. "I can't tell you until I see the ground first and select a crossing, but I think I can make it in two hours," Mason replied. Mason asked for and received force of a thousand men and forty to fifty wagons. He commenced construction at two o'clock in the afternoon while other Engineer Officers arrived and made drawings. They returned at six o'clock with their handiwork and told Mason he could now build the bridge. "What have we here-pictures?" asked Mason. "You can take these pretty drawings back with you Gentlemen! the bridge was completed two hours ago - finished in just two hours just as I told the General it would be," he added. Served as QM of Augusta Raid Guard or Home Guard 11/63. Cut the road for Gen. Lee from the wilderness to Spotslyvania CH. Engineer in charge of repairs on Virginia Central RR 1864-65. Lost everything in the war. Completed repairs of Va. Central 1866. Contractor on Covington & Ohio RR 1870. Completed Jerry's Run fill and Lewis Tunnel in Greenbrier Co. that had broken 4 other contractors, 1872. Built RR lines in Tenn., Ky., Pa Ohio 1870s-80s. Illiterate, but could copy his name. Could walk or ride over a line of survey and determine readily how long it would take a certain number of men to build the road and what it would cost. His principle tool was a twelve-inch rule. He could manipulate this rule vertically and horizontally and determine the cubic content of earth in a mountain. He could calculate (mentally) the interest on a note as quickly as an expert accountant. When challenged to a competition with a UVA Math Professor, he would stare off in the distance and then give the correct answer before the professor could work it on paper. Died Swoope 1/12/85. Bur. Thornrose Cem. The Augusta Co. UDC Chapter was named in his honor.

MATHENY, WESLEY H.: Pvt., Co. H.B. Rockbridge Co. 3/13/37. Laborer, age 23, 1st Dist., Augusta Co. 1860 census. Pvt., Co. E., 93rd Va. Militia. Enl. Staunton 7/23/61 age 24. Present 11/61-4/62. Reenl. 5/1/62. WIA (head) Port Republic 6/9/62. Ab. wounded through 8/31/62. Present 1-12/63. WIA (left leg) Spotslyvania CH 5/12/64. Ab. wounded through 12/31/64. Paroled Staunton 5/1/65. Age 25, 5'8", dark complexion, dark hair, black eyes. Shoemaker, Augusta Co. Died Staunton 11/27/78. Bur. Thornrose Cem.

MATHEWS, ABRAHAM: Pvt., Co. A.B. Augusta Co. 4/1/42. Enl. Staunton 6/23/63. Issued clothing 4/22/64. WIA (forearm) and cap. Bethesda Ch. 5/30/64. Sent to Lincoln Hospital, Washington, DC. Age 18. Sent to Old Capitol Prison. Transf. to Elmira and Point Lookout. Exchanged 11/15/64, Savannah, Ga. NFR. Farmer, near Koiner's Store, Augusta Co. 1888. Died there 6/30/91. Bur. Hildebrand Mennonite Ch. Cem., Madrid.

MATHEWS, DAVID: Pvt., Co. A.B. Augusta Co. 4/4/40. Farmhand, age 26, Burkes Mill Dist., Augusta Co. 1860 census. Enl. Staunton 7/16/61. Present 11/61-4/62. Reenl. 5/1/62. Cap. Front Royal 5/30/62. Exchanged 8/5/62. Present 8/31-12/31/62. Present 1-12/63. Issued clothing 2/20 and 4/27/64. Cap. Bethesda Ch. 5/30/64. Sent to Point Lookout. Transf. Elmira. Exchanged 3/14/65. In hospital Richmond 3/18-19/65. Illiterate. NFR. Farmer, Augusta Co. Died near Waynesboro 9/9/96. Bur. Hildebrand Mennonite Ch. Cem., Madrid.

MATTHEWS, HENRY R.: Pvt., C. C. B. Loudon Co. 12/11/22. Not on muster rolls. Deserted to the enemy Clarksburg, W.Va. 10/13/63. Took oath and sent north. Carpenter, age 47, Staunton 1870 census. Died Staunton 3/21/99. Bur. Thornrose Cem.

MATTHEWS, J.W.: Pvt., A. Not on muster rolls. Cap. date and place unknown. Died of disease Elmira, NY. 5/23/65. Bur. Woodlawn National Cem., there.

MATTHEWS, JOHN M.: Pvt., Co. A. Enl. Staunton 7/16/61. Present 11/61-4/62. Reenl. 5/1/62. Present 8/31-12/31/62. Present 1-4/63, fined $30.00 for arrest as deserter. Present 4/30-12/63. WIA and cap. Bethesda Ch. 5/30/64. Sent to Elmira. Died of disease there 7/23/65. Bur. Woodlawn National Cem., there.

MAUPIN, ISAAC SIMMS: 1stSgt., Co A.B. Va. 1844? Student, age 16, Burkes Mill Dist., Augusta Co. 1860 census. Enl. Staunton 7/16/61 age 17 as 2ndSgt., sustitute for his father, Logan I. Maupin, by order of Gov. Letcher. Ab. sick 11/20/61-1/5/62. Present 11/61-4/62. Reenl. 5/1/62. Present 8/31-12/31/62, rank shown as 1stSgt. Present 1-9/63. Sent to Staunton to collect clothing for the Regt. 9/30-10/63. Present 11/12-63. Issued clothing 1/28 and 2/27/64. KIA Carter's Farm near Winchester 7/20/64, age 21. Bur. Stonewall Cem., Winchester as "Sgt. J.S. MAUPIN," age 20. "No nobler youth than he has fallen in the struggle for Independence...he stood, cool and undaunted, on the hard rough field of Port Republic, returning when wounded with coolness and deliberation the galing fire of the enemy...The roar of the battle reached his home..."

MAUPIN, JAMES HENRY: 4thSgt., Co. D.B. Augusta Co. 3/15/41. Farmhand, age 19, Burkes Mill Dist., Augusta Co. 1860 census. Enl. Staunton 7/16/61 as Pvt. Present 7/16/61-4/62. Present 5/1-6/8/62. AWOL 6/9-20/62, fined $4.03. Present 6/21-8/31/62. Present 1-4/63, promoted 4thSgt 4/15/63. Present 4/30-12/63. Issued clothing 3/31, 4/22 and 6/12/64. WIA Winchester 9/19/64. Ab. wounded through 12/31/64. Paroled Staunton 5/19/65. Age 24, 5'10", light complexion, light hair, grey eyes. Bricklayer, Augusta Co. 1866. Moved to Middletown, Indiana, 1870.

MAUPIN, JAMES THOMAS: Pvt., Co. H.B. Va. 1/24/32. Enl. Staunton 10/25/64. Present 11-12/64. Paroled Staunton 5/20/65. Age 30, 5'8", fair complexion, dark hair, blue eyes. Farmer, age 38, Fishersville PO, Augusta Co. 1870 census. Died Stuart's Draft 3/2/22. Bur. Calvary Methodist Ch. Cem., Stuart's Draft.

MAUPIN, JOHN S.: Pvt., Co. B(2nd). B. Va. 1837? Farmhand, age 23, Burkes Mill Dist., Augusta Co. 1860 census. Not on muster rolls. Cap. Beverly, W.Va. 5/8/65. Paroled for exchange at Clarksburg 5/11/65. Age 24, 5'10", florid complexion, blue eyes, dark hair, farmer. Resident of Albermarle Co., illiterate.

MAUPIN, THOMAS G.: Pvt., Co. B(2nd). B. Albermarle Co. 1843?. Laborer, Augusta Co. 1863. Enl.Orange CH 2/17/64. Deserted 4/20/64. Can be found in the Scottsville area, Albermarle Co. on deserters notice 2/4/65. NFR.

MAYER, J.E.: 2ndLt., Co. K. Not on muster rolls. Resigned 1/15/65. NFR. Could be John H. Mayers, b. Highland Co. Fell from wagon and broke his neck near Stover, Augusta Co. 6/31/96. Farmer, age 54 years, 9 months and 11 days or John W. Mayser, Commissioner of Revenue, Bath Co. 1897.

MAYFIELD, ANDREW J.: Pvt., Co. F.B. Augusta Co. 1832? Enl. Staunton 7/31/61. Present 11/61-4/62. Reenl. 5/1/62. WIA (arm broken) Fredericksburg 12/13/62. Ab. wounded through 12/63. Present 3-4/64. AWOL 4/20-10/31/64. WIA Petersburg 12/8/64. Ab. wounded 12/9-31/64. Paroled Staunton 5/15/65. Age 33, 5'11", light complexion, light hair, blue eyes. Blacksmith, age 38. Pastures Dist., Augusta Co. 1870 census. Member Stonewall Jackson's Camp C-V Staunton. Died Augusta Co. 1/26/09. Bur. Thornrose Cem., Staunton.

MAYS, CYRUS, B.: Pvt., C. I.B. Va. 1842? Farmhand, age 18, Summerdean, Augusta Co. 1860 census. Enl. Staunton 7/16/61 age 19. Ab. sick 11/61-2/62. Present 3-4/62. Reenl. 5/1/62. WIA (shoulder) McDowell 5/8/62. Present 8/30-9/14/62. AWOL 9/15/62-1/20/63, fined $45.83. Present 2-6/63. Deserted to the enemy 7/21/63. Took oath and released. Farmer, age 29. Riverheads Dist., Augusta Co. 1870 census. Died Augusta Co. 6/30/04. Bur. Mt. Carmel Pres. Ch. Cem., Steele's Tavern.

MAYS, ISAIAH M.: Pvt., Co. I.B. Va. 1838? Laborer, age 22, 1st Dist., Augusta Co. 1860 census. Pvt., Co. I, 93rd Va Militia. Enl. Staunton 3/29/62 age 23. Present 3-4/62. WIA (left side) McDowell 5/8/62. WIA (leg) 2nd Manassas 8/29/62. Ab. wounded through 12/31/64. Served in Capt. John Avis's Co., Provost Guard, Staunton 9/1-12/31/64. NFR. Laborer, age 32, Riverheads Dist., Augusta Co. 1870 census. Living in Augusta Co. 1900.

MAYSE, THOMAS O.: Pvt., Co. C.B. Augusta Co. 1819? Farmer, Augusta Co. Enl. Staunton 7/16/61 age 42. Ab. sick 10/20/61-4/62. Discharged for general bad health and "odernutous continia" 7/21/62. Age 43, 5'6", dark complexion, grey eyes, black hair.

MEEKS, JOHN P.: Pvt., Co. B(1st). B. Rockbridge Co. 1833? Wagoner, age 28, 1st Dist., Augusta Co. 1860 census. Enl. Staunton 8/1/61 Present until transf. 9/28/61.

MESSERLY, JOHN P.: 1stSgt., Co. H. B. Va. 1843? Farmhand, age 17, Northern Dist., Augusta Co. 1860 census. Enl. Staunton 7/23/61 age 19, as Pvt. Present 11/61-4/62, rank shown as Cpl. Reenl. and promoted 3rdSgt 5/1/62. Present 5/1-8/31/62. Promoted 2ndSgt 9/1/62. Present 1/1-8/31/63, promoted 1stSgt 5/1/63. Ab. on detached service for 15 days 10/30/63. Present 11-12/63, and 4/30-6/18/64. Ab. sick 6/19-12/31/64. NFR. Supposed to have been KIA 1865? "Was a noble soldier."

MICHAEL, ABSOLOM: Pvt., Co. D. B. Augusta Co. 1837? Farmhand, age 23, North Mt., Augusta Co. 1860 census. Enl. Staunton 7/16/61. Ab. on leave 7/16-10/31/61. AWOL 11/1/61-2/25/62, fined $11.00 by CM. Present 2/26-4/62. AWOL 5/1-8/31/62. Present in arrest 1-2/63. Died of fever in camp near Fredericksburg 4/4/63.

MICHAEL, ALBERT: Pvt., Co. F.B. Rockingham Co. 8/23/18. Farmer, age 42, Northern Dist., Augusta Co. 1860 census. Not on muster rolls. Enl. Valley Mills 4/62. Put in substitute and discharged 7-8/62. Died North River Dist., Augusta Co. 10/23/75. Bur. Old Salem Ch. Cem., 3 miles west of Mt. Sidney.

MICHAEL, DANIEL: Pvt., Co. D.B. North River Dist., Augusta Co. 8/4/27. Farmhand, age 30, Burkes Mill Dist., Augusta Co. 1860 census. Enl. Staunton 7/16/61. Present 7/16/61-4½62. Reenl. 5/1/62. Present 5/1-6/1/62. AWOL 6/2-20/62 and 7/20-9/1/62, fined $21.26. Detailed as Wagoner 1-4/63. AWOL 7/20-11/6/63. Present in arrest 11/7-12/63. Court Martialed 3/7/64. NFR. Farmhand, age 42, Mt. Sidney PO, Augusta Co. 1870 census. Farmer, North River Dist., Augusta Co. Died there 1/1/85. Bur. Old Emanuel Ch. Cem., near Mt. Solon.

MICHAEL, HUDSON STAUBUS: Pvt., Co. D. B. near Mt. Solon, Augusta Co. 9/8/37. Resident, North Mt., Augusta Co. Enl. Staunton 7/16/61. Ab. on leave 7/16-10/31/61. Present 11/61-2/62, detailed to make candles for Commissary 2/25/62. Present 3-4/62, and 5/1-6/8/62. AWOL 6/9-19 and 7/18-8/6/62, fined $10.26. Deserted 11/24/62. Enl. Co. C. 18thVaCav. Farmhand, age 30, Mt. Sidney PO, Augusta Co. 1870 census. Farmer and carpenter, Stokesville, Augusta Co. Died 1923. Bur. St. Paul's Lutheran Ch. Cem., near Mt. Solon.

MICHAEL, JAMES: Pvt., Co. D.B. Augusta Co. 12/24/34. Farmhand, age 26, Northern Dist., Augusta Co. 1860 census. Not on muster rolls. Deserted to the enemy Tucker Co., W. Va. 5/5/64. Age 29, 5'7", fair complexion, blue eyes, dark hair. Farmer, age 35, Mt. Sidney PO, Augusta Co. 1870 census. Farmer and Miller, Dorcas, Augusta Co. 1897. Died near Mt. Zion Ch., Augusta Co. 2/10/14. Bur. Mt. Zion Ch. Cem.

MICHAEL, JOHN: Pvt., Co. D.B. Augusta Co. 4/15/15. Farmer, age 45, Northern Dist., Augusta Co. 1860 census. Enl. Staunton 7/16/61. Ab. on leave 7/16-10/31/61. Present 11/61-3/62. AWOL 4/24-5/1/62. Present 5/2-8/31/62, detailed as Wagoner 7/20/62. AWOL 11/24-12/31/62. Present in arrest 1-2/63. Ab. in Castle Godwin, Richmond, sentenced to 6 months labor with ball and chain and forfeiture of pay. Present 4/30-8/31/63, pardoned by the President 8/1/63. Present 9-12/63, and 4/30-12/31/64. NFR. Died Long Glade, Augusta Co. 10/27/75. Bur. Old Salem Lutheran Ch. Cem., 3 miles west of Mt. Sidney.

MILLER, ARCHIBALD O.: Pvt., Co. I. On postwar roster. Farmer, Kiracofe. Augusta Co. 1897.

MILLER, BENJAMIN F.: Pvt., Co. B(2nd). B. Augusta Co. 1843? Farmhand, age 17, 1st Dist., Augusta Co. 1860 census. Enl. Waynesboro 7/15/61. Present 11/61-4/62. WIA (hand) McDowell 5/8/62. Present 5/62-3/63. Ab. sick in hospital Lynchburg 4/16-12/31/63. WIA Wilderness 5/6/64. WIA (right cheek and groin) and cap. Winchester 9/19/64. Sent to Point Lookout. Exchanged 10/30/64. Paroled Staunton 5/15/65. Age 22, 5'8", dark complexion, light hair, dark eyes. Farmer, Stonewall, Augusta Co. 1897.

MILLER, BENJAMIN, FRANKLIN: Pvt., Co. E.B. on Whistle Creek. Rockbridge Co. 10/8/43. Res., age 16, 4th Dist., Rockbridge Co. 1860 census. Enl. Camp Shennandoah 4/12/62. Present 4/12-30/62. WIA (head) McDowell 5/8/62. WIA Gaines Mill 6/27/62. WIA (leg, tibua fractured) Cedar Run 8/9/62. In hospital Charlottesville 8/11/62-1/27/63. In hospital Lynchburg 3-7/63. Transf. Co. C. 1stVaCav. 2/18/64. Wheelwright and Wagonmaker, Natural Bridge District, Rockbridge Co. 1870 census. Wagonmaker, Sangersville, Augusta Co. 1897. Died Staunton 10/23/20, resident of Fordwick Augusta Co. Brother of Joseph C. Miller.

MILLER, CHARLES L.: Pvt., Co. H.B. Va. 1844? Resident of Port Republic. Enl. Staunton 7/23/61 age 17. Present 11/61-4/62. Deserted 6/8/62. NFR. Member, Stonewall Jackson Camp CV, Staunton. Carpenter, age 76, Waynesboro, 1910 census. Died there 7/14/17. Bur. there.

MILLER, CLEMENT H.: Pvt., Co. H.B. Augusta Co. 4/22/42. Resident, age 14, Burkes Mill Dist., Augusta Co. 1860 census. Enl. Staunton 7/23/61 age 17. Present 11/61-4/62. Reenl. 5/1/62. Present 5/1-7/17/62. AWOL 7/18-8/12/62. Present 8/13-8/31/62. Discharged for underage near Liberty Ch., Caroline Co. 12/9/62. Age 16, 5'4", fair complexion, dark eyes, dark hair, farmer. Col. Baldwin's Orderly 1-2/62. Died Augusta Co. 12/22/05. Bur. Pleasant Valley Ch. of the Brethren, near Weyer's Cave.

MILLER, E. S.: Pvt., Co. B. (1st). On postwar roster.
MILLER, GEORGE C.: Pvt., Co. A. B. Va. 1822? Farmer, age 38, 1st Dist., Augusta Co. 1860 census. Enl. Staunton 7/9/61 age 38. Present 11/61-4/62. Reenl. 5/1/62. WIA (arm) Gaines Mill 6/27/62. Ab. wounded through 12/63. Detailed for light duty as hosp. guard Richmond and Charlottesville 2/2-12/31/64. Paroled Staunton 5/15/65. Age 42, 5'8", dk. complexion, grey hair, hazel eyes. D. near Ft. Defiance, Augusta Co. 12/6/92.
MILLER, GEORGE HENRY: Pvt., Co. B (2nd). B. Va. 1824. Farmhand, age 36, 1st Dist., Augusta Co. 1860 census. Enl. Waynesboro 7/15/61. Present 11/61-4/62, non-conscript. AWOL 4/18-9/24/62. Present 9/25/62-12/63. KIA Bethesda Ch. 5/30/64. Bur. Mt. Tabor Lutheran Ch. Cem., near Middlebrook.
MILLER, JAMES D.: Pvt., Co. G. Not on muster rolls. Listed as des. 2/63. Resident of Staunton. NFR.
MILLER, JOHN: Capt., Co. B (1st). B. Princeton, N. J. 1819? Pres. Minister, age 41, Lexington, Rockbridge Co. 1860 census. Pastor, Old Oxford Pres. Ch. Enl. Fairfield 6/16/61. Present until transf. 9/28/61.
MILLER, JOHN F.: Pvt., Co. I. B. Va. 1835? Laborer, age 25, North Mt., Augusta Co. 1860 census. Enl. Staunton 7/16/61 age 32. Ab. sick 11/61-1/15/62. Present 1/16-4/62. Reenl. 5/1/62. WIA (hand) Gaines Mill 6/27/62. Present 8/30-10/31/62. AWOL 11/23/62-1/19/63, fined $20.53 by CM. Present 1/20-12/63. Capt. Bethesda Ch. 5/30/64. Sent to Point Lookout, Transf. Elmira. D. of pneumonia there 10/2/64. Bur. Woodlawn National Cem. Brother of Martin M. Miller, Jr.
MILLER, JOHN M.: Pvt., Co. K. B. Bath Co. 1835? Farmhand, age 25, Cleeks Mill PO, Bath Co. 1860 census. Enl. Shenandoah Mt. 4/9/62. Present 4/9-30/62. NFR on muster rolls. Listed as des. 5/1/63. Des. to the enemy Clarksburg, W. Va. 4/8/65. Took oath and released. Resident of Back Creek, Bath Co. Farmer, age 35, Williamsville Dist., Bath Co. 1870 census. Member, Gen. Pegram Camp CV, Valley Head, W. Va. 1915.
MILLER, JOHN PRESSLY: Pvt., Co. E. B. Rockbridge Co. 8/15/37. Enl. Staunton 8/1/61. Present 11/61-4/62. Reenl. 5/1/62. WIA (neck) Sharpsburg 9/17/62. Ab. wounded through 10/31/62. Present 1-12/63. WIA (shoulder) Cedar Creek 10/19/64. Ab. wounded through 12/31/64. Surrendered Appomattox 4/9/65. Laborer, age 33, Natural Bridge Dist., Rockbridge Co. 1870 census. Farmer, Canal Engineer, and Carpenter. D. near Lexington 8/12/16. Bur. Broad Creek Pres. Ch. Cem., near Fancy Hill.
MILLER, JOSEPH ANDREW: Pvt., Co. E. B. Rockbridge Co. 2/4/23. Shoemaker, Rockbridge Co. Enl. Staunton 8/1/61. Present 11/61-2/62. Ab. sick 3-4/62. Disch. for varicose conditions of the legs at camp near Gordonsville 7/20/62. Age 39, 5'8½", fair complexion, blue eyes, light hair. Farmer, age 44, Natural Bridge Dist., Rockbridge Co. 1870 census. Lived in Oregon 9/76. D. Alone, Rockbridge Co. 1/13/12. Bur. Stonewall Jackson Cem., Lexington.
MILLER, JOSEPH CLOYD.: Pvt., Co. E. B. Rockbridge Co. 4/9/41. Farmhand, age 19, Lexington Dist., Rockbridge Co. 1860 census. Enl. Staunton 8/1/61. Present 11/61-2/62. Ab. sick 3/4/62. Died of typhoid fever in hospital Lynchburg, 5/20/62. Age 22. Brother of Benjamin Franklin Miller.
MILLER, MARTIN MANSFIELD, JR.: 4thSgt., Co. I. B. Augusta Co. 3/25/39. Farmer, age 21, 1st Dist., Augusta Co. 1860 census. Enl. Staunton 7/16/61 age 22 as 3rdCpl. Present 11/61-4/62, appointed 5thSgt. 2/5/62. WIA 2nd Manassas 8/28/62. Presetn 8/30-10/31/62. Ab. sick 11/1/62-10/63. Detailed in QM Dept. 12/11/63-10/27/64. Present 10/28-12/31/64, appointed 4thSgt. 10/3/64. WIA Ft. Stedman 3/25/65, "turned back to aid James A. McClure, finding him mortally wounded, and being almost surrounded by Union soldiers and asked to surrender, he chose to attempt to escape, knowing his brother John had been captured and died of diease at Elmira, and was shot down as a result." DOWs 3/26/65. Bur. Mt. Carmel Pres. Ch. Cem, near Steele's Tavern, Augusta Co. Brother of John F. Miller.
MILLER, PETER A.: Pvt., Co. D. B. Rockingham Co. 2/12/28. Mechanic, age 32, Northern Dist., Augusta Co. 1860 census. Enl. Staunton 7/16/61. Discharged by order of the Secretary of War 8/16/61. papermaker, Mt. Solon. Farmer, Age 42, Mt. Sidney PO, Augusta Co. 1870 census. Farmer, Kiracofe, Augusta Co. 1897. Died Mt. Solon 1/7/04.
MILLER, SAMUEL M.: Cpl., Co. A. B. Augusta Co. 2/18/62. Farmhand, age 18, Burkes Mill Dist., Augusta Co. 1860 census. Enl. Staunton 7/9/61 as Pvt. Present 11/61-4/62. Reenl. 5/1/62. Promoted Cpl. WIA Gaines Mill 6/27/62. Ab. wounded through 2/63. Transf. Co. E., 1st Va. Cav. 3/10/63. WIA Gettysburg twice, KIA near Berryville 9/2/64. Bur. Zion Lutheran Ch. Com., Augusta Co.
MILLER, WILLIAM G.: Pvt., Co. H. B. Augusta Co. 1801? Enl. Staunton 7/23/61 age 60. Present 11/61-4/62. Died of dysentery near Scottsville 7/23/62, age 62. Left widow.
MILLER, WILLIAM R.: Pvt., Co. C. B. Rockbridge Co. 7/10/32. Resident, Waynesboro. Pvt., Capt. Paris's Co., 93rd Va. Militia 3/62, exempt for scrofula. Not on muster rolls. Listed as deserter 2/63. Died South River Dist., Augusta Co. 9/5/16. Bur. Mt. Tabor Lutheran Ch. Cem., near Middlebrook.
MILSTEAD, GIDEON: Pvt., Co. A. B. 1837? Resident of New Hope, Augusta Co. Enl. Staunton 7/9/61 age 23. Present 11/61-4/62. Renl. 5/1/62. Deserted 7/4/62. NFR. Died before 5/03.
MISNER, ADDISON L.: Pvt., Co. F. B. Augusta Co. 1838? Farmhand, age 22, Northern Dist., Augusta Co. 1860 census. Enl. Staunton 7/15/61. Present 11/61-4/62. Reenl. 5/1/62. Present 8/31/62-7/26/63. AWOL 7/27-8/8/63, and pay deducted. Present 8/31/62-7/26/63. AWOL 7/27-8/8/63, and pay deducted. Present 8/9-12/63. Detailed in Pioneer Corps 1-10/64. Deserted 11/1/64. NFR. Unofficial rosters list him as being WIA, date and place not indicated.
MISNER, JACOB HARVEY: Pvt., Co. A. B. Augusta Co. 1840? Carpenter, Augusta Co. 1861 Married Augusta Co. 7/11/61. Enl. Staunton 7/16/61. Ab. sick 12/11-31/61. died of typhoid fever in hospital Staunton 1/2/62. Left widow.
MISNER, M. L.: Pvt., Co. F. On postwar roster. Enl. 7/61. NFR.
MITCHELL, ROBERT H.: Pvt., Co. A. B. Augusta Co. 1841? Farmhand, age 19, 1st Dist., Augusta Co. 1860 census. Enl. Staunton 7/15/61. died at home near Staunton of typhoid pneumonia 11/10/61, age 22.
MITCHELL, TULEY JOSEPH: Pvt., Co. A. B. Augusta Co. 1841. Attended Princeton Col. '60-61. Enl. 6/61 and discharged for underage. Enl. 5th Va. Inf. and appointed Capt., and Commissary Officer of Regt., served 1 year. Enl. Co. H, 18th Va. Cav. WIA (head) and cap. New Hope 6/5/64. Sent to Camp Morton. Transf. Johnson's Island, Ft. McHenry, Point Lookout and Ft. Delaware. Farmer, Fauquier Co. 1865-92. Real Estate Agent, Roanoke until his death in 1911.
MOFFETT, JAMES SIDNEY: Pvt., Co. H. B. Augusta Co. 1829? Attended Washington Col. '47-48. Farmer, Augusta Co. Staunton 10/25/64. Present 11-12/64. WIA Hatcher's Run (shoulder) 2/8/65. Admitted to hospital Richmond 2/9/65. Transf. hospital Staunton 2/13/65. NFR. Farmer, age 40, Fishersville PO, Augusta Co. 1870 census. Moved to Ky. and then to Ft. Scott, Kan. Died there 3/24/88 in 59th year.
MONEYMAKER, DAVID C.: Pvt., Co. H. B. Augusta Co. 1841. Farmhand, age 19, Walker's Creek Dist., Rockbridge Co. 1860 census. Enl. Staunton 7/23/61 age 22. Present 11/61-2/62 AWOL 3/5-22/62. Present 3/23-4/62. Reenl. 5/1/62. WIA (ankle) Port Republic 6/9/62. Ab. wounded through 10/63. Detailed in hospital Lynchburg 12/14-31/63. Deserted Salisbury, N. C. 7/28/64. NFR. Farmer, Augusta Co., 1870.

MONTGOMERY, ————: KIA Fredericksburg 5/5/63, unofficial report.

MONTGOMERY, JAMES H.: Pvt., Co. H. B. Augusta Co. 1834. Farmhand, age 22, Summerdean, Augusta Co. 1860 census. Enl. Staunton 7/23/61 age 22. Ab. sick 12/15/61-4/62. AWOL 6/18-8/31/62, and 11/28/62-4/22/63. Sentenced to two years in penitentiary Richmond by CM for desertion and pay forfeited to 8/1/63. In hospital with "scorbutus" 6/13-28/63. Returned to Castle Thunder. Present 11-12/64, charged with bayonet scabbard, haversack, knapsack, canteen, gun sling, 40 rounds ammunition. WIA (left shoulder) Hatcher's Run 2/8/65. Furloughed from hospital Richmond 3/22/65, illiterate. NFR. Farmhand, age 32, Pastures Dist., Augusta Co. 1870 census. Matress Maker, Middlebrook, 1873. Died Deerfield 3/31/87. Bur. Rocky Springs Ch. Cem., Deerfield.

MONTHGOMERY, WILLIAM: Pvt. Co. H. B. Augusta Co. 1835? Farmhand, age 25, 1st Dist., Augusta Co. 1860 census. Enl. Staunton 7/23/61 age 23. Present 11-12/61, fined $11.00 by CM. AWOL 1/5-12/62 and 3/5-4/62. Died 4/30/62 place and cause unknown.

MOONEY, ABNER KIRKPATRICK: Pvt., Co. E. B. Harrisonburg, Rockingham Co. 1830? Shoemaker, age 30, Lexington Dist., Rockbridge Co. 1860 census. Shoemaker, Fancy Hill, Rockbridge Co. 2/64, dark complexion, dark eyes, dark hair, 5' 3½". Enl. Rockbridge Co. 10/31/64. Present 11-12/64. Cap. Petersburg 4/1/65. Sent to Washington, D. C. Took oath and transportation furnished to Wheeling. Shoemaker, age 41, Natural Bridge Dist., Rockbridge Co. 1870 census.

MOONEY, JOHN D: Pvt., Co. A. B. Va. 1833? Carpenter, age 27, New Hope, Augusta Co. 1860 census. Enl. Staunton 7/15/61. Deserted in Staunton 8/21/61. NFR. Alive 1903. Brother of Richard Mooney.

MOONEY, RICHARD: Pvt., Co. A. B. Va. 1841? Farmer, age 19, Burkes Mill Dist., Augusta Co. 1860 census. Enl. Staunton 7/9/61 age 19. Present 11/61-4/62. Reenl. 5/1/62. AWOL 5/20/62-10/9/63. Present in arrest 10/10-12/27/63. Escaped from arrest 12/28/63. Ab. sentenced to 12 months hard labor by CM 4/30-12/31/64. Deserted to the enemy New Creek, W. Va. 2/12/65. Took oath and sent north. Resident of New Hope, Augusta Co., age 23, 5' 5", florid complexion, grey eyes, brown hair. Alive 1903. Brother of John D. Mooney.

MOONEY, WILLIAM WEST: Pvt., Co. C. B. Fauquier Co. 5/7/34. Carpenter, age 25, Waynesboro, Augusta Co. 1860 census. Enl. Staunton 7/16/61 age 26. Present 11/61-4/62. Reenl. 5/1/62. AWOL 5/1/62-10/1/63. Present in arrest for desertion 10/63. Ab. escaped from custody of Regt'l guard on the night of 12/27/63. Ab. confined in Castle Thunder by sentence of GCM for 12 months and pay stopped since 2/1/63, on rolls 4-12/64. Deserted to the enemy New Creek, W. Va. 2/12/65. Resident of Waynesboro, age 30, 5' 7", florid complexion, grey eyes, brown hair, farmer. D. O'Niel, Tenn. 6/1/95.

MOORE, BENJAMIN FRANKLIN: Cpl., Co. E. B. Rockbridge Co. 10/8/43. Farmhand, age 22, Lexington Dist., Rockbridge Co. 1860 census. Serving in the Militia 6/61. Enl. Staunton 8/1/61. Present 11-12/61, was Cpl. to 11/1/61, exchanged with J. A. Wilkinson. Present 1-4/62. Reenl. 5/1/62. Ab. on leave 5/1-10/31/62. Present 1-12/62. WIA (right hip) Wilderness 5/6/64. Present 5-12/64. Surrendered Appomattox 4/9/65. Died Rockbridge Co. 10/23/20. Bur. Stonewall Jackson Cem., Lexington.

MOORE, GEORGE WILLIAM: 3rdLt., Co. B (2nd). B. Va. 1840? Cabinetmaker, age 20, Burkes Mill Dist., Augusta Co. 1860 census. Enl. Waynesboro 7/15/61 as Pvt. Present 11/61-4/62. Present 4/30-10/31/62, Promoted 3rd Sgt. 7/15/62 and 2nd Sgt. 10/2/62. Present 11/62-5/63, promoted 3rd Lt. 3/15/63. Presence or absence not stated 6-12/63. KIA Wilderness 5/6/64. May be Lt. G. W. Moore (Va.) bur. Fredericksburg Confederate Cem.

MOORE, HAMLET W.: Pvt., Co. D. B. Va. 6/27/29. Resident, Mossy Creek, Augusta Co. Enl. near Winchester 5/28/62 as substitute for Erasmus A. Samuels. Deserted 6/1/62. NFR. Died Mt. Sidney 2/6/03. Bur. Pleasant Grove Methodist Ch. Cem., near Mt. Crawford, Rockingham Co.

MOORE, I. K.: Pvt., Co. B. (1st). on postwar roster. Served 4 years.

MORRIS, ANCIL: Pvt., Co. C. B. Va. 1815? Farmhand, age 45, Burkes Mill Dist., Augusta Co. 1860 census. Enl. Staunton 8/1/61. Present 11/61-2/62, detailed to assist in bakery. Present 3-4/62. Ab. sick 6/15-8/15/62. Discharged overage of conscription at camp 10/16/62. Farmhand, age 55, Mt. Sidney PO, Augusta Co. 1870 census.

MORRIS, KENE, SR.: Musician, Co. C. Enl. Staunton 3/18/64 as Pvt. Issued clothing 4/1 and 21/64, with rank of Musician. Presen 4/30-6/64. AWOL 10/7/64 and 12/8/64. Admitted to hospital Petersburg 12/23/64 with remarks "1/7/65 Capt. Davidson". Deserted to the Army of the Potomac near Petersburg 3/8/65. Took oath and transportation furnished to Redding, Pa.

MORRIS, KENE, JR.: Drummer, Co. A. Enl. Staunton 7/9/61 age 11. "mustered into service as drummer by leave of his father." Ab. sick 11/7/61-1/4/62. Present 5/1-4/62. Discharged 12/31/62. Blind. Alive 1903.

MORRIS, LAYTON J.: Pvt., Co. A. B. Va. 1838? Blacksmith, age 22, Northern Dist., Augusta Co. 1860 census. Enl. Staunton 7/9/61 age 23. Ab. sick 12/28-31/61. Died of typhoid fever on way to hospital at McDowell 1/16/62.

MORRIS, LATIN J.: Pvt., Co. C. B. Va. 1839? Farmhand, age 21, Northern Dist., Augusta Co. 1860 census. Enl. Staunton 7/16/61 age 23. Present 11/61-4/62. Reenl. 5/1/62. AWOL 5/18-8/30/62. Present 8/31/62-3/63. Ab. sick 4/25-8/31/63. Present 9-12/63, and 4/30-8/64. AWOL 9-12/64. deserted to the enemy New Creek, W. Va. 3/10/65. Took oath and sent to Ohio.

MORRIS, ROBERT B.: Pvt., Co. B (2nd). B. Augusta Co. 1835? Enl. Waynesboro 7/15/61. Present 11/61-2/62, fined $11.00 for 16 days AWOL. Present 3-4/62. Reenl. 5/1/62. AWOL 5/17-9/22/62. Ab. sick 10/22/62-2/63, and 4/22-30/63. AWOL 4/30-12/63. Present 4/30/64. WIA '64, admitted to hospital Charlottesville 7/11/64. Returned to duty 8/30/64. AWOL 9/28-12/31/64. Can be found near Batesville, Albermarle Co. on deserter notice dated 2/4/65. NFR. Died Riverheads Dist., Augusta Co. 1/19/95

MORRIS, SAMUEL S.: Pvt., Co. B (2nd). B. Va. 1826? Carpenter, age 34, Burkes Mill Dist., Augusta Co. 1860 census. Pvt., Long Meadow Co., Augusta Co. militia 3/62, exempt reason not shown. Not on muster rolls. Deserted to the enemy Clarksburg, W. Va. 8/3/65. Took oath and released. Illiterate. Carpenter, age 50, Mt. Sidney PO, Augusta Co. 1870 census.

MORRIS, SIMEON WHITFIELD: Pvt., Co. B (2nd). B. Greene Co. 1844? Enl. Waynesboro 7/15/61. Ab. sick 11-12/61. Present 1-4/62. Reenl. 5/1/62. WIA Port Republic 6/9/62. Ab. wounded through 2/63. Present 3-4/63. AWOL 5/17-12/31/63. Dropped as deserter. Laborer, Augusta Co. Died of snake bite on Blue Ridge Mts, Augusta Co. 8/9/74.

MORTON, JAMES P.: Pvt., Co. K. B. Bath Co. 11/21/44. Resident, age 15, Cleeks Mill PO, Bath Co. 1860 census. Enl. Shenandoah Mt. 4/9/62. Present 4/9-30/62. KIA McDowell 5/8/62.

MOSES, NORVELL: Pvt., Co. B (2nd). Enl. Camp Winder 5/1/63. Deserted 6/30/63. NFR.

MOSES, SAMUEL M.: Pvt., Co. B (2nd). B., Va. 1840? Farmer, Sherando, Augusta Co. Enl. Waynesboro 7/15/61. Cap. Alleghany Mt. 12/13/61. Exchanged 9/11/62. AWOL ——— to 4/4/63. Present 4/5-7/30/63. AWOL 7/31-10/63. Deserted to the enemy Clarksburg, W. Va. 10/24/63. Took oath and sent north. Age 23, 5' 11½", dark complexion, dark eyes, dark hair. Coal Miner, age 70, South River Dist., Augusta Co. 1910 census.

MOYERS, SAMUEL F.: Pvt., Co. B (2nd). B. Va. 1831. Overseer, age 29, Burkes Mill Dist., Augusta Co. 1860 census. Not on muster rolls. Cap. Beverly, W. Va. 5/8/65. Paroled for exchange 5/11/65. Age 22, 5' 7½", cark complexion, dark eyes, dark hair, farmer, illiterate.

MULLEN, DANIEL WADE: Pvt., Co. G. B. Augusta Co. 9/30/40. Farmhand, age 20, Burkes Mill Dist., Augusta Co. 1860 census. Enl. Staunton 8/2/61. Died of typhoid pneumonia Camp Alleghany 11/19/61. Brother of James D. Mullen.

MULLEN, JAMES D.: Pvt., Co. G. B. Va. 1834? Farmhand, age 26, Burkes Mill Dist., Augusta Co. 1860 census. Enl. Staunton 8/2/61. Ab. sick 9/1-12/61. Present 1-4/62. Reenl. 5/1/62. AWOL 1 month fined $11.00, on rolls 4/30-8/31/62. Present 9/1-10/31/62, deserted to the enemy near Slaughter's Mt. 8/9/62. Exchanged 9/21/62. NFR. Dropped as deserter. Brother of Daniel W. Mullen.

MULLEN, JOHN: Pvt., Co. G. B. Ireland 1830? On postwar roster. Farmer, age 40, 6th Dist., Augusta Co. 1870 census.

MURPHY, JOHN: Pvt., Co. B. B. Ireland 1825? Laborer, Augusta Co. Enl. Bunker Hill 10/9/62 as substitute. Deserted 11/16/62. Resident of Fairfax Co.

MURRAY, WILLIAM: Pvt., Co. H. B. Augusta Co. 1801? Farmer, Augusta Co. Enl. Staunton 7/23/61 age 59 as substitute for John Firebaugh. Ab. sick 11/1/61-2/62. Discharged for general bad health by order Col. Baldwin 4/28/62. Age 59, 5' 6", fair complexion, blue eyes, dark hair. died before 1885.

MURRY, P. G.: Pvt., Co. B (2nd). Not on muster rolls. Pension application indicates disabled for service. Living Pittsylvania Co. 1888.

MUTERSPAW, DANIEL JAMES: Pvt., Co. E. B. Frederick Co. Farmhand, age 30, Collierstown PO, Rockbridge Co. 1860 census. On postwar roster.

MUTERSPAW, JOHN: Pvt., Co. E. B. Va. 1842. Farmer, age 18, Lexington Dist., Rockbridge Co. 1860 census. Enl. Staunton 8/1/61. Died of Erysipelas Camp Alleghany 12/7/61.

MYERS, ASBURY: Pvt., Co. B (2nd). B. Rockbridge Co. Carpenter, age 25, Christian's Creek, Augusta Co. 1860 census. Enl. Waynesboro 7/15/61. Ab. sick 11/61-2/62. Present 3-4/62. Reenl. 5/1/62. Present 4/30-6/8/62. AWOL 6/9-23/62, 15 days pay deducted. Detailed in hospital Staunton 6/23/62-10/63. Printed a retraction for being called a deserter 12/30/62 in Staunton paper. "... had been admitted to hospital Staunton 24 June. Remained to 11/16/62. Granted 60 days leave. I have never attempted to shirk my duty and to be absent from my post, and never will unless sickness should compel me to do so. When I volunteered I did so, feeling it a duty, and I would **Scorn** deserting the standard of my country in this hour of her struggle to freedom and independence ... I intend to do my duty as long as I am able to bear arms. I can be found at my residence on Christian's Creek, 7 miles east of Staunton." Present 11-12/63. WIA (right thigh) and cap. Bethesda Ch. 5/30/64. Admitted to 3rd 5th Army Corps Hospital 5/30/64. DOWs 6/4/64.

MYERS, HENRY K.: Pvt., Co. C. B. Germany 1823? Farmhand, age 37, Burkes Mill Dist., Augusta Co. 1860 census. Enl. Bunker Hill 10/10/62. Present 10/10/62-2/28/63. Discharged for disability 4/2/63. Age 47, 5' 10", fair complexion, dark eyes, dark hair, barber. Farmer, age 48, Pastures Dist., Augusta Co. 1870 census. Died Westview, Augusta Co. 1/27/91.

MYERS, JOHN ELLIS: Pvt., Co. F. Not on muster rolls. Enl. 12/62. KIA Bethesda Ch. 5/30/64. Brother of John F. Myers.

MYERS, JOHN F.: 4th Sgt., Co. F. B. Va. 1844? Resident, age 16, 1st Dist., Augusta Co. 1860 census. Enl. Staunton 7/31/61 as Pvt. Present 11/61-4/62, promoted 4th Sgt. 1/28/62. Reenl. 5/1/62. Present 8/31-12/31/62, rank shown as Pvt. Died of diphtheria at Camp of 52d Va. Rgt. near Port Royal, Caroline Co. 2/1/63. "This brave and faithful soldier continued in service from the organization of this regiment to the day of his death without ever having been absent from his colors a single day—endued without complaint the fatigue and hardship encountered ... —Participated in every battle and skirmish ... always behaving with gallantry unsurpassed ..." Brother of John E. Myers.

MYERS, JOHN S.: 2nd Lt., Co. B (2nd). B. Va. 11/16/33. Doctor, age 27, 1st. Dist., Augusta Co. 1860 census. Enl. Waynesboro 7/15/61 as Pvt. Present 11-12/61. Elected 2nd Lt. 2/62. Ab. on leave 3-4/62. Not reelected 5/1/62. NFR. Doctor, age 36, Mt. Sidney PO, Augusta Co. 1870 census. Died Augusta Co. 3/5/97. Bur. Riverview Cem., Waynesboro.

MYERS, JOSHUA E.: Pvt., Co. F. Enl. Staunton 11/15/62. Present 11/62-12/63. Present 3-4/64. Cap. Bethesda Ch. 5/30/64. Sent to Point Lookout. Transf. Elmira. Died of chronic diarrhea there 6/29/65. Bur. grave #2829, Woodlawn National Cem.

MYERS, SAMUEL CRUMER: Pvt., Co. G. B. near New Hope, Augusta Co. 1827? Farmer, age 33, Burkes Mill Dist., Augusta Co. 1860 census. Enl. Staunton 8/2/61. Present 11/61-3/62. Died of pneumonia Camp Shenandoah 4/14/62, age 34, Left widow.

MYRTLE, GEORGE F.: Pvt., Co. C. B. Augusta Co. 10/7/45. Laborer, age 15, Burkes Mill Dist., Augusta Co. 1860 census. Enl. Staunton 3/20/64. AWOL 8/1-12/31/64. Dropped as deserter. Iron Miner, age 22, 6th Dist., Augusta Co. 1870 census. Died Augusta Co. 12/19/93. Bur. Mt. Carmel Pres. Ch. Cem., Steele's Tavern. Brother of William R. Myrtle (1st).

MYRTLE, WILLIAM R.(1st): Pvt., Co. C. B. Va. 1843? Laborer, age 17, Burkes Mill Dist., Augusta Co. 1860 census. Enl. Staunton 7/16/61 age 16. Present 11/61-4/62. Reenl. 5/1/62. AWOL 6/26-8/31/62. Present 9/1/62-12/63. WIA (face) Bethesda Ch. 5/30/64. Sent hospital Staunton 6/5/64. Deserted to the enemy New Creek, W. Va. 2/12/65. Age 20, 5' 7", florid complexion, grey eyes, fair hair, farmer. Brother of George F. Myrtle.

MYRTLE, WILLIAM R. (2nd): Pvt., Co. B. Va. 1845? Laborer, age 15, Burkes Mill Dist., Augusta Co. 1860 census. Not on muster rolls. WIA (head) Bethesda Ch. 5/30/64. Brought from cars in dying condition 5/30/64. DOWs in Wayside Hospital, Richmond 6/3/64.

MacMANAMY, WILLIAM C.: Pvt., Co. E. B. Rockbridge Co. 2/1/46. Enl. Staunton 8/1/61. Ab. on leave 11-12/61. Present 1-2/62. AWOL 3-4/62. Died as a result of an accidental discharge of a rifle Newtown 5/28/62.

McALLISTER, RICHARD: Pvt., Co. H. B. Albemarle Co. 1833? Enl. Albemarle Artillery. Not on muster rolls of 52nd. Cap. in hospital Richmond 4/3/65. NFR. Farmer, age 27, Mt. Sidney PO, Augusta Co. 1870 census. Member Stonewall Jackson Camp, CV, Staunton 1898.

McCAMBRIDGE, WILLIAM: Pvt., Co. K. B. Ireland 1806? Enl. Shenandoah Mt. 4/9/62. Present 4/9-30/62, and 1-12/63. Detailed in Chimborazo Hospital Richmond 4/30-12/31/64. Cap. there 4/3/65. Sent to Newport News. Released 7/1/65. 5' 6", fair complexion, dark hair, blue eyes. Laborer, age 54, Pastures Dist., Augusta Co. 1870 census.

McCHESNEY, WILLIAM STEELE: Surgeon, F&S. B. Staunton 9/6/27. Attended Washington Col. '43-44. Doctor, age 32, Staunton, 1860 census. Assistant Surgeon 27th Va. Inf. 9/13/61. Transf. hospital duty Staunton. Appointed Surgeon date unknown. NFR. Doctor, age 42, Staunton 1870 census. Died Staunton 3/17/84. Bur. Thornrose Cem.

McCLINTIC, JAMES M.: Pvt., Co. K. B. Bath Co. 6/3/18. Farmer, age 42, Mountain Grove Dist., Bath Co. 1860 census. Pvt., Capt. Davis' Co., 31st Va. Militia. Enl. Shenandoah Mt. 4/9/62. Present 4/9-30/62. Discharged for overage of conscription date unknown. Enl. Co. E, 10th Va. Bn. Reserves 4/16/64 as 1st Lt. 5' 9", dark complexion, dark hair, blue eyes, whiskers. Elected Capt. 8/23/64. Present 12/64. Farmer, age 52, Bath CH PO, Bath Co. 1870 census. Died Bath Co. 12/8/99. Bur. McClintic Cem., on Jackson River on Rt. 603, 3 miles above Bolar, Bath Co. (Moved to Warm Springs Cem.)

McCLUNG, CORNELIUS T.: Pvt., Co. K. B. Bath Co. 1843? Resident, age 17, Cleeks Mill PO, Bath Co. 1860 census. Enl. Shenandoah Mt. 4/9/62. Present 4/9-30/62. WIA (breast) 2nd Manassas 8/29/62. Present in arrest 1-2/63. Present 3-4/63. WIA (hip) Fredericksburg 5/5/63. DOWs 5/6/63.

McCLUNG, WILLIAM T.: Pvt., Co. K. B. Bath Co. 1840? Not on muster rolls. Postwar roster only. Farmhand, age 20, Williamsville Dist., Bath Co. 1870 census. Farmer, McClung, Bath Co. 1897. Alive 1917.

McCLURE, CHARLES H.: Cpl., Co. H. B. Va. 1844? Student, age 16, 1st Dist., Augusta Co. 1860 census. Enl. Staunton 7/23/61 as Pvt. Present 11/61-4/62. Reenl. 5/1/62. Present 5/1-8/31/62, and 1-4/63, promoted Cpl. 4/15/63. WIA Gettysburg 7/3/63. Returned to duty 9/30/63. Present 10-12/63 and 4/30-12/31/64. NFR.

McCLURE, GEORGE WASHINGTON: Pvt., Co. H. B. Churchville, Augusta Co. 1/1/22. Farmer, age 38, Spottswood, Augusta Co. 1860 census. Pvt., Co. C, 93rd Va. Militia 3/62 furnished substitute. Enl. Staunton 2/28/64. Detailed as Agriculturist, Augusta Co. Present 11-12/64. WIA (left knee) and cap. Ft. Stedman 3/25/65. "Ordered to charge a battery they reached the guns, one brother to die (James A. McClure) and the other wounded and taken prisoner. He was given permission to carry his brother off the field." Admitted to Lincoln Hospital, Washington, D. C. Sent to Old Capitol Prison. Transf. Elmira. Released 6/27/65. Age 45, 6', florid complexion, auburn hair, blue eyes. Farmer, age 47, Fishersville, Augusta Co. 1870 census. Farmer, Raphine, Rockbridge Co. 1885. Died near Spottswood, 12/11/90. bur. Bethel Pres. Ch. Cem. Brother of James A. and Matthew T. McClure.

McCLURE, JAMES ALEXANDER "Jimmy": Pvt., Co. H. B. Augusta Co. 10/21/28. Farmer, age 36, Spottswood, Augusta Co. 1860 census. Pvt., Co. B, 93rd Va. Militia 3/63 furnished substitute. Enl. Staunton 2/28/64. Present 11-12/64. WIA (left thigh broken and ball lodged in side and left hand) and cap. Ft. Stedman 3/25/65. Brother George and Pvt., Parley Carpenter of a Pa. Regt. removed him to hospital 2nd Div., 9th Army Corps, Army of the Potomac. DOWs 3/26/65. Bur. City Point Cem. Bible and blanket sent to widow by Pvt. Carpenter. Union Chaplain put his name in bottle at head of grave. Body removed to Bethel Pres. Ch. Cem. 10/65. Left widow and 2 children. Brother of George W. and Matthew T. McClure. "When duty called he hesitated not to go, thr' it were to the mouth of the cannon where he fell."

McCLURE, MATTHEW THOMPSON "Tom": Commissary Sgt., F&S. B. Augusta Co. 7/23/34. Farmhand, age 26, 1st Dist., Augusta Co. 1860 census. Enl. Staunton 7/16/61 as Pvt., Co. H, age 27. Present 11-12/61, Commissary Sgt. since 12/1/61, elected 3rd Lt. 11/23/61. Present 1-4/62, acting Commissary of Rgt. Not reelected but reenl. as Pvt. 5/1/62. Present 8/30-10/31/62, detailed in Commissary Dept. 5/1/62. Present 1-8/63, appointed Commissary Sgt. 6/3/63. Present 9/1-12/14/63. Ab. on leave 12/15-31/63, 5' 9", fair complexion, blue eyes, light hair. Present 4/30-12/31/64, detailed as Commissary Sgt. for Brig. 10/28/64. Promoted 1st Lt. '65 but never received commission. Present 3/25/65. Surrendered Appomattox 4/9/65. Entitled to take with him one private mule. Rode home with an old blanket as a saddle (his saddle was stolen the night before) with Capt. S. W. Paxton and Surgeon John Gibson. Spent the first night at Lynchburg. Proceeded up the canal and spent the second night at Riverside, Rockbridge Co. with the Shields family. Spent the third night with Capt. Paxton near Fairfield, reaching home on the afternoon of the 14th of April, 1865. Farmer, Spottswood. Died near there 1/19/22. Bur. Bethel Pres. Ch. Cem. Brother of George W. and James A. McClure.

McCORD, JAMES A.: Pvt., Co. B (2nd). B. Albermarle Co. Enl. Waynesboro 7/15/61. Present 11-12/61. Died of typhoid fever in hospital Lynchburg 4/24/62. Bur. No. 1, Row 5, Confederate Section Lynchburg City Cem.

McCORMICK, THOMAS NIMROD: 1st Cpl., Co. B(1st). B. Va. 1837? Farmhand, age 23, South River Dist., Rockbridge Co. 1860 census. Enl. Fairfield 7/10/61. Ab. sick when transf. 9/28/61.

McCOWN, JOHN JAMES: Pvt., Co. B(1st). B. Rockbridge Co. 1837? On muster rolls 1861 with no enlistment date.

McCRAY, JOHN L.: 2nd Cpl., Co. G. B. Va. 1936? Farmhand, age 26, Walker's Creek Dist., Rockbridge Co. 1860 census. Enl. Staunton 8/2/61 as 4th Cpl. Ab. sick 12/14/61-3/4/62. Present 3/5-4/62. Cap. Front Royal 5/30/62. Exchanged 8/5/62. Present 12/9/62 as 2nd Cpl. Present 1-12/63. WIA (face and eyes) Spotsylvania CH 5/12/64. Ab. wounded through 12/31/64. Cap. 4/12/65. Took oath and released 4/14/65.

McCRAY, WILLIAM A.: Pvt., Co. B (2nd). B. Augusta Co. 1846? On pension rolls only. Blacksmith, Staunton 1879-88. Farmer, Hot Springs, Bath Co., 1897. Carpenter, age 68, Cedar Creek Dist., Bath Co. 1910 census.

McCRORY, EDWARD HUGHES: Pvt., Co. B (1st). B. Va. 8/13/24. Farmhand, age 35, South River Dist., Rockbridge Co. 1860 census. Enl. Fairfield 7/16/61. Present until transf. 9/28/61.

McCRORY, WILLIAM THOMAS: Pvt., Co. B (1st). B. Va. 1830? Farmhand, age 30, South River Dist., Rockbridge Co. 1860 census. Enl. Staunton 8/1/61. Present until transf. 9/28/61.

McCUNE, ALEXANDER GIVENS: Adjutant, F&S. B. near Fishersville, Augusta Co. 4/2/34. Farmer, age 25, Burkes Mill Dist., Augusta Co. 1860 census. Capt. J. K. Koiner's Co. 32nd Va. Militia exempt as overseer of father's farm 3/21/62. Not on muster rolls. Postwar roster and obit. indicates served as Adjutant 4 yrs. Farmer, age 34, Mount Sidney PO, Augusta Co. 1870 census. D. near Mount Zion Ch. 11/5/14. Bur. Mount Zion Lutheran Ch. Cem. near Fishersville. Brother of Samuel H. McCune.

McCUNE, SAMUEL HOUSTON: Capt., Co. G. B. near Fishersville, Augusta Co. 2/23/19. Farmer, age 40, Burkes Mill Dist., Augusta Co. 1860 census. Col., 32nd Va. Militia. Enl. Staunton 8/2/61. Present 11/61-4/62, Not reelected 5/1/62. Capt. Co. D, 3rd Bn. Valley Reserves. Lt. promoted Major 8/9/64. Capt. Co. D 10-12/64. NFR. Farmer, age 50, Mount Sidney PO, Augusta Co. 1870 census. D. near Mount Zion Ch. 9/29/89. Bur. Mount Zion Lutheran Ch. Cem., near Fishersville. Brother of Alexander G. McCune.

McCUTCHAN, JOHN HOWARD: Pvt., Co. I. B. 1826? Enl. Staunton 7/16/61 age 35. Present 11/61-4/62. Reenlisted 5/1/62. WIA (leg) 2nd Manassas 8/29/62. DOWs Middleburg 10/1/62.

McCUTCHAN, WILLIAM HAMILTON: Pvt., Co. I. B. Augusta Co. 1835? Farmhand, age 25, 1st Dist. Augusta Co. 1860 census. Enl. Staunton 7/16/61 age 27. Present 11/61-4-62. WIA (through the lungs) McDowell 5/8/62. Reported Discharged 9/6/62, however he deserted before it occured. Ab. wounded through 4/63. Ab. sick 7/27-10/63, Ab. permanently disabled, detailed as hosp. guard 12/21/63-8/26/64. Hosp. detail 8/27-2/5/65. Dropped as deserter. Enl. Co. C, 14th Va. Cav. Farmer, Augusta Co. D. Raphine, Rockbridge Co. 1/17/25, age 78.

McDANIEL, DORSEY: Pvt., Co. I. B. Nelson Co. 1840? Farmer. Enl. Waynesboro 7/15/61. Ab. sick 11-12/61. D. in hosp. Staunton 1/10/62. Age 20, 6' 1¼", fair complexion, blue eyes, light hair. Bur. Thornrose Cem., Staunton, grave #880.

McDANIEL, ROBERT M.: Pvt., Co. E. Enl. Staunton 8/1/61. AWOL 9/26/61-2/11/62. Present 2/12-4/62. Reenlisted 5/1/62. Captured Front Royal 5/30/62. Exchanged 8/5/62. AWOL 9/1/62-7/1/63. Present 7/2-12/5/63. AWOL 6-12/31/63. Present 4/30-12/31/64. Joined from desertion 2/22/64. Sentenced to death for desertion 4/27/64. Pardoned by the Sec. of War 7/30/64. Deserted to the Army of the Potomac near Petersburg 3/8/65. Took oath and transportation furnished to Green Co., Ohio.

McDANIEL SCOREIM: Pvt., Co. C. B. Va. 1831? Blacksmith, age 29, Burkes Mill Dist., Augusta Co. 1860 census. Enl. Staunton 7/16/61 age 42. Present 11/61-2/62, detailed as blacksmith. Present 3-4/62, fined $11.00 for days absent. Ab. detailed as blacksmith on rolls 5/1-8/31/62. Discharged 11/18/62, over age of conscription.

McDONALDSON, STEPHEN GERALD: Pvt., Co. B (2nd). B. Rockingham Co. 12/4/35. Miller, Harrisonburg, Rockingham Co. Enl. Bunker Hill 9/24/62. Present 9/24-10/31/62. AWOL 12/6/62-2/20/63. Present in arrest in Regt'l Guard House 2/63. Present 4-12/63. WIA (right arm) Spotsylvania CH 5/12/64. Arm amputated 4 inches from shoulder joint. Ab. wounded through 12/31/64. NFR. D. Keezletown, Rockingham Co. 1/11/04. Bur. Keezletown Cem.

McFADDEN, JOSEPH: Pvt., Co. B (1st). Enl. 1861. Furnished Substitute. Not on Muster rolls. Enl. Co. I, 4th Va. Inf. 10/28/64.

McFARLAND, ROBERT P.: 2nd Lt. Co. I. B. Augusta Co. 1839? Farmer, Augusta Co. 1st Lt., Co. E, Mint Springs, 93rd Va. Militia. Enl. Staunton 7/16/61 as 1st Sgt. age 22. Ab. sick 12/15/61-10/31/62, but reelected 1st Sgt. 5/1/62. Present 1-4/63, elected 2nd Lt. 3/15/63. Present 4/30-12/63. KIA Bethesda Ch. 5/30/64, age 25. Body found covered with earth by Dr. Samuel B. Morrison on 6/6/64 and reburied and grave marked. Removed 6/66 to Bethel Pres. Ch. Cem. near Middlebrook. "A brave, Christian soldier."

McGILL, JOHN: Chaplain, F&S. B. Jefferson, Md. 1840. Appointed 9/62. Captured Gettysburg 7/4/63. Sent to Johnson's Island. Exchanged 10/14/63. Submitted casualty reports and requests for clothing and books to the "Staunton Vindicator" '64-65. Applied for retirement 1/23/65. Approved 1/25/65. Ab. sick 2/26/65. NFR. Rector of various Episcopal Churches in Va., the last one at The Plains, Faquier Co. D. Washington, D. C. 12/27/25. Bur. The Plains Episcopal Ch. Cem.

McGUFFIN, JOHN T.: Pvt., Co. B (2nd). B. 7/3/48. Enl. Waynesboro 7/15/61. Discharged 8/22/61. Underage. Farmer, Alone, Rockbridge Co. 1897. D. Rockbridge Co. 2/21/30. Bur. Bethany Lutheran Ch. Cem., near Lexington.

McGUFFIN, JOHN WESLEY: Pvt., Co. E. B. Augusta Co. 8/15/44. Tombstone only record. Enl. Co.E. 5th Va. Inf. 11/17/63. Farmer, age 66, Greenville Riverheads Dist., Augusta Co. 1910 census. D. 8/5/27. Bur. Calvary Meth. Ch. Cem., Stuart's Draft.

McGUFFIN, SAMUEL R.: Pvt., Co. B(1st). Born Augusta Co. 2/17/41. Farmhand, age 19, South River Dist., Rockbridge Co. 1860 census. Enl. Fairfield 7/10/61. Ab. detailed Staunton when transf. 9/28/61.

McGUFFIN, WILLIAM W.: Pvt., Co. B(1st). B. Rockbridge Co. 3/8/38. Enl. Fairfield 7/10/61. Present until transf. 9/28/61.

McKEE, JAMES MOFFETT: Pvt., Co. I. B. Augusta Co. 1832? Enl. Staunton 7/16/61 age 39. Ab. sick 11-12/62. Present 1-4/62. Reenlisted 5/1/62. Present 4/30-7/26/63. WIA (ball passed through penis paralizing testicles and penis and imparing use of right leg) McDowell 5/8/62. Present 8/30-10/31/62. NFR. Carpenter Zack, Rockbridge Co. 1888.

McKEE, JOHN F.: Pvt., Co. I. On postwar roster.

McMANAMA, THOMAS PRESTON.: Pvt., Co. B (1st). B. Rockbridge Co. 1822? Carpenter, age 38, South River Dist., Rockbridge Co. 1860 census. Enl. Fairfield 7/10/61. Present until transf. 9/28/61.

McMANAWAY, JOHN: Pvt., Co. C. B. Augusta Co. 1843? Enl. Staunton 7/16/61 age 18. Present 11/61-4/62. Reenlisted 5/1/62. KIA Port Republic 6/9/62.

McMULLEN, JAMES W.: Pvt., Co. B. Bath Co. 1827? Wagoner, age 33, Staunton PO, Augusta Co. 1860 census. Not on muster rolls. Deserted to the enemy Beverly, W. Va. 3/1/64. Took oath and released. Age 36, 6' 1", fair complexion, blue eyes, black hair, shoemaker.

McMULLEN, JOHN: Pvt., Co. K. Born Bath Co. 1843. Enl. Shenandoah Mtn. 4/9/62. Present 4/9-30/62. Ab. sick 5/28/62 until died of typhoid fever in hosp. Staunton 6/8/62, age 19. Bur. Thornrose Cem. grave #172.

McNETT, DAVID FRANKLIN: 2nd Cpl., Co. D. B. Ross Co., Ohio 5/1/37. Millwright, age 22, Burkes Mill Dist., Augusta Co. 1860 census. Enl. Staunton 7/16/61, as Pvt. Present 11/61-4/62. Reenlisted and promoted 2nd Sgt. 5/1/62. Present 5/1-8/31/62. WIA (foot) Port Republic 6/9/62. Ab. Wounded through 8/31/62. Present 1-4/63. WIA and captured Gettysburg 7/3/63. Sent to Ft. Delaware. Took oath and joined 3rd Md (U. S.) Cav. 8/30/63. Millwright, age 33, Mount Sidney PO, Augusta Co. 1870 census. Undertaker and Furniture Vender, Cross Keys, 1872. Millwright, Mount Solon, 1897. D. Augusta Co. 1/27/99. Bur. Old Emanuel Ch. Cem., near Mount Solon. Brother of Martin and William H. McNett.

McNETT, MARTIN: Pvt., Co. D. B. Ross Co., Ohio 1842. Carpenter, age 19, Burkes Mill Dist., Augusta Co. 1860 census. Enl. near Martinsburg 9/22/62. Detailed as wagoner 1-2/63. Captured Port Royal 4/20/63. Took oath. "His father moved to Va. about 15 yrs. ago, was a carpenter. He himself is a carpenter . . . Friends in Ohio-Cincinnati." Moved west after the war. Brother of David F. and William H. McNett.

McNETT, WILLIAM HENRY: 3rd Sgt., Co. D. B. Ross Co., Ohio 5/6/39. Laborer, age 21, Sangersville, Augusta Co. 1860 census. Enl. Staunton 7/16/61 as Pvt. Present 7/16/61-4/62. Reenlisted and promoted 3rd Sgt. 5/1/62. Present 5/1-8/31/62. Present in arrest 1-2/63. Present 3-4/63. Fined by CM 3 mos. pay in addition to 1 mo. and 1 day AWOL $50.56, and reduced to Pvt. Deserted Gettysburg 7/5/63. Sent to Ft. Delaware. Took oath and joined 3rd Md. Cav. U. S. Army, 9/22/63. Farmhand, age 31, Mount Sidney PO, Augusta Co. 1870 census. Millwright, Sangersville. D. 9/8/06. In flyleaf of his Bible "William H. McNett Comp. D 52 Va. Inf. received at General Jackson's headquarters April the 12th 1862 year of our Lord." General Jackson's horse was in some danger and he moved it and Gen. Jackson gave him the Bible. Brother of David F. and Martin McNett.

McNUTT, ROBERT: Pvt., Co. B (1st). B. Augusta Co. 12/23/23. Farmer, age 36, 1st Dist., Augusta Co. 1860 census. Enl. Staunton 8/1/61. Present until transf. 9/28/61.

NASH, JONATHAN R.: 2nd Lt., Co. F. B. Va. 1831? Farmhand, age 29, Northern Dist., Augusta Co. 1860 census. Enl. Staunton 7/19/61. Present 11/61-4/62. Not reelected 5/1/62. Reenlisted 8/23/62 as Pvt. Present 8-12/62. Present 1-2/63. Issued clothing 5/19/63, rank shown as 3rd Sgt. Present 4/30-7/26/63. AWOL 7/27-8/14/63 and pay deducted. Present 9-12/63, and 3-4/63. KIA Bethesda Ch. 5/30/64.

NATINI, ———: Pvt., Co. F. (Italian) Unofficial roster indicates enl. 5/62 as sub. and deserted a few days later. NFR.

NEILSON, THOMAS HALL: 2ndLt., Co. A. B. Richmond, Henrico Co. 1845. Enl. Staunton 7/16/61 as Pvt. Present 11/61-2/62. Elected 2ndLt. 2/3/62. Not reelected 5/1/62. Enl. Co. D, 62nd Va. Inf. and promoted color bearer. WIA 4 times and captured twice, the last time Beverly W. Va. 11/64. Sent to Camp Chase. Released on parole and never returned. Lawyer, Member, New York City Bar; Member, New York City Chapter, Confederate Veterans. D. in Masonic Home, Elizabethtown, Pa. 10/18/21.

NEWCOMB, JAMES H.: Pvt., Co. C. B. Va. 1840. Laborer, age 20, Burkes Mill Dist., Augusta Co. 1860 census. Enl. Staunton 5/1/63. Present 5-12/63. WIA (right arm) Bethesda Ch. 5/30/64. Returned to duty 6/20/64. AWOL 6/21-10/31/64 fined 6 mos. and 15 days pay by CM, $114.85. Present 11-12/64. Deserted to the Army of the Potomac near Petersburg 3/8/65. Took oath and transportation furnished to Redding, Pa.

NEWCOMER, WILLIAM S.: Cpl., Co. E. B. Rockbridge Co. 1818? Blacksmith, Rockbridge Co. Enl. Staunton 8/1/61. Present 11/61-4/62. Reenlisted 5/1/62. Present 5/1-6/30/62. WIA (head) Gaines Mill 6/27/62. Ab. wounded through 4/63. Discharged over age of conscription 5/1/63. Reenlisted as Pvt., Richmond '64. Deserted from camp near Richmond 6/6/64. Deserted to the enemy Buckhannon, W. Va. 1/26/65. Took oath and sent to Ohio. Age 47, 5' 11", fair complexion, grey eyes, brown hair. Farmer, age 47, Bath CH PO, Bath Co. 1870 census. Resident of Big Island, Bedford Co. 1902.

NISEWANDER, JACOB: Pvt., Co. F. B. Va. 1834? Machinist, age 26, Northern Dist., Augusta Co. 1860 census. Enl. Valley Mills 4/20/62. Deserted 5/7/62. Resident of Harrisonburg. NFR.

NUTTY, ARCHIBALD: Pvt., Co. I. B. Augusta Co. 1825? Farmhand, age 35, Staunton PO, Augusta Co. 1860 census. Enl. Staunton 7/16/61 age 40. Present 11/61-4/62. Reenlisted 5/1/62. Ab. sick 5/6/62-2/28/63. AWOL 3/1-11/26/63, fined all pay and allowances by CM. Present 11/27-12/63. Present 4/30-6/1/64. WIA (leg) Bethesda Ch. 5/30/64. AWOL 6/2-12/31/64. Deserted to the enemy Clarksburg, W. Va. 4/14/65. Took oath and released. Age 48, 5' 10", fair complexion, blue eyes, light hair, farmer, resident of Rockbridge Co. D. Beverly Manor Dist., Augusta Co. 12/21/93. Bur. Shemariah Ch. Cem. on Rt. #602 near Junction #667, near Middlebrook.

O'BRIEN, JOHN: Pvt. Co. K. B. Ireland 1834? Laborer age 26, Bath CH PO, Bath Co. 1860 census. Enl. near Sheperdstown 9/23/62. Present 1/62. Detailed as nurse in hosp. Lynchburg 2/18/63-12/31/64. Apprehended for desertion 2/21/65. NFR. D. Augusta Co. 8/17/99. Bur. Thornrose Cem., Staunton.

O'BRIEN, LAFAYETTE: Pvt., Co. B (1st).B. Va. 1835? Forgeman, age 25, 5th Dist., Rockbridge Co. 1860 census. Enl. Fairfield 7/16/61. Present until transf. 9/28/61.

O'BRIEN, MICHAEL T.: Pvt., Co. I. B. 1830? Enl. Staunton 7/16/61. age 31. Present 11/61-3/62. Ab. sick in hosp. Lynchburg 4/4-30/62. Present 8/30-10/31/62. Ab. sick 11/21/62-8/31/63. Ab. detailed in American Hosp. Staunton 9/31/63-12/31/64. Hosp. Guard Richmond 3/5/65. NFR. Bur. Riverview Cem., Waynesboro.

OCHILTREE, DAVID LEECH: Pvt., Co. E. B. Rockbridge Co. 19/13/40. Enl. Staunton 8/1/61. Present 11/61-2/62. Ab. sick 3-4/62. D. of pneumonia in hosp. Lynchburg 4/27/62, age 21. Bur. High Bridge Pres. Ch. Cem., near Natural Bridge. Brother of Thomas A. Ochiltree.

OCHILTREE, JOHN M: Pvt., Co.H. B. Augusta Co. 1844? Farmhand, age 16, Burkes Mill Dist., Augusta Co. 1860 census. Enl. Staunton 7/23/61 age 17. Ab. on detached service 11-12/61. Present 1-2/62. D. of diphtheria Camp Alleghany 3/22/62. Age 17, 5' 5", fair complexion, blue eyes, dark hair.

OCHILTREE, THOMAS ALEXANDER: Pvt., Co. E. B. Rockbridge Co. 3/1/39. Farmer, Rockbridge Co. Enl. Camp Alleghany 3/20/62. Ab. sick 3-4/62. D. of typhoid fever in hosp. Lynchburg 7/24/62. Age 20, 5' 5", fair complexion, grey eyes, light red hair. Bur. High Bridge Pres. Ch. Cem., near Natural Bridge. Brother of David L. Ochiltree.

OFFLIGHTER, HARRISON: Pvt., Co. I. Not on muster rolls. Listed on unofficial roster died before 8/1/63, cause unknown.

OFFLIGHTER, WILLIAM H.: Pvt., Co. B (2nd). B. Nelson Co. 1828? Carpenter age 32, Sherando, Augusta Co. 1860 census. Enl. Waynesboro 7/15/61. Present 11/61-4/62, non-conscirpt. AWOL 5/23-7/7/62, and 7/18-10/9/62. AWOL 11/23/62-3/13/63, sentenced to one year in penitentiary by CM. Released by Presidential Proclamation 8/1/63. AWOL 8/1/63 until deserted to the enemy Clarksburg, W. Va. 10/25/63. Took oath and sent north. Age 39, 5/6", dark complexion, brown eyes., dark hair, resident of Nelson Co.

OFFLIGHTER, WILLIAM T.: Pvt., Co. I. B. Augusta Co. 1841? Enl. Staunton 7/16/61 age 20. Present 11/61-4/62. Ab. sick in hospial Charlottesville. 7/22-8/11/62 "acute deaMohea". Died of diphteria at the home of his father in Brownsburg, Rockbridge Co. 11/20/62, age 22.

OREBURN, WILLIAM H.: Pvt., Co, B (1st). B. Rockbridge Co. 1831? Farmer, age 29, South River Dist., Rockbridge Co. 1860 census. Enl. Fairfield 7/16/61. Ab. on detail in Staunton when transf. 9/28/61.

OTT, DAVID ALEXANDER: Pvt., Co. B. (1st). B. near Fairfield, Rockbridge Co. 1843? Farmhand, age 17, 7th Dist., Rockbridge Co. 1860 census. Enl. Fairfield 7/16/61. Present until transf. 9/28/61.

OTT, FREDERICK C.: Pvt., Co. H. B. Frederick Co., Md. 1835? Farmhand, age 25, Burkes Mill Dist., Augusta Co. 1860 census. Pvt., Capt. Koiner's Long Meadow Co., 32 Va. Militia exempt 3/62, reason not shown. Enl. Staunton 10/28/64. Present 11-12/64. Surrendered Appomattox 4/9/65. Farmer, age 45, Mt. Sidney PO, Augusta Co. 1870 census. Died near Stribling Springs, Augusta Co. 4/19/10. Bur. Thornrose Cem., Staunton.

PADGETT, JAMES W.: Pvt., Co. B (2nd). B. Augusta Co. 1833? Farmer, age 27, 1st. Dist., Augusta Co. 1860 census. Enl. Waynesbro 7/15/61. Present 11/61-4/62. Reenl. 5/1/62. AWOL 5/17-6/20/62. Present 6/21-7/12/62. AWOL 7/13-10/7/62. Present 11/8-62-2/63. Transf. Co. F. in exchange for Robert W. Pannell 3/1/63. AWOL 3/4/63 until he deserted to the enemy Clarksburg, W. Va. 1025/63. Took oath and sent north. Age 33, 6', fair complexion, blue eyes, dark hair, farmer, age 37, Fishersville PO, Augusta Co. 1870 census. Farmer, Lynhurst, 1897. Died near Sherando 1/7/08.

PAINE, JULIUS.: Pvt., .Co. B (2nd). B. Va. 1842? Farmhand, age 18, Burkes Mill Dist., Augusta Co. 1860 census. Enl. Waynesboro 7/16/61. Present 11/61-4/62. Reenl. 5/1/62. AWOL 7/17-8/1/62. WIA Sharpsburg 9/17/62. DOWs in hospital Winchester 10/6/62. Bur. Stonewall Cem., Winchester as "Job. Pain."

PAINTER, JAMES H.: Pvt., Co. I. B. Botetourt Co. 1834? Farmhand, age 26, 1st Dist., Augusta Co. 1860 census. Enl. Camp. Alleghany 2/28/62. Present 2/28-4/62, and 8/30-9/28/62. Ab. sick 9/29-10/31/62. WIA (right hand) Fredericksburg 12/13/62. Ab. wounded through 10/63. Present 11-12/63. Issued clothing 4/21 and 29/64 and 5/16/64. AWOL 6/21-12/31/64. Dropped as deserter. Pension application states "Came home sick and wounded and while home surrender took place." Illiterate. Farmhand, age 30, Fishersville PO, Augusta Co. 1870 census. Laborer, near Craigsville, 1909. Died Lofton, Augusta Co. 12/22/11.

PAINTER, JOHN FITCH: Pvt., Co. I. B. Nelson Co. 1841. Farmhand, age 19, 1st Dist., Augusta Co. 1860 census. Enl. Staunton 7/16/61 age 21. Present 11/61-4/62. Reenl. 5/1/62. WIA (arm) McDowell 5/8/62. Present 8/30-10/31/62. WIA Fredericksburg 12/13/62. Ab. wounded through 10/63. Present 11-12/63. Issued clothing 4/27/64. AWOL since 5/21/64 on rolls 4/30-12/31/64, however issued clothing 6/12/64. Dropped as deserter. Farmhand, age 25, Fishersville PO, Augusta Co. 1870 census. Died Newport, Augusta Co. 1911. Bur. Mt. Joy Pres. Ch. Cem.

PAINTER, WILIAM J.: 1stSgt., Co. C. B. Va. 1841? Teacher, age 19, Northern Dist., Augusta Co. 1860 census. Enl. Staunton 7/30/61 age 20, as Pvt. Present 11-12/61, detailed to assist in bakery. Present 1-2/62, promoted 1stSgt. 1/28/62. Present 3-4/62. Reenl. 5/1/62. Present 5/1/62-4/63, reduced to Pvt. 12/13/62. AWOL 7/27/63 until deserted to the enemy Hampshire Co. 9/10/63. Age 22, 6', florid complexion, dark eyes, black hair. Sent to Camp Chase. Transf Rock Island. Took oath and released 8/5/64.

PALMER, JACOB S.: Pvt., Co. C. B. Va. 1835. Farmhand, age 25, Burkes Mill Dist., Augusta Co. 1860 census. Enl. Staunton 7/17/61 age 26. Present 11/61-4/62. Reenl. 5/1/62. WIA (shoulder) McDowell 5/8/62. WIA Port Republic 6/9/2. Present 7/62-12/63. KIA Bethesda Ch. 5/30/64.

PALMER, JAMES WALLACE: Pvt., Co. G. B. Augusta Co. 1827? Farmhand, age 33, Burkes Mill Dist., Augusta Co. 1860 census. Enl. Staunton 8/2/61. Present 11/61-2/62. Ab. sick 3-4/62. Reenl. 5/1/62. Present 5/1-8/31/62. Present 9-10/62,, AWOL 28 days fined $10.30. Present 11/62-6/63. Detailed as nurse Jordan Springs hospital 6/18-7/23/63. Present 7-12/63, and 4/30-12/31/64. Surrendered Appomattox 4/9/65. Illiterate. Farmer, Fishersville. Died Graham 6/25/19. Bur. Port Republic.

PALMER, JOHN: Pvt., Co. I. B. Va. 10/6/27. Laborer, age 33, Staunton PO, Augusta Co. 1860 census. Enl. Staunton 7/1/61 age 36. Present 11/61-4/62. Reenl. 5/1/62. Present 8/30-10/31/62. AWOL 12/7/62-2/14/63, fined $24.56. Present 2/15-12/31/63. WIA 9/29/64. Ab. wounded through 10/31/64. AWOL 12/8-31/64. Dropped as deserter. Paroled Richmond 5/4/65.

PALMER, SAMUEL C.: Pvt., Co. C. B. Augusta Co. 1832? Farmhand, age 28, Burkes Mill Dist., Augusta Co. 1860 census. Pvt., Capt. Koiner's Long Meadow Co., 32nd Va. Militia 3/62 exempt but volunteered. Enl. Shenandoah Mt. 4/11/62 age 26. Present 4/11-30/62. Reenl. 5/1/62. WIA (shoulder) McDowell 5/8/62. DOWs 5/18/62. Age 26, 5' 10", dark complexion, dark eyes, dark hair. Left widow.

PALMER, WILLIAM FRANKLIN: Pvt., Co. C. B. Augusta Co. 1822? Farmhand, age 38, Burkes Mill Dist., Augusta Co. 1860 censu. Pvt., Capt. J. K. Koiner's Long Meadow Co., 32nd Va. Militia 3/62 exempt but volunteered, age 43. Enl. Shenandoah Mt. 4/11/62. Present 4/11-30/62. Died of disease in Augusta Co. 5/29/62. Age 40, 6', fair complexion, blue eyes, dark hair, farmer.

PALMER, WILLIAM GREAVER: 3rdCpl., Co. I. B. Arbor Hill, Augusta Co. 12/8/39. Blacksmith, age 20, Staunton 1860 census. Enl. Staunton 7/16/61 age 22. Present 11/61-4/62. Reenl. 5t/1/62. WIA (arm) Port Republic 6/9/62. Ab. wounded through 8/31/63. Detailed as nurse in General Hospital Staunton 9/30/63-12/31/64. Retired to Invalid Corps 3/3/65. Paroled Staunton 5/1/65. Age 26, 5' 10", fair complexion, brown hair, blue eyes. Died near Greenville 1/21/71. Bur. St. John's Ch. Cem., near Middlebrook.

PANNELL, ADAM SAMUEL: Pvt., Co. B (2nd). B. Augusta Co. 8/24/41. Pvt., Capt. Hall's Co., 32nd Va. Militia 3/62 exempt as shoemaker. Joined from Carpteneter's Battery 2/17/62 in exchange for George F. Arey. Present 2/17-4/63 AWOL 7/31-9/25/63. Present in arrest in Regt'l Guard House 10-12/63. Present 4/30-1/31/64, however, appears on rolls Co. A, Ward's Bn. C. S. Prisoners recently released from confinement for their participation in the defense of Lynchburg, pardoned by the President 7/30/64. Paroled Staunton 5/15/65. Age 25, 5' 9", light complexion, light hair, grey eyes. Illiterate. Shoemaker, age 30, Fishersville PO, Augusta Co. 1870 census. Died near Fishersville 1/14/14. Bur. Rhodes Ch. near Afton, Nelson Co.

PANNELL, GEORGE W.: Pvt., Co. B (2nd). Tombstone only record. Bur. Riverview Cem., Waynesboro.

PANNELL, JOHN: Pvt., Co B (2nd). B. Botetourt Co. 1810? Farmer, age 50, 1st Dist., Augusta Co. 1860 census. Enl. Waynesboro 7/15/61. Present 11/61-4/62. WIA (thigh) McDowell 5/8/62. Present 5-7/62. AWOL 8/7-10/3/62. Present 10/4-12/5/62. AWOL 12/6/62-1/17/63. Present in arrest in Regt'l Guard House 1/18-2/63, fined 6 months pay in addition to time AWOL by CM. Present 4/30/-7/30/63. AWOL 7/31-10/63, and 12/1-31/63. Present 4/30-12/31/64, however listed on rolls Co. A, Ward's Bn C. S. Prisoners release 7/30/64 by Presidential pardon for their participation in the defense of Lynchburg. WIA Ft. Stedman 3/25/65. Cap. in hospital Richmond 4/3/65. Paroled 5/18/65.

PANNELL, WILLIAM: Pvt., Co. B. (2nd). B. Nelson Co. 1835? Enl. Waynesboro 7/15/61. Present 11/61-4/62. WIA (thigh) McDowell 5/8/62. DOWs 7/25/62, age about 27 years.

PARR, ANANIAS: Pvt., Co. A. B. Va. 4/2/32. Resident, Burkes Mill Dist., Augusta Co. 1860 census. Enl. Staunton 7/9/61 age 24. Present 11/61-4/62. Reenl. 5/1/62. Present 8/31-11/22/62. AWOL 11/23/62-1/16/63. Present in arrest 1/17-2/82/63, and pay deducted for time AWOL. Present 3-12/63. Issued clothing 1/30, 4/27 and 6/12/64. WIA (left leg) near Winchester 8/11/64. Ab. wounded through 12/31/64. Paroled Charlottesville 5/29/65. Illiterate. Died Batesville, Albermarle 6/24.

PARR, FRANCIS M.: Pvt., Co. A. B. Augusta Co. 1838? Enl. Staunton 2/11/64. WIA (finger shot off) near Spotsylvania CH 5/20/64. Ab. wounded through 12/31/64. Paroled Charlottesville 5/29/65. Illiterate. Farmer and Lead Miner, Oakridge, Nelson Co. 1906.

PARR, WILLIAM R.: Pvt., Co. A. B. Va. 1832? Farmhand, age 28, Burkes Mill Dist., Augusta Co. 1860 census. Enl. Staunton 7/9/61 age 29. Present 11/61-3/62. Ab. sick 4/23-30/62, and 7/27/62-8/31/63. Present 9-12/63. AWOL 5/22-12/31/64. Dropped as deserter. Paroled Staunton 5/1/65. Age 35, ' 7", dark complexion, black hair, grey eyes. Illiterate. Farmhand, age 38, Mt. Sidney PO, Augusta Co. 1870 census. Died before 5/03.

PARRENT, WILLIAM L.: Pvt., Co B (2nd). B. Va. 1842? Student, age 18, Burkes Mill Dist., Augusta Co. 1860 census. Enl. Waynesboro 7/15/61. Present 11/61-4/62. Reenl. 5/1/62. Present 5/1-7/14/62. AWOL 7/15-8/6/62 and 9/3-10/4/62. Present 10/5/62-12/63. Issued clothing 1/8, 4/1 and 24 and 6/12/64. WIA 8/18/64. Ab. wounded through 12/31/64. Paroled Staunton 5/1/65. Age 23, 5' 7", dark complexion, auburn hair, grey eyes. Moved to Illinois after the war. Living in Los Angeles, Calif. 3/18/09.

PARRISH, ANDREW JULIUS: Pvt., Co. A. B. Va. 5/30/42. Farmhand, age 17, Burkes Mill Dist., Augusta Co. 1860 census. Enl. Staunton 7/15/61. Present 11/61-2/4/62. Ab. sick 2/5-4/28/62. Reenl. 5/1/62. Cap. Front Royal 5/30/62. Exchanged 8/5/62. Present 8/31/62-4/63. Deserted 7/23/63. Deserted to the enemy in W. Va. 1/15/64. Took oath and sent north. Age 20, 5' 8", dark complexion, grey eyes, dark hair. Carpenter, Annex, 1897. Grocer, age 67, South River Dist., Augusta Co. 1910 census. Died Augusta Co. 10/18/15. Bur. Asbury Methodist Ch. Cem., on Rt 795, Augusta Co.

PARRISH, JOHN FREEMAN: SgtMajor, F&S. B. Staunton 9/18/39. Grocer, age 21, Burkes Mill Dist., Augusta Co. 1860 census. Enl. Staunton 8/2/61 as Pvt. Co. G. Present 11/61-4/62. Reenl. 5/1/62, and promoted 3rdCpl. Present 5/1-8/31/62, reduced to Pvt. 7/20/62. Present 9-10/62, rank shown as 3rdCpl. Present 11-12/62., rank shown as 4thSgt. Present 1-4/63, rank shown as 3rdSgt. Present 4/30-12/63, promoted SgtMajor 9/16/63. WIA (left thigh) Spotsylvania CH 5/12/64. Ab. wounded through 10/31/64. Present 11-12/64, however issued clothing in hospital Staunton 12/14/64. Surrendered Appomattox 5/9/65. Farmer, age 31, Mt. Sidney PO, Augusta Co. 1870 census. Store Clerk, age 68, Staunton, 1910 census. Member, Stonewall Jackson Camp CV, Staunton. Died Staunton 12/30/18. Bur. Thornrose Cem.

PARRISH, JOHN WILLIAM: Pvt., Co. G. B. Augusta Co. 1846? Resident age 14, Burkes Mill Dist., Augusta Co. 1860 census. Enl. Sommerville Ford '64. Cap. near Spotsylvania CH 5/23/64. Sent to Point Lookout, Exchanged 11/15/64. Ab. through 12/31/64. Paroled Staunton 5/1/65. Age 19, 5' 9", fair complexion, brown hair, hazel eyes. Farmhand, age 23, 1st Dist., Augusta Co. 1870 census. Died Staunton 2/9/01.

PARSONS, WILLIAM A.: 3rdCpl., Co. E. B. Va. 1842? Farmhand, age 18, Nautural Bridge Dist., Rockbridge Co. 1860 census. Enl. Staunton 8/1/61 as Pvt. Present 11/61-4/62. Reenl. 5/1/62. Present 5/1-10/31/62 and 1-12/63. Promoted 3rdCpl '64. WIA (hip) Spotsylvania CH 5/19/64. DOWs in hospital Lynchburg 5/29/64, "pyemia", age 21. "An Excellant Soldier."

PATTERSON, JAMES ARTEMUS, SR.: Pvt., Co. H. B. Va. 9/3/17. Merchant, age 40, Staunton 1860 census. Pvt., Capt. Shannon's Co. 32nd Va. Militia exempt 3/62 bruised leg, age 44 years 6 months. Enl. Staunton 10/28/64. Conscripted by Capt. Mathews and assigned Camp Lee. Handed over to Civil authority on writ of Habeus Corpus by order of Gen. Early. Discharged by Judge Thompson on writ of Habeus Corpus. Died Sherando, Augusta Co. 8/7/82. Bur. Mt. Horeb Pres. Ch. Cem. Father of James A. and Samuel N. Patterson.

PATTERSON, JAMES ARTEMUS, JR.: 4thSgt., Co. A. B. Waynesboro, Augusta Co. 8/15/43. Clerk, age 16, Staunton 1860 census. Enl. Staunton 7/9/61 age 18 as Pvt. Present 11/61-4/62. Reenl. 5/1/62. WIA (arm) McDowell 5/8/62. WIA 2nd Manassas 8/29/62. WIA (arm) Sharpsburg 9/17/2. Arm amputated. Rank shown as 3rd Sgt. Ab. wounded through 12/31/64. When sufficently recovered, appointed 1st Assistant Doorkeeper of Confederate Congress. Appointed Major by Gen. Breckenridge and detailed to recruit for the cavalry in the Valley and served to the end of the war. Merchant, Miller, Mayor of Waynesboro 6 years; VP, Elastic Wheel and Manufacturing Co., Waynesboro 1888; Member, Waynesboro City Council; Member, Stonewall Jackson Camp CV, Staunton. Died Waynesboro 4/27/18. Bur. Riverview Cem., there. Brother of Samuel N. Patterson and son of James A. Patterson, Sr. "brave efficient and fearless."

PATTERSON, JOHN M.: Pvt., Co. B (1st) and Co. E. B. Fairfield, Rockbridge Co. 1822? Farmer, Rockbridge Co. Served in Militia. Enl. Stautnon 8/1/61. Present until transf. 9/28/61. Joined Co. E. by transf. from 2nd Rockbridge Artillery in exchange for David L. Drain 3/20/62. Present 3-4/62. WIA Cross Keys 6/8/62. Ab. wounded through 12/31/62. AWOL 1/1-12/31/63. Dropped as deserter. age 41, 5' 8", fair complexion, grey eyes, sandy hair. Paid Staunton 2/28/63. NFR. Farmer, age 45, Buffalo Dist., Rockbridge Co. 1870 census, and in 1902. Died Rockbridge Co. 1903.

PATTERSON, JOHN VERNON: 3rdCpl., Co. B. (2nd). B. Va. 11/22/39. Laborer, 35, 6th. Dist., Rockbridge Co. 1860 census. Enl. Waynesboro 7/15/61 as Pvt., Ab. sick 11/61-3/3/62. Present 3/4-4/2, non-conscript. Reenl. 5/1/62. Present 5/1/62-4/62, promoted 4th Cpl. 3/15/63. Ab. sick in hospital Staunton with rheumatism 7/1-8/31/63. Present 9-12/63, rank shown as 3rd Cpl. WIA Winchester 9/19/64. Ab. wounded through 12/31/64. Paroled Staunton 5/22/65. Age 40, 5' 9", fair complexion, dark hair, blue eyes.

PATTERSON, JOSEPH A.: 2ndSgt., Co. B. (2nd). B. Va. 1838? Foundry Worker, age 22, 1st Dist., Augusta Co. 1860 census. Enl. Waynesboro 7/15/61 as 1stCpl. Ab. on leave 1-2/62. Present 3-4/62. WIA McDowell 5/8/62. Promoted 4thSgt. 7/15/62. WIA (arm and side) 2nd Manassas 8/29/62. Ab. wounded through 4/63. Promoted 2ndSgt. 10/2/62. Present 4/21-8/31/63. Ab. detailed as Ward Master Staunton hospital 9/13/63-1/31/64. WIA (hip) near Spotsylvania CH 5/19/4. Returned to duty 6/24/64. Present 6/24-12/31/64. NFR. Carpenter, age 31, Mt. Sidney PO, Augusta Co. 1870 census. Living in Albemarle Co. 1888.

PATTERSON, SAMUEL NEWTON: Cpl., Co. A. B. Waynesboro, Augusta Co. 11/26/40. Farmer, age 19, 1st Dist., Augusta Co. 1860 census. Enl. Staunton 7/9/61 age 19. Present 11/61-4/62. Reenl. 5/1/62. WIA (left leg) 2nd Manassas 8/28/62. Leg amputated. Resident of Mt. Meridian. Requested artificial limb 11/25/64. Carried on rolls as Ab. wounded through 12/31/64. Farmer and Revenue Commissioner, New Hope 1885. Merchant, Harriston, Member, Stonewall Jackson Camp CV, Staunton. Resident of Crimora. Died Newport News 10/5/08. Bur. Mt. Horeb Pres. Ch. Cem., near Grottoes. Son of James A. Patterson, Sr., and brother of James A. Patterson, Jr.

PATTERSON, WILLIAM A.: Pvt., Co. B (1st). B. Va. 11/22/39. Farmer, Rockbridge Co. Enl. Fairfield 7/10/61. Present until transf. 9/28/61.

PATTERSON, WILLIAM BROWN: Pvt., Co. H. B. Augusta Co. 3/16/37. Attended Washington Co. '55-56. Farmer, age 22, 1st Dist., Augusta Co. 1860 census. Enl. Co. E, 1st Va. Cav. Furnished substitute and discharged. Enl. Staunton 10/25/64. Present 11-12/64. WIA Ft. Stedman 3/25/65. Admitted to hospital Richmond 4/3/65. Transf. to Provost Marshal 4/29/65. Farmer, age 32, Fishersville PO, Augusta Co. 1870 census. Farmer and Stockraiser, Fishersville 1885. Died Barterbrook, Augusta Co. 8/21/90. Bur. Tinkling Spring Pres. Ch. Cem.

PATTERSON, WILLIAM DELANIE: Pvt., Co. B (1st). B. Marlbrook, Rockbridge Co. 3/14/39. Farmhand, age 22, South River Dist., Rockbridge Co. 1860 census. Enl. Fairfield 7/10/61. Ab. on detail Staunton when transf. 9/28/61.

PAUL, JAMES ANDREW JACKSON: 2ndSgt., Co. B (1st). B. Rockbridge Co. 1830. Enl. Fairfield 7/10/61. Present until tranf. 9/28/61.

PAUL, SAMUEL: 1st. Lt., Co. D. B. Rockingham Co. 9/154/24. Merchant, age 34, Northern Dist., Augusta Co. 1860 census. Enl. Staunton 7/16/61 as QM of Co. Present 7/16-61-4/62. Elected 1stLt. 5/1/62. WIA (leg shattered by piece of shell) Cross Keys 6/8/62. Foot amputated. Ab. wounded through 4/62. Detailed as Enrolling Officer, Augusta Co. 4/7-12/31/63. Elected Sheriff Augusta Co. 1864 and resigned 1/27/64. Merchant, age 40, Northern Dist., Augusta Co. 1870 census. Treasurer of Augusta Co. 1881. Died Mt. Solon 10/7/82. Bur. Mossy Creek Old Cem.

PAULEY, ALBERT G.: Pvt., Co. B (2nd). B. Augusta Co. 1836? Wagoner, age 24, 1st Dist., Augusta Co. 1860 Census. Enl. Waynesboro 7/15/61. Ab. sick 11/61-3/10/62, pay stopped for 1 musket lost. Died of diphtheria in hospital Camp Alleghany 4/62, age 22.

PAXTON, JAMES POAGUE: Pvt., Co. B (1st). B. Rockbridge Co. 3/11/42. Farmer, age 22, South River Dist., Rockbridge Co. 1860 census. Enl. Fairfield 7/16/61. Present until transf. 9/28/61.

PAXTON, JAMES THOMAS: Pvt., Co. B (1st). B near Fairfield, Rockbridge Co. 3/11/42. Teacher age 18, 7th Dist., Rockbridge Co. 1860 census. Enl. Fairfield 7/10/61. Present until transf. 9/28/61.

PAXTON, JOHN W.: Pvt., Co. B (1st). B. Va. 1824? Farmer, age 36, South River Dist., Rockbridge Co. 1860 census. Enl. Fairfield 7/16/61. Present until transf. 9/28/61. Brother of Thomas N. and William A. Paxton.

PAXTON, JOSEPH SAMUEL: 1stLt., Co.E. B. on James River, Rockbridge Co. 1/1/42. Farmer, age 19, South River Dist., Rockbridge Co. 1860 census. Enl. Staunton 8/1/61 as 1stLt. Present 11/61-2/62. Ab. sick 3-4/62. Reelected 2ndLt. 5/1/62. Presence or absence not stated 5-9/62. Present 10/62 and 1-10/63. Promoted 1stLt. 11/3/63. Present 12/31/63. WIA Spotsylvania CH 5/12/64. WIA (thigh) Bethesda Ch. 5/30/64. Ab. wounded 6/64. Present 7-12/64 commanding Co. K. 10/31/64. Ab. on leave 2/19-28/65. Admitted to hospital Farmville with "eryesificlas" 3/17/65. Transf. to hospital Lynchburg 3/28/65. NFR. Capt., Co E. 108th Va. Militia 1866. Farmer, Stockraiser and Merchant, Sherwood; Clerk of Road Board and Commissioner of Revenue, Natural Bridge Dist. 1888. Member, Lee-Jackson Camp CV, Lexington. Died Buena Vista 12/25/06. Bur. Falling Springs Pres. Ch. Cem. Brother of Martin L. Paxton.

PAXTON, MARTIN LUTHER: 2ndSgt., Co. E. B. Rockbridge Co. 9/22/44. Student, age 16, Glenwood Dist., Rockbridge Co. 1860 census. Enl. Camp Alleghany 3/8/62 as Pvt. Present 3/8-4/62. WIA (arm) McDowell 5/8/62. WIA (hand) Gaines Mill 6/27/62. Present 5-10/62. Present 1-4/3, promoted 2ndSgt. 4/12/63. Present 4/31-11/63. Ab. on special service 12/18-31/63. KIA Carter's Farm near Winchester 7/20/64 age 20. Bur. Stonewall Cem., Winchester (Wrong date on marker). Brother Joseph S. Paxton. "A splendid officer."

PAXTON, SAMUEL WASHINGTON: Capt., Co. E. B. Rockbridge Co. 12/20/30. Attended U. Va. Enl. Staunton 8/1/61 as 1stSgt. Present 11/61-4/62. Elected 1stLt. 5/1/62, presence or absence not stated 5/1-10/31/62, however present 6/18 and 10/62. Present 1-4/63. Commanding Co. 4/30-12/31/63. Promoted Capt. 11/3/63. WIA (right shoulder) Spotsylvania CH 5/12/64. Ab. wounded through 8/30/64. Present 8/31-12/31/64. WIA (thigh) Ft. Stedman 3/25/65. Commanding Regt. at surrender Appomattox 4/9/65. Farmer, age 38, Riverside, South River Dist., Rockbridge Co. 1870 census. Justice of the Peace and Farmer, South River Dist., 1875-1897. Member, Lee-Jackson Camp CV, Lexington. Died South River 7/8/09. Bur. Neriah Baptist Ch. Cem., South River.

PAXTON, THOMAS N.: Pvt., Co. B (1st). B. Fairfield, Rockbridge Co. 1844? Wagonmaker, age 16, South River Dist., Rockbridge Co. 1860 census. Enl. Fairfield 7/10/61. Present until transf. 9/28/61. Brother of John W. and William A. Paxton.

PAXTON, WILLIAM A.: Pvt., Co. B (1st). B. near Fairfield, Rockbridge Co. 9/19/32. Farmhand, age 30, South River Dist., Rockbridge Co. 1860 census. Enl. Fairfield 7/16/61. Present until transf. 9/28/61. Brother of John W. and Thomas N. Paxton.

PAXTON, WILLIAM L.: Pvt., Co. E. B. Va. 1836? Plasterer, 24, Lexington, Rockbridge Co. 1860 census. Unofficial roster indicates WIA Gaines Mill 6/27/62 and WIA Winchester 9/19/64. NFR. Died Rockbridge Co. 4/6/95. Bur. Falling Springs Pres. Ch. Cem.

PELTER, SAMPSON: Pvt., Co. B (2nd) B. 1 mile west of Waynesboro Augusta Co. 9/12/33. Stage Driver, age 27, Burkes Mill Dist., Augusta Co. 1860 census. Enl. Waynesboro 7/15/61. Ab. sick 11/61-4/62. Discharged for epilepsey 7/15/62. age 23, 5' 9¾", fair complexion, pale blue eyes, dark hair, farmer, resident of Waynesboro. Enl. Capt. Hall's Co., 32nd Va. Militia. Enl. Co. E, 1st Va. Cav. WIA Gordonsville, present Appomattox. Farmer and Cattleman, Augusta Co. Died Nitro, W. Va. 1/19/20. Bur. Riverside Cem., Waynesboro.

PETERS, GEORGE WILLIAM: Pvt., Co. A. B. Va. 10/3/43. Farmhand, age 16, Burkes Mill Dist., Augusta Co. 1860 census. Enl. Staunton 7/9/61 age 17, afterwards claimed by his father, William Peters, as substitute and accepted by Major Harman. Present 11/61-4/62. Present 8/31/62-4/63. Cap. Chambersburg, Pa. 7/6/63. Sent to Point Lookout. Exchanged 3/17/64. Issued clothing 3/24/64. Ab. on rolls 4/30-12/31/64. NFR. Farmer, New Hope, 1872. Living Washington, D. C. 1/11/96. Died Centennial, Augusta Co. 5/1/21. Bur. Middle River Ch. of the Brethren.

PETERSON, JOSEPH: Pvt., Co. D. B. Rockingham Co. 1821? Collier, age 39, Northern Dist., Augusta Co. 1860 census. Enl. Staunton 7/16/61. Present 7/16-12/31/61. Discharged by order Sec. of War 1/19/62. Iron Worker, Forrer's Furnace, Mossy Creek. Age 39, 6' 1½", fair complexion, light hair, grey eyes. Collier, age 49, Northern Dist., Augusta Co. 1870 census.

PEUSHA, JULIUS: Pvt., Co. A. Enl. on the march 5/21/62. Deserted 5/30/62. NFR.

PEYTON, COLLIS: Pvt., Co. B (2nd). B. Va. 1830? Laborer, age 30, Burkes Mill Dist., Augusta Co. 1860. Enl. Staunton 9/62. Ab. on rolls 9/62-12/63, "this man is a conscript and was sent from Staunton here to be assigned to this company in September 1862 but before reaching the company he was assigned to duty as a teamster in QM Dept. in which service he remained until 9/25/63 and there are no means of ascertaining when or by whom he was enlisted and has been AWOL since 10/6/63. Deserted to the enemy 10/6/63. Sent to Old Capitol Prison. Took oath and sent north. 5' 10", dark complexion, bown hair, grey eyes, illiterate. Died Parnassus, Augusta Co. 6/1/93.

PEYTON, JOHN W.: Pvt., Co. A. B. 1820? Enl. Staunton 7/9/61 age 41. Present 11/61-4/62, detailed as wagoner since 8/26/61. Discharge over age of conscription at camp near Berryville 10/27/62. Died before 5/03.

PHILLIPS, JOSEPH SOLOMON: Pvt., Co. E. B. Augusta Co. 2/2/33. Farmhand age 25, Collierstown PO, Rockbridge Co. 1860 census. Enl. Staunton 8/1/61. Present 11/61-4/62. Reenl. 5/1/62. Ab. sick in hospital Charlottesville with "Morb. var." 8/11-22/62. WIA (head and arm) Sharpsburg 9/17/62. Ab. wounded through 2/63. Present 3-12/63. Deserted 5/5/64 while on the march from camp near Sommerville Ford to the battle of the Wilderness. NFR. Farmer, age 35, Natural Bridge District., Rockbridge Co. 1870 census. Died Collierstown 1/21/14. Bur. Collierstown Pres. Ch. Cem.

PHILLIPS, LEWIS G.: Pvt., Co. B (2nd). B. Augusta Co. 11/8/35. Laborer, age 18, 1st Dist., Augusta Co. 1860 census. Enl. Waynesboro 7/15/61. Present 11/61-4/62. Reenl. 5/1/62. AWOL 5/17-6/28/62, 7/18-9/29/62, and 10/24/62-4/19/63. Present 4/20-12/3. WIA (hip) Wilderness 5/6/64. Ab. wounded through 10/31/64. AWOL 11-12/64. WIA Hatcher's Run 2/8/65. Surrendered Appomattox 4/9/65. Ironworker, age 29, Fishersville PO, Augusta Co. 1870 census. Farmer, Love Mountain Top, Augusta Co. 1906. Illiterate. Died there 2/10/07. Bur. Pines Chapel Cem.

PHILLIPS, RICHARD HENRY: Chaplain, F&S. B. Spotsylvania Co. 1813? President, age 48, Staunton Female Academy, Staunton 1860 census. Enl. Staunton 8/19/61. Resigned 9/19/62. Served as QM, Staunton and cap. Nelson Co. 6/64. Sent to Camp Chase. NFR. President, age 58, Staunton Female Acadmey 1870 census, and 1880. Died Norfolk 4/7/90. Bur. Thornrose Cem., Staunton.

PINER, JAMES M.: Pvt., Co. G. Enl. Staunton 9/23/62. Present 10/62-4/63. WIA Winchester 76/13/63. Hospitalized Jordan Springs. Ab. wounded through 8/31/63. Present 9-12/63. KIA near Spotsylvania CH 5/19/64.

PITTWAY, T. P.: Pvt., Co. K. Not on muster roll. Bur. Thornrose Cem., Staunton 10/27/62, grave 212.

PLECKER, JOSIAH: Pvt., Co. F. B. Augusta Co. 6/4/37. Farmhand, age 24, Burkes Mill Dist., Augusta Co. 1860 census. Enl. Staunton 7/31/61. Present 11/61-4/62. Reenl. 5/1/62. KIA Port Republic 6/9/62. Age 27, 5' 11", fair complexion, blue eyes, fair hair. Left widow and child. Bur. St. Michaels Reformed Ch. Cem.

PLYBON, JOHN: Pvt., Co. E. Enl. Camp Shenandoah 4/19/62. Present 4/62. Ab. on leave 5/1-10/31/62. NFR. Unofficial report cap. date and place unknown. Sent to Ft. Delaware. Died of disease there.

POTTER, SAMUEL M.: Pvt., Co. B (1st). Enl. Fairfield 7/10/61. Present until transf. 9/28/61.

PRICE, DABNEY S.: Pvt., Co. B (2nd). Exempt 3/29/62 reason unknown. Not on muster rolls. Cap. in hospital Petersburg 4/3/65. Transf. 4/8/65. NFR.

PRICE, DAVID: Pvt., Co. I. B. Augusta Co. 1833. Farmer, age 27, 1st Dist., Augusta Co. 1860 census. Enl. 9/62. Present 9-10/62. Died of diphtheria Winchester 10/29/62, age 33. Left widow.

PRICE, JACOB S.: Pvt., Co. D. B. Va. 1817? Farmer, age 43, 1st Dist., Augusta Co. 1860 census. Enl. near Martinsburg 9/22/62. Present 1-12/63. WIA near Spotsylvania CH 5/19/64. Furloughed from hospital for 40 days 6/3/64. Present 8-12/64. WIA (fractured left arm and in right breast) Ft. Stedman 3/25/65. Cap. in hospital Petersburg 5/8/65. Transf. Camp Hamilton Military Prison, Newport News. Released 6/18/65. Farmer, age 52, Fishersville PO, Augusta Co. 1860 census. Died Arbor Hill 8/12/00. Bur. St. John's Ch. Cem.

PRITT, JAMES: Pvt., Co. K. B. Bath Co. 1830? Farmer, age 30, Bath CH PO, Bath Co. 1860 census. Enl. Mt. Zion Ch. 5/18/62. WIA (shoulder) Cross Keys 6/8/62. Ab. wounded through 4/63. Deserted 5/1/63. Enl. Co. G, 18th Va. Cav. Cap. and died of disease Pt. Lookout 2/26-27/64. Bur. POW Cem. there.

PROFFITT, LINDSAY M: Pvt., Co. A. B. Fluvanna Co. 1830? Miller's Apprentice, age 30, Northern Dist., Augusta Co. 1860 census. Pvt., Co. H, 93rd Va. Militia 3/62 detailed as Miller, age 32. Enl. Staunton 3/1/64. WIIA (neck and shoulder) Bethesda Ch. 5/30/64. In hospital Richmond 6/1-2/64 and paid 6/8/64. Dropped as deserter 12/31/64. NFR. Farmhand, age 39, Mt. Sidney PO, Augusta Co. 1870 census. Died Moffett's Creek, Rockbridg Co. 4/4/84.

PULLINS, JOHN B.: Pvt., Co. G. B. Va. 9/3/37. Farmhand, age 29, Northern Dist., Augusta Co. 1860 census. Enl. West View 4/27/62. Ab. sick 6/9/62-8/31/63. Present 9-12/63. Deserted Staunton 6/27/64. NFR. Died Augusta Co. 9/25/04. Bur. Old Salem Lutheran Ch. Cem., 3 miles west of Mt. Sidney.

PULLINS, WILLIAM M.: Pvt., Co. G. Enl. Staunton 8/2/61. Present 11/61-4/62. Reenl. 5/1/62. Present 5/1-10/31/62, AWOL 21 days, fined $7.78. Present 11/62-4/63, fined 3 months pay by sentence of CM. WIA (right ankle) Gettysburg 7/3/63. Ab. wounded through 8/31/63. Present 9-12/63. KIA near Mechanicsville 6/2/64. Possibly William Pullins bur. Clover Creek Cem., near Churchville, Augusta Co.

RALSTON, JESSE: Pvt., Co. D. B. Va. 1824? Farmer, age 36, Northern Dist., Augusta Co. 1860 census. Enl. Staunton 7/16/61. Present 7/16/61-4/62, and 5/1-8/31/62, AWOL 3 months and 26 days. Present in arrest 1-/63, fined 3 months pay in addition to time ab. Present 3-12/63. Cap. Winchester 9/19/4. Sent to Point Lookout. Exchanged 3/15/65. Paroled Staunton 6/3/65. Age 41, 5' 11½", fair complexion, light hair, blue eyes. Farmer, age 46, Mt. Solon, Augusta Co. 1870 census. Farmer, Stribling Springs 1897. Died Augusta Co. 11/20/00. Bur. Mt. Olivet. Cem.

RALSTON, JOHN M.: Pvt., Co. K. B. Augusta Co. 12/19/21. Farmer, age 36, Cleeks Mill PO, Bath Co. 1860 census. Enl. Shenandoah Mt. 4/9/62. AWOL 4/9-30/62. WIA (shoulder) Cross Keys 6/8/62. DOWS in UVa. hospital, Charlottesville 6/13/6, age 39. Left wife and 4 children.

RAMSEY, BROWN CLAYTOR: Rank and Co. unknown. B. Va. 1845. Enl. as Subs. in 14th Va. Cav. Served in 52nd Va. Inf. Enl. 4th Bn. Va. Cav. Capt. near Staunton '64. Sent to Camp Chase.

RAMSEY, WILLIAM MELTON: Pvt., Co. G. B. Va. 5/41. Tanner's Apprentice, age 18, Burkes Mill Dist., Augusta Co. 1860 census. Enl. Staunton 8/2/61. Died of fever Camp Alleghany 11/29/61, age 19.

RANDOLPH, BLAKEY J.: Pvt., Co. B. Rockingham Co. 1832. Enl. Staunton 7/16/61. Present 7/16-12/31/61. Ab. 1-2/62. AWOL 2/16-28/62, fined $11.00 by CM. Present 3-5/62. AWOL 6/6-8/13/62, fined $25.30. Present 8/14-10/31/62. Deserted 1/1/63. NFR. Pension application claims 3½ years service, sent to hospital. Resident Melrose, Rockingham Co. 1899. Died Harrisonburg 2/1/05.

RANKIN, JAMES W.: Pvt., Co. G. B. Va. 1819? Laborer, age 41, Burkes Mill Dist., Augusta Co. 1860 census. Enl. Staunton 8/2/61.Present 11/61-4/62 non-conscript over 35. WIA (mouth) 2nd Manassas 8/29/62. Hospitalized Charlottesville 9/12/62 until deserted 1/20/63. Ab. wounded through 10/31/64. Present 11-12/64. Cap. Amelia CH 4/6/65. Sent to Point Lookout. Released 6/17/65. 5' 8", dark complexion, brown hair, blue eyes.

RANKINS, JOHN B., SR.: Musician, Co. H. B. Augusta Co. 1798? Farmer, age 62, Burkes Mill Dist., Augusta Co. 1860 census. Enl. Staunton 7/23/61 age 60 as substitute for William G. Andes. Drummer, present 11/61-2/62. Ab. sick 3/28-4/62. Discharged for advanced age 7/24/62. Age 63, 6' 2", fair complexion, blue eyes, grey hair. Millwright. Died before 1885.

RANKIN, THOMAS: Pvt., Co. G. B. New Hope, Augusta Co. 12/15/42. Farmhand, age 17, Burkes Mill Dist., Augusta Co. 1860 census. Enl. Staunton 8/2/61. Present 11/61-12/63. Cap. Bethesda Ch. 5/30/64. Sent to Point Lookout. Transf. Elmira. Released 6/30/65. Resident of Staunton, 5' 8", sallow complexion, dark hair, blue eyes. Pension application indicates "was reduced to almost a skeleton by being starved and exposed in prison." Farmhand, age 25, Mt. Sidney PO, Augusta Co. 1870 census. Farmer, New Hope. D. there 6/28/15. Bur. Middle River Ch. of the Brethren Cem.

RANSON, THOMAS DAVIS: 2ndLt., Co. I.B. "Sycamore" near Charlestown, Jefferson Co. 5/19/43. Enl. Capt. Bott's Co., 2nd Va. Inf. at Charlestown 4/21/61. Transf. Co. I, 52nd Va. 8/16/61 as Pvt., Present 11/61-4/62, appointed SgtMajor 9/2/61. Elected 2ndLt. Co. I 5/1/62. WIA (right knee) Cross Keys 6/8/62. Present 8/30-10/31/62. Resigned because of inability to keep up on the march due to wound to knee joint 11/6/62. Enl. Co. B, 12th Va. Cav. as Pvt. Served as Aide to Gen's William L. Jackson and Edward Johnson. Promoted Capt., Co. I, 12th Va. Cav. In charge of Scouts in Secret Service Dept., reporting directly to Gen's J. E. B. Stuart and R. E. Lee. Cap. Tom's Brook 10/10/64 and held until 6 weeks after the surrender for refusal to take the oath. Grad. Washington Col. '67BL with high honors. Grad. UVa. Law School and U. Pa. Lawyer, Staunton. Law partner of Col. John B. Baldwin and Hon. A. H. H. Stuart. First President Va. Bar Association; President, Staunton Chamber of Commerce; Chairman, State organization for International Arbitration; Trustee, W&LU; Trustee, UVa.; Chairman, State Committee on work among colored people; President, Staunton YMCA; Lt. Commander, Grand Camp Confederate Veterans of Va.; Commander, Stonewall Jackson Camp CV, Staunton. Died "Oakenwold", Staunton 7/21/18. Bur. Thornrose Cem. Col. Baldwin endorsed his resignation: "He is one of the best young soldiers I ever saw and his faithful efficient service I think entitles him to the favor of being allowed to go down in rank and to be transferred to the Cavalry.".

RATLIFF, A. P.: Pvt. Co. K. On postwar roster.

RATLIFF, JAMES P.: Pvt., Co. K. B. Bath Co. 1833? Farmer, age 27, Bath CH PO, Bath Co. 1860 census. Enl. Shenandoah Mt. 4/9/62. Present 4/9-30/62. Ab. sick 10/23/62-4/63. AWOL 5/1-12/31/63. Dropped as a deserter. Deserted to the enemy Clarksburg, W. Va. 4/8/65. Took oath and released. Farmer, age 37, Williamsville Dist., Bath Co. 1870 census. Farmer, Green Valley, 1897. Died before 8/17.

RATLIFFE, WARRICK C.: Pvt., Co. K. B. Back Creek, Bath Co. 1842? Farmhand, age 18, Bath CH PO, Bath Co. 1860 census. Enl. Valley Mills 4/26/62. Present 4/26-30/62. Ab. sick 5/27/62. Cap. Front Royal 5/30/62. Sent to Ft. Delaware. Died there of diphtheria 7/28/62 age 19. Bur. Finn's Point National Cem., N. J.

REA, JESSE S.: Pvt., Co. D. B. 1835? Resident of Sangersville. Enl. Staunton 7/16/61. Present 7/16/61-4/62. Deserted 6/10/62. NFR. Died Millboro, Bath Co. 6/16. Bur. Pres. Ch. Cem.

READON, JOHN W.: Pvt. Co. H. Resident of Charlottesville, Enl. Bunker Hill 10/12/62. Deserted 11/29/62. NFR.

REED, ALEXANDER S.: Pvt., Co. I. B. Va. 1833? Farmhand, age 27, Northern Dist., Augusta Co. 1860 census. Pvt., Co., D. 44th Va. Milita rejected for leg injury 3/62. Enl. Port Republic 7/15/62. WIA (foot) Sharpsburg 9/17/62. Ab. wounded through 10/63. Present 11-12/63. Cap. Spotsylvania CH 5/12/64. Sent to Ft. Delaware. Released 6/21/65. Resident of Rockbridge Co. 5' 8", sallow complexion, dark hair, dark eyes. Age 58, Walker's Creek Dist., Rockbridge Co. 1898.

REED, GEORGE FRANKLIN: Pvt., Co. G. B. Augusta Co. 1838? Farmhand, age 22, Burkes Mill Dist., Augusta Co. 1860 census. Enl. Staunton 8/2/61, Ab. sick 11/61-2/62. Present 3-4/62.Ab. sick 5-10/62. Discharged for pulmonallis and injury to spinal column 11/20/62. Age 26, 5' 9", fair complexion, blue eyes, dark hair.

REED, WILLIAM B.: Pvt., Co. I. B. Nelson Co. 1842? Enl. Staunton 7/16/61 age 19. Present 11/61-4/62. Reenl. 5/1/62. WIA (side) and cap. Sharpsburg 9/17/62. Exchanged 9/24/62. Ab. wounded through 2/63. Present 3-12/63. WIA (jaw) near Spotsylvania CH 5/19/64. Ab. wounded through 12/31/64. Paroled Staunton 5/17/65. Age 22, 5' 6", fair complexion, dark hair, blue eyes.

REED, WILLIAM H.: Pvt., Co. C. B. Augusta Co. 1836? Pvt., Capt. Hall's Co., 32nd Va. Militia 3/62 age 34, Enl. Shenandoah Mt. 4/18/62. Present 4/18-30/62. WIA (arm) McDowell 5/8/62. Ab. wounded through 2/63. Present 3-5/63. Ab. Sick 6/14-8/31/63, served as nurse Jordan Springs hospital 6/18-7/23/63. Present 9/25-11/29/63. Ab. in hospital with rheumatism 11/30-1/23/63. Present 12/24-31/63. KIA Bethesda Ch. 5/30/64, age 28. Left widow.

REESE, EDWARD PETER: Pvt., Co. C. B. Augusta Co. 1/4/30. Enl. Staunton 7/16/16 age 31. Present 11/61-4/62, detailed as Wagoner. Reenl. 5/1/62. Ab. detailed as Teamster in Brig. train since 9/10/61 through 12/31/64. Surrendered Appomattox 4/9/65. Illiterate. Carpenter, age 37, Mt. Sidney PO, Augsta Co. 1870 census. Died Verona 4/28/16. Bur. Verona Methodist Ch. Cem.

REESE, JOHN A.: Pvt., Co. E. Enl. Bunker Hill 10/1/62. Present 10/62, and 1-12/63. Issued clothing 1/28, 4/27 and 5/19/64. WIA Cedar Creek 10/19/64. Ab. wounded through 12/31/64. NFR.

REEVES, PEACHY H.: Pvt., Co. D. B. Va. 1830? Laborer, age 30, Burkes Mill Dist., Augusta Co. 1860 census. Enl. Staunton 7/16/61. Present 7/16-4/62. Reenl. 5/1/62. Present 5/1-6/8/62. AWOL 6/9-20/62, fined $4.03. Present 6/21-8/31/62. Present in arrest 1-2/63, fined pay for 7 months and 3 days in addition to 12 days AWOL by CM. Present 3-4/63. Accidentally wounded in the wrist Gettysburg 7/1/63. Cap. there 7/5/63. Sent to Ft. Delaware. Took oath and joined 3rd Md. Cav. (U.S) 9/22/63.

REEVES, ROBERT F.: Pvt., Co. D. B. Va. 1843? Laborer, age 17, Mossy Creek PO, Augusta Co. 1860 census. Enl. Alleghany Mt. 3/13/62. Present 3/13-7/16/62. AWOL 7/17-8/1/62, fined $5.50. Present 8/2-31/62. Present in arrest 1-2/63, fined $83.60 by CM for 16 days AWOL and 7 months and 3 days in addition. Deserted Williamsport, Md. 7/14/63. Sent to Ft. Delaware. Took oath and joined 3rd Md. Cav. (U.S.) 9/22/63. Brother of Thomas H. Reeves. D. Mossy Creek 12/18/91.

REEVES, THOMAS H.: Pvt., Co. D. B. Va. 1839? Laborer, age 21, Mossy Creek PO, Augusta Co. 1860 census. Enl. Staunton 7/16/61. Present 7/16/61-7/16/62. AWOL 7/17-8/1/62, fined $.50. Deserted 8/25/62. Ab. in arrest 3-4/63. Ab. wounded 4/2/63, (supposed to be self inflicted) through 12/31/64, however sent to Castle Thunder from hospital Richmond 7/9/63. Dropped as deserter, NFR. Brother of Robert F. Reeves.

REID, JAMES ALEXANDER: Pvt., Co. E. B. Va. 1845? Resident, age 15, Glenwood Dist., Rockbridge Co. 1860 census. Enl. Orange CH 9/15/63. Present 9/15-12/31/63. WIA (right side) and cap. Winchester 9/19/64. Sent to Point Lookout. Exchange 11/15/64. Ab. wounded 11-12/64. Illiterate. NFR. Laborer, age 24, Buffalo Township, Rockbridge Co. 1870 census.

REID, JOHN WILLIAM: Pvt., Co. C. B. Frederick Co. 6/28/33. Enl. Staunton 7/16/61. Deserted 8/15/61. Possibly J. William Reed who enl. Co. H or C, 18th Va. Cav. 9/25/62. Obit. states "WIA twice, one ball still lodged against spine at death." Farmer, Ft. Defiance, 1897. Died Grottoes 10/6/99. Bur. Old Salem Lutheran Ch. Cem., 3 miles west of Mt. Sidney.

REID, WILLIAM HENRY: Pvt., Co. E. B. Rockbridge Co. 12/22/35. Farmhand, age 25, Lexington Dist., Rockbridge Co. 1860 census. Enl. Staunton 8/1/61. Present 11/61-4/62. Present on rolls 5/30/62-10/31/62 "AWOL 1 mo." "Deserted 1 Aug. 62." Delivered to PM Staunton 10/21/62. Illiterate. Disch. Front Royal 11/18/62. NFR. Farmer, Rockbridge Co. 1868. D. South Buffalo Dist., Rockbridge Co. 3/3/85. Bur. Rapp's Ch.

REID, WILLIAM N.: Pvt., Co. E. B. Va. 1841. Farmhand, age 19, Glenwood Dist., Rockbridge Co. 1860 census. Enl. Staunton 8/1/61. Des. the same day and joined another company before mustered into service. Resident of Fancy Hill.

REYNOLDS, GEORGE WASHINGTON: 2ndCpl., Co. A. B. Augusta Co. 1839? Enl. Staunton 7/9/61 as Pvt. age 19. Present 11/61-4/62. Reenl. 5/1/62. Present 8/31/62-4/63, appointed 4thCpl. 4/14/63. Present 4/30-12/31/63, rank shown as 3rd Cpl. 11-12/63. Issued clothing 1/28, 2/13 and 27, 4/29, 6/12 and 27/64. Cap. Cedar Creek 10/19/64, rank shown as 2ndCpl. Sent to Point Lookout. Released 6/17/65. 5' 6", fair complexion, blue eyes, auburn hair. Farmhand, age 25, Fishersville PO, Augusta Co. 1870 census. Died Staunton 5/2/23. Bur. Green Hill Cem., Churchville.

REYNOLDS, JASPER M.: Pvt., Co. A. Enl. Sommerville Ford 2/29/64. Present 4/30-12/31/64 hospitalized with typhoid pneumonia 5/23-6/30/64 and furloughed for 30 days. Capt. Farmville and paroled 4/11-21/65.

REYNOLDS, SAMUEL: Pvt., Co. F. Not on muster rolls. Paroled Staunton 5/22/65. Age 19, 6', fair complexion, light hair, brown eyes.

REYNOLDS, SILAS: Pvt., Co. F. Enl. Staunton 9/3/62. Discharged 10/15/62. unofficial report indicates he was a young boy.

RHEA, JOHN SHAW: Pvt., Co. K. B. 1836. On postwar roster. Enl. Co. G, 18th Va. Cav. 12/3/63. Paroled Staunton 5/25/65. 5' 11", dark complexion, black hair, blue eyes. Living Millboro, Bath Co. 9/15/07.

RICE, WILLIAM C. W.: Pvt., Co. B. (2nd). B. Albemarle Co. 1842? Farmer, Augusta Co. Enl. Waynesboro 7/15/61. Died of disease at his father's house in Augusta Co. 12/9/61. Age 19, 5' 4½", dark complexion, pale blue eyes, light hair.

RICKETTS, AUGUSTUS: Pvt., Co. C. On postwar roster.

RIDDLE, JOHN: Pvt., Co. B (2nd) B. 1840? Farmhand, Augusta Co. Not on muster rolls. Tombstone only record. Died 11/69. Bur. Riverview Cem., Waynesboro.

RIDDLE, ZEBIDEE: Pvt., Co. C. B. Va. 1843? Farmhand, age 17, Northern Dist., Augusta Co. 1860 census. Enl. Staunton 7/16/61 age 19. ab sick 12/15/61-4/62. Deserted near Staunton 6/30/62. Resident of Mossy Creek. NFR.

RIDDLEBERGER, ELIAS: Rank and Co. unknown. B. Shenandoah Co. 7/8/28. Not on muster rolls Served in 5th Va. Inf. also. Listed service in 52nd Va. on application to join Stonewall Jackson Camp, CV, Staunton 3/24/04. Farmer, near Bridgewater. Died Augusta Co. 2/17/27. Bur. Mt. Pisgah Meth. Ch. Cem. near Ft. Defiance.

RIDER, JACOB M.: Pvt., Co. K. B. Va. 1836? Farmhand, age 24, Bath CH PO, Bath Co. 1860 census. Enl. Shenandoah Mt. 4/9/62. Deserted 4/15/62. Resident of Back Creek. Died Bath Co. 11/22/67. Bur. Mountain Grove. Cem.

RIFE, ARCHIBALD: Pvt., Co. C. B. Augusta Co. 5/22/32. Distiller, age 28, Burkes Mill Dist., Augusta Co. 1860 census. 1stLt., South River Co., 32nd Va. Militia. Enl. Staunton 7/16/61 age 39. Ab. sick 12/15/61-2/62. Present 3-4/62. Reenl. 5/1/62. WIA (left thigh) 2nd Manassas 8/28/62. Ab. wounded through 2/18/63. Ab. on leave 2/19-3/18/63. Ab. detailed in hospitals Palmyra, Richmond and Farmville 3/19/63-12/31/64. Paroled Staunton 5/17/65. age 33, 5' 9", light complexion, light hair, blue eyes. Farmer, age 38, Mt. Sidney PO, Augusta Co. 1870 census. Farmer, Crimora Station 1897. Died Middle River Dist. 10/27/17 with bullet still in leg. Bur. Laurel Hill Baptist Ch. Cem.

RIFE, DAVID M: Pvt., Co. C. B. Va. 1827? Carpenter, age 33, Burkes Mill Dist., Augusta Co. 1860 census. Enl. Staunton 7/16/61 ag 34. Present 11/61-4/62. Reenl. 5/1/62. Present 5/1-8/14/62. Ab. sick 8/16-31/62. Present 9/62-2/9/63. Ab. sick 2/10-8/31/63. Present 9-12/63. WIA (left hand) Bethesda Ch. 5/2/64, age 37. Ab. wounded through 12/31/64. Deserted to the enemy New Creek, W. Va. 3/12/65. Took oath and released. Age 38, 5' 10", fair complexion, grey eyes, dark hair. Miller, age 43, Mt. Sidney PO, Augusta Co. 1870 census. Millwright, Rivanna, Albemarle Co. 1897. Living Rockingham Co. 1916.

RILEY, PATRICK H.: Pvt., Co. E. B. Va. 1830? Not on muster rolls. Surrendered Appomattox 4/9/65. Farmhand, age 40, Bath CH Bath Co. 1870 census.

RINER, DAVID W.: Pvt., Co. G. B. Greene Co. 1829? Farmer, Enl. Staunton 8/2/61. Present 11/61. Ab. sick 12/17/61-4/62. WIA (hand and thigh) Cross Keys 6/8/62. Ab. wounded through 2/63. Ab. detailed on hospital duty 2/19-12/31/63. WIA (right arm) Cedar Creek 10/19/64. Arm amputated at shoulder joint. Retired for disability 2/10/65. Age 36, 5' 8", fair complexion, hazel eyes, dark hair.

RISK, JAMES PAXTON: Pvt., Co. B. (1st) B. Va. 1844? Farmhand, age 16, Walker's Creek Dist., Rockbridge Co. 1860 census. Enl. Fairfield 7/16/61. Present until transf. 9/28/61.

RITCHIE, HENRY F., JR.: Pvt., Co. D. B. Rockingham Co. 1836? Carpenter, Augusta Co. Enl. Staunton 7/16/61. Present 7/16-10/31/61. Died of camp fever in hospital Yager's Alleghany Mt. 11/29/61. Bur. there. Posessions 2 blankets and 1 knapsack with clothing sent to family. Age 25, 6", dark complexion, dark hair, grey eyes. Left widow.

RIVERCOMB, JOHN CONNELL: 1stSgt., Co. D. B. near Miller's Iron Works, Rockingham Co. 1/4/19. Enl. Staunton 7/16/61. Present 7/16/1-3/11/62. Ab. sick 3/12/62-12/63, reduced to Pvt. 5/1/62. NFR. Cabinetmaker, Sangersville and Bridgewater. Died Bridgewater 10/24/08. Bur. Greenwood Cem., Rockingham Co.

ROARKE, CHARLES K. S.: Pvt., Co.K. B. Bath Co. 1826? Pvt., Capt. Hamilton's Co., 81st Va. Militia. Enl. Shenandoah Mt. 4/9/62. Present 4/9-30/62 and 5/26/62 over 35. Ab. sick 6/19/61-4/62, rheumatism and general debility of long standing. Ab. on leave 6/3-7/2/63. AWOL 7/3-12/31/63. Ab. on detail in government works 3/17-12/31/64. NFR. Farmer, age 44, Warm Springs Dist., Bath Co. 1870 census. Died near Mountain Grove 7/29/96.

ROBERTSON, DAVID: Pvt., Co. B (2nd). B. 1833? Farmer, Augusta Co. Enl. Waynesboro 7/15/61. Present 11/61-3/27/62. Ab. sick 3/28/62-6/63. Deserted to the enemy Beverly, W. Va. 6/16/63. Took oath and sent to Ohio. Age 30, 5' 10¾", florid complexion, blue eyes, light hair.

ROBERTSON, JOSHUA: Pvt., Co. B (2nd). B. Nelson Co. 1828? Farmer, Nelson Co. Enl. Waynesboro 7/15/61. Present 11/61-4/62. Reenl. 5/1/62. AB. sick 5/5/62-2/63. Present 3-5/63. Ab. sick 6/4-10/63. Deserted to the enemy Clarksburg, W. Va. 10/25/63. Took oath and sent north. Age 35, 5' 11", light complexion, dark eyes, fair hair. Farmer, Lynhurst, Augusta Co. 1897, and 1910 census.

ROBERSON, WILLIAM J.: 3rdSgt., Co. I. B. Augusta Co. 1833? Enl. Staunton 7/16/61 age 28, as 1stCpl. Present 11/61-4/62. Reenl. 5/1/62. WIA (thigh) Port Repulic 6/9/62. Present 8/30-10/31/62, and 1-4/63, appointed 4thSgt. 4/1/63. Present 4/30-12/63. WIA (arm) near Spotsylvania CH 5/19/64. Furloughed from hospital Richmond for 60 days 5/29/64. Cap. Staunton 9/25/64, rank shown as 3rdSgt. Exchanged 3/19/65. NFR. Living New Hope 1894. Farmer, Swoope, 1897.

ROBINSON, ANDREW D.: Pvt., Co. E. B. Va. 1846? Resident, age 14, Collierstown PO, Rockbridge Co. 1860 census. Enl. Staunton 8/1/61. AWOL 9/6-12/61. Present 1-4/62. Reenl. 5/1/62. WIA and capt. Sharpsburg 9/17/62. Exchanged 11/10/62. Furloughed for 60 days. Present 1-5/63. Ab. sick 6/4-12/31/63. Enl. Co. C. 1st Va. Cav. D. Augusta Co. 8/14/96.

ROBSON, JOHN S.: 2ndCpl., Co. D. B. Rappahannock Co. 3/26/44. Student, Charlottesville. Enl. Staunton 7/16/61 as 3rdCpl. Present 7/16-61-4/62. Reenl. 5/1/62. Present 5/1-12/62 and 1-12/63, took Regt'l Colors from dying Color Bearer 5/5/63. Promoted 2nd Cpl. 8/30/64. Present 4/30-10/64. WIA (leg) Cedar Creek 10/19/64 while acting as Courier for Gen. Pegram. Leg amputated. Ab. wounded 11-12/64. NFR. Wrote "How A One Legged Rebel Lives" 1876. Farmer, Mossy Creek, Augusta Co. 1889-1902. Died Amissville, Rappahannock Co. 6/20.

ROGERS, GEORGE ANDREW: Pvt., Co. C. B. Augusta Co. 1843? Carpenter, age 17, Deerfield PO, Augusta Co. 1860 census. Enl. Staunton 8/21/61. Ab. sick 12/15/61-3/63. Deserted near Staunton 1/1/62 on rolls 9-10/63. NFR. Died McDowell, Highland Co. 6/40. Bur. Thornrose Cem., Staunton. Last member of 52nd to die.

ROSEN, DAVID FRANKLIN.: Pvt., Co. I. B. Augusta Co. 3/29/39. Farmhand, age 21, 1st Dist., Augusta Co. 1860 census. Pvt., Co E, 93rd Va. Militia. Enl. Staunton 7/16/61 age 22. Present 11/61-4/62. Reenl. 5/1/62. Present 5/1-7/17/62. AWOL 7/18-8/5/62, fined $4.77. Present 8/6-10/31/62 and 1-4/63. Ab. sick with debility from pneumonia 5/16/63-12/31/63. Returned to duty 1/7/65. NFR. Farmer, Augusta Co. Died Greenville 10/24/02. Bur. Bethel Pres. Ch. Cem. Brother of Jacob W. Rosen.

ROSEN, JACOB W.: Pvt., Co.I. B. Augusta Co. 10/29/26. Carpenter, age 32, 1st Dist., Augusta Co. 1860 census. Pvt., Co. E, 93rd Va. Militia. Enl. Staunton 7/16/61. Ab. sick 11/61-2/1/62. Present 2/13-4/62. Reenl. 5/1/62. Present 8/30-10/31/62. Ab. on leave 2/19-28/63. Present 3-4/63. WIA (face and jaw broken) Fredericksburg 5/5/63. Ab. wounded through 12/31/64. Obit. states ". . .detailed in ambulance corps the duty of which often times called him into the thickest and hottest of the battle. Like a hero he served at his post ever faithfully, when at last in the fight . . .a piece of the enemies shell penetrated his jawbone, from the effects of which he never entirely recovered." Surrendered Appomattox 4/9/65. Paroled Staunton 5/17/65. Age 37, 5' 9" dark complexion, brown hair, brown eyes. Carpenter, age 42, 1st Dist., Augusta Co. 1870 census. Farmer, Augusta Co. Died near Mint Spring 11/3/78. Bur. St. John's Reformed Ch. Cem. Brother of David F. Rosen.

SAMPSON, L. A.: Pvt., Co. K. Pension application only record. Enl. '62. Served 3 years. Living Madison Co. 1888.

SAMPSON, MITCHELL A.: 4thCpl., Co. K. B. Va. 1840? Farmhand, age 20, Green Valley Dist., Bath Co. 1860 census. Enl. Co. G. 25th Va. Inf. 1861. Enl. Shenandoah Mt. 4/9/62, as Pvt. Present 4/9-30/62. WIA (neck) Cross Keys 6/8/62. WIA (neck) Gaines Mill 6/27/62. Returned to duty 8/26/62. Ab. sick 11/23/62-2/63. Present 3-4/63, appointed 4thCpl. 4/15/63. WIA (right hip) Gettysburg 7/3/63. Ab. wounded through 8/31/63. Present 9-12/63, and 4/30-12/31/64. Cap. Ft. Stedman 3/25/65. Sent to Point Lookout. Released 6/19/65. 5' 8½", dark complexion, dark brown hair, blue eyes.

SAMUELS, ERASMUS A.: 1stSgt., Co. D. B. Va. 1837? Farmhand, age 23, Burkes Mill Dist., Augusta Co. 1860 census. Enl. Staunton 7/16/61 as 2ndSgt. Present 7/16/61-4/62. Reenl. and appointed 1stSgt. 5/1/62. Furnished substitute and discharged 5/28/62. Listed as POW Elmira 1/65, resident of Sangersville. NFR. Brother of Hiram G. and Joseph M. Samuels.

SAMUELS, HIRAM GREENBERRY: Cpl., Co. D. B. Augusta Co. 1839? Enl. Staunton 7/16/61. Present 7/16-11/61. Died of fever at home in Augusta Co. 12/21/61 age 23. Brother of Erasmus A. and Joseph M. Samuels.

SAMUELS, JOSEPH M.: 3rdSgt., Co. D. B. Augusta Co. 1835? Teacher, age 25, Burkes Mill Dist., Augusta Co. 1860 census. Enl. Staunton 7/16/61. Present 7/16-10/31/61. Ab. sick 11/5/61-2/62, at Augusta Springs. Present 3-4/62. WIA (groin) McDowell 5/8/62. Ab. wounded until discharged for disability 4/16/63. Age 32, 5' 9", fair complexion, black eyes, black hair. Carpenter, age 39, Mt. Sidney PO, Augusta Co. 1870 census. Brother of Erasmus A. and Hiram G. Samuels.

SANDES, WILLIAM: Pvt., Co. G. B. Va. 1817. Farmer, age 43, Northern Dist., Augusta Co. 1860 census. Not on muster rolls. Died Staunton 4/7/62. Bur. Thornrose Cem., grave #922. Left widow and 6 children.

SANGER, SIMON: Pvt., Co. F. B. 8/2/40. Resident of Rockingham Co. Not on muster rolls. Died of disease Rockingham Co. 8/25/62. Bur. Linville Creek Ch. of the Brethren Cem., near Broadway, Rockingham Co.

SAUFLEY, JAMES M.: Pvt., Co. G. B. 11/13/43. Enl. Staunton 10/3/62. Present 10/62-6/63. AWOL 7/15-12/31/63. Dropped as deserter. NFR. Died near Mt. Crawford, Rockingham Co. 3/7/20. Bur. Frieden's Ch. Cem., on Rt 275, east of Harrisonburg.

SCOTT, CHARLES A.: Pvt., Co. E. B. Va. 1840? Farmhand, age 20, Collierstown PO, Rockbridge Co. 1860 census. Served in Militia. Enl. Staunton 8/1/61. Present 11/61-4/62. Reenl. 5/1/62. WIA (shoulder) 2nd Manassas 8/29/62. Ab. wounded through 6/5/62. Deserted near Richmond 6/6/64. NFR. Died Rockbridge Co. 1885. Brother of Thomas F. Scott.

SCOTT, THOMAS F.: Pvt., Co. E. B. Rockbridge Co. 1836? Farmhand, age 24, Collierstown PO, Rockbridge Co. 1860 census. Served in Militia. Enl. Staunton 8/1/61. Present 11/61-4/62. Reenl. 5/1/62. Ab on leave 5/1-10/31/62. Died of chronic diarrhea hospital Mt. Jackson 11/6/62. bur. Mt. Jackson Cem. Age 26, 5' 6", dark complexion, dark hair, dark eyes. Brother of Charles A. Scott. Left widow and 4 children.

SEE, GEORGE, JR.: Pvt., Co. G. On postwar roster.

SEE, GEORGE BAXTER.: Pvt., Co. G. B. 2/27/29. Enl. Staunton 8/2/61, non-conscript over 35. AWOL 4/30-8/31/62. Present 9-12/62 and 1-12/63. Died of typhoid pneumonia Orange CH 2/22/64. $10.00 in effects turned over to QM, CSA. Bur. Tinkling Spring Pres. Ch. Cem., near Fishersville.

SEE, JACOB W.: Pvt., Co. G. B. Va. 1839? Farmhand, age 21, Burkes Mill Dist., Augusta Co. 1860 census. Enl. Staunton 8/2/61. Present 11/61-3/62. Discharged 4/1/62. Died 5/26/62. Bur. Tinkling Spring Pres. Ch. Cem., near Fishersville.

SELPH, BENJAMIN DANIEL: Pvt., Co. E. B. Rockbridge Co. 4/22/41. Farmhand, age 18, Lexington Dist., Rockbridge Co. 1860 census. Enl. Staunton 8/1/61. Present 11-12/61. Ab. sick 1-4/62. Reenl. 5/1/62. Present 5/1-7/18/62. Ab. sick in hospital with debility 7/21-8/8/62. Present 8-10/62, and 1-12/63. WIA (thorax) and cap. Bethesda Ch. 5/30/64. DOWs 6/3/64, age 22. "A good soldier."

SELPH, WILLIAM JACKSON: Pvt., Co. E. B. Rockbridge Co. 7/26/36. Farmhand, age 23, Lexington Dist., Rockbridge Co. 1860 census. Pvt., Co. C, 8th Va. Militia 3/62 disqualified for hernia. Enl. Bunker Hill 9/22/62. Present 9/22-10/31/62, and 1-4/63. Ab. sick 6/9-8/31/63. Ab. detailed in hospital Farmville 9-12/63. Discharged for disability 3/19/64, PO Summers, Rockbridge Co. Age 27, 6' 2½", fair complexion, blue eyes, light hair, farmer, illiterate. Conscripted again 5/64 and assigned to Regt. 5/23/64. Ab. sick with hernia 6/25-10/31/64. Present 11-12/64. Surrendered Appomattox 4/9/65. Resident of Timber Ridge, Rockbridge Co. 1902. Died Buena Vista 8/12/11. Bur. there.

SHAFFER, JEREMIAH HENRY "Jessie": 2ndCpl. Co. E. B. Va. 2/14/25. Farmer, age 35, Glenwood Dist., Rockbridge Co. 1860 census. Enl. Staunton 8/1/61. Ab. on leave 11-12/61. Present 1-4/62. Reenl. 5/1/62. AWOL 5/1-10/31/62, rank shown as Pvt. Present in arrest 1-2/63. Present 3-4/63, fined 9 months and 11 days pay by CM. Discharged 5/25/63, reason not stated. Resident of Fancy Hill. Unofficial report indicates detailed for home duty. Farmer, age 45, Natural Bridge Dist., Rockbridge Co. 1870 census. Postmaster, Sherwood, Rockbridge Co. 1897. Died 2/14/02. Bur. High Bridge Pres. Ch. Cem., near Natural Bridge.

SHAFER, ROBERT P. G.: 1stCpl., Co. E. B. Va. 1842. Farmhand, age 18, Glenwood Dist., Rockbridge Co. 1860 census. Enl. Staunton 8/1/61 as Pvt. Present 11/61-4/62. Reenl. 5/1/62. Present 5/1-10/31/62. WIA Cedar Run 8/9/62. WIA (2nd and 3rd fingers right hand amputated) Fredericksburg 12/13/62. Present 1-4/63, promoted 1stCpl. 4/15/63. WIA (left shoulder) Gettysburg 7/3/63. AB. wounded through 12/63. WIA Wilderness 5/6/64. Ab. wounded through 10/31/64. Present 11-2/64. Surrendered Appomattox 4/9/65. Resident, Rapp's Mills, Rockbridge Co. 1900-08. Laborer, age 68, Buena Vista, 1910 census. Died there 1/9/22.

SHAFER, SAMUEL JACKSON: Pvt., Co. E. B. near Natural Bridge, Rockbridge Co. 2/21/34. Enl. Bunker Hill 9/22/62. Present 9/22-10/31/62, and 1-12/63, serving as sharpshooter. WIA (ball entered left side of chest and was cut out of back) Winchester 9/19/64. Cap. Harrisonburg 9/25/64. Sent to West's Building Hospital, Baltimore. Transf. Point Lookout. Exchanged 10/30/64. Ab. on leave 11-12/64. NFR. Pension application indicates WIA twice. Farmer, age 39, Buffalo Dist., Rockbridge Co. 1870 census. Resident Rapp's Mill 1900. Died near Natural Bridge 5/28/07. Bur. Rapp's Mill Ch. Cem.

SHAFER, JOHN FREDERICK W.: Pvt., Co. D. B. Augusta Co. 1825? Farmer, Mt. Solon, Augusta Co. Enl. Staunton 7/16/61. Present 7/16/61-4/62. WIA (groin) Port Republic 6/9/62. DOWs hospital White Hall 6/14/62. Age 30, dark complexion, dark eyes, 5' 11", left widow.

SHANNON, JACOB ALEXANDRIA: Pvt., Co. B (2nd). B. Va. 1849? Not on muster rolls. Issued clothing 3/5/64. Illiterate. NFR. Contractor in southwest U. S.; Builder, Sante Fe Railroad. Resident, El Paso, Tex. Died Roanoke, Va. 6/22/23.

SHANER, JACOB HAINER: Pvt., Co. G. B. Rockbridge Co. 11/35. Enl. Staunton 8/1/61 Present 11/61-4/62. AWOL 5/21/62 until deserted to the enemy Clarksburg W. Va. 10/14/63. Took oath and sent north Age 22, 6'. Cooper, Augusta Co. 1870. May have served in Co. H, 14th Va. Cav. Farmer, near Staunton 1897. Died on Greenville Road, 3 miles from Staunton 5/28/09. Bur. Thornrose Cem.

SHAVER, GEORGE WASHINGTON: Pvt., Co. C. B. Va. 1839? Enl. Staunton 3/20/63. Present 3-4/63. AWOL 6/12-12/31/63. Dropped as deserter. NFR. Farmer, age 31, Mt. Sidney PO, Augusta Co. 1870 census.

SHEETS, HENRY C.: Pvt., Co. I. B. Augusta Co. 1825? Carpenter, Augusta Co. Listed on postwar roster. Farmer, Augusta Co. Died near Grove Ch. 4/16/89.

SHEETS, JACOB SAMUEL: Pvt., Co. F. B. Augusta Co. 7/25/42. Farmhand, age 18, Burkes Mill Dist., Augusta Co. 1860 census. Enl. Staunton 7/31/61. Present 11/61-4/62. Reenl. 5/1/62. Present 8-9/62. WIA (arm) Sharpsburg 9/17/62. Present 10-12/62. Ab. on leave 2/19-28/63. Present 4/30-12/63, and 3-4/64. Issued clothing 5/19/64. WIA (compound fracture left tibia) and cap. Cedar Creek 10/19/64. Lower left leg amputated. Sent to West Buildings hospital, Baltimore. Transf. Ft. McHenry and Point Lookout. Exchanged 2/20/65. In hospital Richmond 3/4-5/65. paroled Staunton 6/5/65. Age 23, 5' 7", fair complexion, light hair, blue eyes. Farmhand, age 28, Mt. Sidney PO, Augusta Co. 1870 census. Farmer, Middle River Dist., Augusta Co. 1910 census. Died Augusta Co. 2/27/22. Bur. Old Salem Lutheran Ch. Cem., 3 miles west of Mt. Sidney. Brother of John D. Sheets.

SHEETS, JOHN D.: Pvt., Co. F. B. Augusta Co. 4/19/34. Farmhand, age 23, Burkes Mill Dist., Augusta Co. 1860 census. Enl. Mt. Jackson 3/17/63. Present 4-12/63, and 3-12/64. In hospital Richmond 1/26-2/10/65. Age 30, 5' 6", fair complexion, dark eyes, dark hair, farmer. Cap. Amelia CH 4/6/65. Sent to Point Lookout. Released 6/19/65. 5' 5½", light complexion, dark brown hair, grey eyes. Farmer, age 36, Mt. Sidney PO, Augusta Co. 1870 census. Farmer, Folly Mills 1897. Died Augusta Co. 4/4/12. Bur. Old Salem Lutheran Ch. Cem., 3 miles west of Mt. Sidney. Brother of Jacob S. Sheets.

SHEETS, WILLIAM HENRY (1st): Pvt., Co. G. B. Va. 5/2/41. Farmhand, age 19, Burkes Mill Dist., Augusta Co. 1860 census. Enl. Staunton 8/2/61. Present 11/61-4/62. Reenl. 5/1/62. Present 5/1-8/31/62. Present 9-10/62, AWOL 4 days fined $1.46. Present 11/62-7/24/63. AWOL 7/25-8/30/63, fined $12.83. Present 9-12/63, and 4/30-12/31/64. NFR. Died Mt. Sidney 2/26/94. Bur. Mt. Sidney Meth. Ch. Cem.

SHEETS, WILLIAM HENRY (2nd): Pvt., Co. I. B. Va. 7/29/44. Resident, age 15, 1st Dist., Augusta Co. 1860 census. Pvt., Co. E, 93rd Va. Militia. Enl. Staunton 7/8/63. Present 8-12/6/63, and 4/30-12/64. Surrendered Appomattox 4/9/65. Farmhand, age 26, 1st Dist., Augusta Co. 1870 census. Died Brands, Augusta Co. 11/26/20. Bur. Olivet Ch. Cem.

SHEETS, WILLIAM J.: Pvt., Co. F. B. Augusta Co. 1837? Farmhand, age 23, Northern Dist., Augusta Co. 1860 census. Enl. Staunton 7/31/63. Present 11/61-1/62. Ab. sick 2/22-4/62. Reenl. 5/1/62. Present 5/1-7/27/62. AWOL 7/28-9/29/62. Present 9/30-12/31/62 and 1-12/31/63. Present 3-4/64. WIA (right side and left arm) and cap. Bethesda Ch. 5/30/64. DOWs in 3rdDiv., 5th Army Corps, Army of Potomac hospital.

SHEET, WILLIAM P.: 1stSgt., F. B. Va. 8/12/36. Carpenter's Apprentice, age 18, Northern Dist., Augusta Co. 1860 census. Enl. Staunton 7/31/61 as 1stCpl. Promoted 1stSgt. Present 11/61-4/62, exempt under Conscript Act. Present 8/31-10/15/62. Ab. in hospital Winchester 10/16/62-2/63. Discharged from hospital Charlottesville reason unknown 5/1/63. Carpenter, age 30, Craigsville PO, Augusta Co. 1870 census. Died near Mt. Pisgah, Augusta Co. 1/7/85.

SHELTON, THOMAS A.: Pvt., Co. E. B. Va. 1839? Farmhand, age 21, Bath CH. PO, Bath Co., 1860 census. Enl. Shenandoah Mtn. 4/18/62. Present 4/18-30/62. WIA (leg) Gaines Mill 6/27/62. Ab. wounded through 8/31/63, while so absent enl. Co. G, 18th Va. Cav. 1/1/63. KIA McConnelsburg, Pa. 6/29/63. Bur. McConnelsburg.

SHEPHERDSON, ALFRED E.: Pvt. Co. E. B. Lexington, Rockbridge Co. 1840? Served in Militia. Enl. Staunton 8/1/61. Present 11/61-4/62. Reenlisted 5/1/62. KIA McDowell 5/8/62, age 19. "A good soldier."

SHERMAN, JOHN WISE: Pvt., Co. C. B. Rockingham Co. 3/26/38. Farmhand, age 23, Northern Dist., Auguusta Co. 1860 census. Enl. Staunton 3/1/63. Ab. Wagoner in Brig. Train 3/1-12/31/64. Surrendered Appomattox 4/9/65. Farmer; Lumber, Ice, and Sorghum Merchant; Manufacturer of spokes and handles, Mt. Crawford, Rockingham Co. D. there 1/7/24. Bur. Mt. Crawford Meth. Ch. Cem.

SHEWEY, FRANKLIN: Pvt., Co. B (1st). B. Va. 1835? Wagonmaker, age 25, Lexington Dist., Rockbridge Co. 1860 census. Enl. Fairfield 7/10/61. Ab. on detail when transf. 9/28/61.

SHIFFLETT, OR SHIPPLET, FRANKLIN: Pvt., Co. F. B. Va. 1823? Farmhand, age 37, Parnassus, Augusta Co. 1860 census. Enl. Staunton 7/31/61. Ab. sick 11/18/61-12/31/61. Present 1-3/62. AWOL 4/26-30/62. Reenlisted 5/1/62. AWOL 6/6-12/21/62. Present in arrest 1-2/63. Discharged for over age 5/21/63. Des. to the enemy New Creek, W. Va. 11/11/64. Sent to Wheeling, W. Va. Arrested as a spy there 12/26/64. Sent to Camp Chase. Age 47, 5' 9½", fair complexion, grey eyes, light hair. Released 6/26/65. Farmhand, age 49, Mt. Sidney PO, Augusta Co. 1870 census.

SHIFFLETT, GIVENS: Pvt., Co. C. B. Albemarle Co. 1824? Farmer, age 36, 1st Dist., Augusta Co. 1860 census. Enl. Staunton 7/16/61. NFR on muster rolls. Capt. Furguson's Co. L, 93rd Va. Militia 3/62 exempt for rupture. Farmer, age 50, Pastures Dist., Augusta Co. 1870 census. D. near Craigsville 1/25/94 in 74th year. Bur. in neighborhood burying ground near his residence.

SHIFFLETT, PEMBERTON C.: Pvt., Co. B (2nd). B. Rockingham Co. 1837? Laborer, age 23, Waynesboro, Augusta Co. 1860 census. Enl. Waynesboro 7/15/61. Captured Alleghany Mt. 12/13/61. Sent to Camp Chase. Transf. to Johnson's Island. D. of disease on trip to Vicksburg, Miss. to be exchange 12/7/62. Age 23, 5' 10½", fair complexion, blue eyes, auburn hair.

SHIFFLETT, SAMUEL B.: Pvt., Co. F. B. Va. 1840? Not on MR. Paroled Staunton 5/22/65. Farmer, age 30, Mt. Sidney PO, augusta Co. 1870 census.

SHIPP, JAMES HENRY: Pvt., Co. C. B. Va. 1835? Blacksmith, age 25, North Mtn., Augusta Co. 1860 census. Enl. Staunton 7/16/61. Present 7/16/61-4/62. Detailed as nurse in hosp. Camp Alleghany 1-2/62. Present 5/1-7/20/62. AWOL 7/21-8/6/62, fined $5.50. Des. 11/23/62. NFR. Farmer, Augusta Co. Died Riverheads Dist., 7/3/84.

SHIREY, JACOB H.: Pvt., Co. G. B. Va. 10/6/22. Farmhand, age 33, Burkes Mill Dist., Augusta Co. 1860 census. Enl. Staunton 8/2/61. Present 11/61-4/62. Reenl. 5/1/62, non-conscript over 35. Present 5/1/62-12/63 and 4/30-12/31/64. Surrendered Appomattox 4/9/65. Obit. states: "Head of Ambulance Corps, a man of remarkable strength and exercised it conspicuously in the discharge of his hospital duties." Farmer, age 46, Mt. Sidney PO, Augusta Co. 1870 census. Farmer and Stockraiser, Waynesboro 1885. Died Annex 5/20/09. Bur. St. James Lutheran Ch. Cem.

SHIREY, WILLIAM D.: Pvt., Co. G. B. Va. 1828? Farmer, age 32, 1st Dist., Augusta Co. 1860 census. Enl. Staunton 8/2/61. Present 11/61-4/62. Reenl. 5/1/62. WIA (leg, foot and both hands) Cross Keys 6/8/62. Present on extra duty Ambulance Driver 8/2/62-12/63. WIA (jaw) Bethesda Ch. 5/30/64. Cap. Fisher's Hill 9/22/64. Sent to Point Lookout. Released 6/9/65. Farmer, age 40, 6th Dist., Augusta Co. 1870 census.

SHOMO, JACOB L.: Pvt., Co. B. (2nd). B. Augusta Co. 1843? Farmer, Augusta Co. Enl. Waynesboro 7/15/61. Present 11/61-4/62. Reenl. 5/1/62. Present 5/1-7/16/62. AWOL 7/17-9/24/62. Present 9/25-12/5/62. AWOL 12/6/62-1/19/63. Present in arrest in Regt'l Guard House 1-2/63. Present 3/1-5/20/63. AWOL 5/21-31/63 and 7/27-12/31/63. KIA Bethesda Ch. 5/30/64. Left widow.

SHOMO, JAMES CRAWFORD: Pvt., Co. B (2nd). B. New Market, Shenandoah Co. 10/6/29. School Teacher. Enl. Waynesboro 7/15/61. Present 11/61-4/62, AWOL 9 days, fined $11.00. WIA Gaines Mill 6/27/62. Ab. wounded through 2/63. Present 3-4/63. AWOL 4 months, 22 days, fined one months pay in addition to time ab. by Regt'l CM. AWOL 5/21-10/1/63. Present in arrest Regt'l Guard House 11-12/63. Ab. undergoing sentence of CM 4/30-10/31/64. AWOL 11-12/64. paroled Staunton 5/15/65. Age 28, 5' 11", dark complexion, dark hair, dark eyes. Farmer, age 32, Fishersville PO, Augusta Co. 1870 census. Died Fishersville 6/7/15. Bur. Fishersville Cem.

SHORT, WILLIAM "Bill": Pvt., Co. E. Mentioned as being present 10/26 and 11/12/61 in correspondence. NFR.

SHOVER, DAVID: Pvt., Co. I. B. Rockingham Co. 2/28/21. Farmer, age 38, 1st Dist., Augusta Co. 1860 census. Enl.Staunton 7/16/61. Present 11/61-4/62. Reenl. 5/1/62. Ab. detailed as Wagoner, Staunton 7/30/61 not with Regt. since 11/1/62 on rolls 8/30/62-12/31/64. NFR. Farmer, age 49, Fishersville PO, Augusta Co. 1870 census. Died near Arbor Hill 2/13/93. Bur. St. John's Ch. Cem., near Middlebrook.

SHOVER, FRANKLIN: Pvt., Co. B (1st). B. Augusta Co., 1833? Farmer, Rockbridge Co. 1860 census. Enl. Staunton 8/1/61. Ab. sick when transf. 9/28/61.

SHOWALTER, J. T.: Pvt., Co. A. On postwar roster. Enl. 3/62. Served 3 years and 2 months.

SHOWALTER, LEONARD: Pvt., Co. G. B. Va. 1841? Laborer, age 19, Burkes Mill Dist., Augusta Co. 1860 census. Enl. Staunton 8/2/61. Present 11-12/61, Wagoner. Present 1/1-6/5/62. AWOL 6/6-10/6/62. AWOL 11/24/62-2/63. Ab. sick and prisoner since 5/25/62, cap. Middletown. Supposed to be dead on rolls 9-10/63. Dropped as deserter 12/31/63. Resident of Staunton. NFR.

SHRECKHISE, DANIEL KEYSER: 3rdSgt., Co. G. B. Augusta 9/17/39. Farmhand age 19, Burkes Mill Dist., Augusta Co. 1860 census. Enl. Co. L, 5th Va. Inf. '61 (disbanded). Enl. Staunton 8/2/61. Present 11/61-4/62. Reenl. 5/1/62. WIA (breast) Cross Keys 6/8/62. Ab. wounded through 8/31/62. Furnished substitute and discharged 10/26/62. Age 22, 5' 9", fair complexion, fair hair, blue eyes. Farmer and Stockraiser, Mt. Sidney 1885.

SHRECKHISE, JAMES D.: 4thSgt., Co. A. B. Augusta Co. 1839? Enl. Staunton 7/12/61 as Pvt. Present 12/11/61, promoted Cpl. 8/7/61. Present 1-4/62. Reenl. 5/1/62. Present 8/31/62-7/63. Ab. sick 8/7-10/63, "Feb. Cotd." Present 11-12/63 and 4/30-12/31/64, promoted 4thSgt. 5/1/64. Surrendered Appomattox 4/9/65. Alive 1903.

SHUE, ABRAHAM WOLFE: Pvt., Co. F. B. Augusta Co. 4/22/33. Blacksmith, age 26, Northern Dist., Augusta Co. 1860 census. Enl. Richmond 8/6/64. Present 8-12/64. Paroled Staunton 5/1/65. Age 34, 5' 5", fair complexion, light hair, blue eyes. Blacksmith, age 37, Mt. Sidney PO, Augusta Co. 1870 census. Resident of Fishersville 1900. Died Augusta Co. 8/5/13. Bur. St. James Lutheran Ch. Cem.

SHUE, JACOB T.: Pvt., Co. D. B. Rockingham Co. 1833? Blacksmith, age 27, Northern Dist., Augusta Co. 1860 census. Enl. Staunton 7/16/61. Present 7/16-10/31/61. Ab. sick 11-12/61. Present 1-2/62. Ab. sick 1-4/62. Discharged for hernia 7/20/62. Age 31, 5' 7", fair complexion, light eyes, light hair. Blacksmith, age 39, Mt. Sidney PO, Augusta Co. 1870 census.

SHULL, DAVID F.: Pvt., Co. D. B. Augusta Co. 1841? Student, age 17, Burkes Mill Dist., Augusta Co. 1860 census. Enl. Staunton 7/16/61. Present 7/16-10/31/61. Ab. sick 11-12/61. Present 1-2/62. Ab. 4/1-5/1/62. WIA Port Republic 6/9/62. Present 6/15-8/31/62, detailed as Wagoner. AWOL 11/24-12/25/62. Present in arrest 1-2/63. Ab. in arrest Castle Godwin, Richmond sentenced by CM to 6 months hard labor with ball and chain for AWOL. Pardoned by the President 8/1/63. Present 8-12/63. WIA (right wrist) Wilderness 5/6/64. KIA Bethesda Ch. 5/30/64. Believed bur. in Emanuel Ch. Cem. in unmarked grave. Brother of Jacob H. Shull.

SHULL, GEORGE, JR.: Pvt., Co. D. B. Augusta Co. 1830? Laborer, age 33, Burkes Mill Dist., Augusta Co. 1860 census. Enl. Staunton 7/16/61. Present 7/16/61-4/62, and 4/30/62-12/31/64, detailed as Wagoner in Div. Train since 7/28/62. NFR. Farmhand, age 40, Mt. Sidney PO, Augusta Co. 1870 census. Believed bur. in Emanuel Ch. Cem in unmarked grave.

SHULL, HENRY C.: Pvt., Co. D. B. Augusta Co. 1838? Farmhand, age 21, Northern Dist., Augusta Co. 1860 census. Enl. Staunton 7/16/61. Present 7/16/61-1/62. AWOL 1/6-2/4/62, fined $11.00 by CM. Present 2/5-4/62. Present 5/1-8/31/62, detailed as Wagoner 7/20/62. Deserted 3/16/63. Enl. Co.B, 23rd Va. Cav. Farmhand, age 29, Mt. Sidney PO, Augusta Co. 1870 census, illiterate. Resident of Valley Mills 1903. Died Jennings Gap, Augusta Co. 10/3/10. Believed bur. Emanuel Ch. Cm. in unmarked grave.

SHULL, JACOB HENRY: Pvt., Co. D. B. Augusta Co. 1841? Farmhand, age 19, Burkes Mill Dist., Augusta Co. 1860 census. Enl. Staunton 7/16/61. Present 7/16/61-2/25/62. AWOL 2/26-28/62. Present 3-4/62. Present 5/1-8/31/62, detailed as Wagoner 6/6/62. AWOL 11/14-12/25/62. Present in arrest 1-2/63. Ab. in arrest Castle Godwin, Richmond. Sentenced by CM to 6 months hard labor with ball and chain and forfeiture of pay. Pardoned by the President 8/1/63. Present 9-12/63. Cap. in hospital Spotsylvania 5/64. Ab. on parole 6-12/64. NFR. Resident of Augusta Co. 1926. Believed bur. Emanuel Ch. Cem. in unmarked grave. Brother of David F. Shull.

SHULL, SAMUEL: Pvt., Co. D. B.Augusta Co. 1837? Enl. Valley Mill 5/1/62. Present 5/1-8/31/62, detailed as Wagoner 5/27/62. Present in arrest 1-2/63. Present 3-12/63. DOWs received Spotsylvania CH 5/12/64. Illiterate. Believed bur. in Emanuel Ch. Cem. in unmarked grave.

SHULL, WILLIAM: Pvt., Co. D. B. Augusta Co.1839. Farmhand, age 22, Northern Dist., Augusta Co. 1860 census. Enl. Staunton 7/16/61. Present 7/16/61-8/31/62, detailed as Wagoner 6/5/62. AWOL 11/23-12/25/62, fined $12.10. Present 1-12/63 and 4/30-12/31/64. NFR. Farmhand, age 29, Mt. Sidney PO, Augusta Co. 1870 census. Died Mt. Solon 7/6/05. Bur. Mt. Olivet Ch. Cem.

SHULTZ, MARTIN: 2ndSgt., Co. B (2nd). B. Adair Co., Ky. 1831. Clerk, age 29, Burkes Mill Dist., Augusta Co. 1860 census. Enl. Waynesboro 7/15/61. Present 11/61-2/62. Ab. on leave 3-4/62. Reenl. 5/1/62. Present 5-9/62, reduced to Pvt. 7/15/62. WIA Sharpsburg 9/17/62. Ab. wounded through 4/63. Detailed hopsital duty 6/1/63-1/31/64. Discharged at camp near Sommerville Ford for chronic laryngitis and "incipient phthiasis pulmonalis of the right lung," 2/28/64. Age 31, 5' 9¾", fair complexion, blue eyes, black hair. Farmer, Greenwood Depot, Albemarle Co. 1897 and 1900.

SHUMATE, THOMAS SUMNER: Pvt., Co. D. B. Albemarle Co. 12/9/21. Pension application only record. Brick Contractor, Augusta Co. Died Augusta Co. 9/26/04. Bur. Calvary Meth. Ch. Cem. Stuart's Draft.

SILLINGS, ARMSTRONG R.: Pvt.,Co. B (2nd). B. Augusta Co. 1839? Painter, age 21, 1st Dist., Augusta Co. 1860 census. Enl. Waynesboro 7/15/61. Captured Alleghany Mtn. 12/13/61. Sent to Camp Chase. Transf. to Johnson's Island. Exchanged 12/8/62. Reported for duty 12/25/62. Ab. on leave 1-2/63. Present 3-4/63. AWOL 7/19/63. Des. to the enemy in W. Va. 11/63. Took oath and sent north. Age 26, 6', lt. complexion, blue eyes, lt. hair, farmer.

SILLINGS, EPHRIAM W.: Pvt., Co. B (2nd). B. Va. 1834? Farmhand age 26, Northern Dist., Augusta Co. 1860 census. Enl. Waynesboro 7/15/61. Present 11/61-4/62. Reenlisted 5/1/62. AWOL 5/17-7/4/62 and 7/18-10/9/62. AWOL 11/26/62 until des. to the enemy Clarksburg, W. Va. 10/25/63. Took oath and sent north. Age 20, 6', dk. hair, dk. eyes, farmer, resident of Sherando, Augusta Co.

SILVEY, JAMES M.: Pvt., Co. E. B. 1839? May have served in 1st Rockbridge Artillery previously. Enl. at Camp of 52nd Va. Regt. 3/3/64. WIA (foot) Spotsylvania CH 5/12/64. Ab. wounded through 12/31/64. Des. to the Army of the Potomac near Petersburg 3/8/65. Sent to Wash., D. C. Took oath and transportation furnished to Green Co., Ohio. Laborer, age 31, Natural Bridge Dist., Rockbridge Co. 1870 census. Age 75, Millboro Dist., Bath Co. 1910 census. Illiterate. D. Rockbridge Co. 5/19/16.

SIMPSON, CHARLES P.: Pvt., Co. E. Resident, age 14, Natural Bridge Dist., Rockbridge Co. 1860 census. Enl. Camp of 52nd Va. Regt. 4/3/64. WIA (face) Wilderness 5/6/64. Ab. wounded through 10/13/64. AWOL 10/14-12/31/64. NFR. Living Lynchburg 1918. Brother of William D. Simpson.

SIMPSON, JOHN J.: Pvt., Co. E. B. Fluvanna Co. 1838? Factory Laborer, age 22, Lexington Dist., Rockbridge Co. 1860 census. Enl. Staunton 8/1/61. Present 11/61-3/62. Died of typhoid fever in hospital Staunton 4/9/62. Bur. Thornrose Cem., grave 926. Left widow and 2 children.

SIMPSON, MICHAEL: Pvt., Co. K. On postwar roster. Also served Co. G, 25th Va. Inf.

SIMPSON, ROBERT MOFFETT: Pvt.,Co. A. B. Loudon Co. 11/26/39. Apprentice Blacksmith, age 20, Northern Dist., Augusta Co. 1860 census. Enl. Staunton 8/9/61 age 22. Present 11/61-4/62. Reenl. 5/1/62. Cap. Front Royal 5/30/62. Exchanged 9/21/62. Present 10-12/62. Ab. on detached service as Blacksmith 1/21/63-12/31/64. Surrendered Appomattox 4/9/65. Blacksmith, age 30, Mt. Sidney PO, Augusta Co. 1870 census. Died Augusta Co. 11/26/09. Bur. New Hope Meth. Ch. Cem.

SIMPSON, WILLIAM DAVID: Pvt., Co.E. B. Va. 1840? Carpenter, age 20, Glenwood Dist., Rockbridge Co. 1860 census. Enl. Staunton 8/1/61. Ab. sick 11/61-2/62. Present 3-4/62. Reenl. 5/1/62. Cap. Strasburg 6/1/62. Exchanged 6/26/62. Ab. on leave 10/31/72-2/63. Present 3-7/63. AWOL 7/26-11/29/63, fined 1 months pay in addition to time absent and clothing allowance by Regt'l CM. WIA (left elbow shattered above the joint) Bethesda Ch. 5/30/64. Ab. wounded until retired to Invalid Corps 10/14/64. Resident, Montreal, Nelson Co. 1888. Died Shipman, Nelson Co. 3/16/13. Brother of Charles P. Simpson.

SIMS, M. B.: Pvt., Co. Unknown. Not on muster rolls. Deserted to the enemy 2/2/65. NFR.

SIPES, HENRY: Pvt., Co. G. B. 1838? Farmhand, age 22, Burkes Mill Dist., Augusta Co. 1860 census. Enl. Staunton 8/2/61. Present 11/61-4/62. Reenl. 5/1/62. Present 5/1-8/31/62. KIA Fredericksburg 5/5/63.

SIPES, SAMUEL: Pvt., Co. G. B. Augusta Co. Enl. Staunton 8/2/61. Present 4/30-10/31/62. "AWOL one month $11.00." Present 11/62-4/30/63. 63. KIA Fredericksburg 5/5/63.

SIRON, SIMON M.: Pvt., Co. E. B. Rockbridge Co. 1841? Enl. Staunton 8/1/61. Ab. on leave 11-12/61. Present 1-4/62. Reenl. 5/1/62. Present 5/11-31/62. Ab. sick 11/20/62-12/31/64, ulcer on foot, detailed in hospital Richmond. Illiterate. NFR. Farmhand, age 29, Buffalo Dist., Rockbridge Co. 1870 census. Farmer, age 79, Buffalo Dist., Rockbridge Co. 1910 census. Died on Short Hill near Hamilton's School house 12/25/11.

SITLINGTON, JOHN MARSHALL: Pvt., Co. K. B. Bath Co. 8/11/22. On postwar roster. Farmer, age 48, Bath CH. Bath Co. 1870 census. Died Bath Alum Springs 4/21/95. Bur. Windy Cove Pres. Ch. Cem., Millboro Springs.

SKINNER, JAMES H.: Col., F&S, B. Norfolk 1/18/26. UVa. '42-46. Lawyer, Staunton. Member Va. Legislature. Capt., Co. B, and Col. 160th Va. Militia. Enl. Staunton 7/9/61 age 35 as Capt., Co. A. Present 11/61-4/62. Elected LtCol. 5/1/62.Commanded left wing of Regt. McDowell; commanded Regt. Cross Keys, Port Republic Winchester, Gaines Mill, Malvern Hill, Cedar Run, Bristoe Station and until WIA (head) by shell fragment 2nd Manassas 8/28/62. Ab. wounded through 10/62. Commanded Regt. Fredericksburg 12/13/62. Ab. 2/63. Present 4-5/63, commanding Regt. Fredericksburg 5/5/63. Commanding Regt. Gettysbug until WIA (blinded for several months by dirt and gravel in eyes) 7/3/63. Promoted Col. 10/24/63. Commanding Regt. Mine Run. Present 12/63-1/64. Commanding Regt. Wilderness and Spotsylvania CH until WIA (in temple by minie ball, ball entered just below the left temple passed through the left eye amd made its exit on right side of nose just below the left corner of the right eye) 5/12/64. Regained sight in right eye. Commanded Post, Staunton 12/64-4/65. Retired for disability 3/4/65. Court of Inquiry found in 1863: At Bristoe Station 8/27/62 — "received an order which under fire of the enemy to withdraw his Regt. to the rear which he did by marching them by the left flank through a thicket of pines in as good order as the nature of the ground would admit of-he leading the way and indicating to the command the direction in which it should march. That at the battle of Fredericksburg . . .he failed in no particular to perform fully his duties as Commander fo the Regt. remaining with his Command and giving all the necessary orders and commands. That at the battles of Gaines Mills, Cedar Mountain, and Fredericksburg being in command of the 52 Va. Regt. did perform his duties in every respect as became an officer in his command giving all necessary orders and commands and evincing coolness and gallantry." Signed "BGen. Harry T. Hays President of the Court" and approved by Gen. R. E. Lee. Lawyer, Staunton; Colonel of Va. Militia 1871; President of the Augusta Memorial Association to improve and maintain the Confederate Section of Thornrose Cem. 1880; Member, Stonewall Jackson Camp CV, Staunton; Commander of Camp 1884. Died "Sailor's Rest", Staunton 5/19/98. Bur. Cedar Grove Cem., Norfolk. "He is a gentleman, in the highest sense of the term; courteous, intelligent and dignified. His moral qualities are exceptional. His habits are excellent. He is a lawyer by education . . .and practiced his profession with success for 14 years-and gifted with fine talents, and studious in his habits, he has acquired reputation and the confidence of the community (Staunton)."

SLOAN, JOHN COOKE: Pvt., Co. B. (1st). B. Lexington, Rockbridge Co. 1841. Attended Washington Col. '59-60. Overseer, age 22, Lexington Dist., Rockbridge Co. 1860 census. Enl. Fairfield 7/10/61. Present until transf. 9/28/61.

SLUSSER, SAMUEL H.: Pvt., Co. C. B. 12/44. Resident of Cross Keys. Enl. Bunker Hill 10/10/62. Deserted near Madison CH 11/2/62. NFR. Died 1/11/16. Bur. Friedan's Lutheran Ch. Cem., near Mt. Crawford, Rockingham Co.

SLY, ADOLPHUS: Pvt., Co. B (1st). On postwar roster.

SLY, ALFRED F.: Pvt., Co. B (1st). B. Va. 1841? Farmhand, age 19, 1st Dist., Augusta Co. 1860 census. Enl. Fairfield 7/10/61. Ab. on detail when transf. 9/28/61.

SMELTZ, FRANKLIN: Pvt., Co. C. B. Va. 1810? Ironworker, age 50, Waynesboro, Augusta Co. 1860 census. Enl. Staunton 7/16/61 age 39. AWOL 8/1/61-4/27/62, fined pay for time ab. Deserted 5/4/62. Joined from desertion 12/20/63. Ab. confined Castle Thunder, Richmond by sentence of GCM 2/1/64 for 12 months and deducted through 10/31/64. Present 11-12/64. Detailed at Farrars and Cosbys Iron Works 1/7/65. NFR.

SMILEY, JOHN B.: Pvt., Co. B (1st). B. Rockbridge Co. 1839. Enl. Fairfield 7/16/61. Ab. sick when transf. 9/28/61.

SMITH, ALEXANDER: Pvt., Co. F. Not on muster rolls. Unofficial roster says died of disease 9/61.

SMITH, ALFRED: Pvt., Co. F. B. Va. 1834? Farmhand, age 26, 1st Dist., Augusta Co. 1860 census. Enl. Staunton 7/16/61. AWOL 11-12/61. Died in hospital Staunton 1/7/62, cause unknown. Bur. Thornrose Cem. grave #877.

SMITH, ANANIAS: Pvt., Co. E. B. Rockbridge Co. 1/3/45. Enl. Staunton 8/1/61. Ab. on leave 11-12/61. Present 1-2/62. Ab. sick 3-4/62. WIA (face) Cross Keys 5/8/62. Ab. wounded until discharged for under age 11/1/62. Age 17, 5' 7", fair complexion, blue eyes, light hair. Enl. Co. G, 58th Va. Inf. Farmer and Postmaster, Denmark, Rockbridge Co. Died near there 4/7/97. Bur. New Monmouth Pres. Ch. Cem.

SMITH, BENJAMIN FRANKLIN: 4thSgt., Co. B (2nd). B. Louisa Co. 2/4/43. Student, age 17, Burkes Mill Dist., Augusta Co. 1860 census. Enl. Waynesboro 7/15/61. Present 11/61-2/62. Furnished substitute and discharged 3/3/62. Age 18, 5' 8½", dark complexion, blue eyes, light hair. Enl. Co. F, 1st Va. Cav. Spring '63 and served to end of war. Moved to Texas and Mo., returned 1871. Farmer and Merchant near Waynesboro 1885. Died Fishersville 12/29/95. Bur. Riverview Cem., Waynesboro.

SMITH, GREENBERRY R.: Pvt., Co. G. B. Va. 2/27/30. Farmhand, age 30, 1st. Dist., Augusta Co. 1860 census. Enl. Staunton 9/15/62. Present 10-12/62 and 1-4/63, on extra duty as Wagoner since 12/12/62. Present 4/30-12/63. WIA (leg) near Spotsylvania CH 5/18/64. Age 34, occupation Wagoner. Ab. wounded through 12/31/64. NFR. Farmer, age 40, Mt. Sidney PO, Augusta Co. 1870 census. Farmer, Churchville 1897. Died Pastures Dist., Augusta Co. 2/11/17. Bur. Green Hill Cem., near Churchville.

SMITH, HARVEY "Harry": Pvt.,Co. G. B. Augusta co. 1827? Farmhand, age 33, 1st Dist., Augusta Co. 1860 census. Pvt., Capt. Hall's Co., 32nd Va. Militia 3/62 age 35. Enl. Staunton 5/4/64. KIA Bethesda Ch. 5/30/64. Left widow and 3 children.

SMITH, JAMES: Pvt., Co. K. B. Va. 1823? Shepherdstown 9/23/62. Ab. sick 2/16-4/63. WIA (knee, side and shoulder) Gettysburg 7/3/63. Ab. wounded until detailed for hospital duty 10/5/63-10/31/64. Present 1-12/64. NFR. Farmhand, age 47, Bath CH PO, Bath Co. 1870 census. Living Yost, Bath Co. 1901.

SMITH, JAMES E.: 5thSgt., Co. B (2nd). B. Louisa Co. 1842? Blacksmith's Apprentice, age 18, Burkes Mill Dist., Augusta Co. 1860 census. Enl. Waynesboro 7/15/61 as Pvt. Present 11/61-4/62. Reenl. 5/1/62. Present 5/1-10/31/62, promoted 2nd Cpl. 7/15/62 and 1stCpl. 10/2/62. Present 11/62-1/63, promoted 5thSgt. 1/1/63. Ab. on leave 2/63. Present 3-12/63, and 4/30-12/31/64. Surrendered Appomattox 4/9/65. Conductor, C&O Railroad, Clifton Forge. Member, Stonewall Jackson Camp CV, Staunton. Died 4/9/16.

SMITH, JAMES M.: Pvt., Co. K. B. Augusta Co. 1822? Blacksmith, age 38, Cleeks Mill PO, Bath Co. 1860 census. Pvt., Capt. Davis's Co., 81st Va. Militia. Enl. Shenandoah Mt. 4/9/62. WIA (back) Cross Keys 6/8/62. Ab. sick 6/19/62 until died of typhoid fever in hospital Staunton 7/1/62. Bur. Thornrose Cem., grave #1084, age 35.

SMITH, JAMES W. (1st): Pvt., Co. G. B. Va. 1842? Farmhand, age 18, Burkes Mill Dist., Augusta Co. 1860 census. Enl. Staunton 9/10/62. Ab. sick 11/1/62-4/63. Present 4/30-12/63. Cap. Bethesda Ch. 5/30/64. Sent to Point Lookout. Died of disease there 7/13/64. Bur. POW Cem. there.

SMITH, JAMES W. (2nd): Pvt., Co. E. B. Va. 10/8/16. Farmer, age 44, 7th Dist., Rockbridge Co. 1860 census. Served in militia. Enl. Staunton 8/1/61. Present 11/61-4/62. Reenl. 5/1/62. Present 5/1-10/31/62. Present in arrest 1-2/63, AWOL 38 days and fined pay for that period. Present 3-6/63. WIA Winchester 6/13/63. Deserted to the enemy 7/21/63. Took oath. Residence near Clark's Distillery, Rockbridge Co. Farmhand, age 54, Buffalo Dist., Rockbridge Co. 1870 census. Died Rockbridge Co. 1/16/04. Bur. Mt. Carmel Pres. Ch. Cem., Steele's Tavern, Augusta Co. Brother of John S. Smith.

SMITH, JAMES W. (3rd): Pvt., Co. E. B. Va. 1835? Laborer, age 25, Lexington, 1860 census. Not on muster rolls. KIA Winchester 5/25/62.

SMITH, JOHN: Pvt., Co. H. B. Va. 1827? Farmhand, age 33, Burkes Mill Dist., Augusta Co. 1860 census. Enl. Staunton 7/23/61. Present 1/62. AWOL 2/9-24/62. Present 3-4/62. Reenl. 5/1/62. Present 5/1-8/1/62 and 12/63. Present in arrest 1/9-2/63, fined all pay to 1/8/63 by CM. Present 4/30-12/63. Issued clothing 4/27/64. WIA (head) Bethesda Ch. 5/30/64. Issued clothing 6/12/64. WIA and cap. Winchester 9/19/64. DOWs.

SMITH, JOHN A.: Pvt., Co. E. B. Va. 1842? Farmhand, age 18, 7th Dist., Rockbridge Co. 1860 census. Enl. Bunker Hill 9/22/62. Present 9-10/62. Ab. sick 2/2-4/30/63, typhoid fever. Present 5-12/63. Issued clothing 4/21 and 6/12/64. WIA (fractured right temporal and panetal bones) and cap. Carter's Farm, near Winchester 7/20/64. DOWs from "Softening of the brain extravacation of blood" in hospital Cumberland, Md. 8/1/64. Resident of Rapp's Mill, Rockbridge Co. Bur. Confederate Lot, Rose Hill Cem., Claysville, Md.

SMITH, JOHN A. (2nd): Pvt., Co. K. B. Rockbridge Co. 3/18/29. Mexican War Veteran. Farmhand, age 32, Green Valley Dist., Bath Co. 1860 census. Enl. Shenandoah Mt. 4/9/62. Present 4/9-30/62, and 1-9/63. Ab. sick 9/19-12/29/63. WIA Winchester 9/19/64. Ab. wounded through 12/31/64. NFR. Farmhand, age 46, Bath CH PO, Bath Co. 1870 census. Cooper, Augusta Co. 1879. D. Rockbridge Co. 12/10/07. Bur. Zollman Cem. Last Mexican War Veteran in Co.

SMITH, JOHN M.: Pvt., Co. K. B. Va. 3/10/28. Laborer, age 32, Green Valley PO, Bath Co. 1860 census. Not on MR. Enl. Shenandoah Mt. 4/9/62. D. Botetourt Co. 6/07. Bur. Galatia Pres. Ch. Cem. near Gala, Botetourt Co.

SMITH, JOHN S.: Pvt., Co. B. B. Va. 1838? Shoemaker, age 22, Burkes Mill Dist., Augusta Co. 1860 census. Enl. Staunton 9/17/62. Ab. sick 11/1/62-8/31/63, and 10-9/31/63. Ab. sick 11/28/63-10/14/64, chronic diarrhea and chronic rheumatism. Present 11-12/64. Paroled Staunton 5/8/65. Age 27, 5' 9", dark complexion, hazel eyes, dark hair. Shoemaker, age 32, Mt. Sidney PO, Augusta Co. 1870 census. Brother of James W. Smith (2nd).

SMITH, JOSEPH: Pvt., Co. E. B. Va. 1838? Farmhand, age 22, Collierstown PO, Rockbridge Co. 1860 census. Enl. Staunton 8/1/61. Present 11-12/61. Ab. sick 1-4/62. Reenl. 5/1/62. KIA Cedar Run 8/9/62.

SMITH, JOSEPH W. (1st): Pvt., Co. G. B. Augusta Co. 1835? Farmer, age 25, Burkes Mill Dist., Augusta Co. 1860 census. Enl. Staunton 8/1/61. Present 11/61-4/62. Ab. on leave 4/30-8/31/62. KIA Sharpsburg 9/17/62. Age 28, 5' 9", fair complexion, blue eyes, auburn hair, left widow and 1 child.

SMITH, JOSEPH W. (2nd): Pvt., Co. G. On postwar roster. "Recruit." Blacksmith, Waynesboro, 1872.

SMITH, PERRY: Pvt., Co. G. Enl. Staunton 9/17/62. NFR.

SMITH, PETER: Pvt., Co. C. B. Va. 1/10/32. Farmhand, age 36, Burkes Mill Dist., Augusta Co. 1860 census. Enl. Staunton 7/16/61 age 42. Ab. sick 12/15/61-2/62. Present 3-4/62. WIA (arm) Port Republic 6/9/62. Ab. wounded through 8/31/62. Paid Staunton 11/15/62. NFR. Farmhand, age 52, Mt. Sidney PO, Augusta Co. 1870 census.

SMITH, POLK: Pvt., Co. G. On postwar roster.

SMITH, ROBERT P.: Pvt., Co. A. B. Rockbridge Co. 1835? Farmhand, age 25, 1st Dist., Augusta Co. 1860 census. Enl. Staunton 7/15/61. Died of typhoid fever Camp Alleghany 12/8/61. Left widow and 3 children.

SMITH, SAMUEL: Pvt. Co. F. B. 1831? Resident of Harrisonburg. Enl. Valley Mills 4/20/62. Cap. and paroled Winchester 6/1/62. Age 31, 5' 10", AWOL 6/6/62-12/63. Dropped as deserter. NFR.

SMITH, SOLOMON L.: Pvt., Co. C. B. Va. 1825? Bricklayer, age 35, Staunton, 1860 census. Pvt., Co. E. 93rd Va. Militia. Enl. Staunton 7/16/61 Age 38, Present 11/61-1/62. AWOL 2/8-28/62. Ab. sick 3/30-4/62, fined $23.46 for AWOL WIA 2nd Manassas 8/28/62. Ab. wounded through 10/63. Deserted to the enemy Clarksburg, W. Va. 10/13/63. Took oath and sent north. However, he was carried ab. wounded through 12/63 and issued clothing 4/21 and 28/64. AWOL 5/20-12/31/64. Dropped as deserter. NFR.

SMITH, THOMAS: Pvt., Co. E. B. Rockbridge Co. 1837? Farmer, age 23, Lexington Dist., Rockbridge Co. 1860 census. Not on muster rolls. Died of fever in hospital Staunton 10/10/61, age 23. Bur. Thornrose Cem., grave 781.

SMITH, THOMAS J.: Pvt., Co. C. B. Va. 1825? Farmhand, age 35, Northern Dist., Augusta Co. 1860 census. Enl. Staunton 7/16/61. age 35. Present 11/61-2/62, fined $11.00 for 9 days AWOL. Present 3-8/62. AWOL 11/27/62-2/63. Ab. 3-4/63, fined $47.66 for 130 days AWOL. Present 4/30-12/63. KIA Cold Harbor 6/3/64. Left widow and 5 children.

SMITH, WILLIAM H.: Pvt., Co. A. B. Frederick Co. 1820? Shoemaker, age 40, Staunton 1860 census. Not on muster rolls. Unofficial report indicates cap. date and place unknown. Sent to Camp Morton. Died of disease there. However, admitted to Confederate Home in Ardmore, Okla. 7/12/11. Died there 11/23/23.

SMITH, WILLIAM HENRY: Pvt., Co. H. B. Augusta Co. 6/24/40. Farmhand, age 18, near Greenville, Augusta Co. 1860 census. Enl. Staunton 7/23/61 age 20. Present 1/61-4/62. Reenl. 5/1/62. WIA Gaines Mill 6/27/62. Ab. wounded through 8/31/62. Present 1-6/63. WIA and cap. Gettysburg 7/3/63. Sent to Ft. McHenry. Transf. Point Lookout and Ft. Delaware. Took oath and joined US Service 1/25/64. Residence near Greenville. Farmhand, age 29, Fishersville PO, Augusta Co. 1870 census. Farmer, age 64, Cedar Creek Dist., Bath Co. 1910 census.

SMITH, WILLIAM R. (1st): Pvt., Co. F. B. Augusta Co. 3/20/32. Farmhand, age 25, Burkes Mill Dist., Augusta Co. 1860 census. Enl. Staunton 7/31/61. Present 11/61-4/62. Reenl. 5/1/62. Discharged for "phlhisis pulmonales" from hospital Staunton 10/7/62. Age 29, 6' 1", dark complexion, dark eyes, dark hair. Farmhand, age 37, Pastures Dist., Augusta Co. 1870 census. Died near Churchville 2/20/17. Bur. Green Hill Cem.

SMITH, WILLIAM R.: (2nd). Cpl., Co. H. B. Fayette Co., Ky. 1815? Tailor, Augusta Co. Enl. Staunton 7/23/61 age 45. Present 11/61-4/62, reduced to Pvt. 7/1/62. Discharged for chronic rheumatism 8/13/62. Age 47, 5' 10", dark complexion, black eyes, dark hair. Tailor, Staunton. Died there 3/18/81. Bur. Thornrose Cem.

SNAPP, CYRUS H.: 1stLt., Co. F. B. Augusta Co. 2/1/39. Farmer, age 24, Burkes Mill Dist., Augusta Co. 1860 census. Enl. Staunton 7/31/61 as 2ndLt. Ab. sick 10/15-12/61. Present 1-4/62. Elected 1stLt. 5/1/62. Submitted resignation 7/9/62 "being of a weak constitution and not being able to stand the fatigues of infantry service and therefore desire to change to someother service." Commanding Co. 7/31/62. Resignation accepted 8/11/62. Paroled Staunton 4/30/65. Ae 28, fair complexion, fair hair, blue eyes. Farmer, age 30 Mount Sidney PO, Augusta Co. 1870 census. Miller and Farmer, Swoope 1885. Died Augusta Co. 3/3/12. Bur. Augusta Stone Ch. Cem.

SNAPP, JOHN R.: Pvt.,Co. C. Enl. Valley Mills 5/1/62. WIA (leg) Port Republic 6/9/62. Arresting deserters for Conscript Officer, Augusta Co. 12/63. Ab. wounded through 6/64. Requested transf. to Cav. 4/2/64. Cap. Nelson Co. 6/11/64. Sent to Camp Chase. Age 20, 6', fair complexion, dark eyes, dark hair, farmer, resident of Augusta Co. Exchanged 3/12/65. Paroled Staunton 4/30/65. Age 20, 6', dark complexion, black hair, grey eyes. Farmer, age 23, Beverly Manor Dist., Augusta Co. 1870 census.

SNAPP, WILLIAM F.: Sgt., Co. K. B. Staunton, Augusta Co. 8/7/36. Farmer and Distiller, age 22, 1st Dist., Augusta Co. 1860 census. Enl. Valley Mills 4/24/62. Ab. sick 5/17-10/62, rank shown as Pvt. Discharged for "ephritis" from hospital Staunton 10/6/62. Age 28, 5' 10", fair complexion, grey eyes, dark hair. Died Staunton 7/23/01. Bur. Thornrose Cem.

SNEAD, ANTHONY MUSTOE: 4thCpl., Co. K. B. Bath Co. 7/11/39. Farmhand, age 20, Mountain Grove PO, Bath Co. 1860 census. Enl. Shenandoah Mt. 4/9/62. Present 4/9-30/62. Ab. sick 11/20/62-463. AWOL 5/1-12/63. Present 4/30-12/31/64, promoted 4thCpl, served as sharpshooter. Enl. Co. E, 46th Bn. Va. Cav. Farmer and Teacher, Bath Co. Died Alleghany Co. 7/7/11. Bur. Mt. Plesant Ch. Cem., 3 miles north of Covington. Brother of John and Samuel J. Snead.

SNEAD, JOHN: Pvt., Co. K. B. Bath Co. 1842? Farmhand, age 18, Mountain Grove PO, Bath Co. 1860 census. Enl. Shenandoah Mt. 4/9/62. Present 4/9-30/62. Died of "catarrhal fever" in hospital Winchester 10/15/62, age 19. Age 19, 6' 2", dark complexion, brown hair, grey eyes. Possessions: 1 blanket and 1 gum blanket. Bur. Stonewall Jackson Cem., Winchester as "J. Speed". Brother of Anthony M. and Samuel J. Snead.

SNEAD, SAMUEL JACKSON: Pvt., Co. K. B. Bath Co. 6/26/26. Farmhand, age 32, Mountain Grove PO, Bath Co. 1860 census. Pvt., Cap., Davis's Co., 81st Va. Militia. Enl. Shenandoah Mt. 4/9/62. Present 4/9-30/62. AWOL 9/22-10/15/63. Present 2/16-12/63. WIA (left hip) near Spotsylvania CH /19/64. Ab. wounded through 10/31/64. Present 11-12/64. Served as sharpshooter. NFR. Farmer, age 45, Bath CH PO, Bath Co. 1870 census. Died Healing Springs 1/24/96. Bur. in family Cem. "Bullet still in hip at death." Brother of Anthony M. and John Snead.

SNEAD, WILLIAM J.: Pvt., Co. K. B. Albemarle Co. 1833? On postwar roster. Enl. Co. F, 11th Va. Cav. Carpenter Albemarle Co. 1867. Died Rockbridge Co. 1922.

SNELL, DAVID FRANKLIN: Pvt., Co. F. B. Augusta Co. 1837? Farmer, Augusta Co. Enl. Staunton 7/31/61. Present 11/61-2/62, AWOL 2 days fined $4.32. Present 3-4/62. Reenl. 5/1/62. Present 5-1/11/23/62. AWOL 11/24-12/18/62. Present in arrest 1-2/63. WIA (leg) Gettysburg 7/3/63. Ab. wounded through 8/31/63. Present 9-12/63 and 3-4/64. WIA (right wrist and fractured small bone in arm) Spotsylvania CH 5/12/64. Ab. wounded through 10/31/64. Presence or absence not stated 11-12/64. In hospital Richmond 1/31/65 and returned to duty 2/1/65. Retired for disability to hand and arm 2/21/65. Age 28, 6' 0", dark complexion, grey eyes, dark hair, farmer. Paroled Staunton 5/10/66. Age 29, 6' 6", dark complexion, dark hair, blue eyes. Died near Bridgewater, Rockingham Co. 2/2/85. Bur. Beaver Creek Cem.

SNITEMAN, JOHN: Pvt., Co. G. B. Augusta Co. 1843? Farmhand, age 17, Burkes Mill Dist., Augusta Co. 1860 census. Enl. Staunton 8/2/61. Present 11/61-1/62. Died of typhoid pneumonia Camp Alleghany or Monterey 2/22/62. Age 18, 5' 11", fair complexion, fair hair, blue eyes. Effects: 2 blankets, 2 pr. pants, one jacket, one dress coat, and 1 pr. boots sent to his father.

SORRELS, GEORGE D.: Pvt., Co. B (1st) B. Rockbridge Co. 1/29/40. Enl. Fairfield 7/10/61. Present until transf. 9/28/61.

SORRELS, JOHN JOSEPH: Pvt., Co. B (1st). B. Nelon Co. 2/4/36. Farmhand, age 33, South River Dist., Rockbridge Co. 1860 census. Enl. Fairfield 7/10/61. Present until transf. 9/28/61.

SPILLMAN, JOHN: Pvt., Co. C. Joined from Co. B. 39th Bn. Va. Cav. in exchange for William S. Kerr 5/1/63. KIA Gettysburg 7/3/63.

SPITLER, HENRY: Pvt., Co. F. B. Va. 1830? Farmer, age 20, Dist No. 2, Augusta Co. 1850 census. On postwar roster. Enl. 7/61. Served to end of war. Died Augusta Co. 11/2/05. Bur. Green Hill Cem., near Churchville, age 75.

SPITZER, NOAH: Pvt., Co. H. B. Augusta Co. 2/27/27. Farmer, age 33, Burkes Mill Dist., Augusta Co. 1860 census. Pvt., Capt. Western's Co., 32nd Va. Militia 3/62, furnished substitute, age 35. Enl. Staunton 11/7/64. Present 11-12/64. WIA Hatcher's Run 2/8/65. Admitted to hospital Richmond 2/9/65 and transf. to hospital Staunton 2/13/65. Paroled Harrisonburg 5/2/65. Age 37, 5' 8", fair complexion, brown hair, hazel eyes. Farmer, age 43, Mt. Sidney PO, Augusta Co. 1870 census. Died Mt. Meridian 7/12/76. Bur. Pleasant Valley Ch. of the Brethern, near Weyer's Cave.

SPROUSE, BENJAMIN H.: Pvt., Co. A. Enl. New Market 10/22/64. Present 10-12/64. Paroled Charlottesville 5/19/65. Illiterate.

SPROUSE, DAVID: Pvt., Co. H. B. Va. 1822? Farmhand, age 38, Northern Dist., Augusta Co. 1860 census. Enl. Staunton 7/23/61 age 39. Present 11/61-1/62. AWOL 2/22-3/4/62, and 4/21-8/8/62. Present 8/9-31/62. AWOL 9/16/62-12/63. Dropped as deserter. Residence near Staunton. NFR. Farmer, Hot Springs, Bath Co. 1897.

SPROUSE, HENRY: Pvt., Co. A. Pension application only record. Living Ivy Depot, Albemarle Co. 1911.

SPROUSE, JOHN: Pvt., Co. H. B. 1818? Enl. Staunton 7/23/61 age 43. Present 11-12/61. AWOL 1/18-30/62 and 4/23-8/8/62. WIA (leg) 2nd Manassas 8/29/62. Ab. wounded through 4/22/63. AWOL 4/23-12/63. Dropped as deserter. NFR.

SPROUSE, JOHN C.: Pvt., Co. C. B.Va. 1841? Farmhand, age 19, Deerfield, Augusta Co. 1860 census. Enl. Staunton 7/16/61 age 22. Deserted 9/12/61. NFR. Laborer, age 66, Warm Springs Dist., Bath Co. 1910 census. Illiterate.

SPROUSE, MARTIN VAN BUREN: Pvt., Co. A. B. Albemarle 1839? Farmhand, age 21, 5th Dist., Rockbridge Co. 1860 census. Enl. Staunton 7/9/61 age 22. Present 11/61-4/62. Reenl. 5/1/62. WIA 2nd Manassas 8/29/62. Ab. wounded through 2/63. Present 3-11/63. WIA Mine Run 11/27/63. In hospital Richmond 12/13/63. Present 4/30-12/64. AWOL 12/8-31/64. Dropped as deserter. "To be found near Charlottesville" on deserter bulletin 2/4/65. Died Batesville, Albemarle Co. 1896.

SPROUSE, PETER: Pvt., Co. H. B. Albemarle Co. 1839? Farmhand, age 21, 1st Dist., Augusta Co. 1860 census. Enl. Staunton 7/23/61 age 22. Present 11/61-2/62, detailed as Wagoner. Ab. sick 3/27-4/62. Reenl. 5/1/62. AWOL 6/18/62-4/22/63. Sentenced to two years in penitentiary and forfeit all pay for desertion by CM. Present in arrest 4/23-8/31/63. Present 9-12/63. Deserted Orange CH 5/5/64. Cap. Piedmont 6/5/64. Resident of Staunton , 5' 6", florid complexion, dark hair, hazel eyes. Sent to Camp Morton. Released 5/22/65. Farmer, Nick, Albemarle Co. 1897. Resident of Terrell, Augusta Co. 1906, illiterate.

SPROUSE, PETER BENJAMIN: Pvt., Co.H. Pension application only record. Died Fordwick, Augusta Co. 5/18/09.

SPROUSE, WALKER H.: Pvt., Co. K. B. Va. 1843? Farmhand, age 17, Hot Springs PO, Bath Co. 1860 census. Enl. Shenandoah Mt. 4/9/62. Present 4/9-30/62, and 1-1/63. WIA (head) Spotsylvania CH 5/12/64. WIA Carter's Farm near Winchester 7/20/64. Ab. wounded through 12/31/64. Deserted to the Army of the Potomac near Petersburg 3/8/65. Sent to Washington, D. C. Took oath and transportation furnished to Green Co., Ohio. Died before 8/17.

SPROUSE, WILLIAM A.: Pvt., Co. K. B. Albemarle Co. 1822? Farmhand age 38, Hot Springs Dist., Bath Co. 1860 census. Pvt., Capt. Davis's Co. 81st Va. Militia. Enl. Co. G, 25th Va. Inf. 1861. Enl. Valley Mills 5/1/62. WIA (ankle) Cedar Run 8/9/62. Ab. wounded through 10/23/63. Present 10/24 until transf. Capt. George J. Davidson's Co., 23rd Va. Cav. 12/15/63. Farmhand, age 38, Bath CH PO, Bath Co. 1870 census. Farmer and Plasterer, Hot Springs 1904. Died before 8/17.

STACK, HENRY F.: Pvt. Co. G. On postwar roster.

STATON, BENJAMIN T.: Pvt., Co. K. B. Amherst Co. 1825? Enl. Shenandoah Mt. 4/18/62. Present 4/18-30/62. KIA Cross Keys 6/8/62 age 37. Left widow.

STAUNTON, ADDISON C.: Pvt., Co. I. Resident of Staunton. Enl. Staunton 7/16/61. Deserted 8/5/61. NFR.

STAUBUS, ALEXANDER FOLDEN: Pvt., Co. B. (2nd). B. Augusta Co. 5/10/34. Farmhand, age 27, Burkes Mill Dist., Augusta Co. 1860 census. Enl. West View 5/1/62. AWOL 5/23-6/23/62. Present 6/24-12/2, detailed as Teamster in Ordnance Train. Present on detail through 12/31/64. Paroled Staunton 5/1/6. Age 30, 5' 9", light complexion, light hair, blue eyes. Moved to Dry Grove, Illinois, 1870 and Rantoul, Illinois 1878. Farmer, Died there 1904. Brother of Augustus W. and Stephen Staubus.

STAUBUS, AUGUSTUS WILLIAM: Pvt., Co. D. B. Augusta Co. 7/22/21. Farmer, age 37, Burkes Mill Dist., Augusta Co. 1860 census. Enl. Staunton 7/16/61. Present 7/16-10/31/61. Ab. sick 11-12/61. Present 1-2/62, fined $11.00 by CM. Present 3-4/62. Detailed in hospital Weyer's Cave 6/9/62 on rolls through 8/31/62. NFR. on muster rolls, however, he was paid Staunton 12/5/62 and commutation from 8/4-11/30/63 and 12/5-19/63 for arresting conscripts for Conscript Officer, Staunton. Age 41, 5' 8", fair complexion, light hair, blue eyes. NFR. Farmer, age 50, Mt. Sidney PO, Augusta Co. 1870 census. Died Western State Hospital, Staunton 11/7/81. Bur. St. Paul's Lutheran Ch. Cem., near Mt. Solon. Brother of Alexander F. and Stephen Staubus.

STAUBUS, JOHN CHRISTIAN: Pvt., Co. D. B. Mt. Solon. Augusta Co. 2/2/43. Farmhand, age 17, Northern Dist., Augusta Co. 1860 census. Enl. Staunton 7/16/61. Present 7/16/61-4/62. Reenl. 5/1/62. Ab. detailed in hospital Gaines Mill 6/27-10/31/62. Present 1-12/63. Cap. in hospital Spotsylvania 5/64 and paroled. Ab. on parole through 12/31/64. NFR. Farmer, Augusta Co. Moved to Ohio and Ottawa, Kansas.

STAUBUS, STEPHEN: Pvt., Co. B (2nd). B. Augusta Co. 2/13/36. Enl. Waynesboro 7/15/61. Ab. sick 11-12/16. Present 1-4/62. Reenl. 5/1/62. Present 6/1-7/16/62. AWOL 7/17-10/11/62. Present 10/12-12/62, detailed as Teamster in Div. Commissary Train. Present on detail through 10/31/64. AWOL 11-12/64. NFR. Moved to Dry Grove Township, McLean Co., Ill. 1865. Farmer, near Bloomington, Ill. Died there 2/16/96. Brother of Alexander F. and Augustus W. Staubus.

STAUBUS, WILLIAM AUGUSTUS: Pvt., Co. D. B. Augusta Co. 3/16/44. On postwar roster. Died Augusta Co. 4/20/63, reason unknown. Bur. St. Paul's Lutheran Ch. Cem., near Mt. Solon.

STEELE, JAMES A.: Pvt., Co. H. B. Augusta Co. 1834? Enl. Staunton 7/23/61 age 22. Present 11/61-4/62. KIA Port Republic 6/9/62, age 26.

STEELE, JAMES E.: Pvt., Co. B (1st). B. Rockbridge Co. 2/7/38. Farmer, age 22, 1st Dist., Augusta Co. 1860 census. Enl. Midway 7/18/61. Present until transf. 9/28/61.

STEVENS, W. C., JR.: Pvt., Co. I. On postwar roster. Enl. '61. Served 4 years.

STICKLEY, THOMAS J.: Co. and rank unknown. B. Shenandoah Co. 11/8/21. Pension application only record. Blacksmith, New Hope, Augusta Co. Died there 1/28/93. Bur. Mountain View Cem., near New Hope.

STOGDALE, ELIAS P.: Pvt., Co. G. B. Augusta Co. 1839? Farmhand, age 21, 1st Dist., Augusta Co. 1860 census. Enl. Staunton 8/2/61. Present 11/61-4/62. Reenl. 5/1/62. Present 5/1-12/62. Ab. on leave 2/19-3/5/63. Present 3/6-12/63. WIA (side) near Spotsylvania CH 5/19/64. DOWs in hospital Lynchburg 6/7/64, age 25. Bur. Confederate Section, No. 8, Row 4, Lynchburg City Cem. Brother of Henry T. Stogdale.

STOGDALE, GEORGE WELLINGTON: Pvt., Co. A. B. Augusta Co. 4/30/42. Farmhand, age 18, Staunton 1860 census. Enl. Staunton 7/16/61. Present 11/61-4/62. Present 8/31-11/14/62. AWOL 11/15-12/31/62. Present 1-7/63. Ab. sick with scabies 7/24-8/31/63. Present 9-10/63. Ab. sick with typhoid fever 11/5-12/63. Issued clothing 4/29/64. Cap. Fisher's Hill 9/22/64. Sent to Point Lookout. Took oath and joined U. S. Service 10/8/64. Farmhand, age 28, Mt. Sidney PO, Augusta Co. 1870 census. Farmer, Harriston 1897. Died there 6/27/10. Bur. Harriston Meth. Ch. Cem.

STOGDALE, HENRY T.: Pvt., Co. G. B. Augusta Co. 1842? Farmhand, age 18, 1st Dist., Augusta Co. 1860 census. Enl. Staunton 8/61. Present 11/61-4/62. Renl. 5/1/62. WIA (face and shoulder) Gaines Mill 6/27/62. Ab. wounded through 4/63. Present 4/30-12/63. Cap. Bethesda Ch. 5/30/64. Sent to Point Lookout. Transf. Elmira. Died of chronic diarrhea there 8/18/64. Bur. Woodlawn National Cem., grave #118. Brother of Elias P. Stogdale.

STOGDALE, ROBERT A.: Pvt., Co. A. B. Augusta Co. 1838? Farmhand, age 22, Smokey Row, 1st Dist., Augusta Co. 1860 census. Enl. Staunton 7/29/61. Died of typhoid fever Camp Alleghany 12/25/61 age 22.

STOGDALE, THOMAS HENRY: Pvt., Co. A. B. Albemarle Co. 1/1/40. Farmhand, age 20, Staunton 1860 census. Enl. Staunton 7/29/61. Present 11/61-4/62. Reenl. 5/1/62. Present 5/1-7/20/62. Ab. sick with "tertia intersittant and rheumatismus" 7/21-12/31/62. Present 1-4/63. WIA (concussion from piece of shell) Fredericksburg 5/5/63., Cap., sick with "remittant fever" Gettysburg 7/3/63. Transf. to PM 9/14/63. Sent to West's Hospital, Baltimore. Exchanged 9/27/63. Admitted to hospital Richmond 9/28/63 and furloughed 11/18/63. Ab. through 12/63. Issued clothing 3/18 and 4/27/64. On rolls of Co. A, Ward's Bn., C. S. Prisoners released from military prison for their participation in the defense of Lynchburg 7/30/64. WIA (arm) Charlestown 8/23/64. Ab. wounded in hospital Staunton through 12/64. Surrendered Appomattox 4/9/65. Farmhand, age 30, Mt. Sidney PO, Augusta Co. 1870 census. Died Augusta Co. 9/6/01.

STOGDALE, WILLIAM J.: Pvt., Co. G. B. Augusta Co. 1829? Farmhand age 31, 1st Dist., Augusta Co. 1860 census. Enl. Staunton 8/61. Present 11/61-4/62. Reenl. 5/1/62. Present 5/1-10/31/62. Ab. sick 12/11/62-12/63. Cap. near Spotsylvania CH 5/19/64. Sent to Point Lookout. Transf. Elmira. Died there of chronic diarrhea 7/9/64. Age 35. Bur. Woodlawn National Cem., grave #2841. Left widow.

STOLANGER, A. H.: Pvt., Co. F. Not on muster rolls. Deserted to the enemy Clarksburg, W. Va. 10/13/63. Took oath and sent north.

STOMBOCK, DAVID BOYD: Pvt., Co. D. B. Page Co. 1831? Farmhand, age 29, Northern Dist., Augusta Co. 1860 census. Enl. Staunton 7/16/61. Present 7/16/61-4/62, detailed to work in bakery 2/6/63. Reenl. 5/1/62. Present 5/1-6/2/62. AWOL 6/3-20/62, fined $6.23. Present in arrest 1-2/63, fined 5 months and 1 day pay in addition to 18 days AWOL $61.60 by CM. Present 4/30-12/63. Issued clothing 3/31 and 4/22/64. Cap. Fisher's Hill 9/22/64. Sent to Point Lookout. Exchanged 3/19/65 and present Camp Lee, Richmond same date. NFR. Farmhand, age 42, Mt. Sidney PO, Augusta Co. 1870 census. Died Mt. Solon 2/14/95.

STONE, HARRISON F. "Henry": Pvt., Co. D. B. King & Queen Co. 2/3/27. Enl. Staunton 7/16/61. Present 7/16/61-4/62. Reenl. 5/1/62. AWOL 5/8-6/14/62, fined $12.20. Present 6/16-3/3/62. Present 1-6/63. Deserted 7/6/63. Sent to Ft. Delaware. Took oath and joined 3rd Md. Cav. U. s. 9/22/63, but married Staunton 11/10/64. Shoemaker, age 41, Fishersville, Augusta Co. 1870 census. Died South River Dist., Augusta Co. 8/7/14. Bur. Tinkling Spring Pres. Ch. Cem.

STOUTAMOYER, WILLIAM HENRY: Pvt., Co. D. B. Augusta Co. 1836? Farmhand, age 24, Burkes Mill Dist., Augusta Co. 1860 census. Enl. Staunton 7/16/61. Present 7/16/61-4/62. KIA McDowell 5/8/62. Age 26, 5' 5", fair complexion, blue eyes, shoemaker. Bur. McDowell.

STOVER, JOHN EMORY: Pvt., Co. F. B. Augusta Co. 1844? Farmhand, age 16, 1st Dist., Augusta Co. 1860 census. Enl. Staunton 7/31/61. Present 11/61-3/62. Died of diphtheria McDowell 4/3/62. Age 18, 5' 9", dark complexion, black hair, black eyes.

STOVER, JOHN HATCH: 1stSgt., Co. F. B. Augusta Co. 2/27/42. Student age 18, Northern Dist., Augusta Co. 1860 census. Attended Hotchkiss Academy, Churchville. Enl. Staunton 8/31/61 as Pvt. Present 11/61-2/62, promoted 2ndCpl. 1/28/62. Present 3-4/62. Reenl. 5/1/62. Present 8/31/62-/63, and 4/30-8/31/63, elected 1stSgt. Present 9-12/63. Ab. on leave 12/24/63-1/8/64. Commanding Co. 4/30 until WIA (bowels) Wilderness 5/6/64. Commanding Co. 6-10/64. WIA Cedar Creek 10/19/64. Commanding Co. 11/8/64-2/4/65. WIA (thigh) Ft. Stedman 3/25/65. Admitted to hospital Richmond 3/26/65. Cap. there 4/3/65. Sent to Newport News. Released 7/1/65. 5' 6", fair complexion, dark hair, grey eyes. Commissioner of Revenue, age 28, Mt. Sidney PO, Augusta Co. 1870 census. Teacher, Notary Public, Farmer and Stockraiser, New Hope 1885. Died Churchville 2/19/07. Bur Green Hill Cem. Brother of William S. Stover. "A brave soldier."

STOVER, JOSHUA H.: Pvt., Co. H. B. Augusta Co. 1839. Carpenter, age 22, 1st Dist., Staunton, 1860 census. Enl. Staunton 7/31/61. Ab. on leave 11/12/61. Present 1/2/62. AWOL since 4/26/62 on rolls 3-4/62. Dropped as deserter. Res. Dayton, Rockingham Co. Carpenter, Mt. Sidney 1870 and Weyer's Cave 1897. Thrown from a horse and killed near Spring Hill 8/30/97. Bur. Pleasant View Lutheran Ch. Cem.

STOVER, WILLIAM S.: Pvt., Co. F. B. Va. 1844? Student, age 16, Northern Dist., Augusta Co. 1860 census. Pvt., Co. L. 160th Va. Militia 3/62 exempt as Blacksmith, age 19. Enl. Valley Mills 4/24/62. Reenl. 5/1/62. WIA (leg) and cap. Sharpsburg 9/17/62. Exchanged 10/17/62. Admitted to hospital Richmond 10/24/62. Ab. wounded through 2/63. Present 4/30/63 until WIA Gettyburg 7/3/63. Returned to duty 9/29/63. Present through 12/63, and 3-12/64. Surrendered Appomattox 4/9/65. Living in Indiana 1913. Attended reunion Staunton 1916.

STRICKLER, ARCHIBALD W.: Pvt., Co. B. (1st) B. Rockbridge Co. 1840? Age 20, South River Dist., Rockbridge Co. 1860 census. Enl. Kerr's Creek 7/10/61. Present until transf. 9/28/61.

STRICKLER, JOHN H: Pvt., Co. I. B. Rockbridge Co. 3/15/30. Farmhand, Middlebrook Augusta Co. Enl. Staunton 7/16/61, age 28. Present 11/61-4/62. Reenl. 5/1/62. WIA Sharpsburg 9/17/62. Ab. sick white swelling of the leg through 2/31/64. NFR. Shoemaker, age 38, Fishersville PO, Augusta Co. 1870 census. Died Augusta Co. 12/2/14. Bur. Old Providence Pres. Ch. Cem., Spottswood.

STRICKLER, WILLIAM MABRY: Assistant Surgeon, F&S. B. near Luray, Page Co. 9/28/38. Graduate Dickinson Col., Pa; UVa. Med. School. Appointed Assistant Surgeon, C.S.A. 2/8/62 and assigned to Hays' La. Brig. Assigned to 52nd Va. Inf. 3/64. Present through 9/64, and 3/18/65. Surrendered Appomattox 4/9/65. Graduate Medical Col. of Va. Moved to Colo. 1869. City and County Physician, Colorado City; Mayor 1888 and 1893. Died Denver 10/26/08. Bur. Colorado Springs.

STRICKLER, WILLIAM N.: Pvt., Co. I. B. Va. 1843? Famhand, age 17, 1st Dist., Augusta Co. 1860 census. Enl. Staunton 7/16/61 age 18. Present 11/61-4/62. Reenl. 5/1/62. Present 5/1-7/17/62. AWOL 7/18-10/28/62, fined $38.56. Present 1-12/63. WIA (shoulder) Bethesda Ch. 5/30/64. DOWs Camp Winder Hospital, Richmond 6/14/64. Bur. Hollywood Cem., grave #632. (incorrectly listed as Co. I, 57th Va.).

STUART, JOHN ANDREW: Drillmaster, F&S. B. Augusta Co. 10/4/43. VMI '65. Served as Drillmaster, Richmond 4-6/61. Enl. Staunton 7/61 as Drillmaster. Served until 2/62 when he returned to VMI. Not on muster rolls. Orderly Sgt. Co. C. Cadet Bn., at New Market and WIA (right leg) 5/15/64. Rejoined Cadet Corps in Richmond 10/64 and remained until disbanded 4/65. Moved to Mo. 1866 and farmed for 3 years. Returned to Va. and farmed at Ladd, Augusta Co. Moved to Lowery, Bedford Co. Farmer, Teacher, Constable, Magistrate; Member, State Board of Agriculture. Died Lowery 2/13/08.

SULLIVAN, JOHN D.: Pvt., Co. D. B. Va. 1842? Farmhand, age 18, Burkes Mill Dist., Augusta Co. 1860 census. Enl. Staunton 7/16/61. Present 7/16/61-1/62. AWOL 2/16-28/62. fined $11.00 by CM. Present 3-4/62, sent to hospital Staunton 4/21/62. Reenl. 5/1/62. Present 5/1-8/31/672. Present in arrest 1-2/63, fined pay for 3 months and 1 day by CM $33.36. Present 3-12/63. Issued clothing 3/31 and 4/1/64. WIA near Spotsylvania CH 5/18/64. DOWs in hospital 6/20/64. Probably John Sullivan (Va.) bur. Fredericksburg Confederate Cem.

SUMMERS, ANDREW JACKSON: Pvt., Co. I. Enl. Greenville 7/19/63. Present 7-12/63. WIA (thigh and arm) Bethesda Ch. 5/30/64. Furloughed from hospital for 30 days 7/1/64. Present 8-12/74. Deserted to the Army of the Potomac near Petersbug 3/26/65. Sent to Washington D. C. Took oath and transportation furnished to Indianapolis, Ind.

SUMMER, GEORGE P.: 1stSgt., Co. I. B. Va. 1844? Farmhand, age 16, 1st Dist., Augusta Co. 1860 census. Enl. Staunton 7/16/61 age 18 as 3rdSgt. Present 11/61-2/62, promoted 2ndSgt. Present 3-4/62. Reenl. 5/1/62. Present 8/30-10/31/62, and 1-4/63, promoted 1st Sgt. 3/5/63. Present 4/30-12/63. WIA "thoratic Par." and cap. Bethesda Ch. 5/30/64. Admitted to 4thDiv., 5th Army Corps Hospital, Army of the Potomac. DOWs at Armory Square Hospital, Washington, D. C. 6/4/64. "dead when brought in gunshot wound right breast." Bur. Washington, D. C. No effects, however 1 testament and 1 diary were sold at auction for 25¢. Next of kin Mrs. Julia Summers, Middlebrook (friend). Brother of John D. Summers.

SUMMERS, JOHN D.: 1st.Lt., Co. I. B. Va. 1833? Farmer, age 23, 1st. Dist., Augusta Co. 1860 census. Capt. Middlebrook Co., 93rd Va. Militia. Enl. Staunton 7/16/61 age 29 as 1stLt. Ab. sick 12/16/61-2/62. Present 3-4/62. Reelected 5/1/62. Present 8/30-10/31/62. Ab. on leave 2/21-28/63. Present 3-12/3, commanding Co. 4/30-8/20/63. WIA (hand) and cap. Spotsylvania CH 5/12/64. Sent to Old Capitol Prison. Transf. Ft. Delaware. Released 6/16/65. 5' 1", florid complexion, dark hair, hazel eyes. Farmer, Arbor Hill, Augusta Co. Brother of George P. Summers.

SUTER, JACOB D.: Pvt., Co. F. B. Berkley Co. 11/24/43. Farmer, Rockingham Co. Enl. Valley Mills 4/20/62. AWOL 5/20-9/20/62. Present 8/21-12/31/63, and 1-12/63. Present 3-6/64. AWOL 6/21 until deserted to the enemy New Creek, W. Va. 10/31/64. Took oath and sent north. Age 21, 5' 8", dark complexion, blue eyes, black hair. Potter, Harrisonburg, 1900. Died New Erection, Rockingham Co. 6/6/11. Bur.Weavers Mennonite Ch.Cem., on Rt. 33, west of Harrisonburg.

SUTLER, GEORGE, W.: Pvt., Co. B. (2nd). B. Nelson Co. 1825? Teacher age 35, 1st Dist., Augusta Co. 1860 census. Enl. Waynesboro 7/15/61. Ab. sick 11/61-4/62. Non-conscript detailed in hospital Staunton 4/30 until discharged for "hypertrophy of heart" 8/11/63. Age 35, 5' 11", fair complexion, dark eyes, dark hair. Beekeeper, age 43, Fishersville PO, Augusta Co. 1870 census. Died Louisa Co. 6/24/06. Bur. Tinkling Spring Pres. Ch. Cem., near Fishersville.

SWAN, WILLIAM D.: 4thCpl., Co. B (2nd). B. Va. 3/21/41. Miller's Apprentice, age 20, Burkes Mill Dist., Augusta Co. 1860 census. Enl. West View 5/1/62 as Pvt. Present 5-12/62 and 1-2/63, promoted 4th Cpl. 1/1/63. Ab. sick debility and "spasmotic stric" 4/17-29/63. Present 4/30-12/63. Capt. Bethesda Ch. 5/30/64. Sent to Point Lookout. Transf. Elmira. Released 6/30/65. 5' 8", florid complexion, auburn hair, blue eyes. Living in Augusta Co. 1922.

SWEARINGIN, JAMES N.: Pvt., Co. K. B. Bath Co. 1843? Farmhand, age 17, Green Valley Dist., Bath Co. 1860 census. Enl. Shenandoah Mt. 4/9/62. Present 4/9-4/30/62 and 1-7/63. WIA (arm) Gettysburg 7/3/63. Ab. wounded until died of typhoid fever Bath Co. 12/26/63. Bur. Bath Co.

SWINK, ERASMUS: Pvt., Co. H. B. Augusta 4/10/28. Farmhand, age 39, Northern Dist., Augusta Co. 1860 census. Pvt., Co. E. 93rd Va. Militia. Enl. Staunton 7/23/61 age 38. NFR. Farmhand, age 48, Mt. Sidney PO, Augusta Co. 1870 census. Died Spring Hill 5/29/98. Father of James W. Swink who was probably a substitute for him. Brother of Mathew J. and William S. Swink.

SWINK, GEORGE WASHINGTON: Pvt., Co. C. B. Augusta Co. 3/14/42. Farmhand, age 18, Northern Dist., Augusta Co. 1860 census. Enl. Staunton 7/16/61 age 20. Present 11-12/61. Ab. sick 1-2/62. Present 3-4/62. Reenl. 5/1/62. WIA (hand) 2nd Manassas 8/28/62. Present 8/31/62-8/31/63. Ab. sick 9/14-10/63. Present 11-12/63. WIA (hip) Spotsylvania CH 5/12/64. Ab. wounded through 12/31/64. NFR. Farmer, age 28, Mt. Sidney PO, Augusta Co. 1870 census. Commissioner of Revenue, Augusta Co. 1897. Brother of John H. and Martin V. B. Swink.

SWINK, JAMES W.: Pvt., Co. H. B. Augusta Co. 1847? Resident, age 13, Northern Dist., Augusta Co. 1860 census. Enl. Staunton 7/23/61. Present 11/61-4/62. Reenl. 5/1/62. Ab. sick 6/18-8/31/62. AWOL 9/10/62-12/63. Died of typhoid fever Orange CH 2/11/64. Bur. Orange CH. Son of Erasmus Swink.

SWINK, JOHN H.: 2ndCpl., Co. C. B. Augusta Co. 1839? Farmhand, age 21, Northern Dist., Augusta Co. 1860 census. Enl. Staunton 7/16/61 age 21. Ab. sick 11-12/61. Present 1-2/62, reduced to Pvt. 2/20/62. Present 3-4/62. Reenl. 5/1/62. WIA Port Republic 6/9/62. Ab. wounded through 8/30/62. Present 8/31/62-4/63. Died of typhoid fever in Crumpton Hospital, Lynchburg 6/18/63. Brother of George W. and Martin V. B. Swink. "A brave soldier and a true Christian."

SWINK, MARTIN VAN BUREN: Pvt., Co. C. B. Augusta Co. 1837? Farmhand, age 23, Northern Dist., Augusta Co. 1860 census. Pvt., Co. I, 160th Va. Militia 3/62 exempt for fits, age 26. Enl. Valley Mills 4/28/62. Ab. sick 6/62 until discharged for epilepsy 11/6/62. Age 26, 5' 9", dark complexion, black eyes, black hair. Farmhand, age 32, Mt. Sidney PO, Augusta Co. 1870 census. Brother of James W. and John H. Swink.

SWINK, MATTHEW J.: Pvt., Co. H. B. Augusta Co. 1827. Farmer age 32, 1st Dist., Augusta Co. 1860 census. Pvt., Co. E, 93rd Va. Militia. Enl. Saunton 7/23/61 age 32. Ab. on leave 2/14-28/62. Present 3-4/62. Reenl. 5/1/62. Present 5/1-6/8/62. AWOL 6/9-8/15/62. Present 8/16-9/20/62. Ab. sick 9/21/62-2/63. Present 3-4/62. Present, Teamster in Brig. Ambulance Train 5/22-12/63. Deserted Staunton 6/26/64. NFR. Farmhand, age 45, Mt. Sidney PO, Augusta Co. 1870 census. Died Augusta Co. 1910. Bur. Stonewall United Meth. Ch. Cem., Staunton. Brother of Erasmus and William S. Swink.

SWINK, WILLIAM SILLINGS: Pvt., Co. H. B. Va. 5/17/32. Farmer, age 32, 1st Dist., Augusta Co. 1860 census. Pvt., Co. E. 93rd Va. Militia. Enl. Staunton 7/23/61 age 32. Ab. on detail 11-12/61. Ab. sick 4-4/62. Returned to duty 4/30/62. Reenl. 5/1/62. WIA (left foot and right thigh) Port Republic 6/9/62. Ab. wounded through 12/31/64. Paroled Staunton 5/1/65. Age 37, 6' 1", fair complexion, dark hair,blue eyes, illiterate. Farmer, age 43, Mt. Sidney PO, Augusta Co. 1870 census. Resident of Moscow, Augusta Co. Died Long Glade 3/22/05. Bur. Pleasant View Lutheran Ch. Cem. Bother of Erasmus and Mathew J. Swink.

SWISHER, GEORGE WASHINGTON: Pvt., Co. F. B. Augusta Co. 11/19/36. Farmhand, age 23, Northern Dist., Augusta Co. 1860 census. Not on muster rolls. Enl. Valley Mills 5/1/62. Died of fever in hospital Staunton 5/19/62. Bur. Old Salem Lutheran Ch. Cem., 3 miles west of Mt. Sidney.

SWISHER, SAMUEL HENRY: Pvt., Co. A. B. Augusta 4/19/34. Farmhand, Rockingham Co. Pvt., Co. G, 160th Va. Militia, not exempt, age 18. Enl. Staunton 9/1/61. AWOL 11/23/62-2/13/63. Present in arrest 2/63, fined pay for period AWOL. Present 3-7/63. Ab. sick 7/16-8/31/63. Deserted to the enemy Clarksburg, W. Va. 10/14/63. Took oath and sent north. Age 20, 5' 3", Farmer, age 39, Mt. Sidney PO, Augusta Co. 1870 census. Died near Springhill, Augusta Co. 7/17/79. Bur. Old Salem Lutheran Ch. Cem., 3 miles west of Mt. Sidney.

SWISHER, WILLIAM F.: Pvt., Co. B (1st). Farmer, age 22, Burkes Mill Dist., Augusta Co. 1860 census. On postwar roster.

SWITZER,CHRISTIAN: Pvt., Co. F. B. Augusta Co. 1841? Farmer, Mt. Sidney Augusta Co. Enl. Staunton 7/31/61. Present 11/61-4/62. Reenl. 5/1/62. Ab. sick 7/14/62. Died of typhoid fever in hospital Richmond 7/30/62. Age 21, 5' 11", fair complexion, dark hair, grey eyes, Bur. Hollywood Cem, Richmond.

SWITZER, JOHN: Pvt., Co. F. B. 1845? Farmer, Dayton, Rockingham Co. Enl. Valley Mills 4/20/62. AWOL 5/20/62-7/63. Deserted to the enemy in Hampshire Co. 7/20/63. Took oath and released. Age 18, 5' 8½", fresh complexion, dark eyes, auburn hair.

SWORTZELL, GEORGE A.: Pvt., Co H. B. Augusta Co. 10/15/41. Farmhand, age 19, Staunton Dist., Augusta Co. 1860 census. Pvt., Co. E. 93rd Va. Militia. Enl. Staunton 7/23/61 age 20. Present 11/61-1/62. Ab. Sick 2/26-28/62. Present 3-4/62. KIA Port Republic 6/9/62. Age 21, 5' 10", fair complexion, dark eyes, dark hair. Bur. St. John's Ch. Cem., near Middlebrook.

TABLER, JASPER N.: Pvt., Co. unknown. B. 3/23/42. Not on muster rolls. Enl. '62. Promoted 2ndLt. Transf. 37th Va. Bn. Cav. Doctor, Royse, Tex. 1915.

TALIFERRO, JOHN M.: Pvt., Co. B. Augusta Co. 5/12/21. Carpenter, age 38. Burkes Mill Dist., Augusta Co. 1860 census. Enl. Staunton 7/29/61 age 39. Ab. detailed as Wagoner in Augusta Co. 9/10/61-2/62. AWOL 2/28-8/12/62. Present 8/13-31/62 WIA (ankle) Sharpsburg 9/17/62. Leg amputated below the knee. Requested artificial leg from Hanger Brothers, Staunton 11/26/64. NFR. Illiterate. Carpenter, age 48, 6th Dist., Augusta Co. 1870 census. Died Jennings Gap, Augusta Co. 12/12/96.

TALLEY, THOMAS JEFFERSON: Pvt., Co. G. B. Va. 1826? Farmhand, age 34, Burkes Mill Dist., Augusta Co. 1860 census. enl. Staunton 8/2/61. Present 11/61-4/62. Reenl. 5/1/62. Present 5/1/62-12/63. WIA and cap. Spotsylvania 5/12/64. DOW aboard the "Baltic Fleet". Bur. at sea. Left wife and 3 children.

TANNER, GEORGE.: Pvt., Co. F. Conscript sent to Regt. 10/62. NFR.

TAYLOR, A. J.:: Pvt., Co. A. B. 1847. Not on muster rolls. WIA (thigh) Spotsylvania 5/12/64. NFR. Farmer, Rockingham Co. Died 1918. Bur. Ottobine Meth. Ch. Cem., Rockingham Co.

TAYLOR, GEORGE W.: Pvt., Co. H. B. Nelson Co. 1836? Farmhand, age 24, 1st Dist., Augusta Co. 1860 census. Enl. Staunton 7/23/61 age 28. Present 11/61/-2/62/. AWOL 3/1-18/62 and 4/21/62-9/25/63. Ab. in arrest 9-10/63. Deserted 11/15/63, resident of Greenville. NFR. Farmhand, age 65, South River Dist., Augusta Co. 1926.

TAYLOR, HENRY C.: Pvt., Co. G. B. Augusta Co. 1804? laborer, age 46, 1st Dist., Staunton 1860 census. Enl. Winchester 5/27/62 as substitue. Deserted 7/8/62. NFR. Railroad worker. Died near Williamsville, Ky. 12/21/74.

TAYLOR, JAMES EDWARD.: Pvt., Co. B (1st). B. Va. 1833? Farmhand, age 27, Kerr's Creek Cist., Rockbridge Co. 1860 census. On postwar roster.

TAYLOR, JOEL A.: Pvt., Co. I. B. Va. 1840? Farmhand, age 20, Staunton PO, Augusta Co. 1860 census. Enl. Staunton 7/16/61 age 21. Present 11/61-2/62, fined $11.00 by CM. Present 3-4/62. Reenl. 5/1/62. WIA (head) Cross Keys 6/8/62. Present 8/30-10/31/62, and 1-7/63. AWOL 7/27-12/31/63. Dropped as deserter. NFR. Brother of William T. Taylor.

TAYLOR, JOSEPH LARKIN: Pvt., Co. B (1st). B. Va. 1823. Farmhand, age 37, South River Dist., Rockbridge Co. 1860 census. Enl. Staunton 8/1/61. Present until transf. 9/28/61.

TAYLOR, ST. CLAIR.: Pvt., Co. A. B. 1836? Resident, Burkes Mill, Augusta Co. Enl. Staunton 7/9/61 age 25. Present 11/61-4/62. Ab. sick 4/23-30/62. Reenl. 5/1/62. Present 8/31-11/22/62 AWOL 11/23/62-1/16/63. Present in arrest 1-2/63, fined pay for AWOL. Present 3-12/63. Ab. sick 4-12/31/63. Present 4/30-10/3/64. AWOL 10/4-12/31/64. NFR. D. before 5/03.

TAYLOR, SILAS A.: Pvt., Co. A. B. Augusta Co. 1825? Plasterer, age 35, 1st Dist., Augusta Co. 1860 census. Enl. Staunton 7/9/61 age 27. Present 11/61-4/62. Reenl. 5/1/62. WIA (wrist) Gaines Mill 6/27/62. Ab. wounded through 12/31/64. Cap. Burkesville 2/1/65. Sent to Ft. Delaware. Released 6/15/65. 5' 9½", light complexion, dark hair, dark eyes, illiterate. Died before 5/03.

TAYLOR, SOLOMON: Pvt., Co. A. B. 1844? Enl. Staunton 7/19/61 age 17. Present 11/61-4/62. Reenl. 5/1/62. Present 8/31/62-4/63, and 4/30-7/26/63. AWOL 7/27-8/10/63, fined $5.13. Present 8/11-12/63. WIA (thigh) Wilderness 5/5/64. AWOL 9/26-10/31/64, fined 1 months pay by CM. Present 11-12/64. NFR. Died before 5/03.

TAYLOR, WILLIAM C.: Pvt., Co. I. B. Augusta Co. 1837? Farmhand, age 23, Staunton PO, Augusta Co. 1860 census. Enl. Staunton 7/16/61 age 23. Present 11/61-3/62. AWOL 4/9-10/25/62, fined $68.20. Present 10/26-31/62 and 1-4/63. Cap. Gettysburg 7/3/63. Sent to Point Lookout. Exchanged 3/20/64. In hospital Richmond 3/25/64. AWOL 12/8-31/64. NFR. Farmer, Lofton, Augusta Co. Died near there 2/15/09. Bur. Riverview Cem., Waynesboro. Brother of Joel A. Taylor.

TEABO, JOHN A.: Pvt., Co. C. B. Va 1831? Carpenter, age 29, Burkes Mill Dist. Augusta Co. 1860 census. Pvt., Capt. J. K. Koiner's Long Meadow Co., 32nd Va. Militia 3/62 exempt but volunteered. Enl. Shenandoah Mt. 4/11/62. Ab. sick 5/6/62-12/63. Deserted to the enemy in W. Va. 1/15/64. Took oath and sent north. Age 26, 5' 10", light complexion, blue eyes, light hair.

TEADMARSH, JAMES: Pvt., Co. G. Enl. Winchester 9/24/62 as substitute. Deserted near Martinsburg the same day. NFR.

TEMPLE, WILLIAM: Pvt., Co. D. B. Va. 1825? Farmer, age 35, Northern Dist., Augusta Co. 1860 census. Enl. Staunton 7/16/61. Present 7/16-12/61. AWOL 1/5-2/15/62, fined $11.00 by CM. Present 8/3-31/62, AWOL 10/29/62-12/63. Discharged on writ of Habias Corpus at camp of 52nd Va. 3/22/64. Age 35, 5' 11", dark complexion, dark eyes, dark hair, farmer.

TEMPLETON, JAMES ALFRED: Pvt., Co. H. B. Fairfield, Rockbridge Co. 12/25/28. Clerk, age 31, Middlebrook, Augusta Co. 1860 census. Enl. Staunton 11/9/64. Present 11-12/64. Surrendered Appomattox 4/9/65. Clerk, age 41. 1st Dist., Staunton 1870 census. Merchant, Staunton 45 years. Member, Stonewall Jackson Camp CV, Staunton. Died there 12/15/10. Bur. Thornrose Cem.

TEMPLETON, WILLIAM PAXTON: Pvt., Co. B(1st). B. Rockbridge Co. 1/26/32. Farmer, age 28, South River Dist., Rockbridge Co. 1860 census. enl. Staunton 8/1/61. Present until transf. 9/28/61.

TERRELL, ANDREW J.: Pvt., Co. B(2nd). B. Va. 1843? Farmhand, age 17, Burkes Mill Dist., Augusta Co. 1860 census. Enl. Waynesboro 7/15/61. Present 11/64-4/62. AWOL 4/17-12/8/62, qualified his act and was paid at hospital Lynchburg. Present 12/9/62-1/14/63. Ab. sick 1/15-2/63. Present 3-7/63. WIA Gettysburg 7/3/63. Ab. wounded through 8/31/63. Present 9-10/63. Ab. sick in hospital Lynchburg 11/8-12/63. Issued clothing 1/28, 3/17 and 4/22/64. DOW received near Spotsylvania CH5/17/64.

TERRELL, JAMES: Pvt., Co. B(2nd). B. Augusta Co. 1818? Brickmoulder, Augusta Co. Enl. Waynesboro 7/15/61. Present 11-12/61. AWOL 1-2/62, fined $11.00 AWOL 10 days. Present 3-4/62, non-conscript. AWOL 5/17/62-1/21/63. Present in confinement Regt'l Guard House 1/22-4/63. Discharged overage of conscription 5/22/63. Age 45, 5' 7½", dark complexion, dark hair, blue eyes, illiterate. Resident of Basic City, Augusta Co. 1896.

TERRELL, JOHN WILLIAM: Pvt., Co. B(2nd). B. Waynesboro, Augusta Co. 1835. Enl. Waynesboro 7/15/61. Ab. sick 11-12/61. Present 1-6/62 AWOL 7/20-9/24/62. Present 9/25/62-7/63. AWOL 7/28-10/1/63. Ab. in arrest Orange CH 10/63. present in arrest 11-12/63. Ab. sick in hospital Richmond 1/23/64. Issued clothing 4/22/64. WIA (arm and right thigh fractured) and cap. Bethesda Ch. 5/30/64, age 24. Sent to hospital Washington, D. C. Transf. Old Capitol Prison and Elmira. Exchanged 2/20/65, Illiterate. In hospital Richmond with chronic diarrhea 3/4/65. Furloughed for 60 days 3/19/65. Paroled Lynchburg 4/15/65. Farmhand, age 23, Mt. Sidney PO, Augusta Co. 1870 census. Brickmoulder, Basic City, Augusta Co. 1913. Member Stonewall Jackson Camp CV, Staunton. Died Basic City 1/31/19. Bur. UDC lot, Riverview Cem., Waynesboro.

TETER, DANIEL W.: 4th Cpl., Co. A. B. Augusta Co. 1843? Carpenter, Augusta Co. Enl. Staunton 7/9/61 age 18 as Pvt. Ab. sick 11/16/61-1/9/62. Present 1/10-4/62. Reenl. 5/1/62. Promoted 4th Cpl 8/31/62. KIA Sharpsburg 9/17/62. Age 19, 5' 9", dark complexion, light hair, hazel eyes.

TETER, JAMES W.: Pvt., Co. A. Enl. Staunton 7/19/61. Ab. sick 11/16/61-1/6/62. Present 1/7-2/62. Ab. sick 3/28-4/62. Deserted 7/62. Deserted to the enemy Clarksburg, W. Va. 10/13/63. Took oath and sent north. Died before 5/03.

THACKER, WILLIAMS S.: Pvt., Co. G. B. Va. 1844? Farmhand, age 16, Northern Dist., Augusta Co. 1860 census. Enl. Staunton 8/2/61. Died of pneumonia Camp Alleghany 12/8/61.

THOMAS, FENDALL: Pvt., Co. E. B. Va. 1821? Farmer, age 39, Glenwood Dist., Rockbridge Co. 1860 census. Conscripted in Botetourt Co. 9/10/64. Cap. Fisher's Hill 9/22/64. Sent to Point Lookout. Died there of chronic diarrhea 1/25/61, age 35. Bur. in prisoners grave yard, #864. Effects given to friends before death. Left widow and 4 children.

THOMAS, JOHN C.: 4thCpl. Co. H. B. 1844? Enl. Staunton 7/23/61 age 17 as Pvt. Present 11/61-4/62. Reenl. 5/1/62 and appointed 4thCpl. WIA (leg) 2nd Manassas 8/29/62. DOWs in Staunton 12/26/62. Bur. Thornrose Cem.

THOMAS, JOHN G.: Pvt., Co. K. B. Alleghany Co. 1818? Millwright, age 42, Hot Springs PO, Bath Co. 1860 census. Enl. Shenandoah Mt. 4/9/62. Present 5/26/62 over 35. Capt. Winchester 6/1/62 over age of conscription. Enl. Co. E, 10th Bn. Va. Res. 4/16/64 age 45, 5' 8", light complexion, black hair, grey eyes. NFR.

THOMAS, JOHN J.: Pvt., Co. K. B. Alleghany Co. 1817? Millwright, age 42, Healing Springs PO, Bath Co. 1860 census. Served militia. Enl. Shenandoah Mt. 4/9/62. Present 4/9-30/62. Cap. and paroled Winchester 6/1/62. Age 45, 5' 7½". Discharged over age of conscription Winchester 6/1/62. Resident of Healing Springs, Bath Co. Age 45, 5' 8", dark complexion, grey eyes, black hair. Enl. Co. E, 10th Bn. Va. Res. 4/16/64. Age 45, 5' 8", light complexion, light hair, blue eyes. Millwright, age 52, Bath CH PO, Bath Co. 1870 census.

THOMASON, JOHN B.: Pvt., Co. I. B. Augusta Co. 1823. Carpenter, age 39, 1st Dist., Augusta Co. 1860 census. Enl. Staunton 7/16/61 age 41, Ab. sick 11/61-12/61. Present 1/13-4/62. Reenl. 5/1/62. Present 5/1-9/11/62. Ab. sick 9/12-10/31/62. Paid Staunton 12/62. Probably discharged for overage of conscription. Entered Lee Old Soldiers Home, Richmond 2/23/86 age 64. Dropped. Died Spottswood, Augusta Co. 8/16/16. Bur. Mt. Carmel Pres. Ch. Cem., near Steele's Tavern.

THOMPSON, ANDREW JAMES: Capt., Co. B(2nd). B. 1/6/32. Enl. Waynesboro 7/15/61 as 1st Lt. Ab. on leave 1-2/62. Reelected 1st Lt. 5/1/62. Elected Capt. 5/8/62. Present 11/62-5/63. WIA (leg) Gettysburg 7/3/63. Ab. wounded until retired for disability 12/8/64. Paroled Staunton 5/20/65. Age 33, 6' 1", fair complexion, dark hair, blue eyes. Died Augusta Co. 3/12/23.

THOMPSON, ROBERT W.: Pvt., Co. B(2nd). Not on muster rolls. Enl. as substitute 1862. Listed as deserter 2/63. NFR. Living Rockingham Co. 1926.

THORNER, M. B.: Pvt., Co. B(2nd). Not on muster rolls. Cap. date and place unknown. Sent to Point Lookout. Died of disease there and bur. in POW cem. there.

THORNTON, ABSALOM: 3rdCpl., Co. I. B. Augusta Co. 1/2/41. Farmhand, age 19, Staunton PO, Augusta Co. 1860 census. Enl. Staunton 7/16/61 age 21 as Pvt. Present 11/61-4/62. Reenl. 5/1/62. Present 8/31-9/62. WIA Sharpsburg 9/17/62. Present 10/62 and 1-4/63, appointed 3rdCpl 4/1/63. Present 4/30-12/63. WIA (left arm pit) Spotsylvania CH 5/12/64. returned to duty 7/64. Cap. Cedar Creek 10/19/64. Sent to Point Lookout. Released 6/20/65. 5' 11", light complexion, red hair, hazel eyes. Farmhand, age 27, Fishersville PO, Augusta Co. 1970 census, Gas House Operator, City of Staunton. Died there 6/17/10. Bur. Thornrose Cem. "A brave soldier."

THORNTON, DAVID BITTLE: Pvt. Co. I. b. 2/25/45. Pvt., Capt. Snapp's Co., 93rd Va. Militia 3/62 exempt, 17 years old. Not on muster rolls. KIA Spotsylvania CH 5/19/64.

THUMA, ROBERT: Pvt., Co. D. B. Rockingham Co. 2/15? Carpenter, age 30, Mt. Solon, Augusta Co. 1860 census. Enl. Staunton 7/16/61. Present 7/16/61-1/62. AWOL 1/5-2/4/62. Reenl. 5/1/62. Present 3-4/62. KIA Port Republic 6/9/62, age 27. Effects 1 pistol and $2.00. Left widow and 3 children. Bur. Silver Creek Cem., Jamestown, O.

TINSLEY, GEORGE R.: Pvt., Co. E. B. Va. 1838? Farmhand, age 22, Lexington Dist., Rockbridge Co. 1860 census. Served in militia. Enl. Staunton 8/1/61. Present 11/61-4/62. Reenl. 5/1/62. Present 5/1-10/31/62, and 1-4/63. Ab. sick in hospital Richmond with rheumatism 5/3-19/63. Ab. on leave 8/21-31/63. Present 9-12/63. KIA Spotsylvania CH 5/12/64. "A capital soldier."

TINSLEY, WILLIAM J.: Pvt., Co. K. B. Va. 9/9/45. Farmhand, age 17, Miller's Mill PO, Bath Co. 1860 census. Enl. Shenandoah Mt. 4/9/62. Present 4/9-30/62 and 5/26/62. In hospital Lynchburg with accute diarrhoea 7/21-8/11/62. Ab. sick with chronic bronchitis 1/7-6/14/63. AWOL 6/15-12/31/63. Ab. sick leave expired 4/26/64 on rolls 4/31-10/31/64. AWOL 11-12/64. Dropped as deserter. Farmhand, age 30, Lexington Dist., Rockbridge Co. 1870 census. Farmer, Augusta Co. 1875. Died Augusta Co. 12/14/34. Bur. Hammond Cem., on Rt. 635, 1.3 miles east of intersection with Rt. 644 in Augusta Co.

TISDALE, JAMES: Pvt., Co. C. B. Augusta Co. 1833? Farmhand, age 27, 1st Dist., Augusta Co. 1860 census. Pvt., Capt. Hall's Co, 32 Va. Militia 3/62 claimed deformed foot but not exempt, age 28. Enl. Shenandoah Mt. 4/18/62. Ab. sick 5/6-8/31/62 and 11/2/62-2/63. Present 3-4/63 Wagoner. Present 4/30-8/31/63, fined $3.30 for 9 days AWOL. Present 9-11/63. Ab. sick in hospitals Gordonsville and Richmond with chronic diarrhea 11/28-12/31/63. KIA Spotsylvania CH 5/12/64. Left widow and child.

TONER, or TONES, HARDEN L.: Pvt., Co. I. Pension application only record. Enl. 7/61. WIA '64. Discharged.

TRAIMER, GEORGE W.: Pvt., Co. B. Va. 1834? Farmhand, age 26, 1st Dist., Augusta Co. 1860 census. Enl. Staunton 7/16/61 age 24. Present 11/61-4/62. Reenl. 5/1/62. WIA (face) Gaines Mill 6/27/62. WIA 2nd Manassas 8/28/62. DOWs 8/30/62, age about 25 years. Left widow and 4 children.

TRENARY, FREDERICK T. JOHN: 4thSgt., Co. C. B. Frederick Co. 1817? Gardner, age 43, 1st Dist., Staunton 1860 census. Enl. Staunton 7/16/61 age 44. Ab. sick 11-12/61. Ab. detailed as nurse in hospital Staunton 1/62-2/63. Discharged for bronchitis 2/10/63. Age 46, 6', sallow complexion, grey eyes, horticulturist. Nurserman, age 52, Staunton 1870 census. Died Beverly Manor Dist., Augusta Co. 2/1/94.

TROUT, ERASMUS STRIBLING: Capt., Co. H. B. Staunton 4/15/44. VMI '65. Drillmaster of Regt. until returned to VMI 1/62. Present as Cadet at McDowell 5/8/62. Reenl. as Pvt., Co. H. 8/62. Appointed Sgt. Major 8/9/62, "for conspicious gallantry at Cedar Mt." by Col. Skinner. Promoted 2nd Lt. Co. H 8/62. Present 2nd Manassas, Sharpsburg. Promoted 1st Lt. 10/62 for "gallantry and general fitness." Present 3-7/63. WIA Gettysburg 7/3/63. Ab. wounded through 10/6/63. Present 10/7-12/31/63. Promoted Capt. 5/30/64. Present 6-12/64. Ab. on leave 1/18-28/65. Submitted resignation 2/15/65. Surrendered Appomattox 4/9/65. Druggist, Staunton. Died of consumption and "effects of military service" Staunton 10/20/66. Br. Thornrose Cem. "A gentleman of habits and deportment."

TROXELL, AMOS:: Pvt., Co. H. B. Va. 1844? Farmhand, age 16, Staunton PO, Augusta Co. 1860 census. Enl. Staunton 7/23/61 age 19. Present 11/61-4/62. reenl. 5/1/62. Present 4/30-8/31/62. AWOL 9/17/62-4/63, but in hospital Staunton 2/6-4/30/63 with rheumatism. Present 5-12/63. Issued clothing 4/21/64. Cap. Spotsylvania CH 5/12/64. Sent to Ft. Delaware. Released 6/15/65. 5' 4", light complexion, dark hair, grey eyes, resident of Newport, Augusta Co. Died Staunton 1865. On postwar roster. Bur. Thornrose Cem. Son of David and brother of John G. Troxell.

TROXELL, DAVID: Pvt., Co. H. B. Augusta Co. 1822? Farmer, age 39, Staunton PO, Augusta Co. 1860 census. enl. Staunton 7/23/61 age 41. Present 11/61-4/62. AWOL 6/9-7/7/62 and 7/18-10/62. Discharged overage of conscription 8/20/62, illiterate. Farmer, age 52, Moffett's Creek, Riverheads Dist., Augusta Co. 1870 census. Died there 8/3/98. Father of Amos and John G. Troxell.

TROXELL, JEREMIAH M. "JERRY": Pvt., Co. H. B. Augusta Co. 1831? Farmhand, age 29, Staunton PO, Augusta Co. 1860 census. Enl. staunton 7/23/61 age 28. Present 11/61-4/62. Reenl. 5/1/62. Present 5/1-8/31/62, and 1-12/63. Cap. Carter's Farm near Winchester 7/20/64. Sent to Camp Chase. Age 38, 5' 10", dark complexion, dark hair, dark eyes. Exchanged 3/10-13/65. Present Appomattox. Farmer, age 42, Fishersville PO, Augusta Co. 1870 census. Alive 1885. "One of the best soldiers in the company."

TROXELL, JOHN GLASSBURNER: Pvt., Co. I. B. Augusta Co. 4/16/45. Farmhand, age 15, Staunton PO, Augusta Co. 1860 census. Not on muster rolls. Deserted to the enemy Beverley, W. Va. 12/6/64. Took oath and released. Age 17, 5' 4", dark complexion, brown eyes, dark hair. Farmhand, age 26, Riverheads Dist., Augusta Co. 1870 census. Son of David and brother of Amos Troxell.

TUCKER, GARRETT L.: Pvt., Co. E. B. Bedford Co. 1838? Enl. Bunker Hill 9/22/62. Present 9-10/62. Ab. sick 2/22-12/63. Cap. Fisher's Hill 9/22/64. Sent to Point Lookout. Exchanged 11/15/64. NFR. Entered Lee Old Soldiers Home, Richmond 11/1/98 age 59 from Rockbridge Co. Died there 1/11/01 age 62. Bur. Hollywood Cem., Richmond.

TURK, JAMES G.: 2ndSgt., Co. C. B. Augusta Co. 1821? Enl. Staunton 7/16/61 age 42 as 3rdCpl. Present 11-12/61, promoted 2ndSgt. 11/26/61. Ab. on special duty Staunton 2/18-28/62. Present 3-4/62. Colorbearer of Regt. at McDowell, Cross Keys and until WIA (hand) Port Republic 6/9/62. Ab. wounded until discharged overage of conscription 2/20/63. Age 42, 6' 2", dark complexion, dark eyes, dark hair, farmer. Arresting conscripts and deserters for Conscript Officer, Staunton 8-1/63. Farmer, age 49, Mt. Sidney PO, Augusta Co. 1870 census. Died on Mossy Creek 1/12/89. Bur. Mossy Creek Ch. Cem. "A brave and gallant soldier."

TUTWILER, GEORGE W.: Pvt., Co. C. B. Rockingham Co. 1824? Shoemaker, Augusta Co. Not on muster rolls. Deserted to the enemy Clarksburg, W. Va. 10/13/63. Took oath and sent north.

TUTWILER, SAMUEL H.: Pvt., Co. A. B. Rockingham Co. 10/14/37. Farmer, Augusta Co. Enl. Staunton 7/9/61 age 23. Ab. sick 11/16/61-2/2/62. Present 2/3-4/62. Reenl. 5/1/62. WIA (hand) Sharpsburg 9/17/62. Ab. wounded through 12/31/64. NFR. Farmer, Rolla, Augusta Co. 1897. Died near Verona 6/10/10. Bur. Salem Lutheran Ch. Cem., near Mt. Sidney.

TUTWILER, WILLIAM HENRY: Pvt., Co. F. B. Va. 1838? Not on muster rolls. Deserted to the enemy Clarksburg, W. Va. 10/13/63. Took oath and sent north. Shoemaker, age 32, Mt. Sidney PO, Augusta Co. 1870 census. Merchant and Farmer, Mt. Crawford, Rockingham Co. "Always wore grey, a bow tie, small white mustach, rawboned, about 6 feet high."

TYGRETT, WILLIAM ROBERT: Pvt., Co. E. B. Arnold's Valley, Rockbridge Co. 2/20/33. Farmhand, age 23, Glenwood Dist., Rockbridge Co. 1860 census. Enl. Staunton 8/1/61. Present 11/61-2/62. Ab. sick 3-4/62. Reenl. 5/1/62. Present 5/1-10/31/62, and 1-7/63. AWOL 7/27-2/63. Issued clothing 2/16/64. WIA Winchester 9/19/64. Cap. Cedar Creek 10/19/64. Sent to Point Lookout. Released 6/20/65. 6' 3¾", light complexion, red hair, blue eyes. Farmer, Natural Bridge Dist., Rockbridge Co. 1880 and Greenlee 1900. Entered Lee Camp Old Soldiers Home, Richmond from Roanoke Co. Died there 12/9/19. Bur. Stonewall Jackson Cem., Lexington. "The 'character' of the Regiment."

TYLER, DANIEL: Cpl., Co.H. B. Augusta Co. 1828? Farmhand, age 32, Pond Gap, Augusta Co. 1860 census. Enl. Staunton 7/23/61 age 36. Present 11/61-2/62, appointed Cpl. 1/1/62. present 3-7/62. AWOL 8/1/62-12/63. Dropped as deserter. Enl. in Cav. Farmer, Augusta Co. Died Craigsville 7/9/85. Bur. Ingram Cem., on Rt. 614, west of US 42, Bell's Valley, Rockbridge Co.

TYREE, LARKIN F.: Pvt., Co. K. B. Va. 1824? Blacksmith, age 36, Millboro Springs PO, Bath Co. 1860 census. Enl. Shenandoah Mt. 4/9/62. Present 4/9-30/62 and 5/26/62. Discharged overage of conscription 1862. Blacksmith, Williamsville, 1872. Alive 1917. Bur. Falling Springs Pres. Ch. Cem., near Covington, Alleghany Co.

VAIDEN, G. W.: Pvt., Co. A. Pension application only record. Served 2 years.

VAIDEN, JACOB: 1stLt., Co. A. Pension aplication only record. Enl. 12/61. Served 3 years.

VAN FOSSEN, W. E.: Pvt., Co. F. Not on muster rolls. WIA (leg) Wilderness 5/6/64. NFR.

VAN FOSSEN, WILLIAM E.: Pvt., Co. F. B. Augusta Co. 2/29/40. Farmhand, age 21, Burkes Mill Dist. Augusta Co. 1860 census. Enl. Staunton 7/31/61. Present 11/61-4/62.. Reenl. 5/1/62. Present 8/31-10/31/62 and 1-7/63. WIA (leg) and cap. Gettysburg 7/3/63. Right leg amputated 6 inches below the knee. Turned over to PM 9/29/63. Sent to West's Hospital, Baltimore. Exchanged 11/17/63. Admitted to hospital Richmond 11/17/63. Transf. to hospital Staunton 9/26/64. Requested artificial leg 7/3/64. Retired to Invalid Corps 12/23/64. Age 24, 5' 10", light complexion, blue eyes, light hair, carpenter. Laborer, age 30, Mt. Sidney PO,Augusta Co. 1870 census. Resident, Rolla Mills, Augusta Co. 1888.

VANHORN, SAMUEL: Pvt., Co. E. Not on muster rolls. WIA (hand) Gaines Mill 6/27/62. NFR.

VARNER, HENRY HARRISON: Pvt., Co. F. B. Pendleton Co. 7/26/38. Farmhand, age 22, Northern Dist., Augusta Co. 1860 census. Enl. Staunton 7/16/61 age 23. Ab. sick 11-12/61. Present 1-4/62. Reenl. 5/1/62. Present 5/1/62-12/63. Ab. on leave 12/24-31/63. Present 4/30-12/31/64. Paroled Staunton 5/26/65. Age 26, 6' 1", fair complexion, light hair, grey eyes. Farmer, age 32, Fishersville PO; Augusta Co. 1870 census. Died near Christian's Creek 4/25/87. Bur. Tinkling Spring Pres. Ch. Cem.

VENABLE, WILLIAM GEORGE: Pvt., Co. K. B. near Ft. Lewis, Bath Co. 5/20/19. Farmhand, age 37, Cleeks Mill PO, Bath Co. 1860 census. Not on muster rolls. Unofficial roster indicates WIA Cedar Run 8/9/62. Trans. Co. G. 11lh Va. Cav. Farmer, age 50, Bath CH PO, Bath Co. 1870 census. Died Near McClung, Bath Co. 7/15/06. Bur. Woodland Ch. Cem., Bath Co.

VESS, WILLIAM H.: Pvt., Co. E. B. Collierstown, Rockbridge Co. 2/24/40. Farmhand, age 18, Collierstown PO, Rockbridge Co. 1860 census. Enl. Staunton 8/1/61. AWOL 11-12/61. Present 1-4/62. Reenl. 5/1/62. Present 4/30-10/31/62. AWOL 2 months. Present in arrest 1-2/63, sentenced by Gen. CM to 20 days in Castle Thunder, Richmond and to forfeit pay for same time. Ab. in arrest 3-12/63. Present 4-6/64. Deserted near Lexington while on the march from Lynchburg to Staunton 6/25/64. NFR. Farmhand, age 32, Buffalo Township, Rockbridge Co. 1870 cnesus. Farmer, Lower Collier's Creek. Died there 6/23/16. Bur. Collierstown Pres. Ch. Cem.

VESS, GEORGE W.: Pvt., Co. K. B. Rockbridge Co. 1830? Farmhand, age 30, Millboro Springs PO, Bath Co. 1860 census. Enl. Shenandoah Mt. 4/9/62. Present 4/9-30/62. KIA Cedar Run 8/9/62. Left widow and 3 children.

VESTAL, ROBERT: Pvt., Co.I. Enl. Staunton 7/16/61 age 18. Name cancelled on roll. NFR.

VIA, WILLIAM A.: Pvt., Co. I. B. Rockbridge Co. 1822? Laborer, age 38, Burkes Mill Dist., Augusta Co. 1860 census. Enl. Staunton 7/15/61. Present 11/61-4/62, fined by CM $11.00 for AWOL. KIA Port Republic 6/9/62. Left widow.

VINES, JAMES H.: Pvt., Co. B (2nd). B. Va. 1828? Saddler, age 22, 1st Dist., Augusta Co. 1860 census. Enl. Waynesboro 7/15/61. Ab. sick 11/61-3/62. Present 3/8-4/62. Reenl. 5/1/62. Present 5/1-7/16/62. AWOL 7/14-9/24/62. Present 9/15-11/30/62. AWOL 12/6/62-1/21/63. Present confined in Regt'l Guard House 1/22-2/63. Ab. sick 4/29-30/63. Cap. Gettysburg 7/3/63. Sent to Ft. Delaware. Died there of chronic diarrhea 10/31/63. Bur. Finn's Point National Cem., N. J.

VINES, MAJOR COLEMAN: Pvt., Co. G. B. Greenville, Augusta Co. 1834. Farmhand, age 25, 1st Dist., Augusta Co. 1860 census. Enl. Staunton 8/2/61. Present 11/61-4/62. Reenl. 5/1/62.WIA (ankle) Port Republic 6/9/62. Ab. wounded through 2/63. Present 3-4/62. AWOL 5/9-20/63. Present in arrest 12/25/63. Ab. under sentence of CM 4/30-10/31/64. Present 11-12/64. Paroled Staunton 5/1/65. Age 25, 6', dark complexion, black hair, grey eyes. Railroad employee, age 35, Mt. Sidney PO, Augusta Co. 1870 census. Railroad Engineer, Huntington, W. Va. Died there 8/25/06. Bur. Thornrose Cem., Staunton.

VINES, WILLIAM JOSEPH: Pvt., Co. B. (2nd). B. Augusta Co. 1841? Farmhand, age 19, 1st Dist., Augusta Co. 1860 census. Enl. Waynesboro 7/15/61. Present 11/61-4/62. Reenl. 5/1/62. Present 5/1-7/16/62. AWOL 7/17-9/24/62. Present 9/25-10/31/62. AWOL 12/6/62-1/19/63. Present confined in Regt'l Guard House 1/20-2/63. Present 3-4/62, fined 6 months pay in addition to time ab. by Brig. CM. Present 4/30-5/20/63. AWOL 5/21-30/63. Present 6/1-12/63, and 4/30-12/31/64. Paroled Staunton 5/19/65. Age 21, 5' 8", dark complexion, dark hair, grey eyes. Died near Elkton, Rockingham Co. 8/31/86.

VINT, JAMES M.: Pvt.,Co. C. B. Augusta Co. 1835. Shoemaker, Deerfield, Augusta Co. Enl. Staunton 7/16/61 age 21, Present 11/61-4/62. Reenl. 5/1/62. WIA Malvern Hill 7/1/62. Ab. wounded through 8/31/62. AWOL 11/27/62-2/63, fined $36.66 for 100 days AWOL. Ab. 3-4/63. AWOL 7/25-8/31/63. Present 9-10/63, fined $11.00 by CM. Present 11-12/63. WIA (lost finger) Spotsylvania CH 5/12/64. Present 6-9/64. AWOL 9/10 until deserted to the enemy in W. Va. 11/17/64. Took oath and released. Age 24, 5' 11", florid complexion, grey eyes, brown hair. Farmhand, age 24, Mt. Sidney PO, Augusta Co. 1870 census. Died 1909. Bur. Mt. Horeb Cem., on Rts 732 and 752 south of Rt 33, Hinton, Rockingham Co.

WADDELL, LIVINGSTON: Surgeon, F&S. B. Louisa Co. 10/11/1799. Grad UPa. MD. Doctor, age 58, Burkes Mill Dist., Augusta Co. 1860 census. Appointed Surgeon 8/19/61. Present 4-8/62. Transf. Hospital duty Staunton 8/15/62. Surgeon, hospital Staunton to end of war. Doctor, Waynesboro and Lexington. Died Lexington 8/17/81. Bur. in unmarked grave beside wife in Riverview Cem., Waynesboro.

WAID, LUTHER R.: Pvt., Co. I. B. Augusta Co. 1839? Farmer, Augusta Co. 1861. Enl. Staunton 7/16/61 age 22. Present 11/61-2/62. AWOL 2/24-28/62, fined $11.00 by CM. Present 3-4/62. Reenl. 5/1/62. Present 5/1-617/62. AWOL 6/18-10/25/62, fined $46.50 AWOL 11/22/62-1/63. In hospital Staunton with rheumatism 2/2/63. Cap. Williamsport, Md. 7/9/63. Sent to Point Lookout. Took oath and joined U. S. Service 1/25/64. "Died of wounds received in battle" in Register of Deaths for Augusta Co. when he died County Poor House 5/5/67, age 26. Left widow.

WAID, WALLACE H.: Pvt., Co. H. B. 1841? Resident, Craigsville, Augusta Co. Enl. Staunton 7/23/61 age 19. Present 11/61-4/62. Reenl. 5/1/62. Present 5/1/62-12/63, detailed as Teamster since 7/23/61. Issued clothing 4/27/64. Cap. Bethesda Ch. 5/30/64. Sent to Point Lookout. Transf. Elmira. Released 6/30/65. 5' 9", florid complexion, dark hair, hazel eyes. Alive 1885.

WALKER, CLINTON M: 2ndSgt., Co. F. B. 1842? Coachmaker, Augusta Co. Enl. Valley Mills 5/1/62, as 4thSgt. WIA (face) Sharpsburg 9/17/62. Ab. wounded through 12/31/62. Present 1-12/63. Ab. on leave 3-4/64. WIA (face) Spotsylvania CH 5/12/64. Ab. wounded through 12/31/64. Deserted to the enemy in W. Va. 3/15/65. Took oath and released. Age 23, 5' 8", florid complexion, dark eyes, dark hair.

WALLACE, ALEXANDER ALLBRIGHT: 3rdCpl., Co. B. (1st). B. Rockbridge Co. 2/23/31. Carpenter, Rockbridge Co. Enl. Fairfield 7/10/61. Present until tansf. 9/28/61.

WALLACE, EDWIN, SR.: Pvt. Co. B. (1st). B. Augusta Co. 1836? Farmer, age 24, South River Dist., Rockbridge Co. 1860 census. Enl. Fairfield 7/10/61. Ab. on detail in Staunton when transf. 9/28/61. Brother of Samuel Wallace.

WALLACE, JAMES WILLIAM M. "Jim": Pvt. Co. E. B. Augusta Co. 6/20/44. Plasterer, age 17, South River Dist., Rockbidge Co. 1860 census. Attended Mr. Pinkerton's School, Midway, now Steele's Tavern. Enl. Staunton 8/1/61. Ab. sick 11/61-4/62. Reenl. 5/1/62. Present 5/2-10/31/62, and 1-3/63.Ab. sick with fever in hospital Richmond and Lynchburg 4/15-12/31/63.NFR. Enl. Capt. Opie's Co., Imboden's Cav. Farmer and Stockraiser, Greenville, Augusta Co. 1885. Died near Spottswood 3/31/23. Bur. Bethel Pres. Ch. Cem.

WALLACE, SAMUEL: Senior 1stLt., Co. B (1st). B. Va. 2/23/34. Farmer, age 26, 6th Dist., Rockbridge Co. 1860 census. Appointed 6/15/61. Present until transf. 9/28/61. Brother of Edwin Wallace, Sr.

WALTON, BENJAMIN T.: Capt., Co. K. B. Alleghany Co. 1822? Farmer, age 38, Cleeks Mill PO, Bath Co. 1860 census. Pvt., Capt. Davis's Co., 81st Va. Militia. Enl. Shenandoah Mt. 4/9/62. Present 4/9-30/62. WIA (thigh) Port Republic 6/9/62. DOWs in hospital Staunton 6/11/62 age 40. Left widow and 7 children.

WALTON, HENRY S.: Pvt. Co. F. On postwar roster. Enl. 1861. Served 4 years.

WALTON, JOHN A.: Pvt., Co. K. B. Bath Co. 2/1/45. Enl. Shenandoah Mt. 4/9/62 as substitute for his father. WIA Port Republic 6/9/62. Enl. Co. G, 18th Va. Cav. 2/4/63. Farmhand, age 25, Bath CH PO, Bath Co. 1870 census. Farmer, Healing Springs, Bath Co. 1897. Died Cedar Creek Dist., Bath Co. 11/7/24. Bur. Jones Cem.

WAMPLER, BENJAMIN F.: Pvt.,Co.C. B. Augusta Co. 1838? Farmhand, age 22, Burkes Mill Dist., Augusta Co. 1860 census. Enl. Staunton 7/16/61 age 23. Ab. sick 11-12/61. Present 1-3/62. AWOL 4/19-30/62. Ab. on detached service as nurse in hospital Staunton 5/1/62-12/31/64. Paroled Staunton 5/22/65. Age 25, 5' 8", fair complexion, light hair, grey eyes, illiterate. Farmer, age 32, Mt. Sidney PO, Augusta Co. 1870 census. Farmer, Weyer's Cave 1897. Farmer, Barterbook, Augusta Co. 1910 census. Died before 7/22. Bur. Union Pres. Ch. Cem. Brother of Simon W. Wampler.

WAMPLER, SIMON WILLIAM: Pvt.,Co.F. B. Augusta Co. 8/10/45. Farmhand, age 18, Burkes Mill Dist., Augusta Co. 1860 census. Enl. Staunton 7/31/61. Present 11/61-4/62. Reenl. 5/1/62. WIA (head) Cross Keys 6/8/62. Present 8/31-12/31/62, and 1-12/63. Present 4/30-12/31/64. Paroled Staunton 5/1/65. Age 22, 5' 9", dark complexion, black hair, dark eyes. Farmer, Barterbrook, Augusta Co. 1910 census. Died Augusta Co. 1921. Bur. Barren Ridge Ch. of the Brethren. Brother of Benjamin F. Wampler.

WARD, LEVI: Pvt., Co. unknown. Not on muster rolls. Cap. Alleghany Mt. 12/13/61. NFR.

WASKEY, ROBERT C.: Pvt., Co. I.B. Va. 1843? Resident, age 17, 1st Dist., Augusta Co. 1860 census. Enl. Staunton 7/16/61 age 19. Present 11/61-4/62.Reenl. 5/1/62. WIA (thigh) McDowell 5/8/62. Ab. wounded through 4/63. Present 4/30-12/63. WIA (left leg) Spotsylvania CH 5/12/64. Ab. wounded through 12/31/64.Paroled Staunton 5/17/65. Age 22, 6', light complexion, light hair, blue eyes.

WASSON, JOSEPH E.: 1stSgt., Co. F. B. Centre Co., Pa. 1831? M. E. Preacher, Augusta Co. 1861. Enl. Staunton 7/31/61. Ab. on leave 12/15-31/61, rank shown as Pvt. Present 1-2/62. AWOL 30 days and fined $2.00 by CM. Present 3-4/62, claimed to be exempt as minister. Cap. Winchester 6/1/62. Exchanged 9/21/62. Discharged 10/30/62. M. E. Peacher.

WATKINS, JOHN K.: 2ndLt., Co. E. B. near Gilmore's Mill, Rockbridge Co. 1837? Clerk, age 23, Lexington, Rockbridge Co. 1860 census. Enl. Co. K, 11th Va. Inf. 5/25/61, age. Elected 2ndLt. this Co. 10/25/62. Present 1-5/63. Presence or absence not shown 6-7/63. Ab. detached to arrest deserters 8/14-31/63. Present 9-12/63, and 4/30-9/5/64. Ab. on leave 9/6-30/64. Commanding Co.'s A and F 11/18/64. Commanding Co. 3/1 until surrendered Appomattox 4/9/65. Clerk in Drygoods Store, age 33, Natural Bridge Dist., Rockbridge Co. 1870 census. Died Rockbrige Co. 1818. Brother of Thomas H. Watkins.

WATKINS, THOMAS H.: LtCol., F&S. B. near Gilmore's Mill, Rockbridge Co. 1839? Manager of Railroad, age 21, Glenwood Dist., Rockbridge Co. 1860 census. Ironmaster, Glenwood Furnace. Enl. Staunton 8/1/62 as Capt., Co. E. Present 11/61-4/62. Reelected 5/1/62. Present 5/1-8/62. WIA McDowell 5/8/62. WIA Gaines Mill 6/27/62. Commanding Regt. 2nd Manassas, Chantilly, Harper's Ferry, and Sharpsburg. Present 10-12/62. Ab. on leave 1-2/63. Present 3-10/63. Promoted Major 10/24/63. Present 11/63-2/64. Promoted LtCol. 3/2/64. Present Wilderness and Spotsylvania CH. Commanding Regt. 5/12/64 after Col Skinner was WIA. Commanding Regt. until "shot from his horse and killed while leading a dashing charge of his Regiment" Bethesda Ch. 5/30/64. Age 24, 6' 2 ", brown hair, grey eyes. Bur. on the field but moved under large walnut tree at the Stark house. Still there 1912. In 1861 he resigned his position as manager of Glenwood Furnace and enlisted in the company being formed in the neighborhood. When the company was organized he was chosen Captain by an almost unanamious vote. He initially declined to accept the position, professing lack of military training and experience. When the company voted again he received every vote but one, doubtlessly his own. He immediately devoted himself to the study of military tactics and the duties of a soldier and an officer and soon became the most accomplished in the command. Col. Baldwin declared him "second to no Captain in the Confederate Army." He possessed an intellect of unusual brightness and owned an energy which was untiring and a sense of duty and devotion to Virginia. Having a symmetrical, well knit, graceful figure, athletic and manly in all his movements. His features regular and striking. "As an officer he won the devotion, affection and confidence of his men by his uniform courtesy and considerable kindness, his care for their wants, their comfort and welfare and yet required from them the same faithful obedience of orders, and conscientious discharge of duty, which he rendered to his superior officers. He shared their hardships and privations, and in every battle he commanded from the front and led the charge." "One of the most gallant and promising officers and noblest men in the Confederate army." Brother of John K. Watkins.

WATTS, WILLIAM A.: Pvt., Co. I. B. Va. 1844? Farmhand, age 16, Northern Dist., Augusta Co. 1860 census. Enl. Staunton 7/16/61 age 17. Ab. sick 11/61-2/6/62. Present 2/7-4/62. Reenl. 5/1/62. WIA (thigh) Port Republic 6/9/62. Present 8/30-10/31/62. Present 17/63. Cap. Fairfield, Pa. 7/6/63. Sent to Ft. Delaware. Took oath and joined 3rd Md. Cav. (U.S.) 9/22/63.

WAY, JAMES S.: Pvt. Co. B. (2nd). B. Alleghany Co. 9/16/36. Stage Driver, age 22, Burkes Mill Dist., Augusta Co. 1860 census. Enl. West View 5/1/62. WIA (thigh, arm and 3 other places) McDowell 5/8/62. Ab. wounded until discharged for disability 6/20/63. Age 24, 5' 10", dark complexion, grey eyes, dark hair, blacksmith. Crippled rest of life. Died Waynesboro 10/29/06. Bur. UDC lot, Riverview Cem., Waynesboro.

WEAVER, JAMES A.: Pvt., Co. G. B. Augusta Co. 10/15/33. Pvt., Capt. Ellis' Co. B. 32nd Va. Militia. Requested discharge 3/15/62 age 27. On post war roster. Farmer, Milneville 1897. Died Grottoes 4/22/14. Bur. Pleasant Valley Ch. of the Brethren Cem.

WEAVER, SAMUEL HENRY: Pvt., Co. I. B. Augusta Co. 1831? Farmer, age 31, 1st Dist., Augusta Co., 1860 census. Enl. Staunton 7/16/61 age 32. Present 11/61-4/62. Reenl. 5/1/62. WIA (leg) 2nd Manassas 8/298/62. Present 8/31-10/31/62. AWOL 12/1/62-4/7/63, fined 6 months pay by CM. Ab. sick 7/16-10/63. Present 11-12/63 and 4/30-12/31/64. Surrendered Appomattox 4/9/65. Farmer and Tanner, Walker's Creek Dist., Rockbridge Co. 1870 census. Died Riverhead Dist., Augusta Co. 10/83.

WEAVER, WILLIAM: Pvt., Co. G. B. Augusta Co. 5/5/44. Enl. Staunton 8/2/61. Present 11/61-4/62. Reenl. 5/1/62. Present 5/1-8/31/62. Ab. sick 10/14/62-2/63. Present 3-4/63. WIA (side) Gettysburg 7/3/63. Ab. wounded through 12/31/64. Paroled Staunton 5/11/65. Age 22, 6', fair complexion, dark hair, hazel eyes, illiterate. Blacksmith, age 25, Western Dist., Augusta Co. 1870 census and 1880. Died Augusta Co. before 1900. Probably buried Pineville Cem., Rockingham Co. next to wife.

WEBB, GEORGE W.: Pvt., Co. H. B. Augusta Co. 1842? Farmhand, age 18, 1st Dist., Augusta Co. 1860 census. Enl. Camp Alleghany 3/21/62. Present 3-4/62. Ab. sick 5/9/62-2/63. Present 3-12/63. WIA Wilderness 5/6/64. DOWs 5/7/64 age 20. Brother of James L. Webb.

WEBB, JAMES L.: Pvt., Co. H. B. Augusta Co. 1838? Farmhand, age 22, 1st Dist., Augusta Co. 1860 census. Enl. Camp Alleghany 3/321/61. Present 3-4/62. Reenl. 5/1/62. Present 5/1-8/31/62. Ab. sick 10/18/61-10/63 bronchitis. Present 11-12/63. Ab. sick 5/4-12/31/64. Paroled Staunton 5/15/65. Age 27, 5' 7", light complexion, light hair, grey eyes. Shoemaker, Augusta Co. 1885. Bur. Mt. Tabor Lutheran Ch. Cem. Brother of George W. Webb.

WEBBER, JOHN: Pvt., Co. H. Enl. Valley Mills 4/24/62 as substitute. Deserted 6/1/62. NFR.

WEIR, W. T.: Pvt., Co. unknown. Not on muster rolls. Bur. Thornrose Cem., Staunton 12/25/61, grave 281.

WELCH, HENRY D.: Pvt., Co. H. B. Va. 10/3/35. Farmhand, age 23, Burkes Mill Dist., Augusta Co. 1860 census. Enl. Staunton 7/23/61 age 27. WIA (leg) on picket Alleghany Mt. 12/12/61. Leg amputated. DOWs there 2/13/62. Age 26, 5' 4", fair complexion, blue eyes, dark hair. Bur. Scutterlee Cem., on Rt 728 2/10's mile east of intersection with Rt 732 at Frank's Mill, Augusta Co.

WELLER, CHARLES LANSTRAM: Capt., Co. C. B. Richmond, Henrico Co. 1844? Enl. Staunton 7/16/61 age 18 as Pvt. Present 11/61-2/62, promoted 4thSgt. 2/10/62. Present 3-4/62, rank shown as 2ndSgt. Ab. sick 8/26-31/62. Present 9/1-6-2-1/9/63, detailed as SgtMajor since 10/62. WIA Fredericksburg 12/13/62. Elected 2ndLt. 3/15/63. WIA (leg) Fredericksburg 5/5/63. Ab. wounded through 8/31/63. Present 9-10/63. Ab. sick 11/3/12/63. Present Wilderness. WIA (left Knee) Spotsylvania CH 5/12/64, promoted 1stLt. same date. Returned to duty 10/23/64. Commanding Co. 11-12/64. Acting Adj. 2/65. Promoted Capt. 2/23/65. Commanding Regt. 3/18/65. Capt. Ft. Stedman 3/25/65. Sent to Old Capitol Prison. Transf. Ft. Delaware. Released 6/17/65. Resident of Richmond, 5' 7", light complexion, light hair, dark eyes. Moved to Staunton 1868. Store Clerk, age 26, Staunton 1870 census. Merchant; Member, City Counciil 8 years. Lt. Commander, Stonewall Jackson Camp CV, Staunton. Living Fishersville 1913. Died Staunton 9/24/15. Bur. Thornrose Cem.

WELLS, CHRISTOPHER C.: Pvt., Co. E. Enl. Camp 52 Va. Regt. 3/31/64. WIA and cap. near Spotsylvania CH 5/19/64. Sent to Point Lookout. Transf. Elmira. Exchanged 11/15/64. In hospital Richmond with chronic diarrhea 12/16/64-2/23/65. Surrendered Appomattox 4/9/65.

WELLS, JAMES HENRY: 4thCpl., Co. E. B. Va. 2/16/32. Farmhand, age 28, Natural Bridge Dist., Rockbridge Co. 1860 census. Enl. Staunton 8/1/61. Present 11/61-4.'62. Reenl. 5/1/62. Present 4/30-10/31/62, AWOL 1 month, rank shown as Pvt. Present in arrest 1-2/63. Ab. in arrest Castle Thunder, Richmond and fined 4 months pay by Gen. CM. Present 4/30-12/63 and 4/30-12/31/64. Surrendered Apopomattox 4/9/65. Living Glasgow, Rockbridge Co. 1913. Died Glen Wilton, Rockbridge Co. 1/22/14.

WENGER, HENRY M.: Pvt., Co. F. B. Edom, Rockingham Co. 10/31/22. Resident of Harrisonburg. Enl. Valley Mills 4/20/62. Cap. and paroled Winchester 6/162. Age 32, 5' 4". Exchanged. Deserted 6/6/62. "Confined in Libby Prison, Richmond for a long time for Union sentiments." Died Augusta, Butler Co., Kansas 11/10/81. Bur. Burlington, Coffey Co., Kan.

WENGER, JACOB.: Pvt., Co. F. B. Rockingham Co. 4/8/43. Resident of Dayton. Enl. Valley Mills 4/20/62. Mennonite. Informed one of his officers of the "no shoot pledge" because of his religious belief. By compulsion he would be obedient so far but that he would harm no one. Cap.. andparoled Winchester 6/1/62. Age 18, 5' 6". Deserted NFR. D. 2/5/25.

WEST, WILLIAM C.: Pvt., Co. E. B. Va. 1847? Student, age 13, Burkes Mill Dist., Augusta Co. 1860 census. On post war roster.

WESTERN, JAMES: 1stLt., Co. G. B. Va. 2/27/21. Farmer, age 29, Burkes Mill Dist., Augusta Co. 1860 census. Capt. of Militia Co. Enl. Staunton 8/2/61. Ab. sick 1-2/62. Present 3-4/62. Not reelected 5/1/62. Farmer, age 48, Mt. Sidney PO, Augusta Co. 1870 census. Died 4 miles east of New Hope 6/9/95. Bur. Mt. Horeb Pres. Ch. Cem., near New Hope.

WHEELER, GEORGE: Pvt., Co. I. Pvt., Capt. G. B. Koiner's Co. Militia 3/62, exempt. WIA (thigh) Mine Run 11/17/63. In hospital Gordonsville 12/2/63. NFR.

WHEELER, JACOB E.: Pvt., Co. G. B. Va. 1841? Student, age 19, Burkes Mill Dist., Augusta Co. 1860 census. Enl. Staunton 8/2/61. Present 11/61-2/62. Ab. sick 3/1-10/3/62, "Febr. Con. Cour." Present 11/62-12/63. Cap. Bethesda Ch. 5/30/64. Sent to Point Lookout. Transf. Elmira. Died there of chronic diarrhea 10/5/64. Bur. Woodlawn National Cem., grave 555.

WHEELER, JAMES DAVID: Pvt., Co. B (2nd). B. Nelson Co. 10/6/43. Farmhand, Augusta Co. Waynesboro 7/15/61. Present 11/6-4/62. WIA (thigh) McDowell 5/8/62. Present 6/62-2/63, detailed as Ambulance Driver since 7/17/62. Present 3-12/63, and 4/30-12/31/64. WIA Ft. Stedman 3/25/65. Cap. in hospital Richmond 4/3/65. Turned over to PM 4/14/65. NFR. Farmhand, age 37, Mt. Sidney PO, Augusta Co. 1870 census. Miller, Riverheads Dist., Augusta Co. Died there 7/92.

WHEELER, JOEL YANCEY: Pvt., Co. B.(2nd). B. Nelson Co. 1836. Farmhand, Augusta Co. Enl. Waynesboro 7/15/61. Present 11/61-4/62. Reenl. 5/1/62. Deserted 7/18/62. NFR. Living Waynesboro 10/6/68. Farmhand, age 34, Mt. Sidney PO, Augusta Co. 1870 census.

WHITE, BENJAMIN F.: Pvt., Co. B (2nd). B. 1837? Enl. Waynesboro 7/15/61. Present 11/61-4/63, detailed as Teamster in Regt'l Train. Ab. detailed as Teamster in QM Dept. Staunton on Va. 115 4/30-12/63 since 7/27/63. Present 4/30-6/64. Ab. sick in hospital with debility 6/14-16/64. AWOL 6/25-12/31/64. Dropped as deserter. Paroled Winchester 4/25/65. Age 28, 5' 8", light complexion, light hair, black eyes, illiterate.

WHITE, GEORGE: Pvt., Co. B (1st). B. Va. 1840? Distiller, age 21, Waslke's Creek Dist., Rockbridge Co. 1860 census. Enl. Staunton 8/1/61. Ab. sick when transf. 9/28/61.

WHITE, ISAAC MATTHEW: Pvt., Co. B. 1st. B. Ireland 1823? Farmhand, Rockbridge Co. Enl. Fairfield 7/16/61. Ab. on detail Staunton when transf. 9/28/61.

WHITE, JAMES A.: 2ndLt., Co. H. B. 1831? Farmer, Waynesboro, Augusta Co. Enl. Staunton 7/23/61 age 28 as 1stSgt. Present 11/61-4/62. Elected 2ndLt. 5/1/62. Present 5-6/62. WIA (shoulder) Port Republic 6/9/62. Present 7-9/62. Deserted to the enemy Hedgesville 9/24/62. Took oath and released. Age 30, 5' 10", sandy complexion, grey eyes, red hair. Unofficial report stated he died of disease in Baltimore during the war.

WHITE, JAMES M.: Pvt., Co. E. B. Va. 1842? Farmhand, age 17, Lexington Dist., Rockbridge Co. 1860 census. Not on muster rolls. Died of measles Staunton 10/2/61 age 19.

WHITE, JOHN N.: Pvt., Co. H. B. Va. 8/19/19. Carpenter, age 39. 1st Dist., Augusta Co. 1860 census. Enl. Staunton 7/23/61 age 44. Present 1/61-2/62. AWOL 3/3-13/63. fined $11.00 by CM. Present 3/14-8/31/62. Detailed in Pioneer Corps 10/62. Discharged for over age of conscription 11/2/62. Died Bell's Valley, Augusta Co. 4/4/02. Bur. Lebanon Pres. Ch. Cem.

WHITE, JOHN P.: Pvt., Co. B. (1st). Enl. Staunton 8/1/61. Present until transf. 9/28/61.

WHITE, ROBERT: Pvt., Co. B (1st). B. Rockbrige Co. 1826. Enl. Staunton 8/1/61. Present until transf. 9/28/61.

WHITE, WEAVER W.: Pvt., Co. E. Enl. Staunton 8/1/61. Present 11/61-4/62. AWOL 5/20-10/31/62. Disch. Staunton 11/27/62, reason unknown. NFR.

WHITESELL, BENJAMIN HARRISON: Pvt., Co. I. B. Va. 6/29/24. Farmhand, age 32, 1st Dist., Augusta Co. 1860 census. Enl. Staunton 7/16/61. Ab. sick 11/61-2/62. Present 3-4/62. Reenl. 5/1/62. Ab. sick 5/6/62-4/63, hernia, however, he served as guard in hospital Staunton 8/1/62-6/63. Present 7-12/63, and 4/30-5/31/63. Ab. sick in hospital Staunton 6/1-12/31/64, illiterate. NFR. Farmhand, age 48, Fishersville PO, Augusta Co. 1870 census. Died Greenville 10/15/10. Bur. St. John's Lutheran Ch. Cem., near Middlebrook.

WHITESELL, DANIEL THOMAS: Pvt., Co. I. B. Augusta Co. 10/28/34. Farmhand, age 24, 1st Dist., Augusta Co. 1860 census. Enl. Staunton 7/16/61 age 28. Present 11/61-4/62. Reenl. 5/1/62. Present 8/30-10/31/62, and 1-4/63. KIA Gettysburg 7/3/63. Age 29, 5' 5", fair complexion, blue eyes, light hair. Left widow.

WHITESELL, JOHN W.: Pvt., Co. B (1st). B. Rockbridge Co. 1833? Farmer, Augusta Co. On postwar roster.

WHITESELL, PETER F.: Pvt., Co. I. B. Augusta Co. 5/7/23. Carpenter, age 36, 1st Dist., Augusta Co. 1860 census. Enl. Staunton 7/16/61 age 36. Ab. sick 11/5/61-10/31/62. Present 1-4/63. Discharged over age of conscription 5/27/63. Age 42, 5' 9", dark complexion, hazel eyes, dark hair. Farmhand, age 48, Fishersville PO, Augusta Co. 1870 census. Died near Mint Springs 9/2/01. Bur. St. John's Lutheran Ch. Cem., near Middlebrook.

WHITLOCK, WILLIAM B.: Pvt., Co. H. B. Va. 1825? Plasterer, age 26, Beverly Manor Dist., Augusta Co., 1850 census. Resident, Augusta Co. On postwar roster. Enl. Staunton 1861 age 35. Discharged Bunker Hill 10/16/62 overage of conscription. Died before 1885.

WHITMORE, FIELDEN: Pvt., Co. F. B. Augusta Co. 3/11/41. Pvt., Co. L, 160th Va. Militia 2/62 age 21. Enl. Valley Mills 4/24/62. Present 8/31/62-2/63. Ab. sick in hospital 5/31/8/31/63. Present 9-10/63. Ab. sick in hospital 10/27-12/28/63. Present 3-4/7/64. WIA (breast) Bethesda Ch. 5/30/64. Cap. Cedar Creek 10/19/64. Sent to Point Lookout. Released 6/19/65. 5' 9½", dark complexion, brown hair, blue eyes. Farmer, age 29, Mt. Sidney PO, Augusta Co. 1870 census. Living West View 1906. D. Staunton 4/19/08. Bur. Thornrose Cem.

WHITMORE, JAMES G.: Pvt., Co. B (2nd). B. Augusta Co. 1834? Carpenter, Waynesboro. Enl. Waynesboro 7/15/61. Present 11/61-4/62. AWOL 6/1/62 until discharged for "phthisis Pulmonalis" 3/5/63. Age 26, 5' 10", fair complexion, brown eyes, dark hair. Laborer, age 33, Williamsville Dist., Bath Co. 1870 census.

WHITMORE, JOSIAH L.: Pvt., Co. F. B. Augusta Co. 12/3/30. Trader, age 29, Northern Dist., Augusta Co. 1860 census. Enl. Staunton 7/31/61. Present 11/61-4/62. Reenl. 5/1/62. WIA Port Republic 6/9/62. DOWs in hospital Staunton 6/20/62. Age 32, 5' 8", fair complexion, light eyes, light hair, farmer. Bur. Parnassus Meth. Ch. Cem. Brother of Samuel M. and Solomon Whitmore.

WHITMORE, SAMUEL MARTIN: Pvt., Co. F. B. Rockingham Co. 5/12/39. Farmhand, age 21, Northern Dist., Augusta Co. 1860 census. Enl. Staunton 7/31/61. Present 11/61-4/62. Reenl. 5/1/62. KIA McDowell 5/8/62. Age 23, 5' 7", fair complexion, grey eyes, light hair, farmer. Bur. Parnassus Meth. Ch. Cem. "Samuel had no superior and few equals of moral worth." Brother of Josiah L. and Solomon Whitmore.

WHITMORE, SOLOMON: Pvt., Co. F. B. Rockingham Co. 4/16/28. Farmhand, age 32, Northern Dist., Augusta Co. 1860 census. Enl. Staunton 8/27/62. WIA Sharpsburg 9/17/62. DOWs and typhoid fever in Augusta Co. 1/17/63. Age 34, 5' 8½", dark complexion, blue eyes, black hair., farmer. Left widow and 4 children. Bur. Parnassus Mehtodist Ch. Cem. Brother of Josiash L. and Samuel M. Whitmore.

WHITTEN, JOHN W.: Pvt., Co. E. B. Rockbridge Co. 12/7/45. Farmhand, age 16, Lexington Dist., Rockbridge Co. 1860 census. Enl. Staunton 8/1/61. Present 11/61-2/62. Ab. sick 3-4/62. Died of typhoid fever in hospital Lynchburg 5/7/62, age 17.

WIKLE, SAMUEL HENRY: Pvt., Co.D. B. Augusta Co. 1834? Farmhand, age 27, Burkes Mill Dist., Augusta co. 1860 census. Enl. Staunton 7/16/61. Present 7/16/61-2/62. AWOL 2/16-28/62. Fined $11.00 by CM. Present 3-4/62. Reenl. 5/1/62. WIA near Strasburg 6/1/62. Ab. wounded through 4/63. Present 4/30-6/83.63. AWOL 6/9-8/5/63, fined $20.90. Present 8/6-12/63. WIA (arm) Bethesda Ch. 5/30/64. Ab. wounded through 12/31/64, illiterate. NFR.

WILDS, WILLIAM: Pvt., Co. unknown. Not on muster rolls. Deserted to the enemy Bermuda Hundred 3/27/65. Sent to Washington, D. C. Took oath and transportation furnished to Harrisburg, Pa.

WILHELM, WILLIAM: Pvt., Co. E. B. Va. 1823? Farmhand, age 36, Collierstown PO, Rockbridge Co. 1860 census. Enl. Staunton 8/1/61. Present 11/61-2/62. Ab. sick 3-4/62. WIA (head) Cross Keys 6/8/62. Ab. wounded through 4/63. Present 4/30-12/63. KIA Bethesda Ch. 5/30/64. left widow and 3 children.

WILKINSON, JOHN ALFRED: 3rdCpl., Co. E. B. Collierstown, Rockbridge Co. 9/20/42. Surveyor, age 17, Collierstown PO, Rockbridge Co. 1860 census. Enl. Staunton 8/1/61 as Pvt. Present 11/61-4/62, appointed 3rd Cpl. 11/1/61. Reenl. 5/1/62. Ab. sick 9/15/62-2/63. Ab. on detached service with Enrolling Officer 11th Congressional Dist., 3-9/63. AWOL 9/15-12/63. Enl. 18th Va. Cav. without authority. Listed as deserter from 52nd Va. Inf. 1/30/64. Attended Washington College 1866. Farmhand, age 28, Buffalo Township, Rockbridge Co. 1870 census. Farmer, School Teacher and Postmaster, Collierstown. Died there 5/16/16. Bur. Collierstown Pres. Ch. Cem.

WILKINSON, WILLIAM WESLEY: Pvt., Co. E. B. 9/19/40. Enl. Staunton 4/12/62. Present 4/62. Ab. sick 10/14/62-12/31/63. Dropped as deserter. Died Rockbridge Co. 3/21/29. Bur. High Bridge Pres. Ch. Cem., near Natural Bridge.

WILLIAMS, ANDREW J.: Pvt., Co. B. (2nd). B. Goochland Co. 1822? Blacksmith, age 37, Burkes Mill Dist., Augusta Co. 1860 census. Enl. Waynesboro 7/15/61.Ab. sick 11-12/61. Ab. on detached service as blacksmith, Staunton 1-12/62, non-conscript. Discharged 1/7/63. Age 40, 6', fair complexion, dark hazel eyes, dark hair. Resident of Waynesboro. Wagonmaker and Blacksmith, age 48, Mt. Sidney PO, Augusta Co. 1870 census. Died Augusta Co. 1926.

WILLIAMS, ELISHA BROWN: Pvt., Co. K. B. Bath Co. 1827? Farmer, age 33, Bath CH PO, Bath Co. 1860 census. Pvt., Capt., Davis's Co., 81st Va. Militia. Enl. Shenandoah Mt. 4/9/62. Present 4/9-30/62, and 5/26/62. Cap. Front Royal 5/30/62. Enl. Co. A., 62nd Va. Inf. 11/15/62 and deserted. Enl. "Bath Cavalry", 14th Va. Cav. 1/1/63 at Williamsville. Capt. Bath Co. 8/25/63. Exchanged '64. Enl. Co. F, 11th Va. Cav. WIA Wilderness 5/64. Farmer, age 31, Bath CH PO, Bath Co. 1870 census, illiterate. Farmer, Healing Springs, Bath Co. 1897. Farmer, Nelson Co. 1902. Died Farina, Albemarle Co. 7/25/06.

WILLIAMS, NOAH: Pvt., Co. G. Listed on postwar roster.

WILLIAMS, THOMAS JEFFERSON: Pvt., Co. K. B. Healing Springs, Bath Co. 5/16/43. Farmhand, age 17, Cleeks Mill PO, Bath Co. 1860 census. Enl. Shenandoah Mt. 4/9/62. Present 4/9-30/62. AWOL 5/25/62-12/63. Enl. Co. F, 11th Va. Cav. 9/2/62. Transf. officially 5/3/64. Farmer, Healing Springs, Bath Co. 1897. Farmer, Bath Co. 1897. Died there 11/21/23. Bur. Healing Springs Cem.

WILSON, JOHN: Pvt., Co. H. B. Va. 1818? Farmer, age 41, 1st Dist., Augusta Co. 1860 census. Pvt., Capt. Snapp's Co., 93rd Va. Militia 3/62 exempt for spinal irritiation, age 44 years, 6 months. Enl. McDowell 4/6/62 as subst. Present 4/62. Ab. sick 6/26-10/1/63 and 10/18/63-12/64. NFR. Died near Swoope's Depot, Augusta Co. 9/27/70.

WILSON, JOHN A.: Pvt., Co. B (1st). B. Va. 6/11/30. Farmer, Age 30, South River Dist., Rockbridge Co. 1860 census. Enl. Fairfield 7/10/61. Present until transf. 9/28/61.

WILSON, JOSEPH B.: Pvt., Co. G. B. Albermarle Co. 7/17/35. Carpenter, age 24, Burkes Mill Dist., Augusta Co. 1860 census. Enl. Staunton 8/2/61. Present 11/61-2/62. Ab. sick 3-4/62. AWOL 4/30 until returned by force 11/6/62. Present in arrest 11/7/62-2/63, fined $66.00 Present 4/63. Present 4/9-30/62, and 5/26/62. Cap. Front 8-12/63. Issued clothing 3/28/64. Ab. on leave 4/30-5/20/64. Dropped as deserter for failure to return. Paroled Staunton 5/1/65. Age 28, 5' 8", dark complexion, dark hair, gray eyes. Carpenter, age 34, Mt. Sidney PO, Augusta Co. 1870 census. Died Augusta Co. 11/6/07. Bur. Belmont Cem., Annex.

WILSON, THOMAS MATHIS: Pvt. Co. B (1st). B. near Fairfield, Rockbridge Co. 2/6/40. Farmer and School Teacher. On postwar roster.

WILSON, WILLIAM: Pvt., Co. K. Enl. Shenandoah Mt. 4/9/62. Present 4/9-30/62. WIA (arm) Port Republic 6/9/62. Ab. wounded through 12/31/64. May have served in Co. E. 3rd Bn. Va. Reserves. Died Bell's Valley, Augusta Co. 11/11/01. Bur. Lebanon Pres. Ch. Cem.

WINDON, CHARLES W.: Pvt., Co. K. Enl. Shenandoah Mt. 4/9/62. Present 4/9-30/62. WIA (face) Sharpsburg 9/17/62. Ab. wounded through 12/63. Present 4/30-12/31/64. Deserted to the enemy Clarksburg, W. Va. 4/8/65. Took oath and released. Illiterate.

WINE, ROBERTSON ERWIN, SR.: Pvt., Co. B (1st). B. Va. 1846. Pvt., Bolivar Mills Co., Rockbridge Co. Militia 6/61. Not on muster rolls.

WISE, REUBEN: Pvt., Co. H. B. 1827? Enl. Staunton 7/23/61 age 33. Ab. sick 11/61-2/62. Present 3-4/62. Ab. sick 4/22/ until he died of chronic diarrhea in hospital Lynchburg 7/19/62. Bur. Confederate Section, Row #1, Lynchburg City Cem.

WISEMAN, ELIJAH MERCHANT: Pvt., Co. B (1st). B. Va. Augusta Co. 7/18/41. Farmhand, age 19, Walker's Creek Dist., Rockbridge Co. 1860 census. Enl. Staunton 8/1/61. Present until transf. 9/29/61.

WISEMAN, HENRY BAXTER: Pvt., Co. H. B. Va. 1844? Student, age 16, 1st Dist., Augusta Co. 1860 census. Enl. Richmond 7/9/62. Deserted 7/25/62. Resident of Moffett's Creek. NFR. Died near Newport, Augusta Co. 2/15/97. Bur. Mt. Hermon Lutheran Ch. Cem.

WISEMAN, JACOB ADAM: Pvt., Co. E. B. Rockbridge Co. 8/24/47, Carpenter, age 23, Walker's Creek Dist., Rockbridge Co. 1860 census. Not on muster rolls. WIA Gaines Mill 6/27/62. NFR. Farmer, Rockbridge Co. 1874. Died Rockbridge Co 9/26/23. Bur. New Providence Pres. Ch. Cem., near Brownsburg.

WISEMAN, JAMES H.: Pvt., Co. I. B. Augusta Co. 5/9/41. Student, age 19, Staunton PO, Augusta Co. 1860 census. Enl. Staunton 7/16/61 age 21. Present 11/61-4/62, and 5/1-6/3/62. AWOL 6/4-8/13/62, fined $30.00 plus forfeited $25.30. Present 8/14-10/31/2, and 1-4/63. Ab. sick "debility" 6/6-12/63. Issued clothing 4/21/64. WIA (side) and cap. Winchester 9/19/64. DOWs at home in Augusta Co.. 12/5/64 age 20.

WISEMAN, JOHN H.: Co. H. B. Va. 1843. Student, age 15, Staunton PO, Augusta Co. 1860 census. Enl. near Richmond 7/7/62. Deserted 7/25/62. NFR. Farmer, Folly Mills, 1897. Farmhand, age 65, Middlebrook Precinct, Greenville Dist., Augusta Co. 1910 census.

WISEMAN, JOHN PETER.: Pvt., Co. E. B. Augusta Co. 7/18/44. Farmhand, age 16, Walker's Creek Dist., Augusta Co. 1860 census. Enl. Bunker Hill 9/22/62. Present 9/22-10/31/62. Ab. sick 2/17-8/31/63. Present 9-12/63. Issued clothing 1/28 and 4/1/64. WIA (right leg) Spotsylvania CH 5/12/64. Ab. wounded through 10/31/64. Present 11-12/64. Paroled Staunton 5/20/65. Age 20, 5' 9", dark complexion, black hair, black eyes. Farmer and Merchant, age 65, Walker's Creek Dist., Rockbridge Co. 1910 census. Died there 5/28/11. Bur. Emanuel Ch. Cem.

WISEMAN, WILLIAM H. B.: Pvt., Co. H. B. Augusta Co. 1841? Student, age 14, Staunton PO, Augusta Co. 1860 census. Enl. 12/4/64. WIA (arm crushed) Hatcher's Run 2/6/65. Admitted to hospital Richmond 2/9/65. Furloughed for 60 days 2/18/65. paroled Staunton 5/11/65. Age 18, 5' 6", light complexion, light hair, black eyes. Carpenter, age 25, Fishersville PO, Augusta Co. 1870 census. Died Staunton 2/19/25. Bur. Thornrose Cem.

WITHERS, CYRUS: Pvt., Co. B (1st). B. Va. 1825? Farmhand, age 35, South River Dist., Rockbridge Co. 1860 census. Enl. Fairfield 7/10/61. Present until transf. 9/28/61.

WOMENDORFF, DANIEL B.: Pvt., Co. B. Augusta Co. 11/29/39. Farmhand, age 20, 1st Dist., Augusta Co. 1860 census. Enl. Staunton 7/16/61 age 21. Ab. sick 11-12/61. AWOL 1-2/62, fined 6 months and 11 days pay by CM. Present 3-4/62. Reenl. 5/1/62. WIA (shoulder) Port Republic 6/9/62. DOWs.

WOMENDORFF, JACOB B.: Pvt., Co. B (1st). B. Rockbridge Co. 2/18/35. Farmhand, age 23, 7th Dist. Rockbridge Co. 1860 census. Enl. Staunton 8/1/61. Ab. sick when transf. 9/28/61.

WOOD, GEORGE GODFREY: Pvt., Co. B (1st). B. Rappahannock Co. 5/16/35. Farmer, age 25, Walker's Creek Dist., Rockbridge Co. 1860 census. Enl. Fairfield 7/10/61. Present until transf. 9/28/61.

WOOD, JAMES A.: Pvt., Co. E. B. Amherst Co. 5/10/34. Farmhand, age 24, Glenwood Dist., Rockbridge Co. 1860 census. Enl. Orange CH 12/1/62. Present 1-6/63. AWOL 6/26-12/14/63. Present in arrest 12/14-31/63 fined pay and clothing allowance for period AWOL in addition to five months and 19 days. Issued clothing 2/20, 4/27, and 5/1/64. WIA (foot) and cap. Bethesda Ch. 5/30/64. Sent to Point Lookout. Exchanged 3/14/65. Illiterate. NFR. Farmer, age 36, Natural Bridge Dist., Rockbridge Co. 1870 census. Living Sherwood, Rockbridge Co. 1902. Died near Natural Bridge Station 6/11/14. Bur. Tygett Cem., Arnold's Valley.

WOOD, PRYOR: Pvt., Co. B (2nd). B. Albermarle Co. 1838? Laborer, age 25, Burkes Mill Dist., Augusta Co. 1860 census. Enl. Waynesboro 7/15/61. Ab. sick 1/61. Died of disease in hospital Staunton 23/23/61. Age 22, 5' 10", fair complexion, grey eyes, light sandy hair.

WOOD, THOMAS H.: Pvt., Co. C. B. Amherst Co. 7/6/46. Not on muster rolls. Cap. Harper's Farm 4/6/65. Sent to Point Lookout. Released 6/19/65. Resident of Augusta Co., 5' 9½", dark complexion, brown hair, light hazel eyes. Died Glasgow, Rockbridge Co. 6/9/17. Bur. Buena Vista.

WOOD, ZACHARIAH T.: Pvt., Co. C. B. Va. 1828? Shoemaker, age 31. Enl. Staunton 7/16/61 age 23. Present 11/61-4/62, detailed as Wagoner. Ab. sick with rheumatism 8/16/61-8/12/63, and hemorrids 8/13-10/63. Ab. sick "debilitas" 11/2-12/3/63. Present 4/30-6/17/64, detailed in Div. Provost Guard. Ab. sick acute dysentary 6/18-8/7/64. Present 8/8-12/31/64, detailed in Div. Provost Guard. Cap. Amelia CH 4/6/65. Sent to Point Lookout. Released 6/19/65. Living in Albemarle Co. 1920.

WOODS, PARMENAS A.: Pvt., Co. K. B. Va. 1824? Farmhand, age 36, Millboro Springs PO, Bath Co. 1860 census. Enl. Shenandoh Mt. 4/9/62. Not on muster rolls remainder of 1862 or 1863. Enl. Co. F, 11th Va. Cav. 3/27/62. Transf. back Co. K. 52nd Va. 5/3/64. WIA Carter's Farm, near Winchester 7/2/64. Ab. wounded through 10/31/64. Present 11-12/64. Reenl. Co. F, 11th Va. Cav. Cap. and died of disease in prison.

WOODDELL, BENJAMIN FRANKLIN: 2ndLt., Co. D. B. near Sangersville, Augusta Co. 1829? Enl. Staunton 7/16/61 as 4thSgt. Present 7/16/61-4/62. Reenl. 5/1/62. Present 5/1-6/8/62. AWOL 6/9-20/62, fined $4.03, rank shown as Pvt. Present 6/21-8/31/62. Elected 2ndLt. 10/2/62. Presence or absence not stated 1-12/63, submitted resignation 3/24/63. Apparently not accepted. WIA Fisher's Hill 9/22/64. Ab. wounded through 12/31/64, but signed roll commanding Co. 12/31/64. Submitted resignation 2/27/65. Commanding Co. Ft. Stedman and WIA 3/25/65. Cap. in hospital Richmond 4/3/65. DOWs there 4/21/65. Bur. Hollywood Cem. Brother of William H. Wooddell.

WOODDELL, WILLIAM HARRISON: 1stLt., Co. D. B. near Sangersville, Augusta Co. 1829? Farmer, age 31, Burkes Mill Dist., Augusta Co. 1860 census. Enl. Staunton 7/16/61. Present 7/16-12/31/61. Ab. on detached service 1-2/62. Present 3-4/62. Not reelected 5/1/62. May have served in 62nd Va. Inf. Farmer, age 41, Mt. Sidney PO, Augusta Co. 1870 census. Auctioneer, Constable, Mechanic, Music Teacher, and Watchmaker, Sangersville. Died there 7/18/86. Bur. Emanuel Ch. Cem.

WOODWARD, JOHN P.: Pvt., Co. H. B. Augusta Co. 11/16/39. Farmhand, age 20, 1st Dist., Augusta Co. 1860 census. Enl. Staunton 7/23/61 age 21. Ab. sick 11/10/60-2/62. Present 3-4/62. Reenl. 5/1/62. WIA (shoulder) Port Republic 6/9/62. Ab. wounded through 2/63. Present on extra duty 3-4/63, AWOL 8/15-31/63. Present 9-12/63. WIA (leg) Bethesda Ch. 5/30/64. Admitted to hospital Richmond 6/2/64. AB. wounded through 12/31/64. Paroled Staunton 5/16/65. Age 23, 5/7". fair complexion, black hair, dark eyes. Farmer, near Craigsville. Died there 3/23/76. Bur. Estaline Pres. Chapel Cem., Estaline Valley, Augusta Co.

WOODWARD, JOHN SAMUEL: Pvt., Co. H. B. Augusta Co. 1838? Farmhand, age 21, 1st Dist., Augusta Co. 1860 census. Enl. Staunton 7/23/61, age 23, Ab. sick 11/10/61-2/62. Present 3-4/62 KIA Malvern Hill 7/1/62. Age 23, 5' 8", fair complexion, gray eyes, dark hair.

WOODY, HENRY: Pvt., Co. E. Enl. Staunton 8/1/61. Present 11/61-4/62. WIA (leg) McDowell 5/8/62. Present 6-10/62. Paid Staunton 11/13/62, illiterate. Unofficial roster indicates transf. 1st Rockbridge Artillery '64. Deserted.

WOODZELL, GEORGE BUFORD: Pvt., Co. K. B. Augusta Co. 9/42. Farmhand, age 21, Bath CH PO, Bath Co. 1860 census. Enl. Shenandoah Mt. 4/9/62. Present 4/9-30/2. Ab. on leave 1-2/63. Present 3-12/63. Present 4/30-12/31/64. Deserted to the Army of the Potomac near Petersburg 3/8/65. Sent to Washington, D.C. Took oath and transportation furnished to Greene Co., Ohio. Farmer, age 27, Williamsville Dist., Bath Co. 1870 census. Died Warrenton, Va. 1/31/73.

WOODZELL, JOHN: Pvt., Co. K. B. Va. 11/30/30. Pvt., Capt. Hamilton's Co., 81st Va. Militia. Enl. near Sheperdstown 9/22/62. Present 1-12/63. Issued clothing 1/28/65. KIA Bethesda Ch. 5/30/64. Left widow.

WOREL, JOHN: Pvt., Co. I. Enl. Staunton 11/16/64. AWOL 11/18-12/31/64. Dropped as deserter.

WRIGHT, BENJAMIN: Pvt., Co. B (2nd). B. Amherst Co. 1819? Laborer, age 40, 1st Dist., Augusta Co. 1860 census. Enl. Waynesboro 7/15/61. Present 11/61-4/62, non conscript. AWOL 5/17-9/29/62. Ab. sick 10/24/62-1/15/63. Present 1/16-2/28/63. Discharged for overage of conscription 3/6/63. Age 42, 6' 4", fair complexion, brown eyes, sandy hair, farmer.

WRIGHT, JAMES W.: 1stSgt., Co. B (2nd). B. Rockingham Co. 1826? Enl. Waynesboro 7/15/61. Present 11/61-4/62. Detailed in hospital Staunton and reduced to Pvt. 7/15/62. Discharged for "nephretis" 8/11/62. Age 35, 5'8", fair complexion, grey eyes, dark hair, merchant.

WRIGHT, JOHN W.: Pvt., Co. D. B. Va. 1838. Carpenter, age 21, Northern Dist., Augusta Co. 1860 census. Enl. Staunton 7/16/61. Present 7/16/61-4/62. WIA (foot) Cross Keys 6/8/62. Returned to duty 1/28/63. Present 1/28-4/63. AWOL 7/21-10/27/63, fined 1 months pay in addition to time AWOL by Reg'l CM. Present in arrest 11-12/63. Issued clothing 3/31 and 4/27/64. Ab. sick with debility 6/14-7/8/64. Cap. Winchester 9/19/64. Sent to Point Lookout. Exchanged 3/18/65. Paroled Staunton 5/25/65. Age 28, 5'8", dark complexion, dark hair, dark eyes, illiterate. Brother of Samuel H. Wright.

WRIGHT, SAMUEL H.: Pvt., Co. C. B. Augusta Co. Enl. near Liberty Mills 8/8/62. KIA 2nd Manassas 8/28/62. Brother of John W. Wright.

WRIGHT, WILLIAM J.: Pvt., Co. C. B. Rockingham Co. 1839? Farmhand, age 19. 1st Dist., Augusta Co. 1860 census. Enl. Staunton 7/16/61 age 21. Ab. sick 11-12/61. Present 1-4/62. Reenl. 5/1/62. WIA 2nd Manassas 8/28/62. Ab. wounded through 4/62. Deserted to the enemy Hampshire Co. 7/20/63. Age 23, 5' 10½", florid complexion, brown eyes, dark hair. Farmer, Ladd, Augusta Co. 1907.

YEAGER, WILLIAM T.: Pvt., Co. H. Enl. Staunton 8/2/61. Present 1-4/62. Transf. 31st Va. Inf. 7/20/62.

YOUNG, ANDREW A.: B. Augusta Co. 1821. Not on muster rolls. Attended reunion Luray 1881. Farmer, Churchville, 1897.

YOUNT, RUDOLPH CHRISTIAN "Doc": B. near Madrid, Augusta Co. 9/22/44. Farmhand, age 17, Burkes Mill Dist., Augusta Co. 1860 census. Enl. Camp Alleghany 3/10/62. Present 3-4/62, and 8/31-12/31/62. Ab. sick 2/18 until he died of typhoid fever at home near Madrid, August Co. 3/14/63. Bur. Trinity Lutheran Ch. Cem.

YOUNT, WILLIAM ISSAC H.: Pvt., Co. G. B. Augusta Co. 5/20/43. Enl. Staunton 8/2/61. Present 11/61-4/62. Reenl. 5/1/62. WIA (ankle) Cross Keys 6/8/62. Foot amputated. Ab. wounded through 12/64. Resident of New Hope. Died Augusta Co. 4/10/77. Believed bur. in unmarked grave next to parents in Barren Ridge Ch. Cem., Augusta Co.

ZIMBRO, JOHN: 1stSgt., Co. H. B. Rockbridge Co. 1839. Farmer, Rockbridge Co. Enl. Staunton 7/23/61 age 22 as Pvt. Present 11/61-4/62. Reenl. and appointed 2ndSgt. 5/1/62,. WIA (leg and hand) Port Republic 6/9/62. Promoted 1stSgt 9/1/62. Reduced to 2ndSgt 5/1/63. WIA (right knee) Paine's Farm 11/27/63. Returned to duty 12/9/63. Ab. detailed as Ambulance Driver 2ndCorps and reduced to Pvt. 4/30-12/31/64. NFR. Miller, age 32, Craigsville PO, Augusta Co. 1870 census. Died Augusta Co. 1911. Bur. Rocky Springs Ch. Cem., Deerfield.

ZIMBRO, ROBERT: Cpl., Co. H. B. Augusta Co. 7/11/43. Enl. Staunton 7/23/61 age 18, as Pvt. Present 11/61-4/62. Reenl. 5/1/62. Ab. sick 6/12-8/31/62. Promoted Cpl. 9/1/62. Present 1-1/63. Cap. Bethesda Ch. 5/30/64. Sent to Point Lookout. Transf. Elmira. Released 5/15/65. 5' 10", sallow complexion, dark hair, hazel eyes, illiterate. Farmhand, age 27, Pastures Dist., Augusta Co. 1870 census. Member, Stonewall Jackson Camp CV, Staunton. Farmer, Craigsville. 1908.

ZIMBRO, WILLIAM T.: Pvt., Co. H. B. Augusta Co. 1830? Laborer, 28, 5th Dist., Rockbridge Co. 1860 census. Enl. Staunton 7/23/61 age 30. Present 11/61-3/62. AWOL 4/22-30/62. Bur. on the field. Present 4/30-8/31/62. KIA Chantilly 9/1/62, age 31. Left widow.

ZIMMERMAN, JACOB H.: 2ndCpl., Co. I. B. Augusta Co. 2/3/26. Farmhand, age 36, 1st Dist., Augusta Co. 1860 census. Enl. Staunton 7/16/61 age 35. Present 11/61-4/62. Reenl. 5/1/62. WIA (right side and bone fractured in right hand) McDowell 5/8/2. Ab. wounded until discharged for disability, liver damaged by wound, 3/10/63. Age 36, 5' 7", florid complexion, blue eyes, black hair, farmer. Paroled Staunton 5/17/65. Age 38, 5' 8", dark complexion, dark hair, blue eyes. Living near Middlebrook 1891. Died Arbor Hill 8/6/94, "ball still in bone." Bur. Mt. Tabor Lutheran Ch. Cem.

ZIRKLE, JOHN J.: Pvt., Co. G. Enl. Staunton 8/2/61. Present 11/61-4/62. Reenl. 5/1/62. Present 5/1-10/31/63, AWOL 1 month fined $11.00. WIA (both thighs) Fredericksburg 12/12/62. Ab. wounded through 2/63. Present 3-7/63. AWOL 8/1/63 until deserted to the enemy Clarksburg, W. Va. 10/13/63. took oath and sent north. Bur. Mountain View Cem., near New Hope, Augusta Co.

ZIRKLE, JOSEPH B.: Pvt., Co. G. B. 6/20/40. Not on muster rolls. Attended reunion Carlisle, Pa. 9/28/81. Died Augusta Co. 11/18/03. Bur. St. James Lutheran Ch. Cem.

ZIRKLE, ROBERT JOSEPH: Pvt., Co. G. B. 1840? Pvt., Capt., Koiner's Co., 32nd Va. Militia 3/62 detailed as Collier for Crawford Iron Works, age 22. Not on muster rolls. Deserted to the enemy New Creek, W. Va. 9/14/63. Took oath and sent north. Attended reunion Luray 1881.

ZOLLMAN, ALEXANDER MCCORKLE: Pvt., Co. E. B. Rockbridge Co. 11/1/45. Enl. Staunton 8/1/61. Present 11/61-10/31/62. Paid Staunton 11/24/62. Transf. Co. C. 14th Va. Cav. Farmhand, age 24, Natural Bridge Dist., Rockbridge Co. 1870 census. Farmer, Postmaster, Merchant and Distiller, Zollman PO, Rockbridge Co. 1897. Died there 3/28/17. Bur. Falling Springs Pres. Ch. Cem.

ZOLLMAN, JOHN WILLIAM: Pvt., Co. E. B. Rockbridge Co. 11/3/39. Farmhand, age 20, Natural Bridge Dist., Rockbridge Co. 1860 census. Served in Militia. Enl. Staunton 8/1/61. Present 11/61-4/62. Reenl. 5/1/62. WIA Cross Keys 6/8/62. WIA (breast and knee) Gaines Mill 6/27/62. Ab. wounded through 4/63. Ab. on detached service arresting conscripts and deserters in Rockbridge Co. 4/30-9/7/63. Present 9-12/63. Transf. Co. C, 1st Va. Cav. 3/8/64. WIA Cannon's Farm '64 and "carried the bullet to his grave." WIA (arm) Farmville 4/65. Farmer, age 29, Lexington Dist., Rockbridge Co. 1870 census. Died at his home 4½ miles west of Lexington 4/29/23. Bur. Zollman Family Cem. on Rt. 251 west of Lexington.

ZOLLMAN, MADISON: Pvt., Co. E. B. Rockbridge Co. 1/31/32. Farmer, age 27, Goshen Bridge PO, Rockbridge Co. 1860 census. Served in militia. Enl. 4/62. Furnished Robert Johnson as substitute and discharged. Enl. Co. C, 1st Va. Cav. Enl. Co. H, 4th Va. Inf. 10/20/64. WIA (leg shattered) Hatcher's Run 2/9/65. DOWs in hospital Richmond 3/20/65. Bur. Stonewall Jackson Cem., Lexington. Left widow and 2 children.

BIBLIOGRAPHY

Manuscripts.

Augusta County Historical Society
  Cemetery Listings
  Confederate Veterans Papers

Richard Armstrong Collection, Millboro, Virginia.
  A. G. Cleek Papers
  James H. McClintic Papers

Bath County Historical Society
  Militia Rosters (1850-   )

Ralph S. Coffman Personal Collection, Mount Sidney, Virginia.
  Alexander S. Coffman Diary
  Roster of Company G, 52nd Virginia Infantry by John Coley, A. S. Coffman and C. B. Coiner

Donald Dorr Personal Collection, Linthicum Heights, Maryland.
  Adam W. Kersh Papers

Dora C. Fechtmann Personal Collection, Ocala, Florida.
  Cline Family Papers

Georgia Department of Archives and History.
  J. N. Wilkinson Letters

Thelma Goodbar Personal Collection, Lexington, Virginia.
  Abner McC. Wilhelm Papers

Forrest Harris Personal Collection, Churchville, Virginia.
  John H. Stover Papers
  Roster of Company F. 52nd Virginia Infantry

Harrisonburg-Rockingham County Historical Society
  Family Bible Information
  Genealogical Files

Jane Hickin Personal Collection, Fairfield, Virginia.
  McClure Family Papers
  Roster of Company I, 52nd Virginia Infantry

John F. Leonard Personal Collection, Phoenix, Arizona.
  Coiner-Koiner Family Papers

Nellie Lilley Personal Collection
    John D. Lilley Diary
    Lilley Family Papers
    Roster of Co. H, 52nd. Virginia Infantry

R. E. Lee Camp of Confederate Veterans Soldiers' Home Burial Ledgers, Hollywood Cemetery, Richmond, Virginia

Rockbridge County Historical Society
    Cemetery Listings
    Confederate Veterans Papers
    List of Participants in the Civil War from Rockbridge
    List of Exempts, Reserved and Substitutes from Rockbridge County
    Muster Rolls First Battalion, Eighth Regiment Virginia Militia 1841-1862

Irvin Rosen Personal Collection, McKinley, Virginia
    James H. Wiseman Papers

Jean Thomas Personal Collection Spottswood, Virginia
    McClure Family Papers
    Roster of Company I, 52nd Virginia Infantry by M. T. McClure and G. P. Lightner

Joseph B. Yount III Collection, Waynesboro, Virginia
    Rudolph C. Yount Papers

United Daughters of the Confederacy, Richmond
    Alphabetical Card File
    Application Forms, Virginia Division, UDC

University of Virginia
    Bowman Family Papers
    Jacob Bumgardner Family Papers
    G. W. Finley Family Papers
    McGuffin Family Papers
    George Q. Peyton Diary 1864
    St. Johns German-Reformed Church Register and Cemetery Records, Augusta County, Virginia

Virginia Military Institute
    Cyrus B. Coiner Papers
    John D. Ross Papers

Virginia State Library
> Compiled Roster of Regiments from Virginia
> Diary of Lieutenant John D. Summers, Company I, 52nd Virginia Infantry
> Seth Bassett French Biographical Sketches
> Muster Rolls, Company G, 52nd Virginia Infantry, January-February 1862, October-December 1862, January-February 1863 and December 31, 1863
> Pension Applications. Acts of 1888, 1900, 1902 from Albemarle, Augusta, Amherst, Bath, Nelson, Rockbridge and Rockingham Counties

Washington and Lee University
> Francis McFarland Diaries 1861-1866
> John D. Ross Papers

West Virginia University
> H. J. Mugler Diary

Public Documents

Augusta County, Virginia Censuses: 1860, 1870, 1910

Bath County, Virginia Censuses: 1860, 1870, 1910

Compiled Service Records of Confederate Soldiers Who Served in Organizations from the State of Virginia, Microfilm No. M324, Rolls 288, 934-941.

Rockbridge County, Virginia Censuses: 1860, 1870, 1910

County and City Records

Augusta County, Virginia
> Muster Rolls of Augusta County 1861-1865 (Actually membership Stonewall Jackson Camp Confederate Veterans, Staunton circa 1920).
> Order Books 1861-1865
> Register of Deaths 1861-1896 and 1912-1917 (incomplete and missing 1863).
> Register of Marriages Augusta County 1854-1880

Bath County, Virginia
> Order Books 1861-1865
> Register of Deaths 1861-1938 (incomplete and missing 1863)
> United Daughters of the Confederacy Applications, Bath County

Rockbridge County, Virginia
> Application for Commutation of Wounds Under the Act of 1884
> Order Books 1861-1865
> Register of Deaths 1861-1873 and 1913-1917 (incomplete and missing 1863)
> Rockbridge County Marriage Bonds 1853-1872
> Roster of Companies of Rockbridge County Men in the War Between the States

Staunton, Augusta County, Virginia
> Register of Burials Thornrose Cemetery
> Register of Deaths 1861-1896 (incomplete and missing 1863)
> Register of Marriages 1853-1882

Periodicals

Confederate Veteran 1893-1932.

Land We Love. Vol. 1-6. 1866-1869.

O'Meara, Ted. ed. *Tracks* Vol. 37, nos. 4-6. Cleveland, Ohio. Chesapeake and Ohio Railway. April-June 1952.

Southern Historical Society Papers. 52 vols. Richmond, 1876-1953.

Newspapers

Augusta County *Argus*
Lexington *Star*
Lexington *Gazette*
Rockingham *Register*
South River *Advertiser*
Staunton *News Leader*
Staunton *Spectator and Vindicator*
Staunton *Vindicator*
The Valley Virginian
Waynesboro *Weekly Tribune*
Yost's Weekly

Published Works

A Descriptive List of the Burial Places of the Remains of Confederate Soldiers Who Fell in the Battles of Antietam, South Mountain, Monocacy. Hagerstown, Maryland, n.d.

A Historical sketch of Michael Keinadt and Margaret Diller, his wife: the history and genealogy of their numerous posterity in the United States up to the year 1893. Staunton, Virginia. Stoneburner & Prufer, Publishers, 1893.

Bailey, Gladys, compiler. *The Descendents of John Huff Sr.* Verona, Virginia, March 1973

Barber, Lucy Harrison Miller, compiler. *Behind the Old Brick Wall.* Richmond, Virginia, Whittet and Sheppherson, 1968.

Barnhart, Nat G. and Katie Rea. *The Barnhart Family History: Augusta County Virginia 1767-1967.* Verona, Virginia, McClure Printing Company, 1967.

Bell, John W. *Memoirs of Governor William Smith.* New York. The Moss Engraving Company. 1891.

Bell, Robert T. *11th Virginia Infantry.* Lynchburg, Va. H. E. Howard, Inc., 1985.

Benson, Eleanor Vanneman. *The Dudley's.* Privately Printed, Greensboro, North Carolina, 1980.

Binford, Alma Fultz, compiler. *A Right To Be Here, Ancestors and Descendents of Dr. George Simeon Fultz, Sr. and Alma Knee Fultz.* Jacksonville, Florida, 1970-77.

Brice, Marshall M. *Conquest of A Valley,* Charlottesville, University of Virginia Press, 1965.

Brock, Robert A., *Hardesty's Historical and Geographical Encyclopedia Illustrated, Special Virginia Edition.* New York, H. H. Hardesty and Company, 1884.

Bruce, Philip A. *History of Virginia.* 5 vols. Chicago, The American Historical Society, 1924.

Bruck, Harry Anthony. *History of Mennonites in Virginia 1727-1900.* Harrisonburg, Virginia, H. A. Bruck Publisher, 1959.

Buck, Captain Samuel D. *With the Old Confeds.* Baltimore, H. E. Houck and Company, 1925.

Buckley, William. *Buckley's History of the Great Reunion of the North and the South and of the Blue and the Gray.* Staunton, Virginia, William Buckley Publisher, 1923.

Burials in Augusta County Cemeteries, Part I. Augusta County Historical Society, 1979.

Burials in Old Graveyard at Augusta Stone Church. n.p. n.d.

Bushong, Millard Kessler. *Old Jube.* Burke, Virginia, Carr Publishing Company, 1955.

Churches and Family Cemeteries Eastern Section, Rockingham County. n.p. n.d.

Churches and Family Cemeteries Western Section, Rockingham County. n.p., 1961-1971.

Churches and Private Cemeteries in Lexington Presbytery. Richmond, Synod of Virginia, 1970.

Cleek, George W. *Early Western Augusta Pioneers: Including the Families of Cleek, Gwin, Lightner and Warwick.* Staunton, Virginia. 1957.

Couper, William. *History of the Shenandoah Valley, Family and Personal Records.* Vol. III. New York, Lewis Historical Publishing Company, Inc., 1942.

_____. *Jackson Memorial Cemetery Survey Complete to 1960.* n.p., n.d.

Culpepper, Ruth Rhodes, *My Heritage.* Harrisonburg, Virginia, Parkview, Press, 1982.

Curry, Charles. *John Brown Baldwin, Lawyer, Soldier, Statesman.* Staunton, Virginia, 1928.

Delauter, Roger E. Jr. *History of The 18th Regiment Virginia Cavalry.* Lynchburg, Va. H. E. Howard, Inc. 1985.

Diehl, George West. *Old Oxford Church and Her Families.* Verona, Virginia, McClure Press, Inc., 1971.

_____ A. Maxim Coppage III, compiler. *Rockbridge County, Virginia Notebook.* Owensboro, Kentucky, McDowell Publishing Company, 1982.

_____ *The Brick Church on Timber Ridge.* Verona, Virginia, McClure Printing Company, Inc., 1975.

Douglas, Henry Kyd. *I Rode with Stonewall.* Chapel Hill, North Carolina, University of North Carolina Press, 1940.

Early, Jubal Anderson. *War Memoirs.* Bloomington, Indiana, Indiana University Press, 1960.

Ebaugh, Thomas A. *Ancestors of George McNett and Susan Armentrout.* New Orleans, Louisiana, Accurate Letter Company, 1961.

Evans, Clement A., ed. *Confederate Military History.* 12 vols. Atlanta, Confederate Publishing Company, Inc., 1898.

Evans, Robert Lee. *History of the Decendents of Jacob Gochenour.* Boyce, Virginia, Carr Publishing Company, Inc., 1977.

Freeman, Douglas Southall. *Lee's Lieutenants: A Study In command.* 3 vols. New York, Charles Scribner's Sons, 1942-1944.

Goddard, Clyde, compiler and publisher. *The Baugher-Schreckhise-Christian Family.* New Boston Missouri, 1973.

Goeller, Mildred S. *The Steeles of Steeles Tavern, Virginia and Related Families.* n.p., 1874.

Gordon, John B. *Reminiscences of the Civil War.* New York, C. Scribner's Sons, 1903.

Hale, Laura V. and Stanley S. Phillips. *History of the Forty-ninth Virginia Infatry, CSA.* Lanham, Maryland, S. S. Phillips and Associates, 1981.

Harman, Col. M. G. *Notice to Members of the 52 Regiment Virginia Volunteers* with casualty lists for June 8-9, 1862. Staunton, June 22, 1862. Broadside.

Harmon, John William, ed. *Harman-Harmon Genealogy and Biography.* Parsons, West Virginia, 1928.

Heatwole, Cornelius J. *History of the Heatwole Family.* Second Edition. North Newton, Kansas, The Mennonite Press, 1970.

Heizer, James Marion. *The Heizer Family American Pioneers.* Louisville, Kentucky, n.d.

Hollywood Memorial Association. *Register of Confederate Dead, Interred in Hollywood Cemetery.* Richmond, Garry, Clemmett and Jones, Printers, 1869.

Holmes, Clay W. *The Elmira Prison Camp.* New York, Putnam, 1912.

Horst, Samuel. *Mennonites In the Confederacy.* Scottsdale, Pennsylvania, Herald Press, 1967.

Hotchkiss, Jedediah. *Historical Atlas of Augusta County, Virginia: Maps from Original Surveys.* Chicago, Waterman, 1880.

Hull, Susan R. *Boy Soldiers of the Confederacy.* New York and Washington, The Neale Publishing Company, 1905.

Johnson, John Lipscomb. *The University Memorial.* Baltimore, Turnbull Brothers, 1871.

Jones J. William. *Christ In the Camp.* Richmond, B. F. Johnson and Company, 1887.

Jordan, Weymouth C. Jr., Compiler. *North Carolina Troops 1861-1865 A Roster.* Vol. 5. Raleigh, North Carolina Division of Archives and History, 1975.

Joyner, Peggy Shomo, compiler. *Ancestors and Decendents of Joseph Shomo (Shammo).* Baltimore, Gateway Press, Inc, 1983.

_____ *Henry Roosen-Rosen to Pennsylvania 1765 German Ancestors-American Decendents.* Norfolk, Virginia, Liskey Lithograph Corporation, 1980.

Kemp, Vernon E. *The Alumni Directory and Service Record of Washington and Lee University.* Lexington, Virginia, the Alumni Inc., 1926

Kershner, Ruth Bownds. *The Kershner Family of Maryland 1731-1977.* Houston, Texas, The Kershner Family Association, 1978.

Krick, Robert K. *Lee's Colonels: A Biographical Register of the Field Officers of the Army of Northern Virginia.* Dayton, Ohio, Press of the Morningside Bookshop, 1979

_____*Roster of the Confederate Dead in the Fredericksburg Confederate Cemetery.* Fredericksburg, Virginia, 1974.

Kurtz, Lucy F. and Benny Ritter. *A Roster of Confederate Soldiers Buried in Stonewall Jackson Cemetery, Winchester, Virginia.* Winchester, Virginia, Farmers and Merchants Bank, 1962.

Lemert, Ann Arnold. *First You Take A Pick and Shovel: The Story of the Mason Companies.* The John Bradford Press, Lexington, Kentucky 1979.

List of Stones, Augusta County, Virginia: Family Cemeteries and Burial Grounds: Abandoned Churches. n.p., n.d.

McChesney, John M. Jr. *The McChesneys of Virginia.* Waynesboro, Virginia, n.d.

McClure, James Alexander. *The McClure Family.* Petersburg, Virginia, Presses of Frank A. Owen, 1914.

McDonald, Archie P., ed. *Make Me A Map of The Valley. The Civil War Journal of Jed Hotchkiss.* Dallas, Southern Methodist University Press, 1973.

Maxwell, Hu. *History of Randolph County, West Virginia.* Morgantown, Acme Publishing Company, 1898.

May, Clarence Edward. *Life Under Four Flags in the North River Basin of Virginia.* Verona, Virginia, McClure Press, 1976.

Mickle, William E. *Well Known Confederate Veterans.* New Orleans, 1907.

Miller, Lula Mae. *Johannes Frederick Kirshof. Early Settler and Patriach of Northern Augusta County.* 2 vols. Verona, Virginia, McClure Publishing Company, 1981-1982.

Moore, Ralph E., compiler. *The Alexander-Carson-Moore Genealogy.* Spottswood, Virginia, Ralph E. Moore, Publisher, 1967.

Morton, Oren F. *Annals of Bath County.* Staunton, Virginia, The McClure Company, Inc., 1917.

_____*History of Rockbridge County, Virginia.* Baltimore, Regional Publication Company, 1973.

Paxton, W. M. *The Paxton Family.* Platte City, Missouri, Landmark Printing, 1903.

Perry, Alfred. *History of Amherst County.* Madison Heights, Virginia, Perry Press, 1961.

Peyton, John Lewis. *History of Augusta County, Virginia.* Staunton, Virginia, S. M. Yost & Son, 1882.

Reese, Margaret C. *Reese-Fox Ancestors. Waynesboro, Virginia, 1981.*

Report of Board of Visitors Lee Camp Soldiers Home Richmond, Virginia. Richmond. Ferguson Printing, 1902, 1905, 1906, 1912, 1913, 1917, 1918, 1919.

Reunion of Ex-Soldiers of the North and South held at Luray, Virginia 21 July 1881 and at Carlisle, Pennsylvania 28 September 1881. Carlisle, Pennsylvania, Herald and Mirror Print, n.d.

Richey, Homer, ed. *Memorial History of the John Bowie Strange Camp, United Confederate Veterans.* Charlottesville, Press of the Michie Company, 1920.

Robson, John S. *How a One-Legged Rebel Lives.* Durham, North Carolina, Educator Company, printers, 1898.

Roster of Confederate Pensioners of Virginia. Richmond, 1912, 1914, 1920, 1926.

Some Cemeteries in Augusta County. n.p., n.d.

Some Cemeteries in Highland County. Staunton, Virginia, 1972.

Strickler, Harry M. *Forerunner of the Stricklers, A History or Genealogy of the Families.* Harrisonburg, Virginia, C. J. Carrier Company, 1977.

Supplement to First Records of Cemeteries East of US Highway #1, Rockingham County, Compiled by Massanutten Chapter of the Daughters of the American Revolution 1965-1975 with additions and corrections to first volumes of records 1974-1975. n.p., n.d.

Suter, Mary Eugenia. *Memories of Yesteryear: A History of the Suter Family.* Waynesboro, Virginia, Charles F. McClung Printer, Inc., 1959.

Swank, J. Robert, compiler. *A record of Burial Places in Rockingham County.* Singers Glen, Virginia, 1957.

Tompkins, Edmund Pendleton, MD. *Rockbridge County, Virginia: An Informal History.* Whittet & Shepperson, Richmond, 1952.

Turner, Dr. Herbert S. *The McFarland Family of Augusta County, Virginia.* Staunton, Virginia, 1957.

Tyler, Lyon G. *Encyclopedia of Virginia Biography.* 5 vols. New York, Lewis Historical Publishing Company, 1915.

University of Virginia.*Students of the University of Virginia, a semi-centennial catalogue, with brief biographical sketches.* Baltimore, 1878.

U. S. War Department. *The War of the Rebellion: A Compilation of the Official Records of the Union and Confederate Armies.* 128 volumes. Government Printing Office, 1880-1901.

Virginia Military Institute. *Virginia Military Institute Register of Former Cadets.* Lexington, Virginia, 1957.

Virginia State Gazetteer and Business Directory 1870-1872. Richmond, J. H. Hill Printing Co. 1872.

Virginia State Gazetteer and Business Directory 1897-1898. Vol. No. 7. Richmond, J. H. Hill Printing Co. 1896.

Waddell, Joseph A. *Annals of Augusta County, Virginia.* Staunton, Virginia, C. R. Caldwell, 1902.

Walker, Charles D. *Memorial, Virginia Military Institute.* Philadelphia, J. B. Lippincott & Company, 1875.

Wallace, Lee A. Jr. *A Guide to Virginia Military Organization, 1861-1865.* Richmond, Virginia Civil War Commission, 1964.

Wayland, John W. *Augusta Bibliographies.* 2 vols. (unfinished manuscript). Bridgewater, Virginia, Bridgewater College, 1960.

Weaver, Dorthy Lee. *The John George Weaver Family.* Locust Grove, Virginia, Privately Printed, 1980.

Welliver, Virginia, compiler. *Index to Augusta County, Virginia Cemeteries.* Staunton, Virginia 1978.

Wenger, Samuel F., ed. *The Wenger Book.* Lancaster, Pennsylvania, Pennsylvania-German History, Inc., 1978.

Wilson, Howard, McKnight. *The Tinkling Spring: Headwater of Freedom.* Fishersville Virginia, The Tinkling Spring and Hermitage Presybyterian Churches, 1954.

Wise, Jennings C. *The Military History of the Virginia Military Institute from 1839-1865.* Lynchburg, Virginia, J. P. Bell Company, Inc., 1915.

Works Progress Administration, Virginia. *Augusta County Cemeteries,* Staunton, Virginia, 1938.

Yowell, Claude Lindsay. *A History of Madison County, Virginia.* Strasburg, Virginia, Shenandoah Publishing House, 1926.

Yost, Samuel M. *Beautiful Thornrose, Memorial Edition.* Staunton, Virginia, The McClure Company, 1921.

52ND VIRGINIA INFANTRY
ADDENDUM FOR THE SECOND EDITION

Page 1 — after Augusta Fencibles (originally)
Page 18 — WIA's Co. A 10 Co. H 10 Total 66
Page 25 — railroad cut in front. .
Page 28 — On September. .
Page 31 — Lieutenant Colonel Skinner wrote the *Vindicator*. .
Page 45 — WIA's add Co. B 1 Co. D 2 Co. F 1 Co. I 4 Total 11
Page 57 — KIA's F&S 2 Total 14
Page 58 — KIA's F&S 2 Total 14
Page 58 — KIA's Co. C 1 Total 11
Page 61 — . . . the regiment had four more men killed and three wounded. . . . defense of Staunton on June 5.
Page 63 — . . . after a hot fight the Yankees . . .
Page 71 — CEDAR CREEK OCTOBER 19, 1864
Page 84 — Fort McHenry vice Henry
Page 87 — . . . until June, 1940 . . DESERTIONS 1864 92 Total 325 186 were killed in action. . . 570 were known. . .
Page 88 — Company B(1st)
Page 89 — Company E Captains Thomas H. Watkins and Samuel Washington Paxton. Lieutenants Mathew Bryan Campbell, William V. Knick, Joseph Samuel Paxton and John K. Watkins.
Page 166 — List of Exempts, Reserves. .
Page 170 — The Steeles of Steeles Tavern . . . 1974.
Page 174 — Virginia State Gazatteer. . . 1897-1898. J. H. Hill Printing Co. 1898.

Caption Corrections

Page 78 — Pvt. John J. McCoun should be Pvt. John J. McCown
Courtesy of Mrs. William Bird, Buena Vista, Va. should be Courtesy of Mrs. Robert E. Bird
Page 79 — Courtesy of John Brane, Greenville, Va. should be Courtesy of John Brake, Greenville, Va.
Page 80 — Adj. John William Lewis caption should read Second Lt. John Addison Carson, Co. D

Col. John B. Baldwin, wartime

Col. John B. Baldwin, postwar

Lt. Col. John Dehart Ross

Lt. Col. John DeHart Ross, 1880

Adjutant Alexander Givens McCune

Lt. Col. John Doak Lilley, circa 1890

Adj. John Wm. Lewis, VMI 1859

Captain and Asst. Comm. of Subsistence Bolivar Christian, 1879

Chaplain John Magill

Ensign William H. Lackey

QM Sgt. Matthew Thompson McClure

Courtesy: Randy Hickin, Spottswood, Va.

QM Sgt. Patrick Maloy

Drillmaster John Andrew Stuart

Tuley Joseph Mitchell, Co. A

Ordnance Sgt. Adam G. Cleek

John Newton Patterson, Co. A

Edward Hughes McCrory

John Timothy Dwight Gisinger, Co. B

Second Lieut. John S. Lipscomb, Co. B, 2nd

Pvt. Sampson Pelter, Co. B, 2nd

Pvt. John Wise Sherman, Co. C

Cpl. Andrew Jackson Kennedy, Co. C

Sgt. James Campbell Lilley, Co. C

Pvt. Stephen Staubus, Co. B, 2nd

James H. Maupin, Co. D

Pvt. James W. Wiseman, Co. C

John Wm. Daggy, Co. C

Samuel Forrer, Co. C

John S. Robson, Co. D

James Michael, Co. D

Cornelius Stanley Knott, Co. D

Pvt. Augustus Wm. Staubus, Co. D

John Alfred Wilkinson, Co. E

Pvt. David Boyd Stomback, Co. D

Pvt. John Pressly Miller, Co. E

James William Marshall, "Cyclone Jim", Co. D

Jacob D. Suter, Co. F

John William Zollman, Co. E

Pvt. Adam. W. Kersh, Co. F

Survivors of Co. F, 1916
Standing: Pvt. Joseph H. Fauver, Cpl. Edward M. Furr, Cpl. John C. Rutherford, Pvt. William S. Stover, 1st Lt. John A. Fauver; **sitting:** Pvt. Jacob S. Sheets, Capt. James Bumgarder Jr., Pvt. Simon W. Wampler.

Capt. James Bumgardner, Jr., Co. F

Edward M. Anderson, Co. F

Pvt. John William Hale, Co. F

Pvt. Samuel Henry Hale, Co. F

Capt. Samuel Houston McCune, Co. G

Captain Cyrus Benton Coiner, Co. G

Alexander S. Coffman, Co. G

John Hatch Stover, Co. F

Captain Clairborne Rice Mason, Co. H

Pvt. James William Hawpe, Co. I

James Alfred Templeton, Co. H

Lt. Thomas D. Ranson, Co. I

Lt. Thomas Davis Ranson, Co. I

Pvt. David Franklin Rosen, Co. I

Pvt. Wm. Harrison Harris, Co. I

Courtesy: Mrs. Virginia Heizer, Middlebrook, Va.

Lt. John Alexander Lindsay, Co. K

Pvt. Paul McNeel Lindsey, Co. K, 52nd; Co. G, 18th Cav.

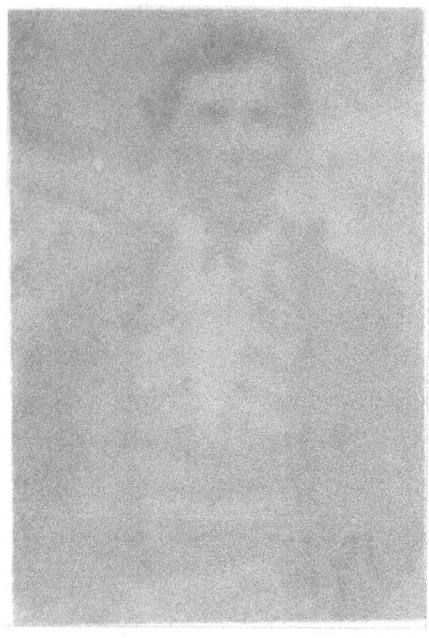

Benjamin A. Deffenbaugh, Co. F

Standing, left to right: Cpl. Gerald E. Crist, Co. I, 52nd; John Joseph Fauber, QM Dept., Staunton; George W. Willson, Co. I, Thomas, Co. C, 33rd; Comm. Sgt. Matthew T. "Tom" McClure, 14th Cav.; Horatio T. "Rush" Wilson, Boys Co. Junior Reserves; 52nd; John Earhart, Co. E, 5th; Edward A. Demastus, Co. F, 5th; John P. Smith, Jr., 2nd Rockbridge Arty.; Qm. Sgt. George Pilson Pvt. James W. M. Wallace, Co. E, 52nd. Front row: Alex G. Brown, Lightner, 52nd; William A. McCorkle, 1st Rockbridge Arty.; Co. G, 27th; J. Martin Harris, Marquis' Boys Bty., Staunton; Pvt. William S. Humphries, Co. E, 5th; James W. Houser, Co. E, 5th; James F. Harris, Co. I, 52nd; Pvt. David H. Rosen, Co. I, 52nd Inf. William J. Berry, Co. C, 52nd; Wm. H. McCutchan, Co. I, 52nd. Standing, front row: Thomas M. Smiley, Co. D, 5th. Second row:

Courtesy: Mrs. Nellie Harris, Raphine, Va.

52ND VIRGINIA INFANTRY
ADDENDUM FOR THE SECOND EDITION

AILOR or AYLOR, JOHN L.: d. Deerfield after 1903.
ALEXANDER, THOMAS WOODWARD: Delete. Served in Co. H. 5th Va. Inf.
ALLEN, CHARLES W.: Bur. Augusta Stone Presb. Ch. Cem.
ALMARODE, JOHN WILLIAM: Enl. Co. I, 5th Va. Inf. 3/23/62. Transf. Co. I, 52nd 12/1/62. Carpenter, Riverheads Dist., Augusta Co. 1870 census.
ANDERSON, EDWARD MANOR: Transf. Co. C, 14th Va. Cav. 7/31/64. Capt. Cedarville 11/12/64. Sent to Point Lookout. Released 6/19/65, 5'8½", florid complexion, brown hair, grey eyes.
AREY, GEORGE F.: Paroled Staunton 5/19/65, 5'9½", blue eyes, light hair, light complexion.
ATKINS, ALEXANDER B.: Present 9-10/64. Ab. sick a few days before the surrender. Farmer, Mt. Solon. d. Roland, Oklahoma 12/20/30. Bur. Salizlau, Okla.
AYERS, STEPHEN P.: d. 1864.

BAILEY, WILLIAM: Pvt., Co. I. b. Augusta Co. circa 1833. Receiving pension Rockingham Co. 1903.
BARTON, J. M.: Pvt., Co. K. Bur. Confederate Cem., UVa., Charlottesville.
BEARD, HUGH S.: b. 3/15/35.
BEARD, JOHN D.: Enl. Co. D, 8th Va. Cav. Monroe Co. 8/12/62. Des. 9/10/64. Moved to Monroe Co. 1867. Farmer, Monroe Co., W.Va. 1916.
BEATY, GEORGE: b. Rockbridge Co. 12/15/39. d. Mingo, Randolph Co., W.Va. 3/19/13. Bur. Mingo Cem. Brother of John Beaty.
BEATY, JOHN: d. Millboro Springs, Bath Co. 9/8/89. Brother of George Beaty.
BELLAMY, ABNER HARRIS: b. Va. circa 1833. Laborer, age 27, Northern Dist., Augusta Co. 1860 census.
BENNETT, J. J.: Pvt., Co. C. KIA 5/19/64.
BERRY, WILLIAM J.: b. Va. 8/30/47. d. Raphine, Rockbridge Co. 9/16/20. Bur. Mt. Carmel Presb. Ch. Cem.
BEVERAGE, WILLIAM E.: Delete. Served in 62nd Va. Inf.
BOSSERMAN, SAMUEL: (1st). b. Va. 2/17/26. d. 5/4/11. Bur. West Augusta Cem.
BOSSERMAN, WILLIAM HENRY: b. 6/20/26.
BOWER (vice BOWES), LEONARD: b. Va. circa 1840. Laborer, age 20, Northern Dist., Augusta Co. 1860 census.
BRIDGE, ALEXANDER: Brother of Jefferson Bridge.
BROOKS, ROBERT T.: Enl. Co. F, 37th Bn. Va. Cav. Staunton 6/20/64. Present through 12/64. NFR.
BROWN, JAMES PERRY: Pvt., Co. unknown. b. Nelson Co. 1837. Claimed service in Marquis's Battery, 5th and 52nd Va. Inf. on pension application. d. Stuart's Draft 3/7/12.
BROWN, JAMES W.: delete, duplicate entry.
BRUCE, ROBERT A.: b. Va. circa 1842. Miller's Apprentice, age 18, Waynesboro, 1860 census.
BRYANT, LORENZO SHAW: b. Va. 1836. Gd. Brownsburg Acad. Att. UVa. 51-52. Enl. 1861. Not on MR.
BULL, WILLIAM HENRY.
BUNCH, WILLIAM W.: Accidentally shot and killed Patterson's Station, Augusta Co. 8/28/81.
BUSH, CHARLES E.: Brother of Daniel L. Bush.
BUSH, DANIEL L.: Enl. Co. C, 5th Va. Inf. 6/9/61 age 15. Ab. 7/23-12/31/61. d. 9/20/65. Bur. Mt. Hebron Cem., Winchester. Brother of Charles E. Bush.
BUSHONG, ISAAC A.: Enl. Co. E, 11th Va. Cav. Brownsburg 3/64.
BUTTERLY, JOHN P.: Pvt., Co. unknown. b. Mass. circa 1834. Trader, age 26, Staunton, 1860 census. Enl. Co. G, 5th Va. Inf. 4/28/61. Transf. 52nd Va. Inf. 9/18/61 as baker. Returned to Co. G, 5th Va. Inf. by 4/62. Absent as courier 10/63. NFR.
BYRD, WILLIAM WALLACE: Bur. Oakland Grove Presb. Ch. Cem., Lowmoor, Alleghany Co.

CAMPBELL, LAWSON P.: d. Staunton 12/25/89.
CAMPBELL, NIMROD McPHERSON: b. Va. 6/16/43.
CARPENTER, WILLIAM JACOB: Enl. Co. H, 12th Va. Cav. 9/15/63. Capt. Mt. Crawford 3/1/65. Sent to Ft. Del. Released 6/7/65, 5'9", light complexion, hazel eyes, brown hair.
CARPENTER, WILLIAM R. N.: Five brothers and sisters died of diptheria at the same time.
CASH, JOHN WESLEY: b. Va. circa 1818.
CAULEY, THOMAS JEFFERSON: Bur. Confederate Cem., U.Va., Charlottesville.
CAULEY, WILLIAM BROWN: Also served in Co. F, 11th Va. Cav.
CHANCELLOR, JAMES EDGAR: Bur. U.Va. Cem.
CHANDLER, WILLIAM: d. Staunton 11/8/15 age 73, Laborer. Bur. Pleasant View Luth. Ch. Cem.
CHAPLAIN, WILLIAM JAMES: Carpenter, age 45, 1st Dist., Augusta Co. 1860 census and age 57, 2nd Dist., Augusta Co. 1870 census.
CHILDRESS, DAVID DILLARA: Also listed Co. A, 25th Va. Inf. and Co. C, 39th Bn. Va. Cav. d. Waynesboro 7/7/25.
CHILDRESS, WILLIAM M.: d. Augusta Co. 8/9/20.
CHRISTMAS, NOAH LEVI: d. Churchville 12/26/16. "A gallant soldier."
CHRISTMAN, THOMAS FRANKLIN: WIA Port Republic 6/9/62.

CLARK, JAMES MADISON C.: Farmhand, age 45, Buffalo Dist., Rockbridge Co. 1870 census.
CLAYTOR, GEORGE WASHINGTON: d. Crimora 6/18/13. Bur. Mt. Bethel Ch. Cem.
CLAYTOR, WILLIAM HENRY HARRISON, JR.: Moved to Rockbridge Co. 1909. d. near Fancy Hill 5/4/12½ Bur. Falling Spring Presb. Ch. Cem.
CLEMENTS, STEPHEN E.: b. Va. 12/20/43. Living Alexandria 1900.
CLEMMER, JAMES W.: Delete. Served in the 62nd Va. Inf.
COCHRAN, ANDREW A.: b. near Greenville 5/18/27.
COFFEY, EDWARD F.: b. circa 1830. Also listed Latham's Va. Battery. d. Stuart's Draft 2/9/06.
COFFEY, ROBERT W.: b. Nelson Co. circa 1831.
COINER, DAVID M., JR.: b. Augusta Co. 6/5/29.
COINER, JOHN CALVIN: Enl. Co. H, 5th Va. Inf. 4/19/61. Hired a subt. and disc. 4/28/62.
COINER, WILLIAM PETER: Delete COINER, WILLIAM PETER. Other information is correct. Enl. Co. E, 1st Va. Cav. Transf. Co. G, 52nd Va. WIA Bethesda Ch. 5/30/64. Brother of David W. Coiner.
COLE, WILLIAM HARRISON.
CONNELL, JAMES HUGH: b. Md. circa 1825. Millwright, age 35, Northern Dist., Augusta Co. 1860 census.
COOKE, JACOB W.: b. Augusta Co. circa 1824. Age 36 on 1860 census. Attended reunion 6/91, res. Ft. Defiance.
COOK, JOHN WILLIAM: Delete. Duplicate entry.
COOK, JOHN WESLEY: b. Nelson Co. 5/14/39. WIA Winchester 9/17/64. "ball entered my head, lodged somewhere in my brain, and has remained ever since." Res. near Ft. Defiance. d. Staunton 2/18/21. Bur. Thornrose Cem.
COOK, SIMON P.: Son of Thomas Cook.
COOK, THOMAS: Farmhand, age 55, 3rd Dist., Augusta Co. 1870 census. Father of Simon P. Cook.
CRENSHAW, D.: d. 3/9/65.
CRESS, JOSHUA T.: Brother of Nicholas Cress.
CRITZER, JAMES ANDREW: Pvt., Co. B(2nd). b. Nelson Co. 4/17/33. Served in Co. E, 1st Va. Cav. On postwar roster. Farmhand, 2nd Dist., Augusta Co. 1870 census. d. Basic City 2/8/13. Bur. Riverview Cem., Waynesboro.
CROSS, GABRIEL: Brother of John A. Cross.
CRUM, JACOB: b. Va. 7/26/19. Farmer, Pastures Dist., Augusta Co. 1870 census.
CUPP, FREDERICK: d. 11/9/73.
CUPP, WILLIAM J.: Brother of John C. Cupp.
CURRIER, ROBERT H.: Brother of John W., Joseph S. and William W. Currier.
CURRY, ALEXANDER S.: vice R. d. Bath Co. circa 1907 on wife's pension application.
CURRY, GEORGE HARVEY ANDERSON: Transf. Co. C, 14th Va. Cav. 11/62. Capt. Greenbrier Co. 11/26/62. Sent to Wheeling. Age 18, 5'6", fair complexion, blue eyes, brown hair. Transf. Camp Chase. Exch. 4/8/63. d. in N.C. Hospital, Petersburg of pneumonia 4/20/63.
CURRY, JOHN M.: d. Blue Grass, Highland Co. 1/1/13.
CURRY, JOHN W.: b. Va. circa 1839. Teacher, age 21, Northern Dist., Augusta Co. 1860 census. Brother of James H. and William L. Curry.
CURRY, JOSEPH: b. Va. circa 1839. Laborer, age 21, Northern Dist., Augusta Co. 1860 census.
CURRY, SAMUEL M.: b. 1827. Bur. Curry family cem. near Frost.

DAGGY, JOHN WILLIAM.
DAVIS, JOHN HENRY: Moved to Ohio 1868. d. West Milton, O. 11/22/16.
DAY, J. T.: Pvt., Co. D. WIA Mine Run 11/19/63. In hospital Gordonsville 12/1/63. Transf. to Richmond hospital 12/2/63. NFR.
DEAKINS, ANDREW J.: b. Ohio. Clerk, age 32, Stribling Springs Hotel, Augusta Co. 1860 census. Enl. Co. L, 5th Va. Inf. 3/17/62. WIA 8/28/62. Surrendered Appomattox 4/9/65. Bookkeeper, Staunton. d. 11/27/83 age 73. Bur. Thornrose Cem.
DEFENBAUGH, BENJAMIN AMI: Deserted to the enemy New Creek 2/8/64. Took oath and sent north.
DEFENBAUGH, HENRY F.: Delete. Served in Co. C, 5th Va. Inf.
DENDEN or DRYDEN, JAMES: Pvt., Co. H. Present 10/8/61. NFR. d. Rockbridge Co. 12/77.
DEPRIEST, JOHN W.: b. Rockbridge Co. d. Augusta Co., 1868.
DONOHOO, JOHN W.: Also served in Marquis's Battery, Junior Reserves 4/64.
DULL, AUGUSTUS B.: b. Augusta Co. 9/2/36. d. Augusta Co. 3/24/00. Bur. Mt. Tabor Luth. Ch. Cem.
DULL, GEORGE LEWIS: Miller, Winston-Salem, N.C. d. there 4/18/32. Bur. Salem Cem.
DULL, WILLIAM HENRY, SR.: Enl. Co. C, 14th Va. Cav. date unknown.
DUNLAP, JOHN J.: Married Augusta Co. 2/4/64. WIA 4/28/64. In hospital Gordonsville 4/29/64. Transf. to Staunton same day. Carpenter.

EAKIN, JAMES MOORE: d. Va. 2/21/30. Stonemason, Rockbridge Co.
EDMONDSON, WILLIAM H.: Bur. Spotsylvania Confederate Cem.
EWING, WILLIAM DAVIS: b. Rockingham Co.

FARROW, EDWARD: Pvt., Co. F. WIA (right thunb) Cold Harbor 6/2/64. NFR.
FAUBER, ANDREW S.: Farmhand, age 32, Walker's Creek Dist., Rockbridge Co. 1870 census.
FENDELL, THOMAS B.: delete. See Fendell E. Thomas.

FERGUSON, JESSE B.: Pvt., Co. C. b. Va. 2/27/27. in hospital Gordonsville 8/19/63 with "Morti Cut." Transf. Lynchburg 8/29/63. NFR. Carpenter, South River Dist., Augusta Co. 1870 census. d. Augusta Co. 6/14/10. Bur. Riverview Cem., Waynesboro.
FISHER, J.: change to FISHARD, J. Delete date of birth and 1860 census information.
FORD, WILLIAM ALEXANDER: b. Rockbridge Co. 3/14/41. Farmhand, 6th Dist., Rockbridge Co. 1860 census.
FORRER, HENRY: vice blank. b. Va. 12/45. Res. Stuart's Draft. d. Rockingham Co. 2/14/37. Bur. Cook's Creek Ch. Cem., near Harrisonburg. Brother of Samuel and William Forrer.
FORRER, JOHN K.: b. Pa. 10/6/30. Att. Mossy Creek Acad. Bur. Moorland Bapt. Ch. Cem., Red Hill.
FRANCISCO, GEORGE MARSHALL: Drillmaster. b. Marshall, Mo. 9/6/43. Moved back to Va. Att. VMI. Drillmaster Richmond and 52nd Va. Enl. Co. I, 14th Va. Cav. 3/10/62. Present Appomattox. Merchant and Real Estate Dealer, Marshall, Mo. d. 10/5/03.
FRASIER, SANDY F.: Res. Sherando, Augusta Co. before 1903.
FRAZIER, L. F.: Delete.
FULWIDER, JAMES W.: Member, Stonewall Jackson Camp, CV, Staunton.

GARBER, DAVID W.: Alive 1870.
GARRISON, JOHN W.: Delete entry on service in 18th Va. Cav.
GIBSON, BURWELL: Enl. Co. B, 23rd Va. Cav. Father of John W. Gibson.
GIBSON, JOHN ST. PIERRE.
GIBSON, JOHN WICKLEY: "Served with distinction." Son of Burwell Gibson.
GILLETT, ANDREW W.: b. Pocahontas Co. 11/10/38. d. Flood, Highland Co. 3/18/21. Bur. Gillett family cem. on Rt. 608, Highland Co.
GILMER, D. C.: Pvt., Co. H. WIA Port Republic 6/9/62. NFR.
GOCHENOUR, MARTIN JOSEPH.
GOODNIGHT, JOHN B.: b. Augusta Co. 10/17/26.
GORDON, THOMAS. Delete.
GRANDSTAFF, ISAAC M.
GRASS, WILLIAM E.: Bur. Sherando Meth. Ch. Cem.
GRAY, DAVID H.: b. Va. circa 1829. Farmer, age 31, 1st Dist., Augusta Co. 1860 census.
GREAVER, JACOB S.: b. Augusta Co. 11/5/39.
GREGER, JOSEPHUS: Sgt. vice Pvt. Promoted Sgt. before 5/30/64.
GRIFFITH, WILLIAM H.: b. Hanover Co.
GRIM, ELIJAH PIPES: Enl. Co. I, 14th Va. Cav. 8/16/62. Bur. McKinley Meth. Ch. Cem.
GROOMS, JOHN S.: b. Augusta Co. 12/4/40. Constable, 2nd Dist., Augusta Co. 1870 census. Bur. Thornrose Cem.
GUINN, WILLIAM C.: Enl. Co. I, 19th Va. Cav. vice 14th Va. Cav. Bur. Rocky Springs Presby. Ch. Cem., Deerfield.

HALL, JOSEPH D.: Carpenter, age 32, South River Dist., Rockbridge Co., 1870 census.
HALL, REUBEN WYMAN: b. Va. 10/37/38. d. Augusta Co. 11/28/85. Bur. Tinkling Spring Presb. Ch. Cem.
HALLL, WILLIAM L. (2nd): b. Va. circa 1833. Laborer, age 27, Northern Dist. Augusta Co. 1860 census.
HAMILTON, JAMES HARVEY: Enl. Co. F, 5th Va. Inf. 4/29/61. Ab. sick 7/25-12/61. AWOL 2/62. NFR.
HARNSBERGER, ROBERT S.: Also listed Co. E, 1st Va. Cav. and Co. A, 10th Va. Cav.
HARRIS, ALEXANDER: b. circa 1825. d. near Middlebrook 6/7/95 age 70. Bur. St. John's Ch. Cem.
HARTMAN, F.: Pvt., Co. F. In hospital Gordonsville with "Feb. Int." 10/28/63. Transf. Lynchburg 10/30/63. NFR.
HARUFF, JACOB A.: b. Bath Co. 6/20/36. d. 9/20/23.
HARSHBARGER, JOHN B.: Bur. Mossy Creek Old Cem.
HAWKINS, JOHN: Ankle broken by falling tree at Alleghany Mt. 1861.
HAYSLETT, ANDREW J.: Res. Bath Co. 1894.
HAYSLETT, ANDREW JACKSON.
HAYSLETT, BRADLEY: Enl. Co. F, 27th Va. Inf. 3/19/62. Des.
HAYSLETT, SARGENT MC DONALD: b. Va. circa 1847. In hospital Gordonsville with chronic rheumatism 6/10/63. Transf. Lynchburg 6/11/63. d. near Lexington 2/19/17 age 60. Bur. Collierstown Presb. Ch. Cem.
HEATON, JACOB HARVEY: b. Augusta Co. 7/29/44.
HEIZER, WILLIAM JAMES: b. Augusta Co. 10/23/32.
HILL, JOHN LYON: b. Augusta Co. 4/28/38. Enl. Co. I, 14th Va. Cav. 4/19/61 as 1st Lt. Transf. back to Co. I, 14th Va. Cav. by 9/62. Paroled Staunton 5/20/65, 5'11", dark complexion, dark hair, gray eyes. Teacher, Surveyor, Farmer. d. near Summerdean 2/27/09. Bur. Bethel Presb. Ch. Cem.
HINKLE, WILLIAM H.: Delete death and cem. information. Retired, age 64, Kerr's Creek Dist., Rockbridge Co. 1910 census.
HINTY, WILLIAM HENRY.
HISE, EMANUEL: b. Wurtenburg, Germany.
HITE, BENJAMIN LEWIS: Bur. Greenville Bapt. Ch. Cem.
HOGSHEAD, NEWTON HENDREN.
HOLBERT, EDWARD A.: d. Augusta Co. 11/16/20.
HOLLINGSWORTH, W. B.: Pvt., Co. F. In hospital Gordonsville with dysentery 9/14/63. Transf. Richmond 9/15/63. NFR.

HOOVER, JACOB A.: Pvt., Co. K.
HOOVER, JOHN J.: b. Va. circa 1830. Plasterer, age 30, Northern Dist. Augusta Co. 1860 census.
HOUSER, JOHN A.: Bur. Confederate Cem., U.Va., Charlottesville.
HUFFMAN, JOHN R.: Alive 1895. Brother of John S., Sylvester F. and William D. Huffman.
HULTZ, DAVID W.: d. Stuart's Draft 7/26/11, age 80.
HULTZ, WILLIAM H.: vice. Holtz.
HUMPHREYS, JAMES G.: b. Rockbridge Co.
HUMPHREYS, JAMES GREEN: Carpenter vice Teacher when enlisted.
HUMPHRIES, JAMES H.: b. Rockbridge Co.
HUNTER, DAVID S.: Enl. Co. C, 14th Va. Cav. 8/25/62. Present through 8/63. Brother of John T. Hunter.
HUPMAN, JOHN ALEXANDER: Bur. Thornrose Cem.

INGRAM, JOSEPH W.: Laborer, age 19, Goshen, Rockbridge Co. 1860 census.

JACKSON, ANDREW GEORGE: Delete 18th Va. Cav. information.
JOHNSON, JOHN: Pvt. Co. unknown. Hired as subt. by Peter N. Long 1/30/63. NFR.
JOHNSON, ROBERT M.: Pvt. Co. E. b. circa 1833. Enl. Camp Shenandoah 4/18/62. Present 4/18-30/62. Ab. on leave 5/1-10/31/62. Present 1-3/63. Ab. sick 4/29/63. Present 4/30-9/63. Ab. sick 10/8-30/63. Present 11-12/63. Issued clothing 1/28, 2/16 and 6/2/64. Des. while on the march from Lynchburg to Staunton 6/25/64. NFR. Mason, Bath Co. 1872.
JONES, CALVIN H.: WIA 11/29/63.
JONES, JOHN G.: b. Bath Co. 1846. Age 14, Northern Dist., Augusta Co. 1860 census. DOW's and diptheria Augusta Co. 8/27/62 age 16. Bur. Thornrose Cem., Staunton, grave #1205.

KASHER, JOHN: Delete. See HASHER, JOHN.
KEATON, JOHN THOMAS: "One of the bravest."
KEBLINGER, WILLIAM J.: d. Old Soldiers Home, Richmond 1/02. Bur. Charlottesville.
KEISTER, L. B.: Pvt., Co. H. In hospital Gordonsville with debilitus 7/29/63. Returned to duty 8/14/63. NFR.
KENNEDY, JOHN H.: d. Danforth, Ill. 3/25/91.
KERR, LORENZO DOW: b. Va. 1830.
KERSHNER, JAMES ADDISON: Disch. Nashville, Tenn. 8/23/65. 5'11", dark complexion, blue eyes. dark hair. Brother of John R. Kershner.
KERSHNER, JOHN ROBINSON: b. Augusta Co. 12/3/38. Enl. McClanahan's Va. Battery before 2/20/64. Paroled Staunton 5/22/65, 6', fair complexion, light hair, blue eyes. Moved to Fairland, Mich. and Red Oak, O. d. Ohio 12/14/01.
KESTERSON, WILLIAM GRASS: "The fight at McDowell was the first time he was under fire and like many other young (and old too) men he was much excited and was loading and firing as rapidly as possible, but shooting straight up in the air. Noticing this his captain, John M. Humphreys, came to him and said "Billy, there are no Yankees up there," which called him to himself and he replied, "No, and I'm afraid there never will be," and began firing where he knew they were."
KIDD, CLIFFORD C.: Pvt., Co. E. b. Nelson Co. 3/9/42. Pension application only record. Mason, Hermitage, Augusta Co. 1904. d. 7/5/24. Bur. Hermitage Cem.
KIDD, JOHN PAXTON: b. near Alexandria.
KINCAID, THOMAS McD.
KIRACOFE, NELSON BITTLE: Enl. McNeil's Rangers 3/10/63. Paroled Staunton 5/18/65, 5'11", dark complexion, dark hair, hazel eyes.
COINER vice KOINER, GEORGE MICHAEL: Res. Marshall, Mo. 1910.
KOINER, JOSEPH: KIA 5/12/64.
CRAFT vice KRAFT, JOHN F.

LACKEY, WILLIAM HARVEY: b. 3/25/42.
LAMBERT, JOHN MOORE: Bur. Bethel Presb. Ch. vice Riverview Cem.
LAMBERT, THOMPSON WILSON: d. Augusta Co. Bur. Bethel Presb. Ch. Cem.
LANDES, DANIEL B.: (1st) b. Va. circa 1838. Laborer, age 23, Northern Dist., Augusta Co. 1860 census. Enl. Co. H, 5th Va. Inf. 3/23/62. Des. 5/6/63. Sawyer, age 32, Mt. Sidney PO, Augusta Co. 1870 census.
LANDES, DANIEL B.: (2nd) b. Va. 1/31/35. Delete all after his parole. d. near Weyer's Cave 7/30/04. Bur. Pleasant Valley Ch. of the Brethren.
LANDES, DAVID: Enl. Co. C, 14th Va. Cav. 8/25/62. Capt. Greenbrier Co. 11/26/62. Sent to Wheeling. Age 23, 5'3", ruddy complexion, blue eyes, brown hair. Transf. Camp Chase and Alton, Ill. d. there of typhoid fever 2/17/63. Bur. Conf. Cem. there.
LANDES, WILLIAM R.: Bur. Thornrose Cem., grave #929.
LANDON, JOHN W.: b. Sullivan Co., Tenn. circa 1834.
LANDRUM, WILLIAM F.: b. Louisa Co. circa 1830. Carpenter, age 30, Staunton, 1860 census.
LEE, JOHN M.: Farmhand, age 50, 2nd Dist., Augusta Co. 1870 census.
LEECH, JAMES GRANVILLE: b. Rockbridge Co. circa 1845.
LEECH, LUCIAN THEODORE: b. Rockbridge Co. 3/23/43.
LEWIS, JOHN: d. Charlottesville 10/5/84. Bur. Maplewood Cem.
LIPSCOMB, OSCAR C.: b. Madison Co. 12/25/38.

LIPSCOMB, ROBERT M.: d. Staunton 5/10/75. Bur. Thornrose Cem.
LONG, PETER N.: Pvt., Co. unknown. Conscripted Rockingham Co. 1/30/63 age 35, 5'9½", fair complexion, brown hair, blue eyes, Farmer. Hired John Johnson as subt. and disch.
LUCAS, PETER, JR.: Detailed in Nitre and Mining Bureau with Robert Craig.
LUCK, L. T.: b. Va. circa 1844. Enl. 1861 on pension application.
LYNN, SAMUEL: b. Va. circa 1836. Brickmason, age 24, Waynesboro, 1860 census. No occ., age 35, 2nd Dist., Augusta Co. 1870 census.

MANN, JOHN A.: b. Albermarle Co. 1844. Delete census information.
MARSHALL, JAMES WILLIAM: d. New Castle 11/27/11. Bur. Roncerverte, W.Va.
MARSHALL, MOTON "Mote": Laborer, age 53, 2nd Dist., Augusta Co. 1870 census.
MARTIN, JOHN J.: b. Augusta Co. Bur. Bethlehem Luth. Ch. Cem.
MEEKS, JOHN P.: b. Rockbridge Co. 6/26/30.
MESSERLY, G. W.: Pvt., Co. H. In hospital Gordonsville with acute rheumatism 5/4/64. Transf. Richmond 5/15/64, NFR.
MICHAEL, CORNELIUS ALBERT vice ALBERT.
MICHAEL, HUDSON STABUS: Enl. Co. C, 18th Va. Cav. 12/15/62. Present 10/31/64. NFR. d. 1/2/23.
MICHAEL, JAMES: Enl. Co. I, 18th Va. Cav. 12/1/62. Present 11/1/63. NFR.
MILLER, BENJAMIN FRANKLIN: "A brave soldier."
MILLER, CLEMENT H.: Enl. McClanahan's Va. Battery. Paroled Winchester 5/4/65, 5'4", dark complexion, dark hair, hazel eyes.
MILLER, JOHN: Gd. College of N. J. 1836. Gd. Princeton Theol. Sem. 1842. Presb. Minister, Fairfield, 1857-59.
MILLER, JOHN M.: Farmer, Mingo, W. Va. 1884. d. Valley Head, W.Va. 1915.
MILLER, MARTIN MANSFIELD, JR.: WIA Mine Run 11/29/63. In hospital Gordonsville 12/1/63. Transf. Richmond 12/2/63.
MILLER, PETER A.: Bur. Old Emanuel Ch. Cem.
MONTGOMERY, WILLIAM.
MOORE, HAMLET W.: Shoemaker, Mt. Sidney PO, Augusta Co. 1870 census.
MOORE, ISAAC K.: vice I. K. b. Va. 8/12/45.
MORRIS, KENE JR. and SR.: change to KENNEY, MORRIS JR. and SR.
MORRIS, LAYTON J.: Bur. McDowell Cem.
MOSES, NORVELL: b. Va. circa 1844. Farmhand, age 16, 1st Dist., Augusta Co. 1860 census. Brother of Samuel M. Moses.
MOYERS, SAMUEL F.: May have served in Co. G, 62nd Va. Inf.
MURRAY, PEACHY G. vice MURRY, P. G. b. Va. circa 1840. Age 20, no occ., 1st Dist., Augusta Co. 1860 census.
MYERS, JOHN ELLIS: b. Va. circa 1845. Laborer, age 15, Northern Dist., Augusta Co. 1860 census.
McCHESNEY, WILLIAM STEELE: Gd. Med. Col., Phila., P. Acting surgeon, Co. D, 5th Va. Inf. '61.
McCORD, JAMES A.: b. Va. circa 1837. Carpenter, age 23, 1st Dist., Augusta Co. 1860 census.
McCUTCHAN, WILLIAM HAMILTON: WIA Mine Run 11/29/63. In hospital Gordonsville 12/1/63. Transf. Richmond 12/2/63. Delete all after deserted.
McDANIEL, SCOREIM JAMES: Blacksmith, Rockbridge Co. d. Fairfield 3/21/73.
McFADDEN, JOSEPH: b. Va. 1/4/24. Farmer, Rockbridge Co. d. 3/31/92. Bur. New Providence Presb. Ch. Cem.
McGUFFIN, WILLIAM W.: b. 3/8/36.
McKEE, JAMES MOFFETT: Cooper, age 49, Riverheads Dist., Augusta Co. 1870 census. d. Walker's Creek 11/11/89.

NIELSON, THOMAS HALL: Also listed McClanahan's Va. Battery. Capt. Beverly 10/29/64. Sent to Camp Chase. 5'10", fair complexion, dark hair, black eyes.

OFFLIGHTER, HARRISON: b. Va. circa 1817. Distiller, age 43, 1st Dist., Augusta Co. 1860 census.

PADGETT, JAMES W.: b. Augusta Co. 7/4/32. Bur. Sherando Metho. Ch. Cem.
PAINE, JAMES P. or F.: 2nd Lt., Co. K. b. circa 1840. Pvt., 1st Co. G, 25th Va. Inf. 1861. Resigned as 2nd Lt. Co. K, 52nd Va. Inf. 2/15/63. Enl. 2nd Co. F, 62nd Va. Inf., 9/29/63. Ab. sick 12/31/64. Res. Rockingham Co. 1910. NFR.
PAINTER, JAMES H.: Delete death information.
PAINTER, JOHN FITCH: d. Lofton 12/29/11.
PALMER, JACOB S.: Brother of James W. Palmer.
PANNELL, ADAM SAMUEL: Enl. Carpenter's Va. Battery 4/3/62, drafted from the militia.
PANNELL, GEORGE W.: May have served in Co. G, 2nd Va. Inf. Res. Lynhurst 1902 age 67. Alive 2/17/06.
PARR, ANANIAS: Bur. Our Holy Cross Episcopal Ch. Cem., on Rt. 635, Alb. Co.
PARRENT, WILLIAM LEWIS: b. Hermitage, Augusta Co. circa 1842. d. Sacramento, Calif. 2/1/23.
PARRISH, JOHN WILLIAM: Bur. Thornrose Cem.
PATERSON, JOHN ARTEMUS, SR.: Enl. Co. G, 5th Va. Inf. 3/23/62. Present to 3/31/62. NFR.
PATTERSON, WILLIAM BROWN: WIA Ft. Stedman 3/25/65 (had thumb shot off, cut knuckles in two and shot off piece of thigh bone).
PAXTON, MARTIN LUTHER: "He is remembered by his old comrades as cool, courageous, effective in battle, and displaying gallantry in warm engagements."

PEGRAM, J. W.: Pvt., Co. K. In hospital Gordonsville with "Feb. Cont." 11/27/63. Transf. Richmond 11/28/63. NFR.

PHILLIPS, RICHARD HENRY: b. Fredericksburg, Spotsylvania Co. 11/19/10.

PLEASANTS, JOHN: Pvt., Co. unknown. Pension application only record. Carpenter, age 55, Staunton 1870 census. Res. Augusta Co. 1902.

POTTER, SAMUEL MARTIN: b. Va. 1/11/43. Farmhand, 6th Dist., Rockbridge Co. 1860 census.

RAMSEY, BROWN CLAYTOR: b. Augusta Co. 1845. WIA (shoulder). Enl. 47th Bn. Va. Cav. Exch. 3/65. Farmer, Bath Co. d. McClung 12/18.

RANKIN, JOHN B., SR.: Enl. Co. C, 5th Va. Inf. 4/17/61. Disch. 6/16/61. d. Springhill 7/7/63 in 64th year. Bur. Burnett-Rankin Cem.

RAWLINS, WILLIAM: Pvt., Co. unknown. KIA Cold Harbor 6/2/64.

REID, WILLIAM H.: vice M.

REID, WILLIAM HENRY: (2nd). Pvt., Co. E. b. Va. 3/24/21. Farmer, 7th Dist., Rockbridge Co. 1860 census. On postwar roster. Farmer, Buffalo Dist., Augusta Co. 1870 census. d. Murat 8/4/89.

REYNOLDS, GEORGE WASHINGTON: Colorbearer. Brother of Jasper M.

REYNOLDS, JASPER M.: b. Augusta Co. circa 1845. Age 12, 1st Dist., Augusta Co. 1860 census.

RHEA, JOHN SHAW: Delete. Served in 62nd Va. Inf.

RIDDLE, JOHN L.: Farmhand, age 19, Waynesboro 1860 census. d. 11/69 age 29.

RIDDLEBERGER, ELIAS: Stage Driver, detailed much of the war on pension application.

RIFE, ARCHIBALD S.

RILEY, PATRICK H.: b. Va. circa 1847. Age 13, Lexington Dist. Rockbridge Co. 1860 census. Delete postwar census information.

ROARKE, CHARLES K. S.: d. 7/22/96.

ROGERS, GEORGE ANDREW: b. Halifax Co. d. 6/45.

ROSS, ABSALOM RICHARD: Pvt., Co. B (2nd). b. Va. circa 1832. Carpenter, Waynesboro. Enl. Waynesboro 7/15/61. Present 11-12/61 and 3-4/62. AWOL 9/17/62-2/63. Pay stopped for AWOL. Ab. sick with "ulcus" 4/29/63. WIA (side) Bethesda Ch. 5/30/64. Ab. wounded thigh 12/31/64. NFR. Farmer, age 38, 2nd Dist., Augusta Co. 1870 census. Entered Old Soldiers Home, Richmond 7/4/98 age 61. Left 6/15/00.

ROSS, JOHN DE HART "Mad Boy": Lt. Colonel. b. "Bel Pae", Culpeper Co. 5/1/40. Gd. VMI 1859. Studied law summer 1860. Assistant Professor of Latin, French and Tactics, VMI 60-61. With 10 cadets escorted gunpowder from VMI to Colonel Jackson at Harpers Ferry 4/21/61. Served briefly as drillmaster there and then on Jackson's staff until ordered back to VMI to instruct young officers being commissioned in the Provisional Army of the Confederate States. Soon ordered to western Virginia as Engineer Officer on staff of Gen. William W. Loring. Appointed Major of the 52nd 10/19/61. Commanded the regiment in the battle of Alleghany Mt. 12/13/61. Ordered to VMI as instructor 1/62. Reelected Major 5/1/62. Present McDowell. WIA (bullet in leg and hand cut in two by shell fragment) Port Republic 6/9/62. Served as Enrolling Officer, Staunton 11/22/62 and Lexington 1863. Promoted Lt. Col. 6/6/63. Present Gettysburg and Mine Run. Resigned 12/19/63 for an "intolerable situation" (he had accused Colonel Skinner of incompetence and he had been tried and acquitted) and his inability to write and was suffering from rheumatism in the wounded hand. Major and Assistant Professor of VMI 1863. Lt. Col. and Assistant Professor of Math 1864-65. Wrote President Davis requesting assignment to the Adjutant General's Office 2/27/64. Served on Colonel Smith's staff at New Market. Served with the cadets in Richmond until disbanded April, 1865. Paroled Staunton 5/23/65. Studied law but never practiced. Farmer, Rockbridge Co. President of Lexington Manufacturing Co. Served on County Board of Supervisors and other minor offices. Member, Stonewall Jackson Camp, CV, Lexington. d. "Sunnyside" near Lexington of Bright's disease 12/12/12. Bur. Stonewall Jackson Cem. Brother of William A. Ross.

ROSS, WILLIAM ALEXANDER: 2nd Lt., Co. C. b. "Bel Pae", Culpeper Co. 11/27/42. Att. U.Va. 60-61. Enl. Port Royal 11/17/62 as Pvt. Present as Sergeant Major of Regiment through 2/28/63. Elected 2nd Lt. 3/16/63. AWOL 8/27-31/63. Present 9-12/63. CM'd 3/4/64. WIA (breast, severe by shell) Bethesda Ch. 5/30/64. DOW's 6/2/64. Bur. Culpeper Co. Brother of John D. Ross.

RUEBUSH, WILLIAM H.: 4th Cpl., Co. E. Enl. Staunton 7/31/61. Present 11/61-4/62. Reenl. 5/1/62. Present 8/31-12/31/62 and 1-4/63. Transf. Co. G, 5th Va. Inf. 5/29/63. NFR. d. near Singers Glen, Rockingham Co. 10/29/91.

RUFF, GEORGE WILLIAM: Pvt., Co. D. b. Augusta Co. 5/26/43. Farmhand, Burke's Mill Dist., Augusta Co. 1860 census. Enl. Staunton 7/16/61. Present 7/31-10/31/61. d. in hospital at Yager's on Alleghany Mt. of fever 11/4/61. Bur. Cedar Hill Cem., Sangersville.

RULEY, JOHN FRANKLIN: 4th Sgt., Co. E. b. Rockbridge Co. 12/23/42. Farmhand, Lexington Dist., Rockbridge Co. 1860 census. Enl. Staunton 8/1/61 as Pvt. Present 11/61-4/62. Reenl. 5/1/62. Present 4-9/62. WIA Sharpsburg 9/17/62. Present as 2nd Cpl. 10/62. Present 1-2/18/63. Ab. on leave 2/19/63 for 16 days. Promoted 4th Sgt. 2/1/63. Present 4-12/63. WIA (left hip) Bethesda Ch. 5/30/64. Ab. wounded through 12/31/64. Captured and paroled Farmville 4/11-21/65. Railroad employee. d. Huntington, W.Va. 12/23/84. Bur. Rapp's Presb. Ch. Cem., Rockbridge Co.

RUNKLE, DAVID: Pvt., Co. I. b. Augusta Co. 9/9/39. Farmhand, Burkes Mill Dist., Augusta Co. 1860 census. Enl. Staunton 7/16/61 age 21. Present 11/61-4/62. WIA (mouth) McDowell 5/8/62. Ab. wounded through 10/31/62. AWOL 11/22/62-2/63. Present 3-5/63. WIA (foot) Fredericksburg 5/5/63. Present 6-12/63 and 4-10/64. Des. to the enemy Beverly 12/28/64. Took oath and transportation furnished to Parkersburg 12/28/64. Age 24, 5'8", fair complexion, grey eyes, fair hair, Farmer. Farmer near Madrid. d. 12/23/96. Bur. Hildebrand Menonite Ch. Cem.

RUSMISELL, GEORGE SHERMAN: Pvt., Co. C. b. Augusta Co. 1843. Farmhand, Northern Dist., Augusta Co. 1860 census. Enl. Staunton 7/16/61 age 19. Present 11/61-4/62. Reenl. 5/1/62. WIA (hand) Malvern Hill 7/1/62. Ab. wounded through 8/31/62. Present 9/62-4/63. WIA Gettysburg 7/3/63. Ab. wounded through 8/31/63. Present 9-12/63. AWOL 9/5-12/31/64. Wrote letter requesting detail as clerk in hospital, Staunton 1/5/65 because of rheumatism. Surrendered Appomattox 4/9/65. Music Teacher, Craigsville PO, Augusta Co. 1870 census. Owner, Iron Mine, Marble Valley 1897. Bur. St. John's Ch. near Middlebrook.

RUSMISELL, WILLIAM HENRY HARRISON: Sgt., Co. C. b. Va. circa 1840. Farmhand, age 20, Northern Dist., Augusta Co. 1860 census. Enl. Staunton 7/16/61 as Pvt. age 21. Present 11/61-4/62. Reenl. 5/1/62. Present through 2/28/63. Present 3-5/63 as 3rd Cpl. Present 5-12/63 and 4-10/64. Promoted Sgt. 10/31/64. Ab. on leave 11-12/64. Paroled Staunton 6/6/65 age 22, 5'4", fair complexion, dark hair, grey eyes. House Painter, Staunton. Member, Stonewall Jackson Camp, CV, Staunton. Entered Old Soldiers Home, Richmond 12/28/10 age 70. d. 1/14/13 age 73. Bur. Hollywood Cem.

RUST, N. A.: Pvt., Co. unknown. Attended reunion Carlisle, Pa. 9/28/81, only record.

RUTHERFORD, JOHN C.: 1st Cpl., Co. F. b. Jefferson Co. 3/10/42. Mail Carrier, Burkes Mill Dist., Augusta Co. 1860 census. Enl. Staunton 7/31/61 as Pvt. Present 11/61-1/62. Ab. sick 2/5-28/62. Present 3-4/62. Reenl. 5/1/62. Present 8/31/62-2/18/63. Ab. on leave 2/19/63. Promoted 1st Cpl. 2/1/63. Present 4-12/63 and 3-4/64. Capt. Bethesda Ch. 5/30/64. Sent to Pt. Lookout. Transf. Elmira. Released 6/30/65, 5'7", dark complexion, dark hair, hazel eyes. WIA twice on postwar roster. Clerk, Staunton, 1870 census. Farmer, Beverly Manor Dist., Augusta Co. 1910 census. Member, Stonewall Jackson Camp, CV, Staunton. d. Staunton 5/5/28. Bur. Thornrose Cem.

RYAN, JOHN M.: Pvt., Co. G. b. Rockingham Co. 4/27/07. Enl. Staunton 9/20/62 as substitute. Present 10/62-4/63. Present 8/31/63, fined two months pay by sentence of CM for AWOL. Present 9-11/63. WIA (left leg) Mine Run 11/29/63. Ab. wounded through 1/64. Detailed in hospital Richmond 1/12/64. NFR. d. Louisa Co. 5/14/90. Bur. Thornrose Cem., Staunton.

SAMUELS, ERASMUS A.: Teacher, age 35, 3rd Dist., Augusta Co. 1870 census.

SAMUELS, JOSEPH M.: Enl. Co. I, 5th Va. Inf. 10/25/64. Surrendered Appomattox 4/9/65.

SANDES, WILLIAM: Delete.

SHAFER, ROBERT PRESTON GEORGE: b. Rockbridge Co. 11/22/41. Bur. Green Hill Cem.

SHANER, JACOB H.: b. Va. circa 1837. Farmhand, age 23, 6th Dist., Rockbridge Co. 1860 census. Delete 1870 census and all following information.

SHEETS, HENRY C.: b. Augusta Co. 1838. Carpenter, age 32, Northern Dist., Augusta Co. 1860 census.

SHEETS, JOHN D.: Enl. Co. I, 5th Va. Inf. 3/23/62. Transf. Co. F, 52nd Va. 3/17/63.

SHEETS, WILLIAM: Pvt., Co. K. b. Va. circa 1839. Wagoner, age 21, Northern Dist., Augusta Co. 1860 census. In hospital Gordonsville with dysentery 9/14/63. Transf. Charlottesville 9/15/63. NFR.

SHEETS, WILLIAM J.: Left widow and one child.

SHIFFLETT, SAMUEL B.: Delete, served in Co. G, 62nd Va. Inf.

SHIPLET, BENJAMIN FRANKLIN vice SHIFFLETT or SHIPLET, FRANKLIN: b. Va. 7/2/22. d. Augusta Co. 2/6/96. Bur. Parnassas Meth. Ch. Cem.

SHIREY, WILLIAM D.: WIA Cold Harbor 6/2/64 vice Bethesda Ch. 5/30/64.

SHRECKHISE, DANIEL KEYSER: Conscripted 2/12/64 2nd Co. I, 62nd Va. Inf. WIA Winchester 7/25/64. Paroled Harrisonburg 5/2/65. d. Mt. Sidney 8/1/08. Bur. Melancthon Chapel Cem.

SHRECKHISE, JAMES D.: Farmhand, age 19, Northern Dist., Augusta Co. 1860 census. d. Washington, D.C. 5/11/01.

SHULL, WILLIAM C.

SHULTZ, MARTIN S.: b. Adair Co., Ky. 1830. d. Albermarle Co. 1897. Bur. Lebanon Presb. Ch. Cem., Albermarle Co.

SIMPSON, CHARLES POWHATAN: b. Va. circa 1846. "ball lodged in lower jawbone. Removed 1894."

SIPES, SAMUEL H.: b. Augusta Co. circa 1825. School Teacher, age 35, Northern Dist., Augusta Co. 1860 census.

SMILEY, JOHN B.: b. Rockbridge Co. 1821. Farmer, age 39, 6th Dist., Rockbridge Co. 1860 census.

SMITH, J. A.: Pvt., Co. K. b. Bath Co. circa 1827. Not on MR. d. of typhoid fever in hospital Staunton 7/1/62 age 35. Bur. Thornrose Cem., grave #1084.

SMITH, JAMES: Living Yost, Highland Co. 1901.

SMITH, JAMES EDWIN: b. Louisa Co. 1844. Bur. Crown Hill Cem., Clifton Forge.

SMITH, JOSEPH W.: (2nd). b. 10/8/16. d. Augusta Co. 1/16/04. Bur. Mt. Carmel Presb. Ch. Cem.

SNAPP, CYRUS H.: d. Richmond.

SNAPP, WILLIAM F.: b. Staunton 9/17/37. d. Staunton 7/24/01.

SNEAD, WILLIAM JAMES: b. North Garden, Albemarle Co. 6/1/42. Railroad employee. Farmer, Walker's Creek Dist., Rockbridge Co. 1910 census. d. near Goshen 3/11/22. Bur. Goshen Bapt. Ch. Cem.

SNELL, DAVID FRANKLIN: b. Augusta Co. 5/31/37.

SPITLER, HENRY: Delete. Served in Co. F, 5th Va. Inf.

SPROUSE, BENJAMIN H.: Married Albemarle Co. 1868.

SPROUSE, PETER: Alive 2/15/16.

STAUBUS, AUGUSTUS WILLIAM: Served in Co. F, Harper's Regiment, Augusta Co. Reserves. WIA Piedmont 6/5/64.

STOCKDALE, JOHN H.: Pvt., Co. E. b. 2/27/34. WIA Cold Harbor 6/2/64. d. near Arbor Hill, Augusta Co. 4/4/32. Bur. Mt. Tabor Luth. Ch. Cem. Obit only record.

STRICKLER, ARCHIBALD W.: b. Augusta Co. 4/18/40. Laborer.

SULLIVAN, JOHN D.: Bur. Spotsylvania Confederate Cem. vice Fredericksburg.

SUMMERS, ANDREW JACKSON: b. Va. circa 1845. Farmhand, age 15, 1st Dist., Augusta Co. 1860 census.

SWINK, GEORGE WASHINGTON: d. Moscow 3/10/04. Bur. Mossy Creek Presb. Ch. Cem.

SWINK, MARTIN VAN BUREN: d. before 1910.

SWINK, MATTHEW J.: b. Augusta Co. 3/17/27.

SWITZER, JOHN D.: Student, age 13, Northern Dist., Augusta Co. 1860 census.

TAYLOR, HENRY C.: b. Augusta Co. 1814.

TERRELL, ANDREW J.: Bur. Spotsylvania Confederate Cem.
TETER, DANIEL W.: Laborer, age 16, Northern Dist., Augusta Co. 1860 census. Brother of James W. Teter.
TETER, JAMES W.: b. Augusta Co. circa 1845. Apprentice Blacksmith, age 15, Northern Dist., Augusta Co. 1860 census. Brother of Daniel W. Teter.
THOMAS, FENDALL E.: d. 1/25/65.
THOMAS, JOHN G.: Disch. 7/16/62.
THOMASON, JOHN B.: b. Augusta Co. 1/27/21.
THOMPSON, ANDREW JAMES: Moved to Clarke Co. Member, J. E. B. Stuart Camp, CV, Clarke Co.
THORNER, M. B.: d. 10/29/63.
THORNTON, ABSALOM: Member, Stonewall Jackson Camp, CV, Staunton.
TINSLEY, WILLIAM H.: vice J. b. Alleghany Co. 9/9/42.
TOMES, HARDEN L. vice TONER or TONES: b. Va. circa 1816. Farmer, age 54, South River Dist., Augusta Co. 1870 census.
TRAINER, GEORGE W. vice TRAIMER.
TUTWILER, GEORGE W.: Delete. Served in Co. C, 5th Va. Inf.
TUTWILER, WILLIAM HENRY: Enl. Co. C, 5th Va. Inf. 4/17/61 age 23, Shoemaker. WIA Chancellorsville 5/3/63 and Gettysburg 7/3/63. Disch. 8/63.

WALLACE, SAMUEL: b. Rockbridge Co. 2/23/34.
WAMPLER, SIMON WILLIAM: d. Augusta Co. 4/8/21.
WARD, LEVI: Resident of Pocahontas Co. Captured 1863 in Randolph Co. Sent to Ft. Delaware. d. of disease there.
WHITE, ISAAC M. vice MATTHEW: b. Va. circa 1836.
WHITE, WEAVER W.: b. Va. circa 1845. Farmhand age 15, Natural Bridge Dist., Rockbridge Co. 1860 census. Brother of James M. White.
WICKLE, SAMUEL HENRY: Farmer, age 36, 3rd Dist., Augusta Co. 1870 census.
WOOD, THOMAS H.: Bur. Green Hill Cem., Buena Vista.
WOODZELL, GEORGE BUFORD: Bur. Warrenton Cem.

YEAGER, WILLIAM ASBURY: b. Pocahontas Co. 11/30/40. Transf. Co. G, 31st Va. Inf. 7/20/62. Had over 17 bullet holes in his clothing at Spotsylvania CH 5/12/64. KIA Hatcher's Run 2/6/65. Over 6'.
YOUNT, RUDOLPH CHRISTIAN "Doc": Pvt., Co. A.
YOUNT, WILLIAM ISAAC H.: Clerk, 2nd Dist. Augusta Co. 1870 census. d. "Woodlawn", Cumberland Co.

ZEIGLER, A. E.: Pvt., Co. B(2nd). Not on MR. d. Batesville. Albermarle Co. 5/4/71. Wife's pension application only record.
ZIMBRO, JOHN: Moved to Bath Co. 1898.
ZIMBRO, ROBERT: d. Augusta Co. 2/5/24.
ZIRKLE, JOHN J.: b. Va. circa 1843. Farmhand, age 17, Northern Dist., Augusta Co. 1860 census. Brother of Robert J. Zirkle.
ZIRKLE, ROBERT JOSEPH: b. Va. circa 1841. Laborer, age 19, Northern Dist., Augusta Co. 1860 census. Brother of John J. Zirkle.
ZOLLMAN, JOHN WILLIAM: WIA Cannon's Farm 5/26/64.

Heritage Books by Robert J. Driver, Jr.:

Augusta County, Virginia Confederate Soldiers

Augusta County, Virginia Confederate Soldiers: Photo Pages

Confederate Sailors, Marines and Signalmen from Virginia and Maryland

Confederates of Prince Edward County and Deaths in Farmville Hospital, 1862–1865

First and Second Maryland Cavalry, C. S. A.

First South Carolina Cavalry

Richmond Local Defense Troops, C.S.A.

The First and Second Maryland Infantry, C.S.A.

The Virginia Regimental Histories Series: 2nd Virginia Cavalry

The Virginia Regimental Histories Series: The 1st and 2nd Rockbridge Artillery, 2nd Edition

The Virginia Regimental Histories Series: 1st Battalion Virginia Infantry, 39th Battalion Virginia Cavalry, 24th Battalion Virginia Partisan Rangers, 1st Edition
Robert J. Driver, Jr. and Kevin C. Ruffner

The Virginia Regimental Histories Series: 1st Virginia Cavalry, 2nd Edition

The Virginia Regimental Histories Series: 5th Virginia Cavalry, 1st Edition

The Virginia Regimental Histories Series: 14th Virginia Cavalry, 2nd edition

The Virginia Regimental Histories Series: 52nd Virginia Infantry, 2nd Edition

The Virginia Regimental Histories Series: 58th Virginia Infantry, 1st Edition

The Virginia Regimental Histories Series: The Staunton Artillery-McClanahan's Battery

Virginia Civil War Battles and Leaders Series: Lexington and Rockbridge County in the Civil War, 2nd Edition

www.ingramcontent.com/pod-product-compliance
Lightning Source LLC
Chambersburg PA
CBHW051924160426
43198CB00012B/2031